Lecture Notes in Computer Science   16611

The series Lecture Notes in Computer Science (LNCS), including its subseries Lecture Notes in Artificial Intelligence (LNAI) and Lecture Notes in Bioinformatics (LNBI), has established itself as a medium for the publication of new developments in computer science and information technology research, teaching, and education.

LNCS enjoys close cooperation with the computer science R & D community, the series counts many renowned academics among its volume editors and paper authors, and collaborates with prestigious societies. Its mission is to serve this international community by providing an invaluable service, mainly focused on the publication of conference and workshop proceedings and postproceedings. LNCS commenced publication in 1973.

Lejla Batina · Ferruh Özbudak

Editors

Arithmetic of Finite Fields

11th International Workshop, WAIFI 2026
Santander, Spain, June 3–5, 2026
Revised Selected Papers

 Springer

Editors
Lejla Batina
Radboud University
Nijmegen, The Netherlands

Ferruh Özbudak
Sabancı University
Istanbul, Türkiye

ISSN 0302-9743   ISSN 1611-3349 (electronic)
Lecture Notes in Computer Science
ISBN 978-3-032-27573-8   ISBN 978-3-032-27574-5 (eBook)
https://doi.org/10.1007/978-3-032-27574-5

Preface

These are the proceedings of WAIFI 2026, the 11th International Workshop on the Arithmetic of Finite Fields, held in Santander, Spain, during June 3–5, 2026. The previous editions of this workshop were held in Madrid, Spain (WAIFI 2007), Siena, Italy (WAIFI 2008), Istanbul, Türkiye (WAIFI 2010), Bochum, Germany (WAIFI 2012), Gebze, Türkiye (WAIFI 2014), Ghent, Belgium (WAIFI 2016), Bergen, Norway (WAIFI 2018), Rennes, France (WAIFI 2020), Chengdu, China (WAIFI 2022), and Ottawa, ON, Canada (WAIFI 2024). Springer published all previous volumes of the WAIFI proceedings in the LNCS series.

Since 2008, WAIFI has been held every even year, bringing together mathematicians, computer scientists, engineers, and physicists conducting research on the theory, applications, and implementations of finite fields.

The scientific program of WAIFI 2026 consisted of three invited talks and 21 contributed papers. The invited speakers were John Sheekey, Violetta Weger, and Delaram Kahrobaei. The contributed papers were selected from 31 submissions. The review process for contributed papers was double blind. Most submissions received three reviews, while a small number received two or four reviews. The final selection was based on the written reports and the discussions of the Program Committee.

The invited papers included in this volume were solicited by the Program Co-chairs and were reviewed by the volume editors for relevance, clarity, and suitability for the proceedings volume. For papers coauthored by committee members, the conflicted members were not involved in the corresponding editorial decisions.

We are very grateful to the members of the Program Committee for their dedication, professionalism, and careful work during the review and selection process. We also sincerely thank the General Chair, the Program Co-chairs, the Poster Session Chair, the Publicity Chair, the Steering Committee, and the members of the Organizing Committee for their support and work in organizing the workshop.

As with the previous volumes, Springer agreed to publish the WAIFI 2026 papers as an LNCS volume. We thank the staff at Springer for making this possible. Finally, and most importantly, we warmly thank all authors who submitted their work to WAIFI 2026 and all participants who contributed to the scientific success of the workshop.

April 2026

Lejla Batina
Ferruh Özbudak

Organization

General Chair

Domingo Gómez Pérez — University of Cantabria, Spain

Program Co-chairs

Lejla Batina — Radboud University, The Netherlands
Ferruh Özbudak — Sabancı University, Türkiye

Poster Session Chair

Francisco Javier Soto — Rey Juan Carlos University, Spain

Publicity Chair

José Luis Imaña — Complutense University of Madrid, Spain

Organizing Committee

Ana Isabel Gómez Pérez — Rey Juan Carlos University, Spain
Edgar Martínez Moro — University of Valladolid, Spain
Daniel Sadornil Renedo — University of Cantabria, Spain

Steering Committee

Lilya Budaghyan — University of Bergen, Norway
Claude Carlet — Universities of Paris VIII, France and Bergen, Norway
Anwar Hasan — University of Waterloo, Canada
José Luis Imaña — Complutense University of Madrid, Spain
Çetin Kaya Koç — University of California, Santa Barbara, USA
Sihem Mesnager — University of Paris VIII, France

Ferruh Özbudak   Sabancı University, Türkiye
Svetla Petkova-Nikova   KU Leuven, Belgium
Francisco Rodríguez-Henríquez   Technology Innovation Institute, UAE
Erkay Savaş   Sabancı University, Türkiye

Program Committee

Herivelto Martins Borges Filho   Universidade de São Paulo, Brazil
Claude Carlet   Universities of Paris VIII, France and Bergen,
                                        Norway
Maria Corte-Real Santos   ENS Lyon, France
Thomas Decru   KU Leuven, Belgium
Sylvain Duquesne   University of Rennes, France
Ana Isabel Gómez Pérez   Rey Juan Carlos University, Spain
Sophie Huczynska   University of St Andrews, Scotland
José Luis Imaña   Complutense University of Madrid, Spain
Jorge Jiménez Urroz   Polytechnic University of Madrid, Spain
Angshuman Karmakar   IIT Kanpur, India
Gohar Kyureghyan   University of Rostock, Germany
Edgar Martínez Moro   University of Valladolid, Spain
Sihem Mesnager   University of Paris VIII, France
Alessandro Neri   University of Naples Federico II, Italy
Svetla Nikova   KU Leuven, Belgium
Daniel Panario   Carleton University, Canada
Hilder Vitor Lima Pereira   State University of Campinas, Brazil
Håvard Raddum   Simula UiB, Norway
Francisco Rodríguez-Henríquez   Technology Innovation Institute, UAE
Ana Salagean   Loughborough University, UK
Amin Sakzad   Monash University, Australia
David Thomson   Carleton University, Canada
Alev Topuzoğlu   Sabancı University, Türkiye
Monika Trimoska   Eindhoven University of Technology, The
                                        Netherlands
Qiang (Steven) Wang   Carleton University, Canada
Violetta Weger   Technical University of Munich, Germany
Nusa Zidaric   Leiden University, The Netherlands

Contents

Cryptography and Boolean Functions

Finite Field Arithmetic, Algorithms, and Implementations

Invited Talks

New Invariants for Rank Metric Codes, with Applications to the Classification of Rank Two Semifields of Order 256

Jack Gilchrist[1], Stefano Lia[2], Arani Paul[1], and John Sheekey[1(✉)]

[1] University College Dublin, Dublin, Ireland
`john.sheekey@ucd.ie`
[2] Umea University, Umea, Sweden
`stefano.lia@umu.se`

Abstract. In this paper we completely classify semifields of order $2^8 = 256$ containing a nucleus of order $2^4 = 16$. We introduce new invariants for semifields, and apply new computational techniques for calculating old invariants. Together these make the computational classification significantly quicker.

Keywords: Semifield · Rank-metric Codes · Invariant

1 Introduction

A *finite semifield* is a finite division algebra in which multiplication is not necessarily associative. Thus they are natural generalisations of finite fields. Dickson [9] showed that there exist non-trivial finite semifields. Albert [1] and Knuth [11] deepened the study of semifields, demonstrating equivalences between semifields and certain types of projective planes and hypercubes (threefold tensors). Since then semifields have proved useful in a variety of settings; we refer to [12] for further background and history.

It is well known that to each semifield $\mathbb{S}$ one can associate a *semifield spread set* $C(\mathbb{S})$; an additively closed set of matrices in which the difference of any pair of matrices is invertible. By an appropriate choice of parameters, a semifield spread set is an $\mathbb{F}_q$-subspace of $M_n(\mathbb{F}_{q^s})$ of dimension ns over $\mathbb{F}_q$ in which every nonzero element is invertible. In this setting we can view $C(\mathbb{S})$ as a *matrix rank metric code*, specifically an *MRD code* in $M_n(\mathbb{F}_{q^s})$ with minimum distance n. Semifields which give rise to spread sets in this space are those who have *right nucleus* containing $\mathbb{F}_{q^s}$.

The natural notion of equivalence for semifields is that of *isotopism*; this notion coincides with the equivalence of the corresponding spread sets as matrix rank metric codes. This fact has been exploited in order to perform computational classifications of semifields of small orders; see [5,6,19,20].

Further advances for general semifields of larger orders appears to be out of reach at present. Therefore attention turns to semifields with added assumptions;

L. Batina and F. Özbudak (Eds.): WAIFI 2026, LNCS 16611, pp. 3–11, 2026.
https://doi.org/10.1007/978-3-032-27574-5_1

for example, symplectic semifields [13,15], or semifields of small dimension over a nucleus [16]. When both these assumptions are combined we consider *rank two commutative semifields*, which have numerous connections to other topics in finite geometry. See [3] for these connections, and a remarkable classification result.

Strong classification results on semifields of dimension 2 over a nucleus $\mathbb{F}_{q^s}$ have been achieved for arbitrary q for $s = 2$ in [4] and $s = 3$ in a series of papers including [10,17].

In this paper we introduce some new invariants for semifields and rank-metric codes which can be used to more efficiently determine inequivalence. As the problem of determining the equivalence of two matrix rank-metric codes is known to be difficult, any efficiently computable equivalence invariant can be used to speed up equivalence classifications, by reducing the number of full equivalence tests necessary. We demonstrate their utility by classifying semifields of order 2^8 containing a right-nucleus of order at least 2^4, or equivalently $\mathbb{F}_2$-linear MRD codes in $M_2(\mathbb{F}_{2^4})$ with minimum distance 2, with the new invariants resulting in reducing the computation time by a factor of 20.

2   Rank Metric Codes and Spread Sets

2.1   Matrix Rank Metric Codes

For the purposes of this paper, a *matrix rank metric code* is an $\mathbb{F}_q$-subspace of $M_{\ell \times n}(\mathbb{F}_{q^s})$, equipped with the distance function

$$d(A, B) := \mathrm{rank}(A - B).$$

We will mostly work with the case $\ell = n$, and write $M_n(\mathbb{F}_{q^s}) := M_{\ell \times n}(\mathbb{F}_{q^s})$.

Two codes C, C' are said to be *equivalent* if there exist invertible matrices $X, Y \in M_n(\mathbb{F}_{q^s})$ and a field automorphism ρ of $\mathbb{F}_{q^s}$ such that

$$C' = \{XA^\rho Y : A \in C\},$$

where A^ρ is the matrix obtained from A by applying the automorphism ρ entry-wise. Throughout we will let σ denote the Frobenius automorphism, that is, the map $\sigma : x \to x^q$.

Each matrix of $M_n(\mathbb{F}_{q^s})$ naturally defines an $\mathbb{F}_{q^s}$-endomorphism from $(\mathbb{F}_{q^s})^n$ to itself. We can also regard such a map as an $\mathbb{F}_q$-endomorphism from $(\mathbb{F}_q)^{ns}$ to itself; we will make this explicit later.

The problem of determining the equivalence or otherwise of two codes is well known to be a difficult problem; indeed, many public key cryptosystems rely on the computational hardness of this problem. Therefore any *invariant*, that is, any (easily computable) function from the set of codes to some target set which takes the same value at any two equivalent codes, holds great value in both cryptographic and classification problems.

2.2 Vector Rank Metric Codes and q-Systems

A *vector rank metric code* is an $\mathbb{F}_{q^s}$-subspace of $(\mathbb{F}_{q^s})^N$ equipped with the distance function

$$d(u, v) := \dim_{\mathbb{F}_q} \langle u_i - v_i : i \rangle_{\mathbb{F}_q}.$$

By picking an $\mathbb{F}_q$-basis for $\mathbb{F}_{q^s}$, we can identify such a code with an $\mathbb{F}_q$-subspace of $M_{s \times N}(\mathbb{F}_q)$, in which case the distance function becomes the rank of the difference of two matrices, as for matrix rank metric codes.

Two vector rank metric codes D, E are said to be *equivalent* if there exists an invertible matrix $Q \in M_N(\mathbb{F}_q)$ and a field automorphism ρ of $\mathbb{F}_{q^s}$ such that

$$E = \{v^\rho Q : v \in D\}.$$

It is much easier to determine the equivalence or otherwise of two vector rank metric codes; see for example [7].

A correspondence between vector rank metric codes and *q-systems* has been developed in recent years, leading to significant outcomes in both areas; see for example [2,14,18]. A q-system is an $\mathbb{F}_q$-subspace of an $\mathbb{F}_{q^s}$-vector space $(\mathbb{F}_{q^s})^k$, with equivalence defined under the natural action of $\Gamma L(k, \mathbb{F}_{q^s})$. Such objects have been studied for the past couple of decades in finite geometry through the *linear sets* they define in $\mathrm{PG}(k - 1, q^s)$. The correspondence is realised by taking a basis for D, creating a matrix G whose rows are this basis (a *generator matrix* for D), and associating to G the $\mathbb{F}_q$-subspace U generated by its columns. Conversely, choosing an $\mathbb{F}_q$-basis for U and forming a matrix G with these basis elements as its columns, we obtain a vector rank metric code as the $\mathbb{F}_{q^s}$-subspace generated by its rows. While some care should be taken around degeneracy (where the dimensions of either D or U may not equal k or N respectively), in general we can pass freely between these two settings.

Consider an $\mathbb{F}_q$-subspace C of $M_{\ell \times n}(\mathbb{F}_{q^s})$ of $\mathbb{F}_q$-dimension k. Then we can define a vector rank metric code D in $(\mathbb{F}_{q^s})^k$ of dimension at most ℓn as follows. We identify $M_{\ell \times n}(\mathbb{F}_{q^s})$ with $(\mathbb{F}_{q^s})^{\ell n}$ as $\mathbb{F}_q$-vector spaces, for example by identifying a matrix A with its vectorisation $\mathrm{vec}(A)$ in column-major order. Then the image of C is naturally a q-system, which defines by the above correspondence a vector rank metric code of length k and dimension at most ℓn.

2.3 Semifield Spread Sets

Let $\mathbb{S}$ be a presemifield with multiplication $\star$. Then the maps

$$L_x : y \mapsto x \star y$$
$$R_y : x \mapsto x \star y$$

are known as the *maps of left-* and *right-multiplication* respectively. Each are additive, and invertible for nonzero x, y.

Definition 1. *The spread set of a semifield $\mathbb{S}$ is denoted by $C(\mathbb{S})$ and defined as*

$$\{R_y : y \in \mathbb{S}\}.$$

It is well-known that there exist fields $\mathbb{F}_q$, $\mathbb{F}_{q^s}$, and a positive integer n such that $C(\mathbb{S})$ can be naturally embedded into $M_n(\mathbb{F}_{q^s})$ as an $\mathbb{F}_q$-subspace in which every nonzero element is invertible. We refer to [8] for a detailed exposition. In this way we see $C(\mathbb{S})$ as an $\mathbb{F}_q$-linear MRD code in $M_n(\mathbb{F}_{q^s})$.

3   Vector Rank-Metric Codes from Spread Sets

As outlined in the previous section, we may construct a matrix G from an $\mathbb{F}_q$-basis of $C(\mathbb{S})$ to view $C(\mathbb{S})$ as a q-system, and hence obtain an $\mathbb{F}_{q^s}$-linear vector rank metric code $D(\mathbb{S})$ of dimension at most n^2 and length ns. Choosing different bases for $C(\mathbb{S})$ leads to equivalent codes, so we abuse notation and omit this choice.

Lemma 1. *If two semifield spread sets $C(\mathbb{S})$ and $C(\mathbb{S}')$ are equivalent, then the vector rank-metric codes $D(\mathbb{S})$ and $D(\mathbb{S}')$ are equivalent.*

Proof. Suppose $C(\mathbb{S}') = XC(\mathbb{S})^\rho Y$ for some $X, Y \in M_n(\mathbb{F}_{q^s})$ invertible. Choose an $\mathbb{F}_q$-basis $\{A_1, \ldots, A_{ns}\}$ of $C(\mathbb{S})$, and form the $\mathbb{F}_q$-basis $\{XA_1^\rho Y, \ldots, XA_{ns}^\rho Y\}$ of $C(\mathbb{S}')$. Then $\mathrm{vec}(XA_i^\rho Y) = (Y^T \otimes X)\mathrm{vec}(A_i)^\rho$, and so the matrices $G(\mathbb{S})$ and $G(\mathbb{S}')$ satisfy $G(\mathbb{S}') = (Y^T \otimes X)G(\mathbb{S})^\rho$, giving that $D(\mathbb{S})$ and $D(\mathbb{S}')$ are equivalent.

Remark 1. Note that the converse is certainly not true in general; two inequivalent spread sets can certainly give rise to equivalent vector rank-metric codes. For example, in [17, Section 3.1] a number of inequivalent semifield spread sets are shown to have equivalent associated vector rank-metric codes (although the result is stated in terms of *linear sets*, which are projective counterparts to q-systems).

However we can utilise the contrapositive of this lemma to act as an invariant of semifields.

Corollary 1. *If the vector rank-metric codes $D(\mathbb{S})$ and $D(\mathbb{S}')$ arising from two semifields $\mathbb{S}$ and $\mathbb{S}'$ are inequivalent, then the semifield spread sets $C(\mathbb{S})$ and $C(\mathbb{S}')$ are inequivalent, and the semifields $\mathbb{S}$ and $\mathbb{S}'$ are not isotopic.*

This approach was in part used in [4,17] to assist in the classification of semifields with $n = 2, s = 2, 3$. There the methods were stated in terms of linear sets rather than q-systems, and certain geometric properties of the q-systems were used rather than the full equivalence of the vector rank-metric codes.

4   Embeddings of Matrix Spaces

It is well-known that we can view matrices over a large field as larger matrices over a subfield; that is, we can create the following embedding.

$$M_n(\mathbb{F}_{q^s}) \hookrightarrow M_{ns}(\mathbb{F}_q) \hookrightarrow M_{ns}(\mathbb{F}_{q^m}).$$

The first embedding can be achieved by replacing elements of $\mathbb{F}_{q^s}$ by their corresponding element of $M_s(\mathbb{F}_q)$ under a fixed regular representation, for example by mapping a primitive element θ to the companion matrix of its minimal polynomial. Concretely, let $\phi : \mathbb{F}_{q^s} \to M_s(\mathbb{F}_q)$ be such a regular representation. Then we define $\overline{\phi} : M_n(\mathbb{F}_{q^s}) \to M_{ns}(\mathbb{F}_q)$ by mapping the matrix $A = (a_{ij})$ to the block matrix $\overline{\phi}(A) := (\phi(a_{ij}))$. The second embedding is simply extending scalars to $\mathbb{F}_{q^m}$, where m is any positive integer.

Definition 2. *Let U be an $\mathbb{F}_q$-subspace of $M_{ns}(\mathbb{F}_q)$. We define the m-th embedded space*
$$E_m(U) = \langle U \rangle_{\mathbb{F}_{q^m}}.$$
For an $\mathbb{F}_q$-subspace C of $M_n(\mathbb{F}_{q^s})$, we define
$$E_m(C) := E_m(\overline{\phi}(C)).$$

Let $\mathbb{S}$ be a semifield of order q^{ns} with a nucleus of order a power of q^s, and $C(\mathbb{S}) \subset M_n(\mathbb{F}_{q^s})$ its spread set. The following is immediate from the definition of equivalence.

Theorem 1. *Suppose $\mathbb{S}$ and $\mathbb{S}'$ are isotopic. Then $E_m(C(\mathbb{S}))$ and $E_m(C(\mathbb{S}'))$ are equivalent for any m, and hence have the same rank distribution.*

In this way we can exploit ideas from (matrix) rank-metric codes which are not useful when studying $C(\mathbb{S})$ directly. The penalty is that we must consider a larger space, which obviously comes with a computational cost.

Definition 3. *Let $\mathbb{S}$ be a semifield of order q^{ns} with $C(\mathbb{S})$ an $\mathbb{F}_q$-subspace of $M_n(\mathbb{F}_{q^s})$, and let $m \in \mathbb{N}$. We define the m-ranks of $\mathbb{S}=$, $\mathrm{ranks}^{(m)}(\mathbb{S})$, as the multiset,*
$$\mathrm{rank}^{(m)}(\mathbb{S}) := \{\mathrm{rank}(A) : A \in E_m(C(\mathbb{S}))\}.$$

We note that the 1-ranks of a semifield is the same for all semifields, as each non-zero element of $E_1(\mathbb{S})$ has full rank. Hence $\mathrm{rank}^{(1)}(\mathbb{S}) = \{0^1, (ns)^{q^{ns}-1}\}$, where the exponents are the multiplicities.

Corollary 2. *If $\mathrm{rank}^{(m)}(\mathbb{S}) \neq \mathrm{rank}^{(m)}(\mathbb{S}')$ for some m, then the semifield spread sets $C(\mathbb{S})$ and $C(\mathbb{S}')$ are inequivalent, and the semifields $\mathbb{S}$ and $\mathbb{S}'$ are not isotopic.*

5 Classifying Semifields

We will use Corollaries 1 and 2 in order to speed up the classification of semifields of order q^{ns} with nucleus containing $\mathbb{F}_{q^s}$. We will assume an equivalence testing algorithm, for example as in [19]. Our advantage comes from reducing the number of times that this algorithm has to be performed.

Let us fix some θ such that $\mathbb{F}_{q^s} = \mathbb{F}_q(\theta)$. We form ns sets S_{i+js} of invertible matrices whose first row is $\theta^i e_j$ for $i = 0, \ldots, s-1$ and $j = 1, \ldots, n$. Note that any semifield spread set in $M_n(\mathbb{F}_{q^s})$ must contain a basis containing precisely one element from each S_k.

For an $\mathbb{F}_q$-subspace C of $M_n(\mathbb{F}_{q^s})$, we define $i(C) = (\mathrm{rank}^{(2)}(\overline{\phi}(C)), D(C))$, where $D(C)$ is a representative for the equivalence class of the vector rank-metric code obtained from C through the process outlined in Sect. 2.2.

We proceed iteratively as in [19]. Suppose we have fully classified up to equivalence all k-dimensional $\mathbb{F}_q$-subspaces of $M_n(\mathbb{F}_{q^s})$ spanned by elements of $S_1, \ldots, S_k$ in which every nonzero element is invertible; let T_k denote a set of representatives. Let T_{k+1} and i_{k+1} be the empty set.

- For $C' \in T_k$, $A \in S_{k+1}$, construct $C = \langle C', A \rangle_{\mathbb{F}_q}$. If every nonzero element of C is invertible, continue.
- Calculate $i(C)$.
  - If $i(C) \notin i_{k+1}$, add C to T_{k+1} and $i(C)$ to i_{k+1}.
  - If $i(C) \in i_{k+1}$, sequentially test if C is equivalent to C'' for all $C'' \in T_{k+1}$ such that $i(C'') = i(C)$. If each of these tests returns false, add C to T_{k+1}.

Repeat this until $k = ns$. We obtain T_{ns} a complete set of representatives for the equivalence classes of MRD codes in $M_n(\mathbb{F}_{q^s})$, and hence a complete set of representatives for the isotopy classes of semifields of order q^{ns} with right nucleus containing $\mathbb{F}_{q^s}$.

5.1   Classification for $q = 2$, $n = 2$, $s = 4$

We implemented the above procedure for $M_2(\mathbb{F}_{2^4})$.

Theorem 2. *The number of isotopy classes of finite semifields of order 2^8 with right nucleus containing $\mathbb{F}_{2^4}$, equivalently the number of equivalence classes of MRD codes in $M_2(\mathbb{F}_{2^4})$ of minimum distance 2, is 757.*

We outline now to what extent the invariants introduced here speed up the calculation.

Amongst the 757 classes there are

- 66 different 2-ranks;
- 17 different equivalence classes of vector rank-metric codes;
- 378 different pairs $i(C)$.

At the final step, there were 530873 spread sets. In order to classify them without utilising the invariants, we would need to calculate roughly 2×10^8 equivalence tests, assuming they are uniformly distributed. With a version of [19] tailored to this case, such a computation would take about 50 days.

Instead, we calculate 2×10^6 invariants, and then many orders of magnitude fewer equivalence tests.

A direct equivalence test takes approx 0.021 seconds; calculating the 2-ranks takes approx 0.08 seconds, and calculating the equivalence class of vector rank-metric codes takes approx 0.03 seconds. Overall, this step of the calculation is performed approximately 20 times faster using these invariants than without using them. As the invariants can be used at each step in creating T_k, we obtain significant time savings.

6   Divisibility Properties of Embedded Spaces

In this section we will demonstrate some divisibility properties of the ranks of embedded spaces. We extend the embedding from the previous section by extending scalars from $\mathbb{F}_q$ to $\mathbb{F}_{q^m}$ to $\mathbb{F}_{q^{\mathrm{lcm}(m,s)}}$ to obtain the following.

$$M_n(\mathbb{F}_{q^s}) \hookrightarrow M_{ns}(\mathbb{F}_q) \hookrightarrow M_{ns}(\mathbb{F}_{q^m}) \hookrightarrow M_{ns}(\mathbb{F}_{q^{\mathrm{lcm}(m,s)}}).$$

Let ϕ be the regular representation from the previous section. It is well-known that we can simultaneously diagonalise the elements of $\mathrm{im}(\phi)$ over $\mathbb{F}_{q^s}$. Explicitly, let Z be an invertible *Moore matrix*, that is,

$$Z = \begin{pmatrix} v_1 & v_2 & \cdots & v_s \\ v_1^\sigma & v_2^\sigma & \cdots & v_s^\sigma \\ \vdots & \vdots & \ddots & \vdots \\ v_1^{\sigma^{s-1}} & v_2^{\sigma^{s-1}} & \cdots & v_s^{\sigma^{s-1}} \end{pmatrix}$$

for some appropriately chosen $\mathbb{F}_q$-basis $\{v_1, \ldots, v_s\}$ of $\mathbb{F}_{q^s}$. Then

$$Z\phi(\alpha)Z^{-1} = \mathrm{diag}(\alpha, \alpha^q, \ldots, \alpha^{q^{s-1}}) =: D_{\alpha x}.$$

Note that $Z\phi(\alpha)Z^{-1}$ is the *Dickson matrix* corresponding to the *linearised polynomial* $f_\alpha(x) = \alpha x$. See for example [21] for details.

We can then perform a similar transformation for the image of $\overline{\phi}$ by defining $\overline{Z} = I_n \otimes Z$, and noting that

$$\overline{Z}\overline{\phi}(A)\overline{Z}^{-1} = (D_{a_{ij}x}).$$

We can then permute rows and columns using an appropriate permutation matrix P so that

$$(P\overline{Z})\overline{\phi}(A)(P\overline{Z})^{-1} = A \oplus A^\sigma \oplus \cdots \oplus A^{\sigma^{s-1}} =: \psi(A),$$

where A^{σ^i} is the matrix obtained by raising each entry of A to the power of q^i. Note that this resulting matrix $\psi(A)$ is an element of $M_{ns}(\mathbb{F}_{q^s})$.

Let C be any $\mathbb{F}_q$-subspace of $M_n(\mathbb{F}_{q^s})$. Define $\overline{\phi}(C)$ and $\psi(C)$ in the natural way. Then the following is straightforward.

Lemma 2. *For any $\mathbb{F}_q$-subspace C of $M_n(\mathbb{F}_{q^s})$, the following multiset equalities hold:*

$$\{s \cdot \mathrm{rank}(x) : x \in C\} = \{\mathrm{rank}(x) : x \in \overline{\phi}(C)\} = \{\mathrm{rank}(x) : x \in \psi(C)\}.$$

Proof. Since P and $\overline{Z}$ are invertible, it is clear that $\mathrm{rank}(\overline{\phi}(A)) = \mathrm{rank}(\psi(A))$. From the direct sum expression for $\psi(A)$, it is clear that $\mathrm{rank}(\psi(A)) = s \cdot \mathrm{rank}(A)$, since the operations $A \mapsto A^\sigma$ preserve the rank function.

We have defined new invariants for $\mathbb{F}_q$-subspaces of matrices by considering the $\mathbb{F}_{q^m}$-span of the elements of $\overline{\phi}(A)$. The previous discussion allows us to calculate and analyse the ranks of these elements of $M_{ns}(\mathbb{F}_{q^m})$ directly from elements of $M_n(\mathbb{F}_{q^{\mathrm{lcm}(m,s)}})$.

Lemma 3. *Let C be an $\mathbb{F}_q$-subspace of $M_n(\mathbb{F}_{q^s})$, with $\mathbb{F}_q$-basis $\{A_1, \ldots, A_k\}$. Let $\alpha_1, \ldots, \alpha_k \in \mathbb{F}_{q^m}$. Then*

$$\mathrm{rank}\left(\sum_{i=1}^k \alpha_i \overline{\phi}(A_i)\right) = \sum_{j=0}^{s-1} \mathrm{rank}\left(\sum_{i=1}^k \alpha_i A_i^{\sigma^j}\right) = \sum_{j=0}^{s-1} \mathrm{rank}\left(\sum_{i=1}^k \alpha_i^{q^{m-j}} A_i\right).$$

If m divides s, then

$$\mathrm{rank}\left(\sum_{i=1}^k \alpha_i \overline{\phi}(A_i)\right) = \frac{s}{m} \sum_{j=0}^{m-1} \mathrm{rank}\left(\sum_{i=1}^k \alpha_i^{q^j} A_i\right).$$

In particular, each element of the $\mathbb{F}_{q^m}$-span of $\overline{\phi}(C)$ has rank divisible by s/m.

Proof. The first equality holds simply by noting again that conjugating by $P\overline{Z}$ does not change the rank of a matrix, regardless of in which field the entries of the matrix lie. The second equality follows from the observation that for any $\alpha_i \in \mathbb{F}_{q^m}$ it holds that $(\sum \alpha_i^{q^{m-j}} A_i)^{\sigma^j} = \sum \alpha_i A_i^{\sigma^m}$, and hence the ranks of $\sum \alpha_i^{q^{m-j}} A_i$ and $\sum \alpha_i A_i^{\sigma^m}$ are equal. The final equation then follows from the fact that $\alpha_i^{q^{rm}} = \alpha_i$ for each i and each integer r, and a reindexing.

Example 1. In the computation of Sect. 5, the 2-ranks consist of even numbers, since $s/m = 4/2 = 2$. In fact, we observe that the ranks of the nonzero elements of $E_2(C(\mathbb{S}))$ are always $4, 6,$ or 8. There exist some semifields where $E_2(C(\mathbb{S}))$ has no elements of rank 6, and some where $E_2(C(\mathbb{S}))$ has no elements of rank 4, but none where all nonzero elements have rank 8.

Disclosure of Interests. The authors have no competing interests to declare that are relevant to the content of this article.

References

1. Albert, A.A.: Finite division algebras and finite planes. In: Proc. Sympos. Appl. Math., vol. 10, pp. 53–70. Amer. Math. Soc., Providence, RI (1960)
2. Alfarano, G.N., Borello, M., Neri, A., Ravagnani, A.: Linear cutting blocking sets and minimal codes in the rank metric. Journal of Combinatorial Theory, Series A **192**, 105658 (2022)
3. Blokhuis, A., Lavrauw, M., Ball, S.: On the classification of semifield flocks. Adv. Math. **180**(1), 104–111 (2003). https://doi.org/10.1016/S0001-8708(02)00084-1
4. Cardinali, I., Polverino, O., Trombetti, R.: Semifield planes of order q^4 with kernel F_{q^2} and center F_q. European J. Combin. **27**(6), 940–961 (2006). https://doi.org/10.1016/j.ejc.2005.04.005

5. Combarro, E.F., Rúa, I.F., Ranilla, J.: New advances in the computational exploration of semifields. Int. J. Comput. Math. **88**(9), 1990–2000 (2011). https://doi.org/10.1080/00207160.2010.548518
6. Combarro, E.F., Rúa, I.F., Ranilla, J.: Finite semifields with 7^4 elements. Int. J. Comput. Math. **89**(13–14), 1865–1878 (2012). https://doi.org/10.1080/00207160.2012.688113
7. Couvreur, A., Debris-Alazard, T., Gaborit, P.: On the hardness of code equivalence problems in rank metric. arXiv preprint arXiv:2011.04611 (2020)
8. de la Cruz, J., Kiermaier, M., Wassermann, A., Willems, W.: Algebraic structures of MRD codes. Adv. Math. Commun. **10**(3), 499–510 (2016). https://doi.org/10.3934/amc.2016021
9. Dickson, L.E.: On commutative linear algebras in which division is always uniquely possible. Trans. Amer. Math. Soc. **7**(4), 514–522 (1906). https://doi.org/10.2307/1986243
10. Johnson, N.L., Marino, G., Polverino, O., Trombetti, R.: Semifields of order q^6 with left nucleus $\mathbb{F}_{q^3}$ and center $\mathbb{F}_q$. Finite Fields Appl. **14**(2), 456–469 (2008). https://doi.org/10.1016/j.ffa.2007.04.005
11. Knuth, D.E.: Finite semifields and projective planes. J. Algebra **2**, 182–217 (1965). https://doi.org/10.1016/0021-8693(65)90018-9
12. Lavrauw, M., Polverino, O.: Finite semifields. In: Storme, L., De Beule, J. (eds.) Current research topics in Galois geometry, pp. 127–155. Mathematics Research Developments, Nova Science (2011)
13. Lavrauw, M., Sheekey, J.: Symplectic 4-dimensional semifields of order 8^4 and 9^4. Des. Codes Cryptogr. **91**(5), 1935–1949 (2023). https://doi.org/10.1007/s10623-023-01183-y
14. Marino, G., Neri, A., Trombetti, R.: Evasive subspaces, generalized rank weights and near mrd codes. Discret. Math. **346**(12), 113605 (2023)
15. Marino, G., Pepe, V.: On symplectic semifield spreads of $PG(5, q^2)$, q odd. Forum Math. **30**(2), 497–512 (2018). https://doi.org/10.1515/forum-2016-0133
16. Marino, G., Polverino, O.: On the nuclei of a finite semifield. In: Theory and applications of finite fields, Contemp. Math., vol. 579, pp. 123–141. Amer. Math. Soc., Providence, RI (2012). https://doi.org/10.1090/conm/579/11525
17. Marino, G., Polverino, O., Trombetti, R.: On $\mathbb{F}_q$-linear sets of $PG(3, q^3)$ and semifields. J. Combin. Theory Ser. A **114**(5), 769–788 (2007). https://doi.org/10.1016/j.jcta.2006.08.012
18. Randrianarisoa, T.H.: A geometric approach to rank metric codes and a classification of constant weight codes. Des. Codes Crypt. **88**, 1331–1348 (2020)
19. Rúa, I.F., Combarro, E.F., Ranilla, J.: Classification of semifields of order 64. J. Algebra **322**(11), 4011–4029 (2009). https://doi.org/10.1016/j.jalgebra.2009.02.020
20. Rúa, I.F., Combarro, E.F., Ranilla, J.: Determination of division algebras with 243 elements. Finite Fields Appl. **18**(6), 1148–1155 (2012). https://doi.org/10.1016/j.ffa.2012.07.001
21. Sheekey, J.: MRD codes: constructions and connections. In: Combinatorics and Finite Fields—difference Sets, Polynomials, Pseudorandomness and Applications, Radon Ser. Comput. Appl. Math., vol. 23, pp. 255–285. De Gruyter, Berlin (2019). https://doi.org/10.1515/9783110642094-013

A Survey on Code Equivalence: The State-of-the-Art and Open Questions

Anna-Lena Horlemann[1] ⓘ, Abhinaba Mazumder[2] ⓘ, Michael Schaller[2], and Violetta Weger[3(✉)] ⓘ

[1] University of St. Gallen, St. Gallen, Switzerland
`anna-lena.horlemann@unisg.ch`
[2] University of Zurich, Zürich, Switzerland
`{abhinaba.mazumder,michael.schaller}@math.uzh.ch`
[3] Technical University of Munich, Munich, Germany
`violetta.weger@tum.de`

Abstract. In this work, we provide a comprehensive survey of the code equivalence problem and its variants. We explain the existing results, highlighting the relationships between different problem formulations, algorithmic techniques, and hardness assumptions. In addition, we systematically review known attacks, identify the parameter regimes in which they are effective, and discuss their limitations. Lastly, we outline several open problems and research directions, with the aim of clarifying the current landscape and guiding future work toward a deeper understanding of the hardness of code equivalence.

Keywords: Coding Theory · Code-based Cryptography · Code Equivalence

1 Introduction

The code equivalence problem asks whether two linear codes are equivalent under a Hamming-metric isometry. Two prominent variants are commonly studied: the Permutation Equivalence Problem (PEP), where the isometry is restricted to permutations, and the Linear Equivalence Problem (LEP), where monomial transformations are allowed.

In recent years, these problems have attracted renewed attention due to their role as hardness assumptions in post-quantum cryptography, where they underpin several cryptographic constructions [2,5,8,11,17,31]. At the same time, a wide range of algorithmic techniques has been developed to attack them, drawing from diverse areas such as combinatorics, algebra, and graph theory. These include approaches based on low-weight codewords [15,20], hull properties [42], graph isomorphism reductions [7], canonical forms [23,37], algebraic methods [40], and Schur product techniques [10,12]. A good reference for code equivalence is the thesis of Saeed [40], where several concepts, such as hull, puncturing, and guessing techniques are explained.

L. Batina and F. Özbudak (Eds.): WAIFI 2026, LNCS 16611, pp. 12–31, 2026.
https://doi.org/10.1007/978-3-032-27574-5_2

Despite this substantial amount of work, our current understanding of the hardness of code equivalence remains incomplete. Different techniques apply in different parameter regimes, often with little overlap, and the resulting picture is far from unified. For instance, several families of weak instances for PEP are known, leading to efficient (often average-case polynomial-time) algorithms [7,12,42]. In contrast, for LEP the situation appears different: apart from specific small-field cases [44], and small dimension cases [10], no broad class of weak instances is known, and generic instances over larger fields are widely conjectured to be hard. This disparity, together with the diversity of existing techniques, raises several fundamental questions. Which variants of code equivalence should be considered secure for cryptographic purposes? How do the different algorithmic approaches compare across parameter regimes? To what extent can hardness results or attacks be transferred between PEP, LEP, and other related problems? At present, answers to these questions are scattered across the literature, making it difficult to obtain a clear and coherent view of the field.

In this survey we address these questions by systematically reviewing existing results and presenting a unified picture of the different variants of code equivalence, their relationships, and their complexity. We conclude with a discussion of several relatively unexplored approaches and a presentation of open problems and questions we have encountered.

2   Preliminaries

We denote by $\mathbb{F}_q$ the finite field with q elements, where q is a prime power and denote by $\mathbb{F}_q^*$ its multiplicative group, i.e., $\mathbb{F}_q \setminus \{0\}$. The identity matrix of size k is denoted by Id_k. By $\mathrm{diag}(d_1,\ldots,d_n)$ we denote the diagonal matrix with diagonal entries $d_1,\ldots,d_n$. Sets are denoted by upper case letters and for a set $S \subseteq I$, we denote by $|S|$ its cardinality and by S^C its complement, i.e., $I \setminus S$. By $\mathrm{GL}_n(\mathbb{F}_q)$ we denote the $n \times n$ invertible matrices over $\mathbb{F}_q$. Finally we denote by S_n the symmetric group on n elements, i.e., the group of permutations of $\{1,\ldots,n\}$. By abuse of notation we will also say that a $n \times n$ permutation matrix corresponding to the permutation $\sigma \in S_n$ is in S_n. For a vector $x \in \mathbb{F}_q^n$ and a set $S \subset \{1,\ldots,n\}$ we denote by x_S the vector consisting of the entries of x indexed by S. Similarly, for a matrix $A \in \mathbb{F}_q^{m \times n}$ and a subset $S \subseteq \{1,\ldots,n\}$ we denote by A_S the matrix consisting of the columns of A indexed by S. Moreover, we denote by $\langle x, y \rangle := \sum_{i=1}^{n} x_i y_i$ the standard inner product. We write $\mathrm{poly}(x)$ for a function that has polynomial growth in x^1.

Definition 1. *Let $1 \leq k \leq n$ be integers. An $[n,k]_q$ linear code $\mathcal{C}$ is a k-dimensional linear subspace of $\mathbb{F}_q^n$. The parameter n is called the* length *of the code, the elements in the code are called* codewords *and $R = k/n$ is called the* rate *of the code. A matrix $G \in \mathbb{F}_q^{k \times n}$ is called a* generator matrix *of an $[n,k]_q$ code $\mathcal{C}$ if $\mathcal{C} = \left\{ xG \mid x \in \mathbb{F}_q^k \right\}$.*

---

[1] where we allow x to be a vector of several variables.

We will often write $\langle G \rangle$ to denote the code generated by G. Similarly, for $x_1, \ldots, x_k \in \mathbb{F}_q^n$, $\langle x_1, \ldots, x_k \rangle$ denotes the subspace generated by these vectors. An $[n, k]_q$ linear code $\mathcal{C}$ has *information set* $I \subset \{1, \ldots, n\}$ of size k, if $|\mathcal{C}| = |\mathcal{C}_I|$, for $\mathcal{C}_I = \{c_I \mid c \in \mathcal{C}\}$. As a consequence, G_I is an invertible $k \times k$ matrix. For any generator matrix $G \in \mathbb{F}_q^{k \times n}$ there exist some $n \times n$ permutation matrix P and some invertible matrix $U \in \mathbb{F}_q^{k \times k}$ that bring G in *systematic form*, i.e., $UGP = \left(\mathrm{Id}_k \; A \right)$, where $A \in \mathbb{F}_q^{k \times (n-k)}$.

Definition 2. *Let $k \leq n$ be positive integers and let $\mathcal{C}$ be an $[n, k]_q$ linear code. The* dual code $\mathcal{C}^\perp$ *is an $[n, n-k]_q$ linear code, defined as*

$$\mathcal{C}^\perp := \{x \in \mathbb{F}_q^n \mid \langle x, y \rangle = 0 \; \forall \; y \in \mathcal{C}\}.$$

A matrix $H \in \mathbb{F}_q^{(n-k) \times n}$ is called a parity-check matrix *of $\mathcal{C}$, if it is a generator matrix of $\mathcal{C}^\perp$.*

Definition 3. *Let $\mathcal{C} \subseteq \mathbb{F}_q^n$ be a linear code. Then the* hull *of $\mathcal{C}$ is defined as $\mathcal{H}(\mathcal{C}) := \mathcal{C} \cap \mathcal{C}^\perp$.*

Note that the dimension of $\mathcal{H}(\mathcal{C})$ is given by $k - \mathrm{rk}(GG^\top)$. We have two extreme cases, which will heavily impact the hardness of code equivalence: If the hull is trivial, i.e., $\mathcal{H}(\mathcal{C}) = \{0\}$, then $\mathcal{C}$ is called a *Linear Complementary Dual* (LCD) code. On the other hand, if $\mathcal{H}(\mathcal{C}) = \mathcal{C}$, i.e., $\mathcal{C} \subseteq \mathcal{C}^\perp$ then $\mathcal{C}$ is called a *self-orthogonal* code. As the next result shows, for large enough q and n, we expect random codes to have a trivial hull.

Theorem 4 [43]. *Let q be a prime power and $k \leq n$ be positive integers. Let $\mathcal{C}$ be a random $[n, k]_q$ linear code. Then $\mathcal{H}(\mathcal{C}) = \{0\}$ with probability $\geq 1 - 1/q - 1/q^2$, for n growing.*

Several metrics can be used in coding-theoretic applications, however, in this work we will focus on the Hamming metric.

Definition 5. *Let n be a positive integer. For $x \in \mathbb{F}_q^n$, the support of x is given by*

$$supp(x) = \{i \in \{1, \ldots, n\} \mid x_i \neq 0\}.$$

The Hamming weight *of x is given by the size of its support, i.e., $wt_H(x) := |supp(x)|$. For $x, y \in \mathbb{F}_q^n$, the* Hamming distance *between x and y is given by the number of positions in which they differ, i.e., $d_H(x, y) := |\{i \in \{1, \ldots, n\} \mid x_i \neq y_i\}|$. The* minimum Hamming distance *of $\mathcal{C} \subseteq \mathbb{F}_q^n$ is denoted by $d_H(\mathcal{C})$ and given by*

$$d_H(\mathcal{C}) := \min\{d_H(x, y) \mid x, y \in \mathcal{C}, \; x \neq y\}.$$

If the minimum distance d of an $[n, k]_q$ linear code $\mathcal{C}$ is known, we write that $\mathcal{C}$ is an $[n, k, d]_q$ linear code. One can generalize the idea of minimum distance to the weight enumerator and weight distribution of a code:

Definition 6. *Let $C \subseteq \mathbb{F}_q^n$ be a linear code. For any $w \in \{0, \ldots, n\}$, let us denote by $A_w(C) := |\{c \in C \mid wt_H(c) = w\}|$ the* weight enumerator *of C. We further denote by $(A_0(C), \ldots, A_n(C))$ the* weight distribution *of C.*

Another generalization is the so-called generalized weight. For this, we need to introduce the support of a code:

Definition 7. *Let C be an $[n, k]_q$ linear code. The* support *of C is defined as*

$$supp_H(C) := \bigcup_{c \in C} supp(c).$$

The weight *of the code C is given by $wt_H(C) := |supp_H(C)|$.[2] Let $r \in \{1, \ldots, k\}$. The rth generalized weight of C is given by $d_r(C) := \min\{wt_H(\mathcal{D}) \mid \mathcal{D} \subset C, dim(\mathcal{D}) = r\}$.*

Note that the first generalized weight $d_1(C)$ is the minimum distance $d_H(C)$.

On several occasions, it is beneficial to project the code to some coordinates, i.e., to *puncture* the code in the other indices:

Definition 8. *Let C be an $[n, k, d]_q$ linear code and $S \subset \{1, \ldots, n\}$ be a subset of size s. The* punctured code *in S is $\mathcal{P}(C, S) := \{(c_i)_{i \notin S} \mid c \in C\}$. If we consider the subcode $C(S) = \{c \in C \mid c_i = 0 \text{ for all } i \in S\}$, then the* shortened code *in S is given by $\mathcal{S}(C, S) := \mathcal{P}(C(S), S) = C(S)_{S^c}$.*

Let us denote by $*$ the componentwise product or *Schur product* between x and y, i.e., $x * y = (x_1 y_1, \ldots, x_n y_n)$ for $x, y \in \mathbb{F}_q^n$. The Schur product has many algebraic properties, in particular, it is symmetric and bilinear.

Definition 9. *Let C_1 be an $[n, k_1]_q$ linear code and C_2 be an $[n, k_2]_q$ linear code. The* Schur product code *of C_1 and C_2 is defined as*

$$C_1 * C_2 := \langle\{c_1 * c_2 \mid c_1 \in C_1, c_2 \in C_2\}\rangle.$$

The ℓth power code *of C_1 is defined as*

$$C_1^{(\ell)} = \langle\{c_1 * \cdots * c_\ell \mid c_1, \ldots, c_\ell \in C_1\}\rangle.$$

For $\ell = 2$ we call this the square code *of C_1.*

Lastly we need the notion of (Hamming) isometries to be able to define code equivalence later on. A map $\varphi : \mathbb{F}_q^n \to \mathbb{F}_q^n$ is called an *isometry* if for all $x, y \in \mathbb{F}_q^n$ we have $d_H(x, y) = d_H(\varphi(x), \varphi(y))$.

Proposition 10 ([38], **Corollary 1.5.14**). *The linear isometries of the Hamming metric consist of monomial transformations, i.e., $(\mathbb{F}_q^\star)^n \rtimes \mathcal{S}_n =: M_{n,q}$ acting on $\mathbb{F}_q^n$ via $((d_1, \ldots, d_n), \sigma) \cdot (x_1, \ldots, x_n) = (d_1 x_{\sigma^{-1}(1)}, \ldots, d_n x_{\sigma^{-1}(n)})$.*

---

[2] Clearly, for a non-degenerate code, the support will be full, i.e., $\{1, \ldots, n\}$, however, as soon as we go to subcodes of C, this will change and therefore become of interest.

3 Code Equivalence

We will now define several types of code equivalence and the corresponding computational problems.

Definition 11. *We say that two codes $C_1, C_2 \subseteq \mathbb{F}_q^n$ are* linearly equivalent, *if there exists a map $\varphi \in (\mathbb{F}_q^\star)^n \rtimes S_n$, such that $\varphi(C_1) = C_2$.*

We say that two codes $C_1, C_2 \subseteq \mathbb{F}_q^n$ are permutation equivalent, *if there exists a permutation of indices, which transforms C_1 into C_2, that is, there exists $\sigma \in S_n$, such that $\sigma(C_1) = C_2$.*

Permutation equivalence is thus a special instance of the more general linear equivalence. We remark that an even more general notion is the one of *semilinear* equivalence, where also field automorphisms are taken into account. However, since the number of such field automorphisms is small (indeed $|\mathrm{Aut}(\mathbb{F}_q)| = \log_p(q)$, where p is the characteristic of $\mathbb{F}_q$), these do not play a role from a cryptographic point of view, and we will hence not consider this variant.

Naturally, we can express code equivalence in terms of matrix operations. For this let G_1, G_2 be the generator matrices of two $[n, k]_q$ linear codes C_1 and C_2. Then there exists a $\varphi = (d, \sigma) \in (\mathbb{F}_q^\star)^n \rtimes S_n$ such that $\varphi(C_1) = C_2$ if and only if there exist a $S \in \mathrm{GL}_k(\mathbb{F}_q)$, $D = \mathrm{diag}(d)$ and $P \in S_n$ such that $G_2 = SG_1DP$.

Both types of equivalence behave well under duality, as the following shows.

Proposition 12. *If C_1 and C_2 are permutation equivalent via $\sigma \in S_n$ then $\sigma(C_1^\perp) = C_2^\perp$. On the other hand, if C_1 and C_2 are linearly equivalent via $\varphi = ((d_1, \ldots, d_n), \sigma)$ then $\varphi' = ((d_1^{-1}, \ldots, d_n^{-1}), \sigma)$ is such that $\varphi'(C_1^\perp) = C_2^\perp$.*

A special case of equivalence arises when a code is equivalent to itself; in this case, we speak of an automorphism of the code:

Definition 13. *Let C be an $[n, k]_q$ linear code. The (monomial)* automorphism *group of C is given by the linear isometries[3] that map C to C:*

$$\mathrm{Aut}(C) := \{\varphi \in (\mathbb{F}_q^\star)^n \rtimes S_n \mid \varphi(C) = C\}.$$

We remark that the automorphism group of a binary random linear code is with high probability trivial [34], i.e., $\mathrm{Aut}(C) = \{\mathrm{id}\}$. While it is believed that for $q > 2$, random codes have automorphism group $\{\lambda\,\mathrm{id} \mid \lambda \in \mathbb{F}_q^\star\}$, a proof is still missing in the literature.

Remark 14. *There are several parameters or properties of equivalent codes which remain invariant under an application of a linear isometry, e.g., the length, dimension and minimum distance. Moreover, the weight enumerator, the generalized weights, and the automorphism group are invariants.[4]*

---

[3] Note that *automorphism* sometimes also refers to any invertible linear map from $\mathbb{F}_q^n \to \mathbb{F}_q^n$ such that $\varphi(C) = C$; however, we will assume that it is an element of $M_{n,q}$.

[4] However, if two codes share the same weight enumerator or the same generalized weights, they are not necessarily equivalent, as can be seen with two Reed-Solomon codes with different evaluation points.

From Proposition 12, we also have that the hull is an invariant of permutation equivalence, i.e., if $\sigma(\mathcal{C}) = \mathcal{C}'$ for some $\sigma \in S_n$, then $\sigma(\mathcal{H}(\mathcal{C})) = \mathcal{H}(\mathcal{C}')$.

Although they are not invariants, power codes also can serve as a tool to check for equivalence, since it was shown in [12] that the power codes of two linearly equivalent codes remain linearly equivalent.

Lemma 1. *If $\mathcal{C}, \mathcal{C}'$ are two $[n,k]_q$ linear codes and $\varphi = ((d_1, \ldots, d_n), \sigma) \in M_{n,q}$ is such that $\varphi(\mathcal{C}) = \mathcal{C}'$, then $\varphi^{(\ell)} = ((d_1^\ell, \ldots, d_n^\ell), \sigma) \in M_{n,q}$ is such that $\varphi^{(\ell)}(\mathcal{C}^{(\ell)}) = \mathcal{C}'^{(\ell)}$.*

Code Equivalence Problems

We now define the various code equivalence problems that appear in cryptographic contexts. For this note that for complexity classifications the decisional versions are used, while for actual cryptosystems we usually turn to the computational versions of these problems. We will now formulate the computational ones, where we start with the Linear and the Permutation Equivalence Problems (LEP and PEP), according to the two types of equivalence from Definition 11.

Problem 1 (LEP). Given two $[n,k]_q$ linear codes $\mathcal{C}, \mathcal{C}'$ find (if it exists) a $\varphi \in (\mathbb{F}_q^\star)^n \rtimes S_n$ such that $\varphi(\mathcal{C}) = \mathcal{C}'$.

Problem 2 (PEP). Given two $[n,k]_q$ linear codes $\mathcal{C}, \mathcal{C}'$ find (if it exists) a $\sigma \in S_n$ such that $\sigma(\mathcal{C}) = \mathcal{C}'$.

As before, PEP is a special instance of LEP, with the scalars all being equal to one. In a similar spirit one could also allow the scalars to be arbitrary, but fix the permutation to be the identity, which would result in the following question: given two generator matrices $G, G' \in \mathbb{F}_q^{k \times n}$ find $d \in (\mathbb{F}_q^\star)^n$ and $S \in \mathrm{GL}_k(\mathbb{F}_q)$ such that $SG\mathrm{diag}(d) = G'$. However, this problem can easily be solved by computing the reduced row echelon forms of both generator matrices and then solving a system of $k(n-k)$ linear equations in $n-k$ variables. Therefore, we do not consider this variant in cryptographic applications.

While PEP and LEP are the main problems considered in the literature, there are several variants, which are captured by the following problem [10]:

Problem 3 (LEP(U)). Let $U \subseteq \mathbb{F}_q^\star$ be a multiplicative subgroup. Given two $[n,k]_q$ linear codes $\mathcal{C}$ and $\mathcal{C}'$ find (if it exists) an isometry $\varphi = (d, \sigma) \in U^n \rtimes S_n$ such that $\varphi(\mathcal{C}) = \mathcal{C}'$.

By setting $U = \mathbb{F}_q^\star$, we recover LEP, for $U = \{1\}$, we get PEP and for $U = \{\pm 1\}$, we get the Signed Permutation Equivalence Problem (SPEP). Note that in the Lee metric, the linear isometries are exactly the signed permutations. For $q = 2$, LEP is PEP, and for $q = 3$, LEP is SPEP.

Another generalization of PEP considers subcodes. It is also known as *Permuted Kernel Problem* (PKP) (see e.g. [41]), and is used in the signature scheme PERK [1]:

Problem 4 (PKP). Let $0 < k' \leq k \leq n$. Given an $[n,k]_q$ code $\mathcal{C}$ and an $[n,k']_q$ code $\mathcal{C}'$, find (if it exists) a $\sigma \in S_n$, such that $\mathcal{C}' \subseteq \sigma(\mathcal{C})$.

While it was shown in [39] that the decisional version of PKP is NP-hard, the (decisional) variants of the code equivalence problems generalized by LEP(U) lie in the complexity class AM∩coAM [39]. As for other isomorphism-type problems (e.g. graph isomorphism), this suggests that they are unlikely to be NP-hard, since NP-hardness would imply a collapse of the polynomial hierarchy. While this is shown in [39] for LEP, the argument extends directly to LEP(U) for any multiplicative subgroup $U \subseteq \mathbb{F}_q^\star$. Despite this, no efficient algorithm is known for solving code equivalence in general.

4   Reductions to Other Problems

One way of solving any of the code equivalence problems is to reduce it to another problem for which a (efficient) solving algorithm is known. In this section we give an overview of various such reductions.

4.1   Linear Equivalence to Permutation Equivalence

There exists a very elegant reduction from linear equivalence to permutation equivalence [43]. For this the authors introduce the *closure* of a code. We describe a generalization from [10].

Definition 15. *Let $U \subseteq \mathbb{F}_q^\star$ be a multiplicative subgroup. Let $\mathcal{C}$ be an $[n,k]_q$ linear code and $\alpha \in \mathbb{F}_q$ be a primitive element. Let $r = |U|$ such that $r \mid (q-1)$. Then U is the subgroup generated by $\alpha^{(q-1)/r}$. Define $\lambda = (1, \alpha^{(q-1)/r}, \ldots, \alpha^{(r-1)(q-1)/r})$. The r-th partial closure of $\mathcal{C}$ is given by the Kronecker product $\lambda \otimes \mathcal{C}$. In the case $U = \mathbb{F}_q^\star$ we just speak of the* closure.

The new code is now of length nr and still of dimension k. In fact, if G is a generator matrix of $\mathcal{C}$, then $\lambda \otimes G$ is a generator matrix of $\lambda \otimes \mathcal{C}$ [43].

Proposition 16 ([10]). *Let $\mathcal{C}_1, \mathcal{C}_2$ be two $[n,k]_q$ linear codes. Then $\mathcal{C}_1$ is equivalent in the LEP(U)-sense to $\mathcal{C}_2$ if and only if $\lambda \otimes \mathcal{C}_1$ is permutation equivalent to $\lambda \otimes \mathcal{C}_2$. In particular we get a reduction from LEP(U) to PEP in time* poly$(n, |U|)$.

This effectively transforms LEP(U) into a PEP instance of length rn. The first case of such a reduction in case $U = \mathbb{F}_q^\star$ was done in [43]. Also the special case $U = \{\pm 1\}$, i.e., the reduction from SPEP to PEP has already been done before in [25]. However, the next proposition which is a generalization of [43] shows that the partial closure is usually self-orthogonal.

Proposition 17. *If $|U| > 2$ then the partial closure $\lambda \otimes \mathcal{C}$ is self-orthogonal. In particular, for $q \geq 4$ the closure is self-orthogonal.*

This result relies on the fact that $\sum_{\beta \in U} \beta^2 = 0$ if squaring defines a nontrivial character.

4.2   Permutation Equivalence and Graph Isomorphism

For this reduction, let us first recall some graph theory. A graph $\mathcal{G}$ consists of vertices V and edges E between the vertices, i.e., $E \subset V \times V$. We will focus on undirected graphs, thus whenever $(u,v) \in E$ also $(v,u) \in E$ and we label the edges with a weight $w(u,v)$. We say that two weighted graphs $\mathcal{G} = (V,E)$ and $\mathcal{G}' = (V',E')$ are isomorphic, if there exists a bijective map $f : V \to V'$ with $(u,v) \in E \leftrightarrow (f(u),f(v)) \in E'$ and $w(u,v) = w(f(u),f(v))$. Thus, we can focus on $V = V' = \{1,\ldots,n\}$ and maps $f = \sigma \in \mathcal{S}_n$.

Problem 5 (Weighted Graph Isomorphism Problem). Given $\mathcal{G},\mathcal{G}'$, find (if it exists) isomorphism $\sigma \in \mathcal{S}_n$ mapping $\mathcal{G}$ to $\mathcal{G}'$.

There are several reductions from graph isomorphism to permutation equivalence or linear equivalence including [13,29,32,39]. However, from here on we are interested in the converse direction.

The *adjacency matrix* of a weighted graph $\mathcal{G}$ is defined as the $n \times n$ matrix A with entries $A_{i,j} = w(i,j)$ if $(i,j) \in E$ and 0 else. Note that two graphs $\mathcal{G}, \mathcal{G}'$ are isomorphic if and only if there exists a permutation matrix P such that $P^\top A P = A'$.

The GI problem has been shown to be quasi-polynomial time by Babai in 2017 [3], that is we can solve GI in time $2^{o(\log(n)^c)}$ for some constant c. This makes the following randomized reduction [7] very interesting: For $\mathcal{C}$ with trivial hull, we define

$$A(G) = G^\top (GG^\top)^{-1} G.$$

Clearly, this matrix can only exist if $GG^\top$ is invertible, i.e., if the hull of $\mathcal{C}$ is trivial. The matrix $A(G) \in \mathbb{F}_q^{n \times n}$ is symmetric, $\langle A(G)\rangle = \mathcal{C}$ and, moreover, it is independent of the choice of G, as $A(SG) = A(G)$ for $S \in \mathrm{GL}_k(\mathbb{F}_q)$.

Theorem 18 [7]. *PEP can be reduced to weighted GI, for codes with trivial hull.*

This reduction of PEP to GI can easily be extended to SPEP. In fact, if $G' = SGDP$, for some $S \in \mathrm{GL}_k(\mathbb{F}_q)$, permutation matrix P and $D = \mathrm{diag}(d)$ and $d \in \{\pm 1\}^n$, then $A(G') = P^\top D A(G) D P$, which is another version of GI, called signed graph isomorphism, which is GI-complete and hence the solver of Babai can also be used to solve SPEP, whenever $GG^\top$ is invertible.

For LEP it would be natural to apply the reduction via the closure to PEP and then to GI. However, this does not work for $q > 3$, due to Proposition 17.

The case $q = 4$ can be handled using the *Hermitian inner product* instead of the standard one [43]. This product exists in general when $q = p^{2m}$ and is defined as $\langle x,y \rangle_H := \sum_{i=1}^n x_i y_i^{p^m}$ for any $x,y \in \mathbb{F}_q^n$. This form is $\mathbb{F}_{p^m}$-linear in each component and satisfies $\langle x,y \rangle_H = (\langle y,x \rangle_H)^{p^m}$.

Hence, we define the Hermitian dual by $\mathcal{C}^\star := \{x \in \mathbb{F}_q^n \mid \langle x,y \rangle_H = 0 \text{ for all } y \in \mathcal{C}\}$. Let $G^{[p^m]}, H^{[p^m]}$ be the matrices given by raising each element to the p^m-th power. If $\mathcal{C} = \langle G \rangle = \ker(H^\top)$, then $\mathcal{C}^\star = \langle H^{[p^m]} \rangle = \ker((G^{[p^m]})^\top)$

and $(\mathcal{C}^\star)^\star = \mathcal{C}$. The *Hermitian hull* is defined as $\mathcal{H}^\star(\mathcal{C}) := \mathcal{C} \cap \mathcal{C}^\star$. If the Hermitian hull is trivial, we can define

$$A^\star = (G^{[p^m]})^\top \left(G(G^{[p^m]})^\top\right)^{-1} G,$$

which is again independent of the generator matrix. As before, random codes have trivial Hermitian hull with high probability [43]. It remains to check, whether closures also have trivial hulls: Since

$$(\lambda \otimes G)((\lambda \otimes G)^{[p^m]})^\top = \lambda(\lambda^{[p^m]})^\top G(G^{[p^m]})^\top,$$

we are left with computing $\lambda(\lambda^{p^m})^\top = \sum_{\beta \in \mathbb{F}_q^*} \beta^{p^m+1}$, which is only non-zero if $(p^{2m} - 1) \mid (p^m + 1)$, which implies $p = 2, m = 1$, i.e., $q = 4$. This shows that this approach cannot be generalized to larger q.

4.3   Search to Decision

While it is easy to see that the decisional version of any code equivalence problem can be reduced to the search (or computational) version, the other direction is not obvious. In [16], the authors prove a search-to-decision reduction for PEP:[5]

Suppose we are given an oracle that decides whether two linear codes are permutation equivalent. The reduction recovers the secret permutation by iteratively determining the image of each coordinate. More precisely, let $\mathcal{C}, \mathcal{C}'$ be two codes and $\sigma \in S_n$ be such that $\sigma(\mathcal{C}) = \mathcal{C}'$. For an index i the reduction tests whether $\sigma(i) = j$ by appending multiple copies of selected columns to the generator matrices of $\mathcal{C}$ and $\mathcal{C}'$. These repetitions enforce that any permutation on the augmented codes must have $\sigma(i) = j$. Querying the oracle on the augmented codes therefore decides whether $\sigma(i) = j$. By repeating this procedure a polynomial number of times, one recovers σ.

A big open question is whether such an (efficient) oracle exists, e.g., an invariant which is characterizing and efficient to compute.

4.4   SPEP to LIP and Additional Reductions

Problem 6 (Lattice Isomorphism Problem (LIP)). Given two lattices Λ_1, Λ_2, find (if it exists) an orthogonal matrix O such that $\Lambda_1 = \Lambda_2 O$.

It is well known that this can be rephrased in terms of quadratic forms. In [14] the authors prove a reduction from Signed Permutation Equivalence for prime fields with cardinality p (SPEP$_p$) to Lattice Isomorphism Problem (LIP).

Theorem 19 ([14]). *Let p be a prime and $n \in \mathbb{Z}_{>0}$. There is a $\mathrm{poly}(n, \log(p))$ reduction from* SPEP$_p$ *on codes with length n and minimum distance at least $p^2 + 1$ to LIP on lattices of rank n.*

---

[5] For LEP this is still an open question.

The basic idea is as follows: From a code $\mathcal{C}$ we can use the Construction A lattice $\mathcal{C}+p\mathbb{Z}^n$ to get a lattice. One can show that the minimum distance requirement ensures that the shortest vectors in this lattice are of the form $\pm pe_i$ where e_i is a standard basis vector. Since orthogonal matrices preserve lengths and are invertible, in this case the isomorphisms between two such Construction A lattices have to be signed permutations.

Additionally, the authors present a reduction from LEP to LIP via several intermediate steps, one of which involves UCE, which we do not introduce here for brevity. In particular, the use of closure in one of these steps renders the reduction less efficient, yielding only a $\mathrm{poly}(n,p)$ complexity. The authors also provide improved reductions for some previously known results. We summarize the relevant reductions below.

- There is a reduction from PEP to LEP that runs in $\mathrm{poly}(n,\log(q))$ time [22]. (With the results from above this implies a reduction from PEP to LIP.)
- There is a reduction from PEP to SPEP in $\mathrm{poly}(n,\log(q))$ time [22].
- There is a reduction from LEP to Matrix Code Equivalence (MCE) that runs in $\mathrm{poly}(n)$ time [24]. For one of the two reductions from LEP to MCE of [24] the authors simply embed a vector in a diagonal matrix.
- Using ideas of [36] and combining the results of [28] and [6, Theorem 5] one gets a probabilistic reduction from PEP and LEP to the problem of finding the order of the (permutation or monomial) automorphism group of a code.

We summarize the known reductions in Fig. 1, where we omit certain reductions that can be derived through alternative approaches.

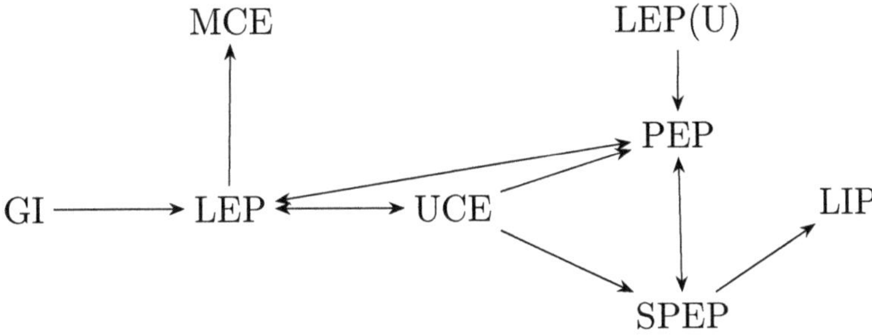

Fig. 1. Known reductions between several equivalence and isomorphism problems.

A natural direction for future work is to extend existing reductions to encompass LEP(U). While some of these extensions may be relatively direct, a systematic treatment remains open. We emphasize that the figure does not reflect the parameter dependencies of the reductions. In particular, some run in time polynomial in n and q, while others are polynomial in n and $\log(q)$. Achieving reductions that uniformly run in time $\mathrm{poly}(n,\log(q))$ would be desirable. Moreover, the reductions to LIP are currently restricted to prime fields, indicating a further limitation. Finally, additional reductions have been established in the literature [26,30,33], but fall beyond the scope of this survey.

5   Solvers

In this section we will give an overview of the known solvers for the different code equivalence problems. The computational complexities will be summarized in Tables 1 and 2 in Sect. 5.6.

5.1   Algebraic Solvers

The algebraic solver goes back to Saeed's thesis [40], where he uses that if $\mathcal{C} = \langle G \rangle = \ker(H^{\top})$ and $\mathcal{C}' = \langle G' \rangle = \ker(H'^{\top})$ are linearly equivalent, i.e., there exist $S \in \mathrm{GL}_k(\mathbb{F}_q), D = \mathrm{diag}(d), P \in S_n$ such that $SGDP = G'$, then $G'H'^{\top} = 0$ implies that $GDPH'^{\top} = 0$. Hence it is enough to solve the linear system $GMH'^{\top} = 0$, for the monomial matrix $M = DP$, with $k(n-k)$ equations and where the entries of M are the n^2 unknowns. The main problem is: how to put the fact that we are only interested in $M = DP$ into this model as equations?

5.2   Combinatorial Solvers Using Invariants

The combinatorial solvers date back to Leon [35], and have been improved by Beullens [15]. This line of combinatorial solvers is often referred to as *codeword search*, to separate it from the canonical form solvers in the next section. In [28] it was shown that for some classes of codes, knowing the automorphism group of $\mathcal{C}$ is enough to solve LEP. This underpins also the idea of Leon [35].

The idea of the codeword-search solvers is to find subsets of codewords $S \subset \mathcal{C}$ and $S' \subset \mathcal{C}'$ which are also invariant under the secret monomial φ, i.e., $\varphi(S) = S'$. For the smaller sets S, S' it will become easier to find an isometry between them. This is usually done through invariants. However, to find the subset S most solvers first need to solve a much harder problem: the problem of finding low weight codewords, which is known to be NP-hard. Thus, their bottleneck is usually the exponential cost of Information Set Decoding (ISD) algorithm.

Leon: The main observation of Leon [35] is that the sought isometry has to map all codewords of weight w in $\mathcal{C}$ to all codewords of weight w in $\mathcal{C}'$. Thus, Leon first constructs the sets

$$S = \{c \in \mathcal{C} \mid \mathrm{wt}_H(c) = w\}, \quad S' = \{c' \in \mathcal{C}' \mid \mathrm{wt}_H(c') = w\}$$

and then searches for $\varphi \in M_{n,q}$ such that $\varphi(S) = S'$. The cost of finding such a map φ is polynomial in the size of S. However, to construct the sets, we rely on ISD.

Beullens' Algorithm for PEP: The main observation of Beullens [15] is that a permutation σ does not only fix the weight of a vector, but also the multiset of its entries. Thus, instead of seeing the elements in S and S' as vectors c respectively c', Beullens' algorithm treats them as multisets and searches for a collision in $S \times S'$, i.e., $c \in S$ has the same multiset as $c' \in S'$. This allows us to store less elements in S. Each found collision is then used to piece-wise reconstruct the permutation: if c, c' have the same multisets and $c_i \neq c'_j$, then we guess $\sigma(i) \neq j$. Again, the bottleneck remains constructing the sets S, S' using ISD.

Beullens' Algorithm for LEP: Beullens also provides an algorithm to solve LEP [15]. In the case of linear equivalence, the multiset of the entries are clearly not preserved. However, for any subcode $\mathcal{D} < \mathcal{C}$ there exists a $\varphi(\mathcal{D}) = \mathcal{D}' < \mathcal{C}'$ with the same dimension and support size. In particular, Beullens proposes to use subcodes of dimension 2, that is: Beullens uses as invariant the second generalized weight. The lists now contain generator matrices in $\mathbb{F}_q^{2 \times n}$:

$$S = \{D \in \mathbb{F}_q^{2 \times n} \mid \langle D \rangle < \mathcal{C}, |\mathrm{supp}(\langle D \rangle)| = s\},$$

and similarly we construct S'. One then again searches for collisions, i.e., an isometry $\varphi(\langle D \rangle) = \langle D' \rangle$.

There is also another combinatorial solver introduced by Sendrier [42], the so-called Support Splitting Algorithm (SSA).

SSA: This algorithm solves PEP in a runtime of $\mathcal{O}(q^{\dim(\mathcal{H}(\mathcal{C}))})$, under heuristic assumptions on the refinement process, and thus not reflecting worst-case complexity. As random codes have with high probability $\dim(\mathcal{H}(\mathcal{C})) = c$ for some small constant c, this algorithm already breaks many instances and suggests that the hardest instance of PEP are self-orthogonal codes.

The idea of SSA is to first puncture the codes $\mathcal{C}$ and $\mathcal{C}'$ in each index $i \in \{1, \ldots, n\}$, getting the lists

$$\mathcal{C}_1 = \mathcal{P}(\mathcal{C}, \{1\}), \ldots, \mathcal{C}_n = \mathcal{P}(\mathcal{C}, \{n\}), \quad \mathcal{C}'_1 = \mathcal{P}(\mathcal{C}', \{1\}), \ldots, \mathcal{C}'_n = \mathcal{P}(\mathcal{C}', \{n\}).$$

The algorithm then heavily relies on the fact that the hull is an invariant for permutation equivalence, which is also the main hindrance in generalizing this algorithm for LEP.

In fact, it then computes the hulls $\mathcal{H}(\mathcal{C}_i)$ and $\mathcal{H}(\mathcal{C}'_j)$ for all $i, j \in \{1, \ldots, n\}$. As the hulls are small, we can even compute their weight distributions and (hopefully) they reveal whether there exists a unique $\sigma \in S_n$ such that $\sigma(\mathcal{H}(\mathcal{C}_i)) = \mathcal{H}(\mathcal{C}'_j)$ for all $i, j \in \{1, \ldots, n\}$ which can also be extended to a $\pi \in S_n$ such that $\pi(\mathcal{C}) = \mathcal{C}'$. If there are, however, different choices, i.e., there exists several permutations which would send i either to j or to another j', we continue the puncturing process, until we get a unique permutation.

In [42], Sendrier also mentions the possibility of extending or shortening the code. For this we recall that shortening the dual is the same as the dual of the punctured code, i.e., $\mathcal{S}(\mathcal{C}^\perp, J) = (\mathcal{P}(\mathcal{C}), J)^\perp$. Hence $\mathcal{H}(\mathcal{P}(\mathcal{C}, J)) = \mathcal{H}(\mathcal{S}(\mathcal{C}^\perp, J))$.

5.3   New Solvers Using Power Codes

While the square code is primarily used in cryptanalysis and for distinguishing structured systems, it also turns out to be a valuable tool for code equivalence.

In fact, in [12], the authors showed that we can also employ the square code to solve (signed) permutation equivalence and in turn attacked the proposal [2].

We can use the square of the hull to solve PEP faster: for this consider two permutation equivalent codes $\mathcal{C}, \mathcal{C}'$, which are self-orthogonal. The idea of [12], is to first puncture $\mathcal{C}$ and $\mathcal{C}'$ in the positions I, respectively I', denoted by $\mathcal{P}(\mathcal{C}, I)$ and $\mathcal{P}(\mathcal{C}', I')$, and then to compute the square codes of their hulls. These are again permutation equivalent.

Thus, a first approach for solving linear equivalence would be to raise $\mathcal{C}$ and $\mathcal{C}'$ to the power $(q-1)/2$. The main problem of this idea is that by computing the $(q-1)/2$-th power code, we get that the dimension quickly reaches the length n, that is $\binom{k+(q-1)/2-1}{(q-1)/2} \geq n$.

However, in the case that $n > \binom{k+(q-1)/2-1}{(q-1)/2}$, the $((q-1)/2)$-th power code approach can be applied and, starting with linearly equivalent codes, Lemma 1 yields power codes that are equivalent with a signed permutation. If the hull of these power codes is small enough, we can use SSA to solve this. Assuming that we found the correct permutation, we can then solve LEP for the original code.

This idea can be improved, as done in [10]. For this, the authors generalize the adjacency-matrix construction from Sect. 4.2 to pairs of codes. Let A, B be generator matrices of linearly equivalent codes $\mathcal{A}, \mathcal{B}$. The authors define auxiliary codes $\mathcal{A}_1, \mathcal{A}_2, \mathcal{B}_1, \mathcal{B}_2$ with generator matrices A_1, A_2 coming from A and B_1, B_2 coming from B. Then they define $\mathrm{Adj}(A_1, A_2) := A_2^\top (A_1 A_2^\top)^{-1} A_1$, and similarly $\mathrm{Adj}(B_1, B_2)$. The auxiliary codes should force the diagonal scalings to cancel. The most basic definition of the auxiliary codes is to use $\mathcal{A}_1 = \mathcal{A}_2 = \mathcal{A}^{(\frac{q-1}{2})}, \mathcal{B}_1 = \mathcal{B}_2 = \mathcal{B}^{(\frac{q-1}{2})}$, as described above. However, we may also use Frobenius powers for non-prime fields. For these auxiliary codes, they compute $\mathrm{Adj}(A_1, A_2), \mathrm{Adj}(B_1, B_2)$, concluding equivalence if the multisets of their diagonal entries are identical. Note that even though it is not guaranteed that this test avoids false positives - it does avoid false negatives - the probability that this happens becomes negligible when $k' \ll n$.

This yields several new classes of weak LEP instances, in particular for small-rate codes, and provides the first attack on LEP exploiting explicitly algebraic properties of the underlying field.

5.4   Canonical Forms

A recent line of work studies LEP using *canonical forms*. Instead of directly searching for the secret monomial transformation, the idea is to associate to each code a canonical representative of its equivalence class. If two codes are equivalent, then their canonical representatives must coincide. This approach was introduced in [23], where the authors also managed to use this new result to get smaller sizes for LESS, and was later improved by [37].

These solvers all stem from the following observation: if I is an information set for $\mathcal{C}$, and we know the secret permutation in I, i.e., σ_I, we can find the whole monomial transformation in polynomial time, see e.g. [20]. Even more, recall that the hardest instances of PEP are self-orthogonal codes. Due to [12], we know that puncturing a self-orthogonal code in an information set results in an LCD code. Thus, given two linearly equivalent codes, one may first compute their closures, then puncture in the corresponding information sets I and $\sigma(I)$, and finally apply the reduction from PEP with trivial hull to GI.

In [23], they take this fact even further and introduce the notion of *Left-Right Linear (LRL) equivalence*: two matrices $G_1 = \left(\mathrm{Id}_k \; A_1\right), G_2 = \left(\mathrm{Id}_k \; A_2\right)$ are LRL-equivalent if there exist monomials $Q_r \in M_{k,q}$ and $Q_c \in M_{n-k,q}$ such that $A_2 = Q_r A_1 Q_c$. Thus the matrices A_1, A_2 are equivalent up to monomial row and column transformations.

A *canonical form function* takes $G = \left(\mathrm{Id}_k \; A\right)$ as input and outputs a canonical representative $G^* = \left(\mathrm{Id}_k \; A^*\right)$ of its LRL-equivalence class together with monomials satisfying $A^* = Q_r A Q_c$, or an error symbol $\bot$. If the algorithm succeeds on two LRL-equivalent inputs, it outputs the same representative. The original canonical form construction achieves constant success probability only for large fields, whereas for constant q, the success probability becomes exponentially small in n. This limitation was addressed in [37], where the author introduced an improved canonical form function with success probability $1 - O(n^{-1})$ for all $q \geq 7$.

Further Improvements. In [13], the authors present asymptotic improvements for several algorithms related to code equivalence. They give a faster deterministic algorithm for LEP, extending Babai's approach from [4] beyond the permutation case, and avoiding the exponential length expansion of the classical closure by handling monomial isometries directly. They also propose a randomized algorithm for LEP and PEP based on a meet-in-the-middle strategy combined with a down-up random walk for near-uniform matroid basis sampling. Finally, they introduce a quantum variant using the BHT collision-finding framework based on Grover's algorithm. The latter two algorithms achieve runtimes comparable to [37], while applying to arbitrary codes and not requiring $q \geq 7$.

5.5   Solvers Using Side Information

Another line of work studies LEP under side information, namely access to pairs of equivalent codewords $(c, c') \in \mathcal{C} \times \mathcal{C}'$ connected through the secret monomial. Such pairs naturally arise when computing lists of low-weight codewords, and earlier approaches required $\Omega(\log n)$ pairs to recover the monomial transformation, while [20] showed that two pairs suffice with high probability by exploiting support intersections and ratio structures. More recently, [19] and [6] demonstrated that even a single pair can be sufficient, achieving the theoretical lower bound of this framework. A complementary direction considers the reuse of the same secret monomial across multiple instances, where several equivalent pairs induce linear constraints on the underlying permutation. As shown in [18], these constraints can be exploited to distinguish shared secrets and partially recover the permutation, highlighting the risks of monomial reuse in cryptosystems.

An open question which remains, is which kind of information can be obtained through side-channel attacks and how will they influence the cost of LEP solvers?

5.6   Summary of the Costs

Tables 1 and 2 provide a short summary on the costs of the previously presented solvers. We use the following notation: H for the binary entropy function. N_w denotes the number of codewords of weight w and C_{ISD} denotes the exponential cost of an ISD algorithm of choice.

Table 1. Complexities of various solvers for PEP.

Algorithm	Complexity	Remarks
Brute Force	$2^{k \log n + o(k \log_2 n)}$	Guess σ on an information set
Via Graph Isomorphism	$2^{n+o(n+q)}$ [13]	Run Babai's algorithm for each information set, i.e., $\binom{n}{k}$ times
Meet-in-the-middle variant	$2^{n/2+o(n+q)}$ [13]	Extension of GI-approach, probabilistic, randomized
Algebraic (Gröbner basis)	$2^{n \log \frac{k(n-k)}{n^2} + o(n \log \frac{k(n-k)}{n^2})}$ [40]	Heuristic
SSA	$\mathcal{O}(n^3 + q^h n^2 \log n)$ [7,43]	Heuristic, h is hull dimension

Table 2. Complexities of various solvers for LEP.

Algorithm	Complexity	Remarks
Brute Force	$2^{k \log n + o(k \log_2 n)}$	Guess σ on an information set (as diagonal can be solved once P is known)
Via Graph Isomorphism	$2^{2(q-1)n+o(n)}$ [13]	Run Babai's algorithm $\binom{n}{k}$ times on the code closure
Meet-in-the-middle variant	$2^{n/2+o(n+q)}$ [13]	Extension of GI-approach, probabilistic
Leon's Algorithm	$\mathcal{O}(\log(N_w)C_{ISD})$ [9]	ISD based
Beullens' Algorithm	$\mathcal{O}(\sqrt{\frac{n \log n}{N_w}} C_{ISD})$ [9]	ISD based
Nowakowski's Algorithm	$2^{n/2 H(k/n)+o(n)}$ [37]	Randomized, $q \geq 7$
SSA extension	$\mathcal{O}(n^3 + q^k n^2 \log n)$ [43]	SSA applied to closure of code

6   New Directions and Open Questions

Every Code is Equivalent to a Code with Trivial Hull: An interesting discovery has been made in [21]: every linear code over $\mathbb{F}_q$ with $q > 3$ is linearly equivalent to a code with trivial hull, i.e., an LCD code. Unfortunately, this cannot be combined with the existing reductions: if we first reduce LEP to PEP by using the closure, we get self-orthogonal codes, which are permutation equivalent. If we now apply [21], we end up with LCD codes, but they are again linearly equivalent and not permutation equivalent.

The Search for New Codes Respecting the Monomial Equivalence: One of the open questions is whether there exists another code $F(\mathcal{C})$, which acts for LEP like the dual code for PEP. That is: for a generator matrix of a linearly equivalent code, $G' = SGDP$ with $S \in \mathrm{GL}_k(\mathbb{F}_q)$, a permutation matrix P and diagonal matrix D, we want that $F(H') = S'F(H)D^{-1}P$, for some $S' \in \mathrm{GL}_k(\mathbb{F}_q)$ and, to use the reduction to GI, we additionally require that $GF(H)^\top$ is invertible. In fact, if such $F(\mathcal{C})$ construction exists, we have that $\mathrm{Adj}(G', F(H')) = P^\top D\mathrm{Adj}(G, F(H))DP$, yet another version of GI, which is still GI-complete and hence, we can apply the quasi-polynomial time solver of Babai to recover (P, D). One such example was the Hermitian hull. Other inner products have, unfortunately, not led to such a reduction.

Rank-Metric Code Equivalence is Easy: In [24] the authors state that the code equivalence problem for $\mathbb{F}_{q^m}$-linear rank-metric codes is either in P or in ZPP, depending on the size of the base field. More precisely, the $\mathbb{F}_{q^m}$-linear rank-metric code equivalence problem is: given two generator matrices $G, G' \in \mathbb{F}_{q^m}^{k \times n}$, find $S \in \mathrm{GL}_k(\mathbb{F}_{q^m})$ and $P \in \mathrm{GL}_n(\mathbb{F}_q)$ such that $SGP = G'$. As this problem is easier than its Hamming-metric counterpart, a natural question is whether we may get a reduction from LEP to the $\mathbb{F}_{q^m}$-linear version of rank-metric equivalence, e.g. along the lines of [27].

Extension Fields: Some New Approaches: If LEP is defined over $\mathbb{F}_{p^s}$ with $s > 1$, a natural idea is to descend to the base field $\mathbb{F}_p$ and try to solve the resulting instance there. Two standard ways to do so are via the *expanded code* and the *trace code*. Fix a basis $\Gamma = \{\gamma_0, \ldots, \gamma_{s-1}\}$ of $\mathbb{F}_{p^s}$ over $\mathbb{F}_p$. Writing each $x \in \mathbb{F}_{p^s}$ as $x = \sum_{i=0}^{s-1} x_i\gamma_i$, we obtain the expansion map

$$\Gamma : \mathbb{F}_{p^s} \to \mathbb{F}_p^s, \ x \mapsto (x_0, \ldots, x_{s-1}),$$

which extends componentwise to vectors and matrices. For a code $\mathcal{C} \subseteq \mathbb{F}_{p^s}^n$, its *expanded code* is $\Gamma(\mathcal{C}) = \{\Gamma(c) \in \mathbb{F}_p^{sn} \mid c \in \mathcal{C}\}$, which is an $[sn, sk]_p$ linear code. If G is a generator matrix of $\mathcal{C}$, then a generator matrix of $\Gamma(\mathcal{C})$ is obtained by expanding each entry of G with respect to Γ. If $G' = SGP$, then $\Gamma(G') = \Gamma(S)\Gamma(G)(\mathrm{Id}_s \otimes P)$, so permutation equivalence is preserved. For linear equivalence, however, the diagonal part causes an obstruction: if $G' = SGDP$ with

$D = \mathrm{diag}(d_1, \ldots, d_n)$, then the induced action on the expanded code is monomial only when all $d_i \in \mathbb{F}_p$. Hence expansion yields an LEP instance over the base field only for LEP(U), with $U \subseteq \mathbb{F}_p^\star$.

Another natural descent is given by the trace map $\mathrm{Tr}_{\mathbb{F}_{p^s}/\mathbb{F}_p}(\alpha) = \sum_{i=0}^{s-1} \alpha^{p^i}$. For a code $\mathcal{C} \subseteq \mathbb{F}_{p^s}^n$, define its *trace code* by $\mathrm{Tr}(\mathcal{C}) = \{(\mathrm{Tr}(c_1), \ldots, \mathrm{Tr}(c_n)) \mid (c_1, \ldots, c_n) \in \mathcal{C}\} \subseteq \mathbb{F}_p^n$. This is an $[n, \le sk]_p$ linear code; moreover, it is the dual of a subfield subcode, i.e., $\mathcal{C} \cap \mathbb{F}_p^n$. If $G' = SGP$, then the corresponding trace codes are again permutation equivalent. However, for trace codes, the diagonal part of a monomial equivalence does not behave well with respect to the trace, since the scalars d_i cannot in general be pulled through the trace coordinatewise. Thus, while both constructions naturally preserve permutation equivalence, neither gives a general reduction of LEP over $\mathbb{F}_{p^s}$ to LEP over $\mathbb{F}_p$.

Connections to Other Incidence Structures: Another natural idea is to exploit the many connections between coding theory and other finite structures, e.g. projective systems, or matroids.

Linear codes admit a geometric interpretation via projective systems. Let $G \in \mathbb{F}_q^{k \times n}$ be a generator matrix of a non-degenerate $[n, k, d]_q$ code $\mathcal{C}$. Considering the one-dimensional subspaces generated by the columns of G yields a multiset of points $\mathcal{M} \subseteq \mathrm{PG}(k-1, q)$, called the *projective system* associated to $\mathcal{C}$. This construction gives a one-to-one correspondence between non-degenerate $[n, k, d]_q$ codes and projective $[n, k, d]_q$ systems. Moreover, column permutations and non-zero column scalings do not change the corresponding points in projective space and hence the projective system. Therefore, linearly equivalent codes give rise to equivalent projective systems.

Another closely related structure is the representable matroid associated to a generator matrix $G \in \mathbb{F}_q^{k \times n}$. Let $E = \{1, \ldots, n\}$ and define the independent sets $I = \{S \subseteq E \mid G_S \text{ has full rank}\}$. This yields a matroid $M(G)$ whose rank function is given by $r(S) = \dim(\langle G_S \rangle)$. If two codes are linearly equivalent, then the linear dependency relations among the columns of their generator matrices are preserved (up to relabeling of the coordinates), i.e., they define isomorphic matroids. This was also noticed in [14]. However, many questions about the relationships of the two problems and the complexity of matroid isomorphism remain open.

Open Questions. Finally, we pose some open problems which we deem fruitful for further research and which might be (partially) in reach using existing techniques.

- Find (efficient) algorithms to find isomorphisms between matroids.
- Improve the complexity of the known reductions by achieving poly($n, \log(q)$) bounds for all reductions presented.
- Are there reductions from LEP to LIP over non-prime fields?
- Introduce an intermediate problem similar to UCE and find more reductions for Fig. 1 including LEP(U).

- Follow [36] to get similar reductions between different versions of code equivalence problems.
- Show that random codes have automorphism group $\{\lambda\,\mathrm{id}|\lambda \in \mathbb{F}_q^\star\}$ with high probability for $q > 2$.
- Give a search to decision reduction for LEP.
- Find an invariant which is characterizing.
- Find linear equations to model that $M \in M_{n,q}$.
- Which information can be found on $\varphi \in M_{n,q}$ through side-channel attacks?

References

1. Aaraj, N., et al.: First round submission to the additional NIST postquantum cryptography call. In: PERK (2023)
2. Albrecht, M.R., Benčina, B., Lai, R.W.F.: Hollow LWE: a new spin. In: Fehr, S., Fouque, P.A. (eds.) Advances in Cryptology - EUROCRYPT 2025, pp. 363–392. Springer, Cham (2025). https://doi.org/10.1007/978-3-031-91101-9_13
3. Babai, L.: Graph isomorphism in quasipolynomial time. In: Proceedings of the Forty-Eighth Annual ACM Symposium on Theory of Computing, pp. 684–697 (2016)
4. Babai, L., Codenotti, P., Grochow, J.A., Qiao, Y.: Code equivalence and group isomorphism. In: Proceedings of the Twenty-Second Annual ACM-SIAM Symposium on Discrete Algorithms, SODA '11, pp. 1395–1408. Society for Industrial and Applied Mathematics (2011)
5. Baldi, M., Battagliola, M., El Mechri, R., Santini, P., Schiavoni, R., De Zuane, D.: SPECK: signatures from permutation equivalence of codes and kernels. Cryptology ePrint Archive (2025)
6. Bardet, M., Brion, C., Otmani, A., Saeed, M., Sendrier, N.: Solving the linear code equivalence problem from single codeword matching. ePrint, Paper 2026/550 (2026). https://eprint.iacr.org/2026/550
7. Bardet, M., Otmani, A., Saeed-Taha, M.: Permutation code equivalence is not harder than graph isomorphism when hulls are trivial. In: 2019 IEEE International Symposium on Information Theory (ISIT), pp. 2464–2468. IEEE (2019)
8. Barenghi, A., Biasse, J.F., Ngo, T., Persichetti, E., Santini, P.: Advanced signature functionalities from the code equivalence problem. Int. J. Comput. Math. Comput. Syst. Theory 7(2), 112–128 (2022)
9. Barenghi, A., Biasse, J.F., Persichetti, E., Santini, P.: On the computational hardness of the code equivalence problem in cryptography. Adv. Math. Commun. 17(1), 23–55 (2023)
10. Battagliola, M., et al.: The power of power codes: new classes of easy instances for the linear equivalence problem (2026). https://arxiv.org/abs/2603.23230
11. Battagliola, M., Mattiuz, L., Meneghetti, A.: VOLE-in-the-head signatures based on the linear code equivalence problem. Cryptology ePrint Archive (2025)
12. Battagliola, M., Mora, R., Santini, P.: Using the Schur product to solve the code equivalence problem. ePrint (2025)
13. Bennett, H., et al.: Asymptotic improvements to provable algorithms for the code equivalence problem. IEEE Trans. Inf. Theory 72(2), 1093–1108 (2025)
14. Bennett, H., Win, Htay, K.M.: Relating code equivalence to other isomorphism problems. Des. Codes Cryptogr. 93(3), 701–723 (2025)

15. Beullens, W.: Not enough LESS: an improved algorithm for solving code equivalence problems over $\mathbb{F}_q$. In: Dunkelman, O., Jacobson, Jr., M.J., O'Flynn, C. (eds.) SAC 2020. LNCS, vol. 12804, pp. 387–403. Springer, Cham (2021). https://doi.org/10.1007/978-3-030-81652-0_15

16. Biasse, J.F., Micheli, G.: A search-to-decision reduction for the permutation code equivalence problem. In: 2023 IEEE International Symposium on Information Theory (ISIT), pp. 602–607. IEEE (2023)

17. Biasse, J.-F., Micheli, G., Persichetti, E., Santini, P.: LESS is more: code-based signatures without syndromes. In: Nitaj, A., Youssef, A. (eds.) AFRICACRYPT 2020. LNCS, vol. 12174, pp. 45–65. Springer, Cham (2020). https://doi.org/10.1007/978-3-030-51938-4_3

18. Budroni, A., Chi-Domínguez, J.J., D'Alconzo, G., Di Scala, A.J., Kulkarni, M.: Don't use it twice! Solving relaxed linear equivalence problems. In: International Conference on the Theory and Application of Cryptology and Information Security, pp. 35–65. Springer, Heidelberg (2024). https://doi.org/10.1007/978-981-96-0944-4_2

19. Budroni, A., Esser, A.: One pair to rule them all: an optimal algorithm for solving code equivalence via codeword search. ePrint (2026)

20. Budroni, A., Esser, A., Franch, E., Natale, A.: Two is all it takes: asymptotic and concrete improvements for solving code equivalence. ePrint (2025)

21. Carlet, C., Mesnager, S., Tang, C., Qi, Y., Pellikaan, R.: Linear codes over $\mathbb{F}_q$ are equivalent to LCD codes for $q > 3$. IEEE Trans. Inf. Theory **64**(4), 3010–3017 (2018)

22. Cheraghchi, M., Shagrithaya, N., Veliche, A.: Reductions between code equivalence problems. In: 2025 IEEE International Symposium on Information Theory (ISIT), pp. 1–6. IEEE (2025)

23. Chou, T., Persichetti, E., Santini, P.: On linear equivalence, canonical forms, and digital signatures. Des. Codes Crypt. 1–43 (2025)

24. Couvreur, A., Debris-Alazard, T., Gaborit, P.: On the hardness of code equivalence problems in rank metric. arXiv:2011.04611 (2020)

25. Ducas, L., Gibbons, S.: Hull attacks on the lattice isomorphism problem. In: IACR International Conference on Public-Key Cryptography, pp. 177–204. Springer, Heidelberg (2023). https://doi.org/10.1007/978-3-031-31368-4_7

26. D'Alconzo, G.: Monomial isomorphism for tensors and applications to code equivalence problems. Des. Codes Crypt. **92**(7), 1961–1982 (2024)

27. Gaborit, P., Zémor, G.: On the hardness of the decoding and the minimum distance problems for rank codes. IEEE Trans. Inf. Theory **62**(12), 7245–7252 (2016)

28. Grochow, J.: Complexity of finding automorphism group of code. Theoretical Computer Science Stack Exchange. https://cstheory.stackexchange.com/q/46298 (version: 2020-02-06)

29. Grochow, J.: Matrix isomorphism of matrix lie algebras. In: Proceedings of the 27th Conference on Computational Complexity, CCC 2012, Porto, Portugal, 26–29 June 2012, pp. 203–213. IEEE Computer Society (2012). https://doi.org/10.1109/CCC.2012.34

30. Grochow, J., Qiao, Y.: On the complexity of isomorphism problems for tensors, groups, and polynomials i: tensor isomorphism-completeness. SIAM J. Comput. **52**(2), 568–617 (2023)

31. Hanzlik, L., Lai, Y.F., Mula, M., Paracucchi, E., Slamanig, D., Tang, G.: Tanuki: new frameworks for (concurrently secure) blind signatures from post-quantum

group actions. In: International Conference on the Theory and Application of Cryptology and Information Security, pp. 35–69. Springer, Heidelberg(2025). https://doi.org/10.1007/978-981-95-5113-2_2

32. Kaski, P., Östergård, P.R.: Classification Algorithms for Codes and Designs, vol. 15. Springer, Heidelberg (2006). https://doi.org/10.1007/3-540-28991-7

33. Kreuzer, M.: Code equivalence, point set equivalence, and polynomial isomorphism. arXiv:2511.06843 (2025)

34. Lefmann, H., Phelps, K.T., Rödl, V.: Rigid linear binary codes. J. Comb. Theory Series A **63**(1), 110–128 (1993)

35. Leon, J.: Computing automorphism groups of error-correcting codes. IEEE Trans. Inf. Theory **28**(3), 496–511 (1982)

36. Mathon, R.: A note on the graph isomorphism counting problem. Inf. Process. Lett. **8**(3), 131–132 (1979). https://doi.org/10.1016/0020-0190(79)90004-8

37. Nowakowski, J.: An improved algorithm for code equivalence. In: International Conference on Post-Quantum Cryptography, pp. 71–103. Springer, Heidelberg (2025). https://doi.org/10.1007/978-3-031-86599-2_3

38. Pellikaan, R., Wu, X.W., Bulygin, S., Jurrius, R.: Codes, Cryptology and Curves with Computer Algebra, vol. 1. Cambridge University Press, Cambridge (2017)

39. Petrank, E., Roth, R.M.: Is code equivalence easy to decide? IEEE Trans. Inf. Theory **43**(5), 1602–1604 (1997)

40. Saeed, M.A.: Algebraic approach for code equivalence. Ph.D. thesis, Normandie Université; University of Khartoum (2017)

41. Santini, P., Baldi, M., Chiaraluce, F.: Computational hardness of the permuted kernel and subcode equivalence problems. IEEE Trans. Inf. Theory (2023)

42. Sendrier, N.: Finding the permutation between equivalent linear codes: the support splitting algorithm. IEEE Trans. Inf. Theory **46**(4), 1193–1203 (2000)

43. Sendrier, N., Simos, D.E.: How easy is code equivalence over $\mathbb{F}_q$? In: International Workshop on Coding and Cryptography-WCC 2013 (2013)

44. Sendrier, N., Simos, D.E.: The hardness of code equivalence over $\mathbb{F}_q$ and its application to code-based cryptography. In: Gaborit, P. (ed.) PQCrypto 2013. LNCS, vol. 7932, pp. 203–216. Springer, Heidelberg (2013). https://doi.org/10.1007/978-3-642-38616-9_14

Eidolon: A Post-Quantum Signature Scheme Based on k-Colorability in the Age of Graph Neural Networks

Asmaa Cherkaoui[1]([✉]) , Ramón Flores[2] , Delaram Kahrobaei[3,4,5,6] ,
and Richard C. Wilson[6]

[1] Laboratory of Mathematical Analysis, Algebra and Applications (LAM2A), Faculty of Sciences Ain Chock (FSAC), University Hassan II, Casablanca, Morocco
`esma1maysan@gmail.com`
[2] Department of Geometry and Topology, Faculty of Mathematics, University of Seville, Seville, Spain
`ramonjflores@us.es`
[3] Departments of Computer Science and Mathematics, Queens College, City University of New York, New York, USA
`delaram.kahrobaei@qc.cuny.edu`
[4] PhD Program in Mathematics, and Initiative for the Theoretical Sciences, Graduate Center, City University of New York, New York, USA
[5] Department of Computer Science and Engineering, Tandon School of Engineering, New York University, New York, USA
[6] Department of Computer Science, University of York, York, UK
`richard.wilson@york.ac.uk`

Abstract. We propose `Eidolon`, a post-quantum signature scheme grounded in the NP-complete k-colorability problem. Our construction generalizes the Goldreich–Micali–Wigderson zero-knowledge protocol to arbitrary $k \geq 3$, applies the Fiat-Shamir transform, and uses Merkle-tree commitments to compress signatures from $O(tn)$ to $O(t \log n)$. We generate instances by planting a coloring while aiming to preserve the statistical profile of random graphs. We present an empirical security analysis of such a scheme against both classical solvers (ILP, DSatur) and a custom graph neural network (GNN) attacker. Experiments show that for $n \geq 60$, neither approach is able to recover a valid coloring matching the planted solution, suggesting that well-engineered k-coloring instances can resist the considered classical and learning-based cryptanalytic approaches. These experiments indicate that the constructed instances resist the attacks considered in our evaluation.

Keywords: Post-quantum cryptography · k-colorability · NP-hard signatures · zero-knowledge proofs · Fiat-Shamir transform · graph neural networks

L. Batina and F. Özbudak (Eds.): WAIFI 2026, LNCS 16611, pp. 32–51, 2026.
https://doi.org/10.1007/978-3-032-27574-5_3

1   Introduction

With the rising threat of quantum computers to traditional cryptography, recent attention has focused on post-quantum cryptographic methods. Based on the current belief that there is no quantum speed-up for NP-complete problems, these problems are potentially a rich source of potential cryptosystems. In particular, graph theory contains several NP-complete problems, including homomorphism, k-colorability, and Hamiltonicity. In this paper, we study the application of k-colorability to a post-quantum signature system.

When considering an NP-hard problem as the basis for a cryptosystem, care must be taken in the concrete realization of a particular instance. This is because many instances can be solved in acceptable time using heuristic algorithms. Indeed, a number of heuristics are available for graph coloring and can quickly solve the problem for many graphs.

In this regard, machine learning has emerged as a particular threat because of its ability to extract powerful heuristics directly from data. In this paper, we explore the issue of instance hardness for k-colorability, from both a theoretical and empirical perspective, to demonstrate the security of the system. In [20], the authors demonstrate how graph neural networks can be used to solve combinatorial optimization problems. Their approach is broadly applicable to canonical NP-hard problems in the form of quadratic unconstrained binary optimization problems, such as maximum cut, minimum vertex cover, maximum independent set, as well as Ising spin glasses and higher-order generalizations thereof in the form of polynomial unconstrained binary optimization problems.

The rapid progress in quantum computing has intensified the search for cryptographic primitives that remain secure in a post-quantum world. While lattice- and code-based schemes currently dominate standardization efforts, combinatorial problems offer an alternative foundation rooted in computational complexity theory. The use of k-colorability in cryptography is limited by the gap between worst-case hardness and typical-instance behavior: many graph instances that are hard in the worst-case sense can still be handled efficiently by heuristic algorithms such as DSatur [4], and more recently by data-driven methods leveraging machine learning.

Recent work by Schuetz, Brubaker, and Katzgraber [20] demonstrates that graph neural networks (GNNs) can effectively approximate solutions to canonical NP-hard problems, including graph coloring, by learning implicit heuristics from data. This motivates examining whether such methods affect the security of schemes based on combinatorial hardness. To date, to the best of our knowledge, no signature scheme based on k-colorability has been proposed that simultaneously offers a complete construction, incorporates instance generation that embeds a secret coloring while aiming to maintain statistical indistinguishability from a random graph, and undergoes empirical validation against both classical and learning-based attacks.

In this paper, we present `Eidolon`, a post-quantum digital signature scheme derived from the k-colorability problem by generalizing the zero-knowledge identification protocol of Goldreich, Micali, and Wigderson (GMW) [8], originally demonstrated for 3-colorability, to arbitrary $k \geq 3$, and then applying the Fiat–Shamir transform. Our design features a planted coloring mechanism, inspired by the "quiet solution" framework of Krzakala and Zdeborová [15], which embeds a secret k-coloring into a k-partite random graph with calibrated edge density, ensuring witness existence while aiming to maintain statistical indistinguishability from an Erdős–Rényi random graph. To reduce the signature size obtained from a direct Fiat–Shamir transformation of the underlying zero-knowledge protocol, we use Merkle-tree vector commitments, compressing the per-round vertex commitments into a single root and thereby reducing the overall signature size from $O(tn)$ to $O(t \log n)$, where t denotes the number of Fiat–Shamir rounds and $n = |V|$.

We evaluate the scheme empirically using two attack strategies: the classical DSatur heuristic and a custom-designed graph neural network (GNN). Our GNN is inspired by the framework of Schuetz, Brubaker, and Katzgraber [20]. Both attackers are tested on the same class of k-partite random graphs. Our experiments show that, for graphs with $n \geq 60$ vertices and our chosen density parameters, neither DSatur nor our GNN is able to recover a coloring matching the planted solution, providing empirical evidence, within the tested regime, that the hardness of our k-colorability instances withstands the considered classical and learning-based cryptanalytic approaches.

In summary, our contributions are threefold: (i) we design `Eidolon`, a Fiat–Shamir signature scheme based on k-colorability with a statistically hidden planted coloring; (ii) we reduce its asymptotic signature size via Merkle-tree vector commitments; and (iii) we provide an empirical hardness study against both classical heuristics and GNN-based attacks. The rest of the paper is organized as follows: Sect. 2 reviews graph coloring and instance generation; Sect. 3 details our signature construction; and Sect. 5 describes our attack models and experimental setup.

2   Graph Instance Generation

Let $G = (V, E)$ be a graph with vertices V and edges E. A k-coloring of a graph is an assignment of one of k colors to each vertex so that no two adjacent vertices share the same color. It is known that determining whether a graph has a k-coloring is NP-complete for $k \geq 3$. The minimum number of colors needed is the *chromatic number* $\chi(G)$, and computing it is NP-hard.

However, worst-case hardness alone is insufficient for cryptography: many concrete instances are easy to solve. For example, if $k = |V|$, a trivial coloring exists; complete graphs have $\chi(G) = |V|$; and heuristic algorithms like DSatur often find valid colorings quickly on structured or sparse graphs. This creates a constraint for signature schemes based on k-colorability: the prover must know a valid k-coloring, which serves as the secret key, while the public graph should not reveal that coloring to the verifier or to an adversary.

Thus, we must *construct* instances where:

– a valid k-coloring is known by design to the prover,

- the graph appears statistically indistinguishable from a random hard instance,
- and recovering the coloring remains infeasible for classical and ML-based attackers.

This ensures the protocol is both *correct* (the prover can always respond) and *secure* (the secret remains hidden).

An (Erdös-Rényi) random graph is a graph $G_R(n, p)$ on n vertices, where each vertex pair is joined uniformly at random with probability p. A number of results are available on the hardness of coloring random graphs. In [7] it is shown that coloring a graph with less than $2\chi(G) - \delta$ is NP-hard. For fixed p, almost every random graph has a chromatic number [2]

$$\chi(G_R(n, p)) = \frac{n}{r}(1 + \epsilon) \tag{1}$$

with $0 \leq \epsilon \leq 3 \log \log n / \log n$ and

$$r = 2\log_d n - \log_d \log_d n + 2\log_d(e/2) + 1. \tag{2}$$

These results hold in the limit of large n. Furthermore, we observe that very sparse and very dense graphs are easy to color, and so we select $p = 1/2$. In this case, it has been shown that $\chi(G_R(n, p)) \sim \log(1/(1 - p))/2\log np$ [17]. However, these results are asymptotic and may not hold for the small graphs under consideration here. We reserve selection of k for the planted coloring until our empirical analysis, noting only that it should be the same size as the expected chromatic number of the equivalent random graph.

2.1   Planted k-Colorable Graph Construction

In order to operate the digital signature scheme above, it is necessary to be able to construct a graph and k-coloring in polynomial time, where the coloring is difficult to discover. We use the natural algorithm for planting a known coloring in a random graph [15]. In particular, we begin by selecting a vertex set V, $|V| = n$ (the problem size) and $k \approx \chi[G_R(n, p)]$. We then follow the following steps.

- V is partitioned into k sets such that $n_i = |V_i|, n_i \in \{\lfloor n/k \rfloor, \lceil n/k \rceil\}$ and $\sum_i n_i = n$
- We iterate through all pairs of vertices (u, v)
- Let P_u, P_v be the partitions of the vertices. We join the vertices with probability p if $P_u \neq P_v$ and probability zero otherwise.

The resulting graph may be colored with k colors simply by assigning one color to each partition. In [15], the authors demonstrate that this is a *quiet solution* which is hidden in the random graph in the sense that the properties of the graph are not much altered by the existence of the known extra solution. In particular, they conjecture that the hardness of the problem is the same as for the original graph.

In general we work with random graphs such that the probability of an edge between two given vertices is a fixed number $p \in [0,1]$, in such a way that if the graph has n vertices, the expected final number of edges is $p\binom{n}{2}$. We consider here a multipartite graph Γ with n vertices and a partition in the set of vertices given by $V = V_1 \cup \ldots V_k$. Given a number $s \in [0,1]$, we compute here the probability p of existence of an edge in Γ such that the expected number of edges in the graph is $s\binom{n}{2}$.

Recall that a complete graph in n_j vertices has $\binom{n_j}{2}$ edges. Hence, the number of forbidden edges in Γ is $S = \sum_{j=1}^{k} \binom{n_j}{2}$. Then, the maximum number of edges of the graph Γ is $\binom{n}{2} - S$, and given a probability p of existence of an edge, the expected number of edges in the graph is $p(\binom{n}{2} - S)$. We have imposed that

$$p\left(\binom{n}{2} - S\right) = s\binom{n}{2},$$

and hence

$$p = \frac{s\binom{n}{2}}{\binom{n}{2} - S}.$$

Observe that $s\binom{n}{2}$ is always bounded in the graph Γ by $\binom{n}{2} - S$, and hence we always obtain $p \leq 1$. Edges in the graph are selected with probability p. We discuss the specific values of p and k in Sect. 5 on our security analysis.

3   Protocol Description

We describe the protocol underlying `Eidolon`, based on graph k-colorability. Let

$$f : \{1, \ldots, k\} \times \{0,1\}^r \longrightarrow \{0,1\}^s$$

be a statistically hiding and computationally binding commitment function, as in [8]. One round of the protocol is as follows. The protocol is then repeated independently T times.

- **Coloring.** The prover holds a valid k-coloring $\phi : V \to \{1, \ldots, k\}$.
- **Permutation.** The prover samples a fresh random permutation $\pi \in S_k$ to mask color labels.
- **Commitment.** For each vertex $i \in V$, the prover samples fresh randomness r_i and commits to the permuted color

$$c_i = f\big(\pi(\phi(i)), r_i\big).$$

  The prover sends all commitments $\{c_i\}_{i \in V}$ to the verifier.
- **Challenge.** The verifier chooses a uniform random edge $(u, v) \in E$ and requests openings for its endpoints.
- **Response.** The prover reveals $\big(\pi(\phi(u)), r_u\big)$ and $\big(\pi(\phi(v)), r_v\big)$.

– **Verification.** The verifier checks

$$f\big(\pi(\phi(u)), r_u\big) \stackrel{?}{=} c_u, \qquad f\big(\pi(\phi(v)), r_v\big) \stackrel{?}{=} c_v,$$

and verifies that

$$(u, v) \in E, \qquad \pi(\phi(u)), \pi(\phi(v)) \in \{1, \ldots, k\}, \qquad \pi(\phi(u)) \neq \pi(\phi(v)).$$

The round is accepted if all checks pass.

Zero-Knowledge Intuition. In each round, the prover uses a fresh permutation π and fresh commitment randomness. The verifier therefore learns only that the challenged edge joins two differently colored vertices under the permuted coloring.

3.1   Soundness

To analyze the soundness of the zero-knowledge proof protocol for k-colorability, consider the case where the graph is *not* k-colorable. In such a scenario, no matter how the prover attempts to simulate a valid coloring, there will inevitably be a set of *violating edges*–edges whose endpoints receive the same color under any attempted coloring. Let t denote the number of these bad edges, and let m be the total number of edges in the graph. Since the verifier selects one edge uniformly at random in each round, the probability that a cheating prover is caught in a single round is at least $\frac{t}{m}$, while the probability of escaping detection is at most $1 - \frac{t}{m}$.

Although this is the general case, we often assume a worst-case scenario where $t \geq 1$ but unknown. In this case, we conservatively lower-bound the detection probability per round by $\frac{1}{m}$. The protocol is repeated independently for m^2 rounds to drive the cheating probability down. Therefore, the probability that a cheating prover escapes detection across all m^2 rounds is:

$$\left(1 - \frac{1}{m}\right)^{m^2} = \left[\left(1 - \frac{1}{m}\right)^m\right]^m.$$

It is well known that:

$$\lim_{m \to \infty} \left(1 - \frac{1}{m}\right)^m = \frac{1}{e},$$

so it follows that:

$$\left(1 - \frac{1}{m}\right)^{m^2} \approx \left(\frac{1}{e}\right)^m = e^{-m}.$$

This quantity becomes *negligibly small* as m increases, for instance, $e^{-40} < 2^{-55}$. Thus, the verifier's probability of accepting a false claim is exponentially small in the number of edges.

The number of rounds is taken to depend on the number of edges m, rather than on the number of vertices n or the number of colors k, because the verifier's challenges are edge-based and violations are detected on edges.

3.2   Identification and Signature Construction

We use a GMW-type identification protocol for graph k-coloring, following the classical approach of Goldreich, Micali, and Wigderson [8]. The prover commits to a randomly permuted coloring of the public graph and, upon receiving a challenge edge, opens only the two commitments corresponding to its endpoints. The verifier then checks that the revealed colors are valid, distinct, and consistent with the commitments (Table 1).

Table 1. One round of the graph (k)-coloring identification protocol

PROVER	**VERIFIER**
Hold a valid coloring $\phi : V \to \{1, \ldots, k\}$	
Sample a fresh permutation $\pi \in S_k$ and fresh r_v for all $v \in V$	
Compute $c_v = f\big(\pi(\phi(v)), r_v\big)$ for all $v \in V$	
Set $C = \{c_v\}_{v \in V}$	

$$\xrightarrow{\quad C \quad}$$

Sample a uniform random edge $(u, v) \in E$

$$\xleftarrow{\quad (u,v) \quad}$$

Reveal $(\pi(\phi(u)), r_u)$ and $(\pi(\phi(v)), r_v)$

$$\xrightarrow{\quad (\pi(\phi(u)),r_u),\ (\pi(\phi(v)),r_v) \quad}$$

Accept if

$$(u, v) \in E$$
$$f(\pi(\phi(u)), r_u) = c_u \quad \text{and}$$
$$f(\pi(\phi(v)), r_v) = c_v$$
$$\pi(\phi(u)), \pi(\phi(v)) \in \{1, \ldots, k\} \quad \text{and}$$
$$\pi(\phi(u)) \neq \pi(\phi(v))$$

3.3   Fiat–Shamir Signature Construction

To obtain a non-interactive signature scheme, we apply the Fiat–Shamir transform [5] to the identification protocol in the random-oracle model. The signer first computes the round commitments using fresh permutations and fresh randomness, then evaluates

$$h = \mathcal{H}(\text{Encode}(G, k, C^{(0)}, \ldots, C^{(t-1)}, M))$$

on the public data, the commitments, and the message M, where $\text{Encode}(\cdot)$ is a canonical encoding procedure. A public hash-to-edges parser deterministically derives the challenge edges from h, and the signer includes the openings corresponding to the endpoints of those edges. Verification recomputes the same hash from (G, k), the commitments, and M, derives the same challenge edges, and checks the openings together with color distinctness.

Challenge Derivation. The challenge edges are derived deterministically from the hash value h. More precisely, we sample t edge indices with replacement so as to match the independent public-coin challenges of the identification protocol. To avoid modulo bias, the mapping from hash output to edge indices is implemented using domain-separated rejection sampling.

Algorithm 1. HashToEdges (h, t, E)

1: $m \leftarrow |E|; \quad b \leftarrow \lambda$ $\triangleright$ λ-bit blocks from $\mathcal{H}$
2: **for** $i = 0$ to $t - 1$ **do**
3:      $j \leftarrow 0$
4:      **repeat**
5:         $B \leftarrow \mathcal{H}(\text{``EdgeDerive-v1''} \parallel h \parallel \langle i \rangle_{32} \parallel \langle j \rangle_{32})$
6:         $x \leftarrow \text{int}(B)$ $\triangleright$ interpret B as a big-endian integer in $[0, 2^{\lambda} - 1]$
7:         $M \leftarrow \lfloor 2^{\lambda}/m \rfloor \cdot m$
8:         $j \leftarrow j + 1$
9:      **until** $x < M$ $\triangleright$ rejection sampling removes modulo bias
10:      $\text{idx}_i \leftarrow x \bmod m; \quad e_i \leftarrow E[\text{idx}_i]$
11: **end for**
12: **return** $(e_0, \ldots, e_{t-1})$

Public Inputs and Secret Key. Public parameters are $(G = (V, E), k, f, \mathcal{H}, \lambda)$, where $V = \{1, \ldots, n\}$, $E \subseteq \{\{u, v\} \mid 1 \leq u < v \leq n\}$, f is a statistically hiding, computationally binding commitment, and $\mathcal{H} : \{0, 1\}^* \to \{0, 1\}^{\lambda}$ is modeled as a random oracle. The *secret key* is a valid k-coloring $\phi : V \to \{1, \ldots, k\}$.

Canonical Encoding with Domain Separation. We follow standard practice for domain separation [1,6] and canonical serialization [14,18], using a fixed domain-separation tag TAG = ``FS-GkColor-v1'' and a canonical serializer Encode($\cdot$):

$$\text{Encode}(G, k, X_0, \ldots, X_{t-1}, M) = \langle \text{TAG} \rangle \parallel \langle n \rangle_{64} \parallel \langle k \rangle_{32} \parallel \langle m \rangle_{64} \parallel \text{Edges}(E)$$
$$\parallel \langle t \rangle_{32} \parallel X_0 \parallel \cdots \parallel X_{t-1} \parallel \langle |M| \rangle_{64} \parallel M.$$

Here $\langle \cdot \rangle_b$ is a b-bit big-endian length/value encoding, and $\parallel$ is concatenation. We fix a vertex order $1 < \cdots < n$ and encode edges as

$$\text{Edges}(E) = \langle m \rangle_{64} \parallel \big\|_{\{u,v\} \in E^{\uparrow}} \big(\langle u \rangle_{\lceil \log_2 n \rceil} \parallel \langle v \rangle_{\lceil \log_2 n \rceil} \big),$$

where $E^{\uparrow}$ lists edges with $u < v$ in lexicographic order. This canonicalization ensures all parties hash identical byte strings.

Eidolon signature scheme Eidolon uses as public parameters a graph $G = (V, E)$, an integer $k \in \mathbb{N}$, a statistically hiding and computationally binding commitment

$$f : \{1, \ldots, k\} \times \{0, 1\}^r \to \{0, 1\}^s,$$

and a hash function $\mathcal{H} : \{0,1\}^* \to \{0,1\}^\lambda$, modeled as a random oracle. The secret key is a valid k-coloring $\phi : V \to \{1,\ldots,k\}$. Let $m = |E|$ and set $t = m^2$.

To sign a message $M \in \{0,1\}^*$, the signer samples, for each $i \in \{0,\ldots,t-1\}$, a fresh permutation $\pi_i \in S_k$ and, for every vertex $v \in V$, fresh randomness $r_v^{(i)}$, and computes

$$c_v^{(i)} = f\big(\pi_i(\phi(v)),\, r_v^{(i)}\big).$$

Let $C^{(i)} = (c_v^{(i)})_{v \in V}$. Then compute

$$h = \mathcal{H}\big(\mathrm{Encode}(G, k, C^{(0)}, \ldots, C^{(t-1)}, M)\big),$$

where $\mathrm{Encode}(\cdot)$ is a fixed canonical encoding with domain separation. The value h is parsed deterministically into a sequence of challenge edges $(e_0,\ldots,e_{t-1})$, where $e_i = (u_i, v_i) \in E$. For each i, the signer includes the openings

$$\big(\pi_i(\phi(u_i)),\, r_{u_i}^{(i)}\big) \qquad \text{and} \qquad \big(\pi_i(\phi(v_i)),\, r_{v_i}^{(i)}\big).$$

The signature is

$$\sigma = \Big(\{C^{(i)}\}_{i=0}^{t-1},\ \{\pi_i(\phi(u_i)), r_{u_i}^{(i)}, \pi_i(\phi(v_i)), r_{v_i}^{(i)}\}_{i=0}^{t-1} \Big).$$

Algorithm 2. EIDOLON.SIGN(G, k, ϕ, M)

1: **Input:** message M, private coloring $\phi : V \to \{1,\ldots,k\}$
2: **for** $i = 0,\ldots,t-1$ **do**
3:     choose a fresh random permutation $\pi_i \in S_k$
4:     **for** each $v \in V$ **do**
5:         choose fresh randomness $r_v^{(i)}$ and compute

$$c_v^{(i)} \leftarrow f(\pi_i(\phi(v)),\, r_v^{(i)}).$$

6:     **end for**
7:     set $C^{(i)} \leftarrow (c_v^{(i)})_{v \in V}$
8: **end for**
9: Compute

$$h \leftarrow \mathcal{H}\big(\mathrm{Encode}(G, k, C^{(0)}, \ldots, C^{(t-1)}, M)\big).$$

10: Parse h deterministically to obtain challenges $e_0,\ldots,e_{t-1}$, where each e_i encodes an edge $(u_i, v_i) \in E$
11: **for** $i = 0,\ldots,t-1$ **do**
12:     set

$$open_i \leftarrow \big(\pi_i(\phi(u_i)),\, r_{u_i}^{(i)},\, \pi_i(\phi(v_i)),\, r_{v_i}^{(i)}\big)$$

13: **end for**
14: Output signature

$$\sigma \leftarrow \Big(\{C^{(i)}\}_{i=0}^{t-1},\ \{open_i\}_{i=0}^{t-1}\Big)$$

Algorithm 3. EIDOLON.VERIFY(G, k, M, σ)

1: **Input:** public parameters (G, k), message M, signature σ
2: Parse σ as $\{C^{(i)}\}_{i=0}^{t-1}$ and openings $\{open_i\}_{i=0}^{t-1}$
3: Recompute
$$h' \leftarrow \mathcal{H}\big(\text{Encode}(G, k, C^{(0)}, \ldots, C^{(t-1)}, M)\big)$$
4: Parse h' into $e'_0, \ldots, e'_{t-1}$, where $e'_i = (u'_i, v'_i)$
5: **for** $i = 0, \ldots, t-1$ **do**
6:      Let $open_i = (\alpha_u, r_u, \alpha_v, r_v)$
7:      Check that
$$f(\alpha_u, r_u) \stackrel{?}{=} c_{u'_i}^{(i)} \qquad \text{and} \qquad f(\alpha_v, r_v) \stackrel{?}{=} c_{v'_i}^{(i)}.$$
8:      **if** either equality fails **then**
9:          **return** Reject
10:     **end if**
11:     **if** $\alpha_u = \alpha_v$ or $\alpha_u, \alpha_v \notin \{1, \ldots, k\}$ **then**
12:         **return** Reject
13:     **end if**
14: **end for**
15: **return** Accept

3.4 Merkle Compression of Commitments

In the plain scheme, each round i includes the full commitment vector

$$C^{(i)} = (c_v^{(i)})_{v \in V}, \qquad c_v^{(i)} = f\big(\pi_i(\phi(v)), r_v^{(i)}\big) \in \{0, 1\}^s.$$

If all t commitment vectors are included in the signature, the size is

$$|\sigma|_{\text{plain}} \approx t\, n\, s + 2t\big(|\alpha| + |r|\big),$$

where $n = |V|$, s is the commitment length, $|\alpha| \leq \lceil \log_2 k \rceil$, and $|r|$ is the commitment randomness length. This yields asymptotic size $O(tn)$.

We compress each vector $C^{(i)}$ to a single Merkle root R_i using a collision-resistant hash function $H : \{0,1\}^* \to \{0,1\}^\lambda$. Let $\text{enc}(v)$ be a fixed-length encoding of the vertex index v, and define

$$L_v^{(i)} = H\big(\text{leaf} \,\|\, \text{enc}(v) \,\|\, c_v^{(i)}\big).$$

The hash function $\mathcal{H}$ is modeled as a random oracle. For concrete instantiations, one should use a standardized hash with output length and security strength matched to the target security level.

A binary Merkle tree is built over the leaves $(L_v^{(i)})_{v \in V}$, and R_i denotes its root. Instead of publishing the whole vector $C^{(i)}$, the signer publishes only R_i. For each challenged vertex, the signature includes the corresponding opening (α, r) together with a Merkle authentication path.

In the Fiat–Shamir transformation, the hash is applied to the roots rather than to the full commitment vectors:

$$h = \mathcal{H}(\mathrm{Encode}(G, k, R_0, \ldots, R_{t-1}, M)).$$

The challenge edges are then derived from h as before. For each round i, if the challenge is $e_i = (u_i, v_i)$, the signer reveals

$$(\alpha_{u_i}, r_{u_i}, \mathsf{path}^{(i)}_{u_i}) \qquad \text{and} \qquad (\alpha_{v_i}, r_{v_i}, \mathsf{path}^{(i)}_{v_i}),$$

where $\alpha_{u_i} = \pi_i(\phi(u_i))$ and $\alpha_{v_i} = \pi_i(\phi(v_i))$.

Lemma 1. *Suppose H is collision-resistant and f is binding. Fix a round i and a vertex index v. Given the root R_i, it is infeasible to produce two distinct valid openings for position v.*

Proof. If two distinct openings $(\alpha, r) \neq (\alpha', r')$ produce the same commitment value, then the binding property of f is violated. Otherwise the corresponding leaves differ. If both leaves authenticate to the same root R_i, this yields a collision in the Merkle tree, contradicting the collision resistance of H.

Each Merkle root contributes λ bits, and each authentication path has length $\lceil \log_2 n \rceil$. The resulting signature size is therefore

$$|\sigma|_{\mathrm{Merkle}} \approx t\lambda + 2t(|\alpha| + |r| + \lambda\lceil \log_2 n \rceil),$$

which gives asymptotic size $O(t \log n)$.

A Python prototype of the Merkle-compressed construction was implemented to validate the encoding, challenge derivation, and signature-size formulas. On a small prototype instance, the measured signature sizes matched the analytic expressions derived above.

4   Security Model and Analysis

We use the standard notion of existential unforgeability under adaptive chosen-message attack (EUF–CMA) for signature schemes, and we analyze the construction in the random-oracle model.

Proposition 1. *Assume that f is computationally binding, that H is collision-resistant, and that $\mathcal{H}$ is modeled as a random oracle. Let*

$$\sigma = (\{R_i\}_{i=0}^{t-1}, \{open_i\}_{i=0}^{t-1})$$

be a signature accepted by the verifier for a message M. Then, except with negligible probability, for every round i, the opening $open_i$ is consistent with the challenged edge derived from

$$h = \mathcal{H}(\mathrm{Encode}(G, k, R_0, \ldots, R_{t-1}, M)),$$

and with the commitments authenticated under the Merkle root R_i. In particular, any successful forgery must either:

1. *produce valid openings for all challenged positions under the corresponding Merkle roots, or*
2. *violate the binding of f or the collision resistance of H.*

Proof. Let

$$h = \mathcal{H}(\mathrm{Encode}(G, k, R_0, \ldots, R_{t-1}, M))$$

be the hash value recomputed by the verifier, and let

$$(e_0, \ldots, e_{t-1})$$

be the sequence of challenged edges derived from h. Since the verifier is deterministic once (G, k, M, σ) is fixed, acceptance means that, for each round i, the verifier has parsed $open_i$, extracted the challenged edge $e_i = (u_i, v_i)$, and accepted all checks attached to that round.

Fix a round i. By construction of the verification algorithm, acceptance implies that the verifier has checked the Merkle authentication paths for the two challenged endpoints u_i and v_i, starting from the corresponding leaf values and ending at the published root R_i. Therefore, if one of these two openings were not consistent with R_i, then either the verifier would reject, or two distinct leaf values would authenticate to the same root. In the latter case, this would contradict the collision resistance of H.

Next, the verifier also checks that the revealed pairs (α_u, r_u) and (α_v, r_v) open the corresponding commitments. Hence, if an accepted opening at one of the challenged positions admitted two distinct valid decommitments, this would violate the binding property of f.

Therefore, except with negligible probability, every accepted round is simultaneously consistent with the challenged edge derived from the hash value, with the openings checked against the commitment function f, and with the Merkle root R_i. Applying the same argument to all rounds yields the claim.

The proposition above only establishes consistency of accepted transcripts with the published commitment structure. A complete EUF–CMA reduction would additionally require a witness-extraction argument for the Fiat–Shamir transform in this setting.

5   Security Analysis and Attack Models

5.1   Exact Attacks

To set appropriate parameters and evaluate the security of **Eidolon**, we analyze the practical difficulty of recovering the secret k-coloring from the public graph using a well-known algorithmic approach. In the context of this paper, we are interested in any coloring of the random graph with k or less colors, as this allows us to impersonate the prover. For a comprehensive analysis of the problem, see

Mann [16]. We use the **exactcolors** algorithm [11], which is based on branch-and-bound with a linear programming solver (Gurobi). This is currently one of the most efficient solvers for the coloring problem on ER-graphs [3].

We begin by confirming the properties of the random graph and the difficulty of finding the chromatic number of a random graph. Figure 1 illustrates the chromatic numbers found using the **exactcolors** solver for graphs drawn from $G_R(n, \frac{1}{2})$ with between 10 and 52 vertices and the minimal coloring found using the DSatur heuristic. We choose $p = 1/2$ as this maximizes the complexity of the graphs and the associated coloring problem. The Bollabas upper and lower bounds are also plotted. It is clear that the actual chromatic numbers are much larger than the asymptotic formula, which does not apply in this range. Instead, the chromatic number seems to follow a power law with best-fit equation $\chi(n, 1/2) = 0.88n^{0.611}$.

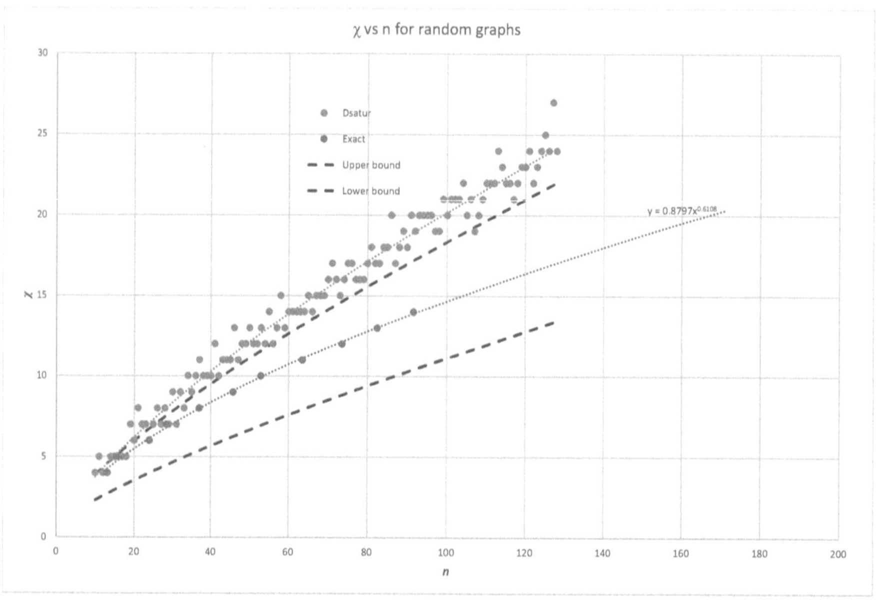

Fig. 1. The observed chromatic number of random graphs using an exact algorithm and the minimal coloring discovered by the DSatur algorithm. The upper and lower bounds of χ due to Bollobás are also plotted.

The secret key is embedded as a planted coloring in the graph. It seems natural to choose χ as the number of colors for the key, but embedding an additional solution may change the difficulty of the problem. To confirm the correct number of colors, we analyzed the solution time to find a coloring of size k embedded in graphs with expected $\chi = 13(n = 72)$, $\chi = 14(n = 82)$ and $\chi = 15(n = 93)$ (Fig. 2. The peak difficulty occurs either at $k = \chi$ or $k = \chi + 1$, and so we choose $k = \chi$.

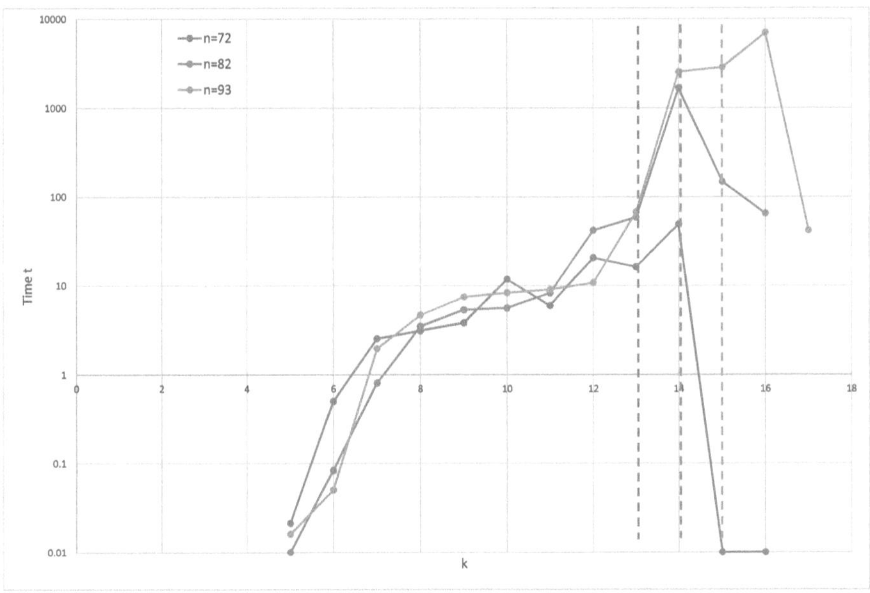

Fig. 2. The time taken to find the minimal coloring of graphs with different planted solution size k for graphs of size 72, 82 and 93. The vertical dashed lines represent the expected chromatic number of the equivalent ER graphs.

Finally, we demonstrate the relationship between n and solution time in Fig. 3. We select k for the planted solution to match the empirical prediction of χ for the equivalent random graph. The solution times appears, empirically, to be super-exponential. These results were produced on a AMD EPYC 7501 at 2.6 GHz with 512 GB of RAM.

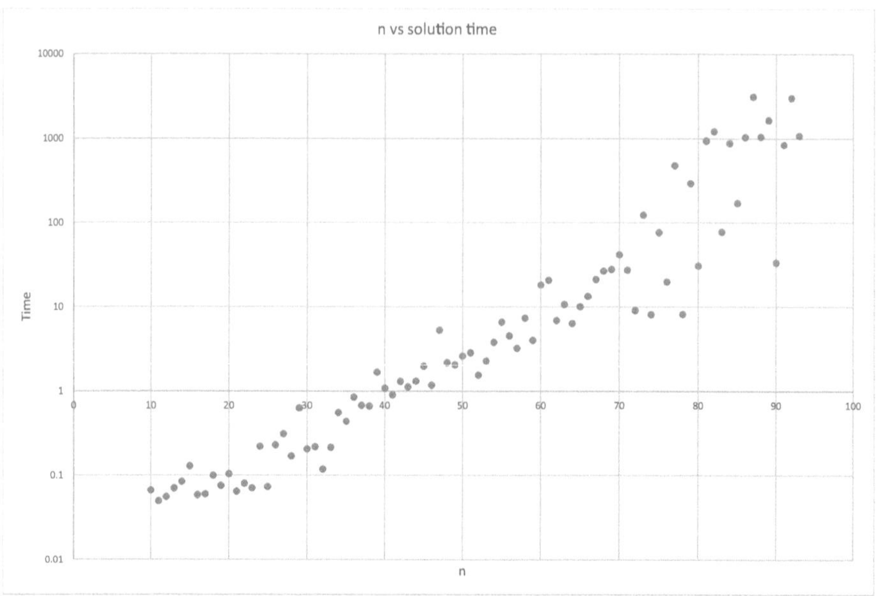

Fig. 3. The time-to-solution for the 'exactcolors' solver on planted graphs with planted k chosen according to the empirical power law.

5.2   Heuristic Attacks

The Dsatur algorithm [4] is a well-known greedy algorithm for coloring graphs. The algorithm sequentially selects vertices for coloring on the basis of the maximum current degree of saturation. It can recover the minimal coloring for small graphs but is known to be limited for larger graphs. We can see this in Fig. 1, where the size of the solution found by Dsatur is shown, along with the chromatic number discovered by the exact algorithm. While the method performs well for small graphs, a gap is evident for larger ones and DSatur cannot recover the minimal coloring for graphs larger than $n = 40$. This result is confirmed in Fig. 4 for planted graphs, where we can see identical behavior to the ER-graphs.

In [10], Gryak, Haralick, Kahrobaei have used machine learning algorithms to solve one of the algorithmic problems, known as conjugacy decision problem for certain classes of groups. This technique has been used for the cryptanalysis of proposed schemes using the conjugacy problem. Machine learning and pattern recognition techniques have been successfully applied to algorithmic problems in free groups. In [10], the authors seek to extend these techniques to finitely presented non-free groups, with a particular emphasis on polycyclic and metabelian groups that are of interest to non-commutative cryptography. As a prototypical example, they utilize supervised learning methods to construct classifiers that can solve the conjugacy decision problem, i.e., determine whether or not a pair of elements from a specified group are conjugate. The accuracies of classifiers created using decision trees, random forests, and N-tuple neural network models are evaluated for several non-free groups. The very high accuracy of these classifiers suggests an underlying mathematical relationship with respect to conjugacy in the tested groups.

Based on these results, we propose a modified version of the Graph Neural Network framework to tackle the GCP. The method is based on that of Schuetz et al. [21]. The method uses a set of random features initially assigned to the vertices of the graph. The GNN is then used to decode these features into color labels for the vertices. The network learns through a cost function to minimize the number of color conflicts in the assignment. We use the GraphSAGE variant described in the paper with one hidden layer. We optimized the parameter settings to find $d_0 = 45, d_1 = 40, \beta = 0.0091$, dropout $= 0.6$, epochs $= 20000$.

We use a modified cost function suggested by Porumbel et al. [19] which weights color violations by the inverse of the degree, favoring violations with a smaller number of neighbors to resolve. If the softmax output of the network is $\mathbf{p}_k$ for each color k, then the standard cost function is given by $\sum_k \mathbf{p}_k^T \mathbf{A} \mathbf{p}_k$. The modified cost function is

$$\sum_k \mathbf{p}_k^T (\mathbf{I} - \mathbf{D}^{-1}) \mathbf{A} \mathbf{p}_k \tag{3}$$

This down-weights the cost of violations on low degree vertices, as these are potentially easier to resolve. [1]

Since the network is fixed, the expected number of colors, k, must be chosen in advance. This is not straightforward, as we found that values of k higher than the known chromatic number of the graph produced better solutions in some cases. However, since the goal is to recover a coloring equivalent to the planted coloring (in order to break the security of the key), we use the same k as the planted coloring.

We found considerable improvement from the network by initializing it with the labels given by the DSatur algorithm. This is achieved by pre-training the network with the DSatur labelling as the target output, using 70 epochs, $\beta = 0.08$ and a cross-entropy loss. Since the DSatur algorithm will typically produce more than k colors, the coloring is reduced by replacing the excess colors with the least-conflicted of the initial k colors before training.

Finally, the network does not usually converge to a non-conflicted configuration (i.e. a graph coloring). We therefore post-process the labelling in a similar fashion to [21], by relabelling conflicted nodes with the smallest number of additional colors.

The results are shown in Fig. 4. The GNN algorithm marginally outperforms the Dsatur algorithm for intermediate sized graphs ($n \in [20, 60]$), but cannot recover the planted coloring for $n \geq 20$. Above $n = 60$, the performance is very similar to Dsatur. This suggests for $n > 60$ it is very challenging to recover the key with known algorithms. The plot also contains results for ER random graphs (lighter lines) showing very similar performance to the planted graphs.

---

[1] Our GNN-based attack code is available at https://github.com/EsmaMaysan/GNN-CSP/blob/main/GNN+CSP.py.

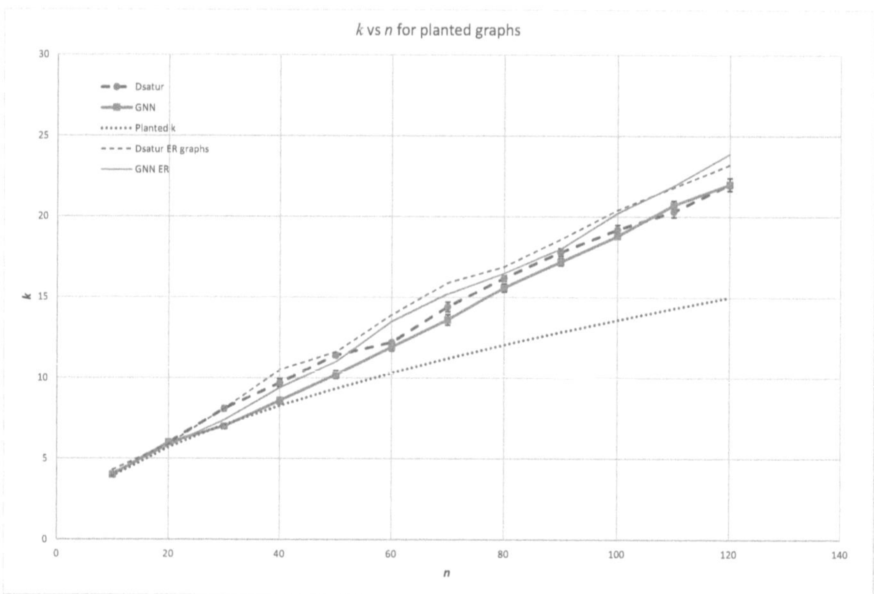

Fig. 4. The size of the recovered colorings for planted graphs using approximate algorithms. The dotted line shows the number of colors k used in the planted graphs. The dashed line gives the number of colors used by DSatur, and the solid line those used by the GNN algorithm. The lighter lines show the results for these algorithms on ER graphs, i.e. random graphs with no planted solution.

6 Conclusion

We have presented `Eidolon`, a practical post-quantum digital signature scheme based on the k-colorability problem, combining a generalized zero-knowledge protocol, Merkle-tree compression, and carefully constructed hard instances via planted colorings.

Our empirical security analysis provides a detailed evaluation of the resistance of the scheme against both exact and heuristic attacks. In particular, the results obtained using the **exactcolors** solver indicate that the time required to recover a valid k-coloring grows super-exponentially with the graph size when parameters are chosen according to the empirical chromatic number. This confirms that recovering the planted coloring becomes computationally infeasible even for moderate values of n.

On the heuristic side, both the DSatur algorithm and the proposed GNN-based approach fail to recover the planted coloring beyond small graph sizes. While the GNN model shows slight improvements over DSatur for intermediate sizes ($n \in [20, 60]$), it does not scale effectively and exhibits similar limitations for larger graphs. In particular, for $n \geq 60$, neither method is able to produce a valid coloring matching the planted solution, and their performance remains close to that observed on random graphs without planted structure.

Overall, these results demonstrate that the hardness of the underlying combinatorial problem is preserved even in the presence of modern heuristic and machine learning techniques. This supports the viability of graph coloring as a foundation for post-quantum cryptographic constructions when parameters are carefully selected.

Future work includes refining parameter selection to align with standardized post-quantum security levels, improving the efficiency of the construction, and further investigating the resilience of the scheme against more advanced attacks, including hybrid and adaptive strategies.

Acknowledgments. The authors acknowledge the support from the Institut Henri Poincaré (UAR 839 CNRS-Sorbonne Université) and LabEx CARMIN (ANR-10-LABX-59-01). This project started from discussions with Farinaz Koushanfar (UCSD) by Delaram Kahrobaei. Consequently restarted during the Trimester on Post-quantum Algebraic Cryptography at IHP. DK conducted this work partially with the support of ONR Grant 62909-24-1-2002. DK thank Institut des Hautes Études Scientifiques - IHES for providing stimulating environment while this project was partially done. DK was supported by the CARMIN fellowship during the completion of this project. RF thanks IMUS-Maria de Maeztu grant CEX2024-001517-M - Apoyo a Unidades de Excelencia María de Maeztu for supporting this research, funded by MICIU/AEI/ 10.13039/501100011033.

Disclosure of Interests. The authors declare that they have no competing interests relevant to the content of this article.

A   Standard Security Definitions

Definition 1 (Probabilistic polynomial-time algorithms and negligible functions [12, Sec. 1.3], [13, Sec. 3.1]). *An algorithm $\mathcal{A}$ is probabilistic polynomial time (PPT) if there exists a polynomial $p(\cdot)$ such that, for every input $x \in \{0,1\}^n$ and every random coin string r, the execution $\mathcal{A}(x; r)$ halts within at most $p(n)$ steps. A function $\mu : \mathbb{N} \to [0,1]$ is* negligible *if, for every polynomial $q(\cdot)$, there exists N such that $\mu(n) < 1/q(n)$ for all $n \geq N$.*

Definition 2 (EUF–CMA security [9, p. 21], [13, Ch. 12]). *Let $\Pi = (\mathsf{Gen}, \mathsf{Sign}, \mathsf{Vfy})$ be a signature scheme with security parameter λ. The EUF–CMA experiment is defined as follows.*

1. Setup: *The challenger samples $(\mathsf{pk}, \mathsf{sk}) \leftarrow \mathsf{Gen}(1^\lambda)$ and gives pk to the adversary $\mathcal{A}$.*
2. Signing queries: *The adversary $\mathcal{A}$ is given adaptive oracle access to $\mathsf{Sign}_{\mathsf{sk}}(\cdot)$. For each queried message m, it receives a signature $\sigma \leftarrow \mathsf{Sign}_{\mathsf{sk}}(m)$. Let Q denote the set of queried messages.*
3. Forgery: *Eventually, $\mathcal{A}$ outputs a pair $(m^\star, \sigma^\star)$.*

The adversary wins if

$$\mathsf{Vfy}_{\mathsf{pk}}(m^\star, \sigma^\star) = 1 \qquad and \qquad m^\star \notin Q.$$

The scheme Π is EUF–CMA secure *if every PPT adversary wins with only negligible probability in* λ.

References

1. Bertoni, G., Daemen, J., Peeters, M., Van Assche, G.: Sha-3 derived functions. Technical report SP 800–185, NIST (2016)
2. Bollobás, B.: The chromatic number of random graphs. Combinatorica **8**(1), 49–55 (1988). https://doi.org/10.1007/BF02122551
3. Brand, T., Faber, D., Held, S., Mutzel, P.: A customized sat-based solver for graph coloring. In: Chowdhury, R., Puglisi, S.J., Ren, B., Veldt, N. (eds.) Proceedings of the 28th Symposium on Algorithm Engineering and Experiments, ALENEX 2026, Vancouver, BC, Canada, January 11–12, 2026, pp. 142–155. SIAM (2026). https://doi.org/10.1137/1.9781611978957.11
4. Br'elaz, D.: New methods to color the vertices of a graph. Commun. ACM **22**(4), 251–256 (1979)
5. Fiat, A., Shamir, A.: How to prove yourself: practical solutions to identification and signature problems. In: Odlyzko, A.M. (ed.) CRYPTO 1986. LNCS, vol. 263, pp. 186–194. Springer, Heidelberg (1987). https://doi.org/10.1007/3-540-47721-7_12
6. Fluhrer, S., et al.: Hashing to elliptic curves. Tech. Rep. RFC 9380, IETF (2023)
7. Garey, M.R., Johnson, D.S.: The complexity of near-optimal graph coloring. J. ACM **23**(1), 43–49 (1976). https://doi.org/10.1145/321921.321926
8. Goldreich, O., Micali, S., Wigderson, A.: Proofs that yield nothing but their validity for all languages in NP have zero-knowledge proof systems. J. ACM **38**(3), 691–729 (1991). https://doi.org/10.1145/116825.116852
9. Goldwasser, S., Micali, S., Rivest, R.L.: A digital signature scheme secure against adaptive chosen-message attacks. SIAM J. Comput. **17**(2), 281–308 (1988). https://doi.org/10.1137/0217017
10. Gryak, J., Haralick, R., Kahrobaei, D.: Solving the conjugacy decision problem via machine learning. Exp. Math. **29**(1), 66–78 (2020)
11. Held, S., Cook, W.J., Sewell, E.C.: Maximum-weight stable sets and safe lower bounds for graph coloring. Math. Program. Comput. **4**(4), 363–381 (2012). https://doi.org/10.1007/S12532-012-0042-3
12. Katz, J.: Comparative book review: Cryptography: An introduction by v. v. yaschenko (american mathematical society, 2002); cryptanalysis of number theoretic ciphers by S.S. wagstaff, jr. (chapman & hall/crc press, 2003); RSA and public-key cryptography by r. a. mollin (chapman & hall/crc press, 2003); foundations of cryptography, vol. 1: Basic tools by o. goldreich, (Cambridge university press, 2001). SIGACT News **36**(2), 14–19 (2005). https://doi.org/10.1145/1067309.1067316
13. Katz, J., Lindell, Y.: Introduction to Modern Cryptography, Second Edition. CRC Press, Boca Raton (2014). https://www.crcpress.com/Introduction-to-Modern-Cryptography-Second-Edition/Katz-Lindell/p/book/9781466570269
14. Krawczyk, H., Eronen, P.: Hmac-based extract-and-expand key derivation function (HKDF). RFC, **5869**, 1–14 (2010). https://doi.org/10.17487/RFC5869

15. Krzakala, F., Zdeborová, L.: Hiding quiet solutions in random constraint satisfaction problems. CoRR abs/0901.2130 (2009). http://arxiv.org/abs/0901.2130
16. Mann, Z.Á.: Complexity of coloring random graphs: an experimental study of the hardest region. ACM J. Exp. Algorithmics **23** (2018). https://doi.org/10.1145/3183350
17. McDiarmid, C.: Colouring random graphs. Ann. Oper. Res. **1**(3), 183–200 (1984). https://doi.org/10.1007/BF01874388
18. Pornin, T.: Deterministic usage of the digital signature algorithm (DSA) and elliptic curve digital signature algorithm (ECDSA). RFC, **6979**, 1–79 (2013). https://doi.org/10.17487/RFC6979
19. Porumbel, D.C., Hao, J.-K., Kuntz, P.: A study of evaluation functions for the graph K-coloring problem. In: Monmarché, N., Talbi, E.-G., Collet, P., Schoenauer, M., Lutton, E. (eds.) EA 2007. LNCS, vol. 4926, pp. 124–135. Springer, Heidelberg (2008). https://doi.org/10.1007/978-3-540-79305-2_11
20. Schuetz, M.J.A., Brubaker, J.K., Katzgraber, H.G.: Combinatorial optimization with physics-inspired graph neural networks. Nat. Mach. Intell. **4**, 367–377 (2022)
21. Schuetz, M.J.A., Brubaker, J.K., Zhu, Z., Katzgraber, H.G.: Graph coloring with physics-inspired graph neural networks. CoRR abs/2202.01606 (2022). https://arxiv.org/abs/2202.01606

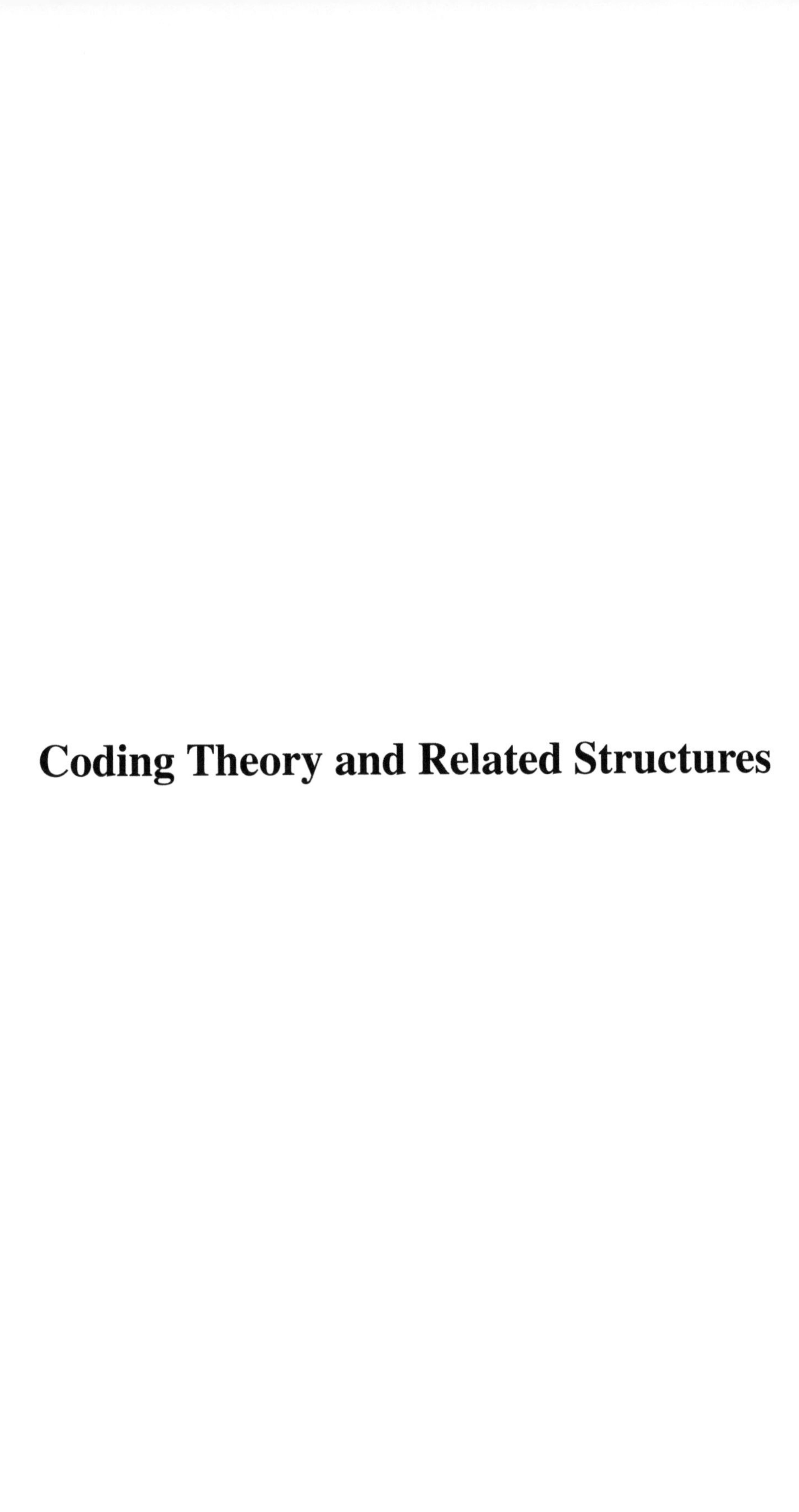

Coding Theory and Related Structures

Output Feedback Transformations for Preserving Observability and Structural Decodability in Convolutional Codes

Noemí DeCastro-García[1,2]($\boxtimes$) (iD), Lucía Mallo-Fernández[1,2] (iD),
and Miguel V. Carriegos[1,2] (iD)

[1] Departamento de Matemáticas, Universidad de León, Campus de Vegazana s/n,
24007 León, Spain
`{ncasg,lmalf,miguel.carriegos}@unileon.es`
[2] Research Institute of Applied Sciences in Cybersecurity, Universidad de León,
Campus de Vegazana s/n, 24007 León, Spain

Abstract. In this paper, we study the properties that an output feedback matrix over a finite field, $K \in \mathbb{F}_q^{k \times (n-k)}$, must satisfy to preserve some structural properties of the associated output feedback system. Our aim is to use these conditions to construct (n, k)-convolutional codes with optimal properties of observability and decoding via their linear associated dynamical systems or input/state/output representations. Thus, we derive explicit algebraic conditions under which output feedback preserves reachability, observability, and invertibility of the state matrix. Our approach is constructive, applying output feedback while simultaneously modifying the external signal space, thereby potentially producing new convolutional codes. Also, we derive explicit lower bounds on the proportion of admissible output feedback matrices that maintain the desired properties.

Keywords: Convolutional codes · ISO representations · Good Decodable Properties · Feedback invariants

1 Introduction

Convolutional codes are a fundamental class of error-correcting codes for communication systems. The construction of such codes with good properties, such as non-propagation of errors, existence of good decoding algorithms, or the distances, is typically addressed algebraically. However, the realization of these codes enables the application of control techniques to modulate the properties that will verify the associated convolutional codes. One of these realizations is the well-known Input/State/Output (ISO) representation, a linear system that models the dynamics of the convolutional code encoder.

In this work, we study how output feedback transformations affect ISO representations of convolutional codes. While output feedback is a classical tool

L. Batina and F. Özbudak (Eds.): WAIFI 2026, LNCS 16611, pp. 55–71, 2026.
https://doi.org/10.1007/978-3-032-27574-5_4

in control theory, its use in coding theory requires additional study. Although minimality is preserved under well-posed feedback, additional properties, such as invertibility of the state matrix, may be lost. Our main goal is to determine explicit algebraic conditions on the feedback matrix $K \in \mathbb{F}_q^{k \times (n-k)}$ that guarantee preservation of determined properties in the ISO representation in order to construct an observable and good decodable (GDP) associated convolutional code. We provide constructive criteria for selecting such feedback matrices and derive quantitative lower bounds on their density inside the parameter space. In particular, we prove that admissible output-feedback gains form a Zariski-open, dense subset, so structural decodability is generically preserved when the field size is sufficiently large.

From a coding-theoretic viewpoint, it is equally important that feedback transformations generate new convolutional codes. If the resulting encoders were equivalent to the original ones, no new algebraic structure would be obtained. We therefore analyze the extent to which output feedback produces non-equivalent codes.

Thus, rather than classifying realizations of a fixed code, we develop a constructive framework for generating new convolutional codes from existing ISO representations while preserving their essential structural and decoding properties.

This article is organized as follows: In Sect. 2, we have include the preliminary results. In Sect. 3, we have developed the main results. Finally, the conclusions and the references are given.

2   Preliminary Results

2.1   Linear Dynamical Systems

Consider a dynamical linear system Σ defined over a finite field $\mathbb{F}_q$, that we denote by $\mathbb{F}$, as follows:

$$\Sigma : \begin{cases} x_{t+1} = Ax_t + Bu_t \\ y_t = Cx_t + Du_t \end{cases} \tag{1}$$

where $A \in \mathbb{F}^{\delta \times \delta}$, $B \in \mathbb{F}^{\delta \times k}$, $C \in \mathbb{F}^{(n-k) \times \delta}$, and $D \in \mathbb{F}^{(n-k) \times k}$.

In the study of linear dynamical systems over finite fields, structural properties such as reachability and observability play a central role. These notions characterize whether the internal state can be influenced by external inputs and whether it can be reconstructed from output measurements. In the context of convolutional codes, these properties are not merely system-theoretic concepts: they directly determine minimality of the associated ISO representation and, consequently, structural properties of the generated code.

Definition 1 (Theorem 2.3, [1]). *Let $\Sigma = (A, B, C, D) \in \mathbb{F}^{\delta \times \delta} \times \mathbb{F}^{\delta \times k} \times \mathbb{F}^{(n-k) \times \delta} \times \mathbb{F}^{(n-k) \times k}$ be a dynamical linear system. Σ is reachable if and only if its controllability matrix*

$$\Phi_\delta(A, B) = \begin{pmatrix} B\ AB\ \dots\ A^{\delta-2}B\ A^{\delta-1}B \end{pmatrix}$$

has full rank; that is, $rank(\Phi_\delta(A,B)) = \delta$.

While reachability guarantees that the state space is fully reachable from the input, observability ensures that no internal state component remains hidden from the output.

Definition 2 (Theorem 2.6, [1]). *Let $\Sigma = (A,B,C,D) \in \mathbb{F}^{\delta \times \delta} \times \mathbb{F}^{\delta \times k} \times \mathbb{F}^{(n-k) \times \delta} \times \mathbb{F}^{(n-k) \times k}$ be a dynamical linear system. Σ is observable if and only if $rank(\Omega_\delta(A,C)) = \delta$, where $\Omega_\delta(A,C)$ is the observability matrix defined by*

$$\Omega_\delta(A,C) = \begin{pmatrix} C \\ CA \\ CA^2 \\ \vdots \\ CA^{\delta-1} \end{pmatrix}.$$

Also, we can consider the Hautus test for observability:

Definition 3 ([1,9]). *Let $\Sigma = (A,B,C,D) \in \mathbb{F}^{\delta \times \delta} \times \mathbb{F}^{\delta \times k} \times \mathbb{F}^{(n-k) \times \delta} \times \mathbb{F}^{(n-k) \times k}$ be a dynamical linear system. Σ is observable if and only if*

$$\mathrm{rk}\begin{pmatrix} \lambda I_\delta - A \\ C \end{pmatrix} = \delta \quad \text{for all } \lambda \in \overline{\mathbb{F}} \text{ where } \overline{\mathbb{F}} \text{ denotes the fixed algebraic closure of } \mathbb{F}.$$

Equivalently, Σ is not observable if and only if there exist $\lambda \in \overline{\mathbb{F}}$ and $x \neq 0$ such that $(A - \lambda I)x = 0$ and $Cx = 0$.

Definition 4 (Ch.10, Definition 6.1 [12]). *Let $\Sigma = (A,B,C,D) \in \mathbb{F}^{\delta \times \delta} \times \mathbb{F}^{\delta \times k} \times \mathbb{F}^{(n-k) \times \delta} \times \mathbb{F}^{(n-k) \times k}$ be a dynamical linear system. Σ is called minimal if it is both reachable and observable.*

Definition 5 (Definition 10 [5]). *Let $\Sigma = (A,B,C,D) \in \mathbb{F}^{\delta \times \delta} \times \mathbb{F}^{\delta \times k} \times \mathbb{F}^{(n-k) \times \delta} \times \mathbb{F}^{(n-k) \times k}$ be a dynamical linear system. For $L \in \mathbb{N}$, define*

$$T_L := [\Omega_{L+1}(A,C) \mid F_L] \in \mathbb{F}^{(L+1)(n-k) \times (\delta + (L+1)k)} = \left(\begin{array}{c|cccc} C & D & & & \\ CA & CB & D & & \\ CA^2 & CAB & CB & D & \\ \vdots & \vdots & & \ddots & \ddots \\ CA^L & CA^{L-1}B & CA^{L-2}B & \cdots & CB & D \end{array} \right).$$

We say that Σ is output observable *if there exists $L \in \mathbb{N}$ such that $\mathrm{rk}(T_L) = \delta + (L+1)k$. That is, T_L has full column rank.*

2.2   ISO Representations of Convolutional Codes over Finite Fields

Definition 6 ([15]). *An (n,k,δ)-convolutional code $\mathcal{C}$ over a finite field is a submodule of $\mathbb{F}[z]^n$ of rank k where n is the number of output symbols, k is the number of input symbols, and δ is the complexity of the code.*

Fix integers $n > k \geq 1$ and $\delta \geq 1$.

Definition 7 ([6])**.** *A convolutional encoder $G(z) \in \mathbb{F}[z]^{n \times k}$ of full column rank over $\mathbb{F}(z)$ represents the transformation of an input sequence $u(z)$ into an output sequence $v(z)$. It is given by $v(z) = G(z) \cdot u(z)$. Equivalently, $\mathcal{C} = \mathrm{Im}_{\mathbb{F}[z]}\, G(z) = \{\, G(z)u(z) : u(z) \in \mathbb{F}[z]^k \,\}$.*

We aim to construct new convolutional codes with good properties from other convolutional codes, ensuring that the new codes are not equivalent to the previous ones. Then, an essential property is the equivalent notion that we will use.

Definition 8 ([15])**.** *Two polynomial matrices $G_1(z), G_2(z) \in \mathbb{F}[z]^{n \times k}$ are said to be unimodularly equivalent if there exists a unimodular matrix $U(z) \in GL_k(\mathbb{F}[z])$ such that $G_2(z) = G_1(z)U(z)$. In this case,*

$$\mathrm{Im}_{\mathbb{F}[z]}\, G_1 = \mathrm{Im}_{\mathbb{F}[z]}\, G_2,$$

and hence both matrices generate the same convolutional code.

Definition 9 (Definition 1.1 [8])**.** *Two convolutional codes $\mathcal{C}_1, \mathcal{C}_2 \subseteq \mathbb{F}[z]^n$ are said to be monomially equivalent if there exists a monomial matrix $M \in GL_n(\mathbb{F})$, that is, a permutation matrix multiplied by a non-singular diagonal matrix such that $\mathcal{C}_2 = M\mathcal{C}_1$. If M is restricted to be a permutation matrix, the codes are said to be permutation equivalent.*

Unimodular equivalence preserves the generated submodule, while monomial equivalence allows for coordinate permutations and scalings of the ambient space. These notions are standard in the theory of convolutional codes, see [6,8,16].

One essential property of convolutional codes is the observability of the code, a characteristic that implies that the code is non-catastrophic.

Definition 10 (Lemma 3.3.2, [21])**.** *Let $G(z)$ be an encoder of an (n, k, δ)-convolutional code $\mathcal{C}$ over $\mathbb{F}$. The code $\mathcal{C}$ is observable if and only if there exists a Forney syndrome ψ such that the following sequence is exact:*

$$0 \longrightarrow \mathbb{F}[z]^k \xrightarrow{\;G(z)\;} \mathbb{F}[z]^n \xrightarrow{\;\psi\;} \mathbb{F}[z]^{n-k} \longrightarrow 0.$$

On the other hand, the construction of convolutional codes with good decoding properties is a central topic in modern coding theory. Let $u(z)$ be an information word, and let $G(z)$ be an encoder for an (n, k, δ)-convolutional code. The codeword $v(z) = G(z) \cdot u(z)$ is transmitted through a channel, where errors may be added or erasures may occur. The objective of the decoding process is to recover $v(z)$ from the erroneous codeword. Thus, an active line of research is the construction of GDP convolutional codes; that is, codes for which efficient decoding algorithms exist. In particular, $\mathrm{GDP}_{\mathrm{noisy}}$ codes guaranty the existence of efficient algebraic decoding algorithms in the presence of substitution errors, whereas $\mathrm{GDP}_{\mathrm{erasure}}$ codes allow efficient recovery under erasure.

The construction of convolutional codes that simultaneously satisfy observability and decoding efficiency conditions constitutes a challenge. One of the best-known approaches to facilitate this construction of optimal convolutional codes is through ISO representations.

Definition 11 ([15]). *Let $C \subset \mathbb{F}[z]^n$ be an (n, k, δ)-convolutional code. An ISO representation of C is a tuple of matrices $\Sigma(C) = (A, B, C, D) \in \mathbb{F}^{\delta \times \delta} \times \mathbb{F}^{\delta \times k} \times \mathbb{F}^{(n-k) \times \delta} \times \mathbb{F}^{(n-k) \times k}$ such that*

$$C = \left\{ v(z) = \begin{pmatrix} y(z) \\ u(z) \end{pmatrix} \in \mathbb{F}[z]^n : \exists\, x(z) \in \mathbb{F}[z]^\delta \text{ such that } \begin{cases} x_{t+1} = Ax_t + Bu_t, \\ y_t = Cx_t + Du_t, \\ v_t = \begin{pmatrix} y_t \\ u_t \end{pmatrix}, \\ x_0 = 0. \end{cases} \right\}.$$

Here $x_t \in \mathbb{F}^\delta$ denotes the state vector, $u_t \in \mathbb{F}^k$ is the information vector, and $y_t \in \mathbb{F}^{n-k}$ is the output vector. The vector v_t is the transmitted code vector.

In order to compute an ISO representation $\Sigma(C) = (A, B, C, D)$ of a convolutional code C starting from a polynomial encoder $G(z)$, it is necessary to pass through a first-order representation. Indeed, first-order representations provide an intermediate linear-algebraic description of the code that makes explicit the relation between polynomial encoders and state-space realizations. Recently, algorithms have been designed to transform the encoder directly into Σ [2]. Nevertheless, the first-order representation remains a fundamental concept for the ISO, as it clarifies the structural constraints underlying its construction.

Let $C \subset \mathbb{F}[z]^n$ be an (n, k, δ)-convolutional code. The matrices $(\mathcal{K}, \mathcal{L}, \mathcal{M})$ are called a first-order representation of the code C if they satisfy the following conditions [15]:

(i) $C = \{v(z) \in \mathbb{F}[z]^n : \exists\, x(z) \in \mathbb{F}[z]^\delta \text{ such that } z\mathcal{K}x(z) + \mathcal{L}x(z) + \mathcal{M}v(z) = 0\}$.
(ii) $\mathcal{K}$ has full rank.
(iii) $(\mathcal{K}, \mathcal{M})$ has full rank.

Moreover, if $(\mathcal{K}, \mathcal{L}, \mathcal{M})$ also satisfies the additional property

(iv) $(z\mathcal{K} + \mathcal{L}, \mathcal{M})$ has full rank,

then the first-order representation is called minimal. From $(\mathcal{K}, \mathcal{L}, \mathcal{M})$, an encoder $G(z)$ can be obtained by computing a minimal basis of the free $\mathbb{F}[z]$-module

$$\ker(z\mathcal{K} + \mathcal{L} \mid \mathcal{M}) = \{v(z) \in \mathbb{F}[z]^n \mid \exists\, x(z) \in \mathbb{F}[z]^\delta :$$

$$(z\mathcal{K} + \mathcal{L})x(z) + \mathcal{M}v(z) = 0\}.$$

Then, from $(\mathcal{K}, \mathcal{L}, \mathcal{M})$, and after suitable similarity transformation and a permutation of the components of the code vector, the representation Σ can be obtained as

$$\mathcal{K}' = \begin{pmatrix} -I_\delta \\ 0 \end{pmatrix}, \quad \mathcal{L}' = \begin{pmatrix} A_{\delta \times \delta} \\ C_{(n-k) \times \delta} \end{pmatrix}, \quad \mathcal{M}' = \begin{pmatrix} 0 & B_{\delta \times k} \\ -I_{n-k} & D_{(n-k) \times k} \end{pmatrix}.$$

From a dynamical systems perspective, the system $\Sigma(\mathcal{C})$ is an reachable system (Lemma 5.3.5, York, 1997).

As we mentioned above, one of the advantages of the ISO representations is that let us use structural properties of the system to construct non-catastrophic and GDP convolutional codes. Thus,

1. If $\Sigma(\mathcal{C})$ is a reachable and observable system, then $\Sigma(\mathcal{C})$ is a minimal ISO representation of an observable/non-catastrophic convolutional code $\mathcal{C}$ [15].
2. If we have an (n, k, δ)-convolutional code $\mathcal{C}(\Sigma) \subseteq \mathbb{F}[z]^n$, any code sequence v_t must satisfy the system dynamics (Proposition 2.6, [14]):

$$\begin{pmatrix} C \\ CA \\ \vdots \\ CA^L \end{pmatrix} x_t + \left(-I_{L+1} \left| \begin{matrix} D & 0 & 0 & \cdots & 0 & 0 \\ CB & D & 0 & \cdots & \vdots & \vdots \\ CAB & CB & D & \cdots & 0 & 0 \\ \vdots & \vdots & \vdots & \ddots & \vdots & \vdots \\ CA^{L-2}B & CA^{L-3}B & CA^{L-4}B & \cdots & D & 0 \\ CA^{L-1}B & CA^{L-2}B & CA^{L-3}B & \cdots & CB & D \end{matrix} \right. \right) \begin{pmatrix} y_t \\ y_{t+1} \\ \vdots \\ y_{t+L} \\ u_t \\ u_{t+1} \\ \vdots \\ u_{t+L} \end{pmatrix} = 0$$

which we denote by $\Omega_{L+1}(C, A) \cdot x_t + (-I_{L+1} \mid F_L) \cdot v_t = 0$, where F_L is a block Toeplitz matrix. Then, if Σ is reachable, observable, and A is invertible, then $\mathcal{C}(\Sigma)$ is observable and GDP_{noisy} [14]. On the other hand, if Σ is reachable, observable, and output observable, then $\mathcal{C}(\Sigma)$ is observable and $GDP_{erasure}$ [19].

In this work, we focus on properties of observability and GDP_{noisy}.

2.3   Output Feedback Transformations

Let $\Sigma = (A, B, C, D) \in \mathbb{F}^{\delta \times \delta} \times \mathbb{F}^{\delta \times k} \times \mathbb{F}^{(n-k) \times \delta} \times \mathbb{F}^{(n-k) \times k}$ be a dynamical linear system over $\mathbb{F}$ such that

$$\Sigma : \begin{cases} x_{t+1} = Ax_t + Bu_t \\ y_t = Cx_t + Du_t \end{cases} \tag{2}$$

where $x_t \in \mathbb{F}^\delta$ is the state, $u_t \in \mathbb{F}^k$ the input, and $y_t \in \mathbb{F}^{n-k}$ the output. We can transform the system Σ by the output feedback. Given an output-feedback matrix $K \in \mathbb{F}^{k \times (n-k)}$, we can consider the static output feedback law

$$u(t) = K\, y(t) + v(t), \tag{3}$$

where again $v_t \in \mathbb{F}^k$ is the new external input. Substituting $y_t = Cx_t + Du_t$ into (3) gives $(I_k - KD)\, u_t = KC\, x_t + v_t$.

Definition 12 ([18]). *We say that the system is* well-posed *if* $I_k - KD \in GL_k(\mathbb{F})$. *Under this condition, define* $\Theta := (I_k - KD)^{-1} \in \mathbb{F}^{k \times k}$.

Then, the transformed system given by applying K, Σ_K, is given by

$$\Sigma_K := (A_K := A + B\Theta KC, B_K := B\Theta, C_K := (I_{n-k} - DK)^{-1}C, D_K := (I_{n-k} - DK)^{-1}D).$$
$$(4)$$

From the perspective of convolutional codes, the output feedback law replaces the original information sequence by a new external input, while the transmitted sequence remains $v_t = \begin{pmatrix} y_t \\ u_t \end{pmatrix}$. Consequently, the ISO representation Σ_K induces a new convolutional code $\mathcal{C}_K(\Sigma_K)$ in the external variables (y, v). Equivalently, in the polynomial domain, if $G(z)$ is an encoder associated with Σ, the output feedback induces a new encoder $G_K(z)$ corresponding to Σ_K, where the input variable is $v(z)$ instead of $u(z)$. Hence, the induced submodule $\mathcal{C}_K(\Sigma_K) = \mathrm{Im}_{\mathbb{F}[z]}G_K(z)$ may differ from $\mathcal{C}(\Sigma)$, even though both arise from ISO representations related by output feedback. Thus, output feedback acts as a code-generating operation rather than as a realization equivalence.

3   Main Results

As mentioned above, the objective of this work is to determine which conditions must be satisfied by the matrices of Σ and by the output-feedback matrix K so that Σ_K can be regarded as an ISO representation of a convolutional code while preserving optimal properties originally possessed by $\mathcal{C}(\Sigma)$. However, the true interest of this approach arises when $\mathcal{C}_K(\Sigma_K)$ and $\mathcal{C}(\Sigma)$ are not equivalent codes. In that case, feedback provides a constructive mechanism for generating new convolutional codes that retain desirable structural properties [4].

The equivalence of convolutional codes and their state-space realizations has been extensively studied in the literature. In particular, [8] characterized all minimal state-space realizations of a fixed encoder and describe the action of the full state feedback group on canonical realizations. Their results show that two minimal realizations generate the same convolutional code if and only if they are related by suitable state transformations together with unimodular input transformations. Thus, their framework concerns different realizations of a fixed submodule of $\mathbb{F}[z]^n$. Related notions of feedback equivalence have also been studied over more finite rings in [3], extending the classical theory beyond the field case. These works analyze when two system representations are equivalent under feedback transformations and generate the same code over finite rings. However, none of the above references address the situation considered in this work, where a feedback transformation is applied to an ISO representation and, simultaneously, the external signal space is modified. In particular, the cited results establish equivalence of realizations of a fixed code, but they do not assert that an arbitrary output feedback transformation preserves the generated submodule when the transmitted variables are changed. Consequently, the feedback constructions studied in this work operate outside the standard equivalence framework of [3,8].

In general, applying output feedback to an ISO representation and then considering the induced code in the corresponding external variable space does not preserve code equivalence (neither permutation nor monomial equivalence). It suffices to exhibit explicit a counterexample. Since unimodular equivalence is stronger than monomial (or permutation) equivalence, non-existence of such $U(z)$ implies the codes are not monomially (hence not permutation) equivalent.

Example 1. Consider the following $(n, k, \delta) = (2, 1, 1)$ system $\Sigma = (0, 1, 1, 0)$ over $\mathbb{F}_2$. Since $B = 1 \neq 0$, Σ is reachable, then Σ can be considered an ISO representation for a convolutional code. Apply output feedback with $K = 1 \in \mathbb{F}_2^{1 \times 1}$ (well-posed since $I - KD = 1$), the output feedback system is $\Sigma_K = (A_K, B_K, C_K, D_K) = (A + BKC,\ B,\ C,\ D) = (1, 1, 1, 0)$. Again $B_K = 1 \neq 0$, hence Σ_K is reachable. Then, we can consider both of them as ISO representations for convolutional codes.

The computation of $\ker(zK + L \mid M)$ provides us an encoder of $\mathcal{C} = \operatorname{Im} G$, $G(z) = \begin{pmatrix} 1 \\ z \end{pmatrix} \in \mathbb{F}_2[z]^{2 \times 1}$. In the same way, we obtain $G_K(z) = \begin{pmatrix} 1 \\ z + 1 \end{pmatrix} \in \mathbb{F}_2[z]^{2 \times 1}$. If $G_K(z) = G(z)U(z)$ for some unimodular $U(z) \in GL_1(\mathbb{F}_2[z])$, then again $U(z) = 1$, and hence we would require

$$\begin{pmatrix} 1 \\ z + 1 \end{pmatrix} = \begin{pmatrix} 1 \\ z \end{pmatrix},$$

which is impossible. Thus no unimodular $U(z)$ exists.

3.1 Preservation of Minimal ISO Representations Under Output Feedback

Proposition 1. *Let $\Sigma = (A, B, C, D) \in \mathbb{F}^{\delta \times \delta} \times \mathbb{F}^{\delta \times k} \times \mathbb{F}^{(n-k) \times \delta} \times \mathbb{F}^{(n-k) \times k}$ be a dynamical linear system. Assume that Σ is reachable. Then, for any well-posed $K \in \mathbb{F}^{k \times (n-k)}$, Σ_K is reachable.*

Proof. This is a classical result in linear systems theory (reachability is invariant under well-posed static output feedback). See, for instance, [11, 20]. $\qquad\square$

Proposition 2. *Let $\Sigma = (A, B, C, D) \in \mathbb{F}^{\delta \times \delta} \times \mathbb{F}^{\delta \times k} \times \mathbb{F}^{(n-k) \times \delta} \times \mathbb{F}^{(n-k) \times k}$ be a dynamical linear system. Assume that Σ is minimal. Then, for any well-posed $K \in \mathbb{F}^{k \times (n-k)}$, the system Σ_K is minimal.*

Proof. Reachability of Σ_K follows from Proposition 1. The invariance of the observability under well-posed static output feedback is classical in linear systems theory. See, for instance, [11, 13]. $\qquad\square$

Theorem 1. *Let $\Sigma = (A, B, C, D) \in \mathbb{F}^{\delta \times \delta} \times \mathbb{F}^{\delta \times k} \times \mathbb{F}^{(n-k) \times \delta} \times \mathbb{F}^{(n-k) \times k}$ be a minimal ISO representation of an observable (n, k, δ)-convolutional code $\mathcal{C}(\Sigma)$. Then, for any well-posed $K \in \mathbb{F}^{k \times (n-k)}$, Σ_K is a minimal ISO representation for an observable (n, k, δ)-convolutional code $\mathcal{C}_K(\Sigma_K)$.*

Proof. By Proposition 1, Σ_K is reachable. Observability of Σ_K follows from Proposition 2. Hence Σ_K is minimal and therefore defines an observable (n, k, δ)-convolutional code $\mathcal{C}_K$.

Example 2. Consider an $(n, k, \delta) = (2, 1, 1)$ dynamical linear system $\Sigma = (A, B, C, D) = (0, 1, 1, 0)$ over $\mathbb{F}_2$. Since $B = 1 \neq 0$, Σ is reachable, and since $C = 1 \neq 0$, Σ is observable. Hence Σ is minimal. Now apply output feedback with $K = 1 \in \mathbb{F}_2$. The output feedback is well-posed because $I_1 - KD = 1 \neq 0$. Moreover $\Theta = (I_1 - KD)^{-1} = 1$ and therefore $\Sigma_K = (1, 1, 1, 0)$, which is again minimal (reachability and observability hold).

Using the first-order representation, we obtain an encoder for $C(\Sigma)$, $G(z) = \begin{pmatrix} 1 \\ z \end{pmatrix} \in \mathbb{F}_2[z]^{2 \times 1}$. Repeating the same procedure for Σ_K yields an encoder for $C_K(\Sigma_K)$, $G_K(z) = \begin{pmatrix} 1 \\ z+1 \end{pmatrix} \in \mathbb{F}_2[z]^{2 \times 1}$. We claim that $G_K(z)$ is not unimodularly equivalent to $G(z)$. Indeed, if $G_K(z) = G(z)U(z)$ for some unimodular $U(z) \in GL_1(\mathbb{F}_2[z])$, then necessarily $U(z) = 1$, and hence we would require $\begin{pmatrix} 1 \\ z+1 \end{pmatrix} = \begin{pmatrix} 1 \\ z \end{pmatrix}$, which is impossible. Therefore, the output feedback with $K = 1$ produces a non-unimodularly equivalent encoder.

From a coding-theoretic perspective, it is natural to quantify how many output-feedback matrices preserve the properties required for an ISO representation to define an observable convolutional code. Since minimality is preserved under well-posed feedback, the first step consists in estimating the number of well-posed output feedback matrices.

Proposition 3. *Let $\Sigma = (A, B, C, D) \in \mathbb{F}^{\delta \times \delta} \times \mathbb{F}^{\delta \times k} \times \mathbb{F}^{(n-k) \times \delta} \times \mathbb{F}^{(n-k) \times k}$ be a dynamical linear system. Set $r := n - k$ and $N := kr$. Define the well-posedness polynomial*

$$P_{\mathrm{wp}}(K) := \det(I_k - KD) \in \mathbb{F}[K_{ij}], \qquad K \in \mathbb{F}^{k \times r} \simeq \mathbb{F}^N.$$

This is not identically zero and $\deg P_{\mathrm{wp}} \leq k$. Consider the set of well-posed output feedback matrices defined as $\mathcal{W} := \{K \in \mathbb{F}^{k \times r} \mid P_{\mathrm{wp}}(K) \neq 0\}$. Then

$$|\mathcal{W}| \geq q^N - k\,q^{N-1} = q^{kr}\left(1 - \frac{k}{q}\right).$$

Equivalently, a uniformly random $K \in \mathbb{F}^{k \times r}$ is well-posed with probability at least $1 - \frac{k}{q}$.

Proof. By definition, K is well-posed if and only if $\det(I_k - KD) \neq 0$. The polynomial P_{wp} is not identically zero since $P_{\mathrm{wp}}(0) = \det(I_k) = 1$. Each entry of $I_k - KD$ is an affine linear form in the variables K_{ij}, so $\deg P_{\mathrm{wp}} \leq k$. By the Schwartz–Zippel bound [17,22], $|\{K \in \mathbb{F}^N \mid P_{\mathrm{wp}}(K) = 0\}| \leq (\deg P_{\mathrm{wp}})\,q^{N-1} \leq kq^{N-1}$, which yields the stated inequality.

Proposition 3 shows that over $\mathbb{F}$ there exist at least $q^{k(n-k)}\left(1 - \frac{k}{q}\right)$ output-feedback matrices $K \in \mathbb{F}^{k \times (n-k)}$ for which the feedback is well posed. In particular, when q is large, almost every matrix K yields a well-defined ISO realization. Combined with Theorem 1, this implies that there are at least $q^{k(n-k)}\left(1 - \frac{k}{q}\right)$

choices of K that produce a minimal ISO representation and therefore generate observable convolutional codes with parameters (n, k, δ).

In the single-input ISO cases $(k = 1)$, unimodular equivalence reduces to multiplication by a nonzero scalar. This allows us to estimate how many well-posed output feedback matrices produce observable convolutional codes that are equivalent to the original one. The following result shows that this set is contained in a proper algebraic subset and therefore is small when the field size is large.

Proposition 4. *Assume* $k = 1$ *and let* $\Sigma = (A, B, C, D) \in \mathbb{F}^{\delta \times \delta} \times \mathbb{F}^{\delta \times 1} \times \mathbb{F}^{(n-1) \times \delta} \times \mathbb{F}^{(n-1) \times 1}$ *be a minimal ISO representation. For each well-posed* $K \in \mathbb{F}^{1 \times (n-1)}$, *let* $G_K(z) \in \mathbb{F}[z]^{n \times 1}$ *denote the encoder obtained from* Σ_K *via a minimal basis of* $\ker(z\mathcal{K}_K + \mathcal{L}_K \mid \mathcal{M}_K)$, *and let* $G(z)$ *be the encoder obtained from* Σ *by the same procedure,* $(\mathcal{K}_K, \mathcal{L}_K, \mathcal{M}_K)$ *being the first order representation of* Σ_K. *Define the set of output feedback matrices producing unimodularly equivalent encoders*

$$\mathcal{E} := \left\{ K \in \mathcal{W} \ \middle| \ \exists \lambda \in \mathbb{F}^\times \ \text{such that} \ G_K(z) = \lambda\, G(z) \right\}.$$

where $\mathbb{F}^\times$ *denotes the multiplicative group of the* $\mathbb{F}$ *defined as* $\mathbb{F}^\times := \mathbb{F} \setminus \{0\}$, *with the multiplication operation. Assume there exists at least one well-posed* K_0 *such that* $G_{K_0}(z)$ *is not unimodularly equivalent to* $G(z)$. *Then*

$$|\mathcal{E}| \ \leq \ (n - 1)(\delta + 1)\, q^{n-2}, \qquad \text{and hence} \qquad \frac{|\mathcal{E}|}{|\mathcal{W}|} \ \leq \ \frac{(n - 1)(\delta + 1)}{q - k}.$$

In particular, for $q > (n - 1)(\delta + 1) + k$, *most well-posed* Ks *produce encoders that are not unimodularly equivalent to* $G(z)$.

Proof. Since $k = 1$, the polynomial module $\ker(z\mathcal{K}_K + \mathcal{L}_K \mid \mathcal{M}_K) \subset \mathbb{F}[z]^n$ is free of rank 1. Hence it admits a minimal basis (in the sense of Forney [6]), consisting of a single column vector $G_K(z) \in \mathbb{F}[z]^{n \times 1}$. Such a generator is unique up to multiplication by a nonzero scalar in $\mathbb{F}^\times$. In order to make this choice canonical, we fix once and for all a normalization convention: namely, we select the unique representative in the scalar equivalence class of minimal generators for which a prescribed leading coefficient. For instance, we can consider the highest-degree coefficient of the last component is equal to 1. With this normalization, the encoder $G_K(z)$ is uniquely determined as a polynomial vector in $\mathbb{F}[z]^n$. Write

$$G(z) = \sum_{t=0}^{\delta} G_t z^t, \qquad G_K(z) = \sum_{t=0}^{\delta} G_{K,t} z^t,$$

with coefficient vectors $G_t, G_{K,t} \in \mathbb{F}^n$. Under the fixed normalization, each coefficient vector $G_{K,t}$ depends algebraically, indeed, polynomially on a Zariski-open set, on the entries of K. Since $k = 1$, unimodular equivalence means multiplication by a nonzero constant: $GL_1(\mathbb{F}[z]) = \mathbb{F}^\times$. Thus $G_K(z) = \lambda G(z)$ for some $\lambda \in \mathbb{F}^\times$ if and only if all 2×2 minors of the $n \times 2$ polynomial matrix $(G(z) \mid G_K(z))$

vanish identically. Equivalently, for each $i \in \{1,\ldots,n-1\}$, the 2×2 minor formed by the i-th and n-th rows must vanish: $g_i(z)g_{K,n}(z) - g_n(z)g_{K,i}(z) = 0$. Expanding coefficients up to degree δ, this yields polynomial equations in the entries of K of total degree at most $(n-1)(\delta+1)$. By assumption, there exists at least one well-posed matrix K_0 for which G_{K_0} is not a scalar multiple of G. Hence the above defining polynomial conditions are not identically zero. Therefore the set $\mathcal{E}$ is contained in the zero set of a nonzero polynomial in $r = n-1$ variables of total degree at most $(n-1)(\delta+1)$. Finally, by the Schwartz–Zippel bound [17,22], this zero set has size at most $(n-1)(\delta+1)\,q^{r-1} = (n-1)(\delta+1)\,q^{n-2}$. Finally, dividing by the lower bound $|\mathcal{W}| \geq q^{n-1} - q^{n-2}k = q^{n-2}(q-k)$ from Proposition 3 yields the proposed estimate.

Corollary 1. *Assume $k = 1$ and let $\Sigma = (A,B,C,D) \in \mathbb{F}^{\delta\times\delta} \times \mathbb{F}^{\delta\times k} \times \mathbb{F}^{(n-k)\times\delta} \times \mathbb{F}^{(n-k)\times k}$ be a minimal ISO representation of an observable $(n,1,\delta)$-convolutional code over $\mathbb{F}$. Let $\mathcal{W}$ be the set of well-posed matrices defined in Proposition 3, and set*

$$E := \{K \in \mathcal{W} : \ \mathcal{C}_K(\Sigma_K) = \mathcal{C}(\Sigma)\}.$$

Then

$$|\{K \in \mathcal{W} : \ \mathcal{C}_K(\Sigma_K) \neq \mathcal{C}(\Sigma)\}| \ \geq \ q^{n-2}\big(q - (n-1)(\delta+1) - 1\big).$$

In particular, if $q > (n-1)(\delta+1)+1$, there exists a well-posed K such that Σ_K is a minimal ISO representation of an observable $(n,1,\delta)$-convolutional code $\mathcal{C}_K(\Sigma_K)$ with $\mathcal{C}_K(\Sigma_K) \neq \mathcal{C}(\Sigma)$. Moreover, among well-posed gains one has

$$\frac{|\{K \in \mathcal{W} : \ \mathcal{C}_K(\Sigma_K) \neq \mathcal{C}(\Sigma)\}|}{|\mathcal{W}|} \geq 1 - \frac{(n-1)(\delta+1)}{q-1}.$$

Proof. By Proposition 3 (with $k = 1$, $r = n-1$), $|\mathcal{W}| \geq q^{n-2}(q-1)$. By Theorem 1, every $K \in \mathcal{W}$ yields a minimal ISO representation Σ_K, hence an observable convolutional code $\mathcal{C}_K(\Sigma_K)$. By Proposition 4, the number of well-posed gains that produce a code equivalent to $\mathcal{C}(\Sigma)$ satisfies

$$|E| \leq (n-1)(\delta+1)\,q^{n-2}.$$

Therefore

$$|\{K \in \mathcal{W} : \ \mathcal{C}_K(\Sigma_K) \neq \mathcal{C}(\Sigma)\}| = |\mathcal{W}| - |E| \geq q^{n-2}(q-1) - (n-1)(\delta+1)q^{n-2},$$

which gives the stated bound.

3.2 Preserving GDP_{noisy} Conditions

Although we have seen that the minimality of the ISO representation of a convolutional code is invariant under output feedback as long as K is well-posed, the additional requirement that A be invertible, necessary for $\mathcal{C}(\Sigma)$ to be

GDP_{noisy}, is not naturally preserved. The requirement that A be invertible is crucial because it guarantees that the state evolution is reversible, so past states can be uniquely reconstructed from future ones. This prevents loss of information in the state dynamics and ensures that the output sequences retain enough structure to enable algebraic decoding in the presence of noise, which is the key feature of GDP_{noisy} codes [14].

Example 3. Consider an $(n, k, \delta) = (2, 1, 1)$ system $\Sigma = (A, B, C, D) = (1, 1, 1, 0)$ over $\mathbb{F}_2$. Since $B \neq 0$ and $C \neq 0$, Σ is a minimal ISO representation for a $(2, 1, 1)$-observable convolutional code $\mathcal{C}(\Sigma)$. Moreover, $A = 1$ is invertible. Let $K = 1 \in \mathbb{F}$. Since $D = 0$, it is well-posed. We have $\Theta = (I_k - KD)^{-1} = 1$, and therefore $A_K = A + B\Theta KC = 1 + 1 \cdot (-1) \cdot 1 = 0$. Hence Σ_K remains reachable and observable, and thus minimal, but the invertibility of A is not preserved under well-posed output feedback.

Proposition 5. *Let $\Sigma = (A, B, C, D) \in \mathbb{F}^{\delta \times \delta} \times \mathbb{F}^{\delta \times k} \times \mathbb{F}^{(n-k) \times \delta} \times \mathbb{F}^{(n-k) \times k}$ be a minimal dynamical linear system. Assume that A is invertible. Let $K \in \mathbb{F}^{k \times (n-k)}$ be a well posed output-feedback. We define $\Theta := (I_k - KD)^{-1}$. Then Σ_K is minimal and A_K is invertible if and only if*

(i) $\det(A + B\Theta KC) \neq 0$.
(ii) $\det(I_{n-k} + CA^{-1}B\Theta K) \neq 0$.
(iv) $\det(I_k + \Theta KCA^{-1}B) \neq 0$.

Proof. Minimality of Σ_K under well-posed output feedback follows from Proposition 2. Now, A_K is invertible $\Leftrightarrow \det(A + B\Theta KC) \neq 0$. The equivalence are given by using the Sylvester's determinant theorem [11] because $\det(A + B\Theta KC) = \det(A) \det(I_{n-k} + CA^{-1}B\Theta K) = \det(I_{n-k} + CA^{-1}B\Theta K) = \det(I_k + \Theta KCA^{-1}B)$.

Proposition 6. *Let $\Sigma = (A, B, C, D) \in \mathbb{F}^{\delta \times \delta} \times \mathbb{F}^{\delta \times k} \times \mathbb{F}^{(n-k) \times \delta} \times \mathbb{F}^{(n-k) \times k}$ be a minimal dynamical linear system. Assume that A is invertible. Consider $K \in \mathbb{F}^{k \times (n-k)}$. Let $M := CA^{-1}B \in \mathbb{F}^{(n-k) \times k}$. If K satisfies $KD = 0$ and $MK = 0$, then Σ_K is minimal, and A_K is invertible.*

Proof. Since $KD = 0$, we have $I_k - KD = I_k$, so Σ_K is well-posed. By Proposition 2 , Σ_K is minimal.

Now, using the Sylvester's determinant theorem [11], $\det(A + BKC) = \det(A) \det(I_{n-k} + CA^{-1}BK) = \det(A) \det(I_{n-k} + MK)$. If $MK = 0$, then $\det(I_{n-k} + MK) = \det(I_{n-k}) = 1$, hence $\det(A_K) = \det(A) \neq 0$ because A is invertible. Therefore, A_K is invertible.

Example 4. Consider the $(n, k, \delta) = (2, 1, 2)$ dynamical linear system $\Sigma = (A, B, C, D)$ over $\mathbb{F}_5$ with

$$A = \begin{pmatrix} 0 & 1 \\ 1 & 0 \end{pmatrix} \in \mathbb{F}_5^{2 \times 2}, \qquad B = \begin{pmatrix} 0 \\ 1 \end{pmatrix} \in \mathbb{F}_5^{2 \times 1}, \qquad C = (0\ 1) \in \mathbb{F}_5^{1 \times 2}, \qquad D = (0) \in \mathbb{F}_5^{1 \times 1}.$$

First, $A \in GL_2(\mathbb{F}_5)$ since $\det(A) = -1 = 4 \neq 0$. Moreover, Σ is minimal. Now compute $M := CA^{-1}B = \begin{pmatrix} 0 & 1 \end{pmatrix} \begin{pmatrix} 1 \\ 0 \end{pmatrix} = 0$. Choose the output-feedback $K = \begin{pmatrix} 1 \end{pmatrix} \in \mathbb{F}_5^{1 \times 1}$ such that $KD = 1 \cdot 0 = 0$ and $MK = 0 \cdot 1 = 0$. The feedback is well-posed and then Σ_K is minimal. Also, $A_K = \begin{pmatrix} 0 & 1 \\ 1 & 1 \end{pmatrix}$, that is invertible.

Proposition 7. *Let* $\Sigma = (A, B, C, D) \in \mathbb{F}^{\delta \times \delta} \times \mathbb{F}^{\delta \times k} \times \mathbb{F}^{(n-k) \times \delta} \times \mathbb{F}^{(n-k) \times k}$ *be a minimal dynamical linear system. Assume that A is invertible. Consider static output feedback $K \in \mathbb{F}^{k \times (n-k)}$. Set $M := CA^{-1}B \in \mathbb{F}^{(n-k) \times k}$. Assume that $\ker(M) \neq \{0\}$ and $\ker(D^{\top}) \neq \{0\}$. Pick any $u \in \ker(M) \setminus \{0\}$ and any $v \in \ker(D^{\top}) \setminus \{0\}$, and define*

$$K := uv^{\top} \in \mathbb{F}^{k \times (n-k)}.$$

Then Σ_K is minimal and A_K is invertible.

Proof. Since $v \in \ker(D^{\top})$, we have $v^{\top}D = 0$ and therefore $KD = u(v^{\top}D) = u \cdot 0 = 0$. Since $u \in \ker(M)$, we have $Mu = 0$ and therefore $MK = M(uv^{\top}) = (Mu)v^{\top} = 0 \cdot v^{\top} = 0$. Now apply Proposition 6.

Proposition 8. *Let* $\Sigma = (A, B, C, D) \in \mathbb{F}^{\delta \times \delta} \times \mathbb{F}^{\delta \times k} \times \mathbb{F}^{(n-k) \times \delta} \times \mathbb{F}^{(n-k) \times k}$ *be a minimal dynamical linear system. Assume that A is invertible. Consider $K \in \mathbb{F}^{k \times (n-k)}$. Let $M := CA^{-1}B$. Assume $\ker(D^{\top}) \neq \{0\}$ and choose any K such that $KD = 0$. Then Σ_K is minimal. Also, A_K is invertible if and only if $\det(I_{n-k} + MK) \neq 0$.*

Proof. If $KD = 0$, then Σ is well-posed and $\Theta = I_k$. Minimality of Σ_K follows from Proposition 2. Also $\det(A_K) = \det(A)\det(I_{n-k} + MK)$. Since $\det(A) \neq 0$, we have $\det(A_K) \neq 0$ if and only if $\det(I_{n-k} + MK) \neq 0$.

Theorem 2. *Let* $\Sigma = (A, B, C, D) \in \mathbb{F}^{\delta \times \delta} \times \mathbb{F}^{\delta \times k} \times \mathbb{F}^{(n-k) \times \delta} \times \mathbb{F}^{(n-k) \times k}$ *be a minimal ISO representation of an observable and $\text{GDP}_{\text{noisy}}$ (n, k, δ)-convolutional code over $\mathbb{F}$, $\mathcal{C}(\Sigma)$. Let $K \in \mathbb{F}^{k \times (n-k)}$ be well-posed and satisfying one of the constructive conditions given in Propositions 6, 7, or 8. Then Σ_K is a minimal ISO representation of an observable and $\text{GDP}_{\text{noisy}}$ (n, k, δ)-convolutional code $\mathcal{C}_K(\Sigma_K)$.*

Proof. Since $\mathcal{C}(\Sigma)$ is $\text{GDP}_{\text{noisy}}$, the ISO system Σ is minimal and A is invertible. By Propositions 6, 7, and 8, we conclude the proof.

Remark 1. The feedback $K \in \mathbb{F}^{k \times (n-k)}$ satisfies $KC = 0$ and $KD = 0$, then $\Sigma_K = \Sigma$. Consequently, all such K produce exactly the same convolutional code $\mathcal{C}(\Sigma)$. Thus, this situation corresponds to a trivial feedback action. Outside this situation, however, different admissible K need not yield unimodularly equivalent convolutional codes.

Algorithm 1: Construction of output-feedback gains preserving minimality and $\mathrm{GDP}_{\mathrm{noisy}}$

Input: A field $\mathbb{F}$ and a minimal ISO system $\Sigma = (A, B, C, D)$ with $A \in GL_\delta(\mathbb{F})$.
Output: Gains K such that Σ_K is minimal and A_K is invertible, and the corresponding systems Σ_K.

Compute A^{-1} and $M := CA^{-1}B$.
Initialize an empty list $\mathcal{K}_{\mathrm{good}}$.

Route A (linear constraints).
Solve the linear systems $KD = 0$ and $MK = 0$.
Let $\mathcal{S} := \{K \in \mathbb{F}^{k \times (n-k)} \mid KD = 0, \ MK = 0\}$.
If $\mathcal{S} \neq \{0\}$, choose $K \in \mathcal{S} \setminus \{0\}$ and add it to $\mathcal{K}_{\mathrm{good}}$.

Route B (rank-one construction).
Compute $\ker(M)$ and $\ker(D^\top)$.
If $\ker(M) \neq \{0\}$ and $\ker(D^\top) \neq \{0\}$, choose $u \in \ker(M) \setminus \{0\}$ and $v \in \ker(D^\top) \setminus \{0\}$,
  set $K := uv^\top$, and add it to $\mathcal{K}_{\mathrm{good}}$.

Route C (reduced test with $KD = 0$).
Solve $KD = 0$ and sample nonzero K satisfying $KD = 0$.
For each such K, if $\det(I_{n-k} + MK) \neq 0$, add K to $\mathcal{K}_{\mathrm{good}}$.

Route D (general random filtering).
Sample random $K \in \mathbb{F}^{k \times (n-k)}$.
If $\det(I_k - KD) = 0$, reject.
Otherwise set $\Theta := (I_k - KD)^{-1}$.
If $\det(I_{n-k} + M\Theta K) = 0$, reject; otherwise add K to $\mathcal{K}_{\mathrm{good}}$.

Closed-loop system.
For each $K \in \mathcal{K}_{\mathrm{good}}$, set $\Theta := (I_k - KD)^{-1}$ and compute

$$A_K := A + B\Theta KC, \quad B_K := B\Theta, \quad C_K := (I_{n-k} - DK)^{-1}C, \quad D_K := (I_{n-k} - DK)^{-1}D.$$

return $\mathcal{K}_{\mathrm{good}}$ and the corresponding Σ_K.

Algorithm 1 always produces output feedback matrices K preserving observability and GDP_{noisy} in the associated convolutional code.

We estimate the computational complexity of Algorithm 1 in terms of the parameters δ, k, and $r := n - k$. All costs are measured in arithmetic operations over $\mathbb{F}$. The inversion of $A \in \mathbb{F}^{\delta \times \delta}$ requires $O(\delta^3)$ operations. The computation of $M = CA^{-1}B$ requires $O(\delta^2 k + r\delta k)$ operations. The route A consists of solving the linear systems $KD = 0$ and $MK = 0$ in the $N := kr$ unknown entries of K. Using Gaussian elimination, the worst-case cost is $O(N^3) = O((kr)^3)$. This method is deterministic and yields the full solution subspace. The route B requires computing $\ker(M)$ and $\ker(D^\top)$. Since $M \in \mathbb{F}^{r \times k}$ and $D^\top \in \mathbb{F}^{k \times r}$, both kernels can be computed in $O((r + k)^3)$ operations. Forming $K = uv^\top$ requires $O(kr)$ operations. Hence the overall cost is dominated by $O((r + k)^3)$. The route C reduces the test with $KD = 0$. After parameterizing the linear solutions of $KD = 0$, each candidate K must satisfy $\det(I_r + MK) \neq 0$. Computing MK requires $O(r^2 k)$ operations, and computing $\det(I_r + MK)$ requires $O(r^3)$ operations. Hence the cost is $O(r^2 k + r^3)$. The rout D uses a randomly chosen $K \in \mathbb{F}^{k \times r}$, well-posedness is tested by computing $\det(I_k - KD)$, which requires

$O(k^2r + k^3)$ operations. If well-posed, one computes $\Theta = (I_k - KD)^{-1}$ and tests $\det(I_r + M\Theta K) \neq 0$. This requires an additional $O(k^2r + r^2k + r^3)$ operations. Hence the total cost is $O(k^2r + k^3 + r^2k + r^3)$. Then, route B is computationally optimal when the required kernels are nontrivial. Route C provides an efficient reduced test when $KD = 0$ is imposed. Route D is the most general approach and is suitable for randomized sampling.

While Theorem 2 provides a constructive criterion for selecting feedback that preserve minimality and invertibility of the state matrix, it is natural to ask how restrictive these algebraic conditions actually are. In other words, given a minimal system over $\mathbb{F}$, how large is the set of feedback matrices $K \in \mathbb{F}^{k \times (n-k)}$ that simultaneously preserves minimality and invertibility conditions.

The output feedback matrices depend rationally on K through the inverse $(I_k - KD)^{-1}$, so the admissible Ks that preserve the conditions of minimality and invertibility conditions must first belong to the well-posed domain $\{K : \det(I_k - KD) \neq 0\}$. The set of ouput feedback matrices preserving minimality and invertibility is the complement of an algebraic variety inside the well-posed domain.

Proposition 9. *Let $\Sigma = (A, B, C, D) \in \mathbb{F}^{\delta \times \delta} \times \mathbb{F}^{\delta \times k} \times \mathbb{F}^{(n-k) \times \delta} \times \mathbb{F}^{(n-k) \times k}$ be a minimal dynamical linear system. Assume that Σ is minimal and that A is invertible. Let $K \in \mathbb{F}^{k \times (n-k)}$. Set $r := n - k$ and $N := kr$. Consider the well-posedness polynomial $P_{\mathrm{wp}}(K) := \det(I_k - KD) \in \mathbb{F}[K_{ij}]$, and let $\Theta = (I_k - KD)^{-1}$ on the set $\{K : P_{\mathrm{wp}}(K) \neq 0\}$. Let $M := CA^{-1}B \in \mathbb{F}^{r \times k}$. Define the polynomial*

$$P_{\mathrm{inv}}(K) := \big(\det(I_k - KD)\big)^r \det\big(I_r + M\Theta K\big) \in \mathbb{F}[K_{ij}].$$

Finally, let

$$\mathcal{U} := \Big\{ K \in \mathbb{F}^{k \times r} : \ P_{\mathrm{wp}}(K) \neq 0 \ and \ P_{\mathrm{inv}}(K) \neq 0 \Big\}.$$

Then:

(i) For every $K \in \mathcal{U}$, then Σ_K is minimal. Also, A_K is invertible.

(ii) The polynomials P_{wp} and P_{inv} are not identically zero, and their degrees satisfy $\deg P_{\mathrm{wp}} \leq k$ and $\deg P_{\mathrm{inv}} \leq kr$.

(iii) Consequently, the number of admissible Ks that preserve the minimality for the system Σ_K and the invertibility of A_K, satisfies the lower bound

$$|\mathcal{U}| \ \geq \ q^N - (\deg P_{\mathrm{wp}} + \deg P_{\mathrm{inv}}) \, q^{N-1} \ \geq \ q^{kr} - k(r+1) \, q^{kr-1} \ = \ q^{kr}\Big(1 - \frac{k(r+1)}{q}\Big),$$

where $r = n - k$. Equivalently, a uniformly random $K \in \mathbb{F}^{k \times (n-k)}$ satisfies both well-posedness and $A_K \in GL_\delta(\mathbb{F})$ with probability at least $1 - k(n - k + 1)/q$.

Proof. (i) By definition, $P_{\mathrm{wp}}(K) \neq 0$ is equivalent to well-posedness, so Θ exists. By Proposition 7, A_K is invertible if and only if $\det(I_r + M\Theta K) \neq 0$, and this is equivalent to $P_{\mathrm{inv}}(K) \neq 0$ on the well-posed set. Minimality is preserved by Proposition 2.

(ii) Since $P_{\mathrm{wp}}(K) = \det(I_k - KD)$ is the determinant of a $k \times k$ matrix whose entries are affine linear forms in the variables K_{ij}, we have $\deg P_{\mathrm{wp}} \leq k$. Moreover, the entries of $\Theta = (I_k - KD)^{-1}$ are ratios of $(k-1)$-degree cofactors by $\det(I_k - KD)$. Thus each entry of $M\Theta K$ is a rational function whose numerator has degree at most k and whose denominator is $\det(I_k - KD)$. Therefore, $\det(I_r + M\Theta K)$ has denominator $\big(\det(I_k - KD)\big)^r$, and multiplying by that factor yields the polynomial $P_{\mathrm{inv}}(K)$. Since the determinant is multilinear in rows/columns and each entry has numerator degree $\leq k$, we obtain $\deg P_{\mathrm{inv}} \leq kr$.

Both polynomials are nonzero: $P_{\mathrm{wp}}(0) = \det(I_k) = 1 \neq 0$, and $P_{\mathrm{inv}}(0) = \det(I_k)^r \det(I_r) = 1 \neq 0$.

(iii) Apply the Schwartz–Zippel bound [17,22] to P_{wp} and P_{inv} in $N = kr$ variables: $|\{K : P(K) = 0\}| \leq (\deg P) q^{N-1}$ for each nonzero polynomial P. A union bound gives $|\mathcal{U}^c| \leq (\deg P_{\mathrm{wp}} + \deg P_{\mathrm{inv}}) q^{N-1}$, yielding the stated lower bound.

Proposition 9 shows that over $\mathbb{F}$ there exist at least $q^{k(n-k)}\left(1 - \frac{k(n-k+1)}{q}\right)$ output-feedback matrices K for which the feedback system Σ_K remains minimal and A_K is invertible. Consequently, for all such K, the induced convolutional code is observable and satisfies the GDP_{noisy} property. In particular, when q is large, almost every $K \in \mathbb{F}^{k \times (n-k)}$ yields an ISO realization generating an observable (n, k, δ)-convolutional code with the same structural parameters, observable and GDP_{noisy}.

4   Conclusions

In this work, we have studied the effect of applying output feedback on ISO representations of observable and GDP_{noisy} convolutional codes over finite fields. In particular, we established algebraic conditions that yield implementable design procedures to maintain good properties of the representations. Moreover, our results show that admissible output-feedback matrices, K_s, form a Zariski-open and dense subset of the parameter space.

The results provide a deterministic, quantitative framework for generating new, observable convolutional codes via output feedback while maintaining their observability and decoding structure, and, in general, producing non-equivalent encoders.

Future work will address whether such feedback transformations preserve additional performance parameters, in particular the code's distance properties, as well as extend the present approach to state-feedback transformations.

Disclosure of Interests. The authors declare that they have no conflict of interest.

References

1. Brewer, J.W., Bunce, J.W., Van Vleck, F.S.: Linear Systems over Commutative Rings. Marcel Dekker, New York (1986)
2. Climent, J., Napp, D., Requena, V.: An algorithm to compute a minimal input-state-output representation of a convolutional code. Linear Algebra Appl. **721**, 715–735 (2025)
3. DeCastro-García, N.: Feedback equivalence of convolutional codes over finite rings. Open Math. **15**(1), 1495–1508 (2017)
4. DeCastro-García, N., Carriegos, M.V., Muñoz Castañeda, A.L.: Construction of observable and MDP convolutional codes with good decodable properties by I/S/O representations. Int. J. Appl. Math. Comput. Sci. **36**(1), 141–154 (2026)
5. Fragouli, C., Wesel, R.D.: Convolutional codes and matrix control theory. In: Proceedings of the 7th International Conference on Advances in Communications and Control, Athens, Greece (1999)
6. Forney, G.: Convolutional codes I: algebraic structure. IEEE Trans. Inf. Theory **16**(6), 720–738 (1970)
7. Gluesing-Luerssen, H., Rosenthal, J., Smarandache, R.: Strongly-MDS convolutional codes. IEEE Trans. Inf. Theory **52**(2), 584–598 (2006)
8. Gluesing-Luerssen, H., Schneider, G.: State space realizations and monomial equivalence for convolutional codes. Linear Algebra Appl. **425**(2–3), 518–533 (2007)
9. Hautus, M.L.J.: Controllability and observability conditions for linear autonomous systems. Indag. Math. **72**(5), 443–448 (1969)
10. Hutchinson, R., Rosenthal, J., Smarandache, R.: Convolutional codes with maximum distance profile. Syst. Control Lett. **54**(1), 53–63 (2005)
11. Kailath, T.: Linear Systems. Prentice-Hall, Englewood Cliffs (1980)
12. Kalman, R.E., Falb, P.L., Arbib, M.A.: Topics in Mathematical System Theory. McGraw-Hill, New York (1969)
13. Rosenbrock, H.H.: State-Space and Multivariable Theory. Nelson, London (1970)
14. Rosenthal, J.: An algebraic decoding algorithm for convolutional codes. In: Picci, G., Gilliam, D.S. (eds.) Dynamical Systems, Control, Coding, Computer Vision. Progress in Systems and Control Theory, vol. 25, pp. 343–360. Birkhäuser, Basel (1999)
15. Rosenthal, J., Schumacher, J.M., York, E.V.: On behaviors and convolutional codes. IEEE Trans. Inf. Theory **42**(6), 1881–1891 (1996)
16. Rosenthal, J., Smarandache, R.: Maximum distance separable convolutional codes. Appl. Algebra Eng. Commun. Comput. **10**(1), 15–32 (1999)
17. Schwartz, J.T.: Fast probabilistic algorithms for verification of polynomial identities. J. ACM **27**(4), 701–717 (1980)
18. Staffans, O.J.: Quadratic optimal control of stable well-posed linear systems. Trans. Am. Math. Soc. **349**(9), 3679–3715 (1997)
19. Tomás Estevan, V.: Complete-MDP convolutional codes over the erasure channel. Ph.D. Thesis, Universidad de Alicante (2010)
20. Wonham, W.: On pole assignment in multi-input controllable linear systems. IEEE Trans. Autom. Control **12**(6), 660–665 (1967)
21. York, E.: Algebraic description and construction of error correcting codes: a linear systems point of view. Ph.D. thesis, Department of Mathematics, University of Notre Dame, Notre Dame, Indiana, USA (1997)
22. Zippel, R.: Probabilistic algorithms for sparse polynomials. In: Ng, E.W. (ed.) Symbolic and Algebraic Computation (EUROSAM 1979). LNCS, vol. 72, pp. 216–226. Springer, Berlin (1979)

The Non-asymptotic Landscape of Abelian Codes: A Framework for (F, G)-Goodness

Beatriz García García$^{(\boxtimes)}$ (iD), Consuelo Martínez López (iD), and Ignacio F. Rúa (iD)

Department of Mathematics, University of Oviedo, 33007 Oviedo, Spain
`garciagbeatriz@uniovi.es`

Abstract. While abelian group codes are classically known to be asymptotically bad—constrained by strict structural limitations such as the Square-Root and Logarithmic barriers—this traditional, infinite-length perspective obscures their rich mathematical behavior at finite block lengths. In this contribution, we move beyond the standard paradigm of asymptotic metrics to investigate the non-asymptotic goodness of abelian codes. We introduce the framework of (F, G)-goodness and a machinery that enables us to rigorously classify abelian codes into distinct performance "tiers".

Keywords: Abelian group codes · Universal reductions · (F, G)-goodness · Non-asymptotic bounds · Group algebras · Permutation equivalence

1 Introduction

Historically, the evaluation of abelian group codes has been overshadowed by a well-known asymptotic failure: they are unequivocally bad as the block length approaches infinity [8]. The literature establishes two fundamental walls depending on the group order. For block lengths of p^m over a field of characteristic p, codes are structurally trapped by the Square-Root Barrier ($d \sim \sqrt{n}$). Conversely, for lengths coprime to p, the semisimple algebra restricts them to the BCH or Logarithmic Barrier ($d \sim n/\log n$) [7].

Establishing exact structural bounds is a critical theoretical necessity, particularly because the order 2 case (binary codes) is the most computationally demanding to explore via brute-force algorithmic search. Rather than searching blindly for good codes, we require an intrinsic algebraic plan to map the feasible parameter space.

In this work, we formalize the non-asymptotic goodness of abelian group codes. In Sect. 2, we provide a rigorous probabilistic proof that the p-elementary group algebra is strictly bounded by the square-root barrier, demonstrating that the "middle slice" of its radical filtration is the unique, Pareto-optimal operating point. In Sect. 3, we introduce the "Bridge Concept" to prove that alternative

L. Batina and F. Özbudak (Eds.): WAIFI 2026, LNCS 16611, pp. 72–88, 2026.
https://doi.org/10.1007/978-3-032-27574-5_5

modular structures—such as cyclic and hierarchical p-groups—are fundamentally equivalent to skewed ideals within the elementary algebra, forcing them into even more severe uniserial distance collapses. We further prove that tensoring with semisimple algebras fails to rescue these modular codes, actively diluting their relative distance.

Finally, in Sect. 4, we abandon explicit polynomial generators in favor of a coordinate-free approach to group codes. We introduce the framework of $\Theta(F, G)$-goodness and the Universal Reduction Machinery. We classify all abelian group codes into a rigorous, five-tier hierarchical taxonomy, proving that asymptotic barriers are inescapably inherited properties of a group's composition factors.

2   Non-asymptotic Goodness of the p-Elementary Group Algebra

To rigorously evaluate the coding-theoretic potential of abelian groups, we will first establish the baseline limitations imposed by the p-elementary group algebra. In this section, we prove that the asymptotic minimum distance of any family of ideals in the modular group algebra $\mathbb{F}_r[C_r^m]$ with non-vanishing rate is upper-bounded by the square root of the block length.

Let $H_{r,m} = (C_r)^m$ be the elementary abelian group of order $n = r^m$, viewed as a vector space of dimension m over $\mathbb{F}_r$. The group algebra $\mathcal{R}_m = \mathbb{F}_r[H_{r,m}]$ is isomorphic to the truncated polynomial ring, $\mathcal{R}_m \cong \mathbb{F}_r[x_1, \ldots, x_m]/\langle x_1^r, \ldots, x_m^r \rangle$. This isomorphism allows us to treat the algebra as a graded vector space, decomposed by the total degree of polynomials:

$$\mathcal{R}_m = \bigoplus_{j=0}^{D_{max}} V_j$$

where V_j is the subspace of homogeneous polynomials of total degree j, and $D_{max} = m(r-1)$ is the maximum degree. The dimensions of these homogeneous subspaces, denoted $h_j = \dim(V_j)$, play a crucial role in our construction.

Definition 1. *Let $\mathcal{R}_m = \bigoplus_{j \geq 0} V_j$ be the graded algebra associated with our construction. We define the irrelevant ideal $\mathfrak{m}$ of $\mathcal{R}_m$ as the ideal generated by the homogeneous elements of positive degree, $\mathfrak{m} := \bigoplus_{j > 0} V_j$. Moreover, we have $\mathfrak{m} = \langle V_1 \rangle$.*

The powers of this ideal provide a natural descending filtration of the algebra $\mathcal{R}_m$, $\mathcal{R}_m = \mathfrak{m}^0 \supseteq \mathfrak{m}^1 \supseteq \mathfrak{m}^2 \supseteq \ldots$, where $\mathfrak{m}^k$ is the subspace spanned by homogeneous elements of degree at least k.

This filtration recovers the graded components V_j of the algebra. V_k is isomorphic to the quotient of consecutive powers of the ideal: $V_k \cong \mathfrak{m}^k/\mathfrak{m}^{k+1}$. This isomorphism allows us to study the asymptotic properties of the dimensions $h_k = \dim(V_k)$ via the algebraic structure of the filtration.

Any linear code $\mathcal{C}$ that is an ideal in $\mathcal{R}_m$ is a subspace satisfying $\mathfrak{m}\mathcal{C} \subseteq \mathcal{C}$. When n tends to infinity, the sequence of dimensions approaches a discrete Gaussian distribution. The generating function for the sequence $\{h_j\}$ is the polynomial:

$$P(z) = \left(\sum_{i=0}^{r-1} z^i\right)^m = (1 + z + \cdots + z^{r-1})^m$$

The coefficient h_k is the number of integer solutions to $e_1 + \cdots + e_m = k$ with $0 \le e_i \le r - 1$. This is equivalent to the distribution of the sum $S_m = \sum_{i=1}^{m} X_i$, where X_i are independent discrete uniform random variables on $\{0, \ldots, r - 1\}$.

Lemma 1. *Let r be fixed. As $m \to \infty$, the distribution of dimensions normalized by the total dimension $n = r^m$ converges to a Gaussian distribution:*

$$\frac{h_k}{n} \approx \frac{1}{\sqrt{2\pi\sigma}} \exp\left(-\frac{(k-\mu)^2}{2\sigma^2}\right)$$

with mean μ and variance σ^2 given by, $\mu = m\frac{r-1}{2}$, $\sigma^2 = m\frac{r^2-1}{12}$.

Proof. The mean and variance of a single uniform variable X_i on $\{0, \ldots, r - 1\}$ are $\mu_X = (r-1)/2$ and $\sigma_X^2 = (r^2-1)/12$. Since the variables are independent, the sum S_m has mean $\mu = m\mu_X$ and variance $\sigma^2 = m\sigma_X^2$. By the Local Limit Theorem for sums of independent lattice variables [4], the probability mass function $P(S_m = k) = h_k/r^m$ converges uniformly to the Gaussian density $\mathcal{N}(\mu, \sigma^2)$. $\square$

This result implies that the "width" of the algebra's layers is concentrated around the middle degree $\mu = D_{max}/2$. The standard deviation scales as $\sigma \propto \sqrt{m}$. Crucially, for any $\epsilon > 0$, the fraction of the total dimension n residing in the "tails" of the distribution vanishes exponentially.

Proposition 1. *Let $V_{tail} = \bigoplus_{k > \mu + \epsilon m} V_k$ be the subspace of $\mathcal{R}_m$ spanned by its homogeneous components of degree strictly greater than $\mu + \epsilon m$, where $\epsilon > 0$ is an arbitrary constant. There exists a strictly positive constant $c > 0$, independent of m, such that, $\dim(V_{tail}) \le n^{1-c}$.*

Consequently, the relative dimension of the tail collapses asymptotically:

$$\lim_{n \to \infty} \frac{\dim(V_{tail})}{n} = 0.$$

Now, we can state the main result of this section.

Theorem 1 (Non-asymptotic Goodness of the p-Elementary Group). *Let $\{\mathcal{C}_n\}_n$ be a sequence of ideals in $\mathbb{F}_r[H_{r,m}]$. If the asymptotic rate is strictly positive, i.e.,*

$$\liminf_{n \to \infty} \frac{\dim(\mathcal{C}_n)}{n} = R > 0,$$

then the minimum distance is bounded by, $d(\mathcal{C}_n) \le \mathcal{O}(\sqrt{n})$.

Proof. We want to establish the minimum distance of the ideal $\mathfrak{m}^k$, which corresponds exactly to a Generalized Reed-Muller code over $\mathbb{F}_r$ [9]. According to the Kasami-Lin-Peterson bound [6], the minimum distance of the ideal $\mathfrak{m}^k$ is precisely $d(\mathfrak{m}^k) = (b+1)r^a$, where $k = a(r-1)+b$ is an integer with $0 \leq b < r-1$.

Assume for contradiction that there exists a family of codes with rate $R > 0$ and minimum distance scaling as n^α with $\alpha > 1/2$.

We define the "bulk" of the algebra as all homogeneous components up to degree $k_{max} = \mu + \epsilon m$, for some $0 < \epsilon \ll (r-1)\left(\alpha - \frac{1}{2}\right)$ (notice that $\alpha > 1/2$). By [6], the maximum possible minimum weight for any ideal contained within this bulk occurs at the upper edge k_{max}. Substituting k_{max} into the distance formula, the weight scales as:

$$d(\mathfrak{m}^{k_{max}}) \approx r^{\frac{\mu+\epsilon m}{r-1}} = r^{\frac{m}{2}+\frac{\epsilon m}{r-1}} = n^{\frac{1}{2}+\frac{\epsilon}{r-1}} < n^\alpha,$$

because of the choice of ϵ. Therefore, any element achieving distance n^α must reside entirely outside the bulk.

But we know, from Proposition 1, that the dimension of this high-degree tail is sub-linear for any fixed $\epsilon > 0$:

$$\lim_{n \to \infty} \frac{\dim(V_{tail})}{n} = 0.$$

However, by hypothesis, the code has a constant, strictly positive rate:

$$\liminf_{n \to \infty} \frac{\dim(\mathcal{C}_n)}{n} = R > 0,$$

so, for sufficiently large n, we have $\dim(\mathcal{C}_n) > \dim(V_{tail})$, a contradiction, since $\mathcal{C}_n$, as a subspace, must be strictly contained in the high-degree tail subspace V_{tail}. $\square$

Corollary 1 (The Square-Root Barrier Discontinuity). *Let $R(\alpha)$ denote the maximum achievable asymptotic information rate for any family of modular abelian group codes $\{\mathcal{C}_n\}$ whose minimum distance scales as $d(\mathcal{C}_n) = \Omega(n^\alpha)$. The trade-off curve $R(\alpha)$ exhibits a strict macroscopic discontinuity at the critical threshold $\alpha = 1/2$.*

2.1   The Optimality of the Square-Root Barrier

In this section, we formalize the trade-off between information rate and error correction capability. We introduce a unified metric, the *Total Decay Metric*, and prove that the construction based on the middle slice of the radical filtration is the unique global optimum within the class of modular abelian group codes.

The Total Decay Metric. Let $\{\mathcal{C}_s\}_s$ be a family of codes each of them in a modular group algebra $\mathbb{F}_r[H_{r,m}]$ of length $n = r^m$, $m = m_s$. We characterize the asymptotic behavior of the code parameters by two decay exponents, $\alpha, \beta \geq 0$:

$$\text{Relative Distance:} \quad \delta_n \sim n^{-\alpha} \quad \implies \quad d_n \sim n^{1-\alpha},$$

$$\text{Rate:} \quad R_n \sim n^{-\beta}.$$

We define the **Total Decay Metric** γ as the sum of these asymptotic penalties: $\gamma(\mathcal{C}_n) = \alpha + \beta$. A value of $\gamma = 0$ corresponds to asymptotically good codes (where $\alpha = 0$ and $\beta = 0$). A value of $\gamma = 1$ corresponds to trivial codes (e.g., repetition codes with $\alpha = 0, \beta = 1$, or single parity checks with $\alpha = 1, \beta = 0$).

Analysis of the Filtration Landscape. We analyze the value of γ as a function of the normalized filtration cut $\lambda \in [0, 1]$, where the code is defined by the ideal $\mathfrak{m}^k$ with $k = \lambda D_{max}$.

The Middle Slice Regime ($\lambda = 1/2$): At the center of the filtration, the dimension is maximal due to the concentration of measure, and the absolute distance is governed by the square-root barrier.

- **Rate:** The dimension is approximately $n/2$, meaning the sequence maintains a strictly positive asymptotic rate $R \sim$ const. Thus $\beta = 0$.
- **Distance:** By the Kasami-Lin-Peterson bound, $d \sim n^{1/2}$, so the relative distance decays as $\delta \sim n^{-1/2}$. Thus $\alpha = 0.5$.
- **Metric:** $\gamma(1/2) = 0.5 + 0 = 0.5$.

The Deep Tail Regime ($\lambda > 1/2$): As we move the cut deeper into the filtration to improve absolute distance, we incur a massive rate penalty. Let $\lambda = 1/2 + \epsilon$ for some $\epsilon > 0$.

- **Distance Gain:** The minimum distance increases, meaning the exponent α decreases from 0.5 towards 0.
- **Rate Loss:** The dimension of the tail is governed by the large deviation principle. The rate decays exponentially in m. Specifically, the rate exponent β grows strictly faster than the distance improvement. The Gaussian nature of the coefficient distribution dictates that the "cost" of distance is strictly convex.
- **Metric Behavior:** For any $\epsilon > 0$, the penalty $\beta(\epsilon)$ heavily dominates the marginal gain in $\alpha(\epsilon)$. $\gamma(1/2 + \epsilon) > 0.5$.

The Shallow Regime ($\lambda < 1/2$): Moving towards the top of the algebra increases the dimension slightly, but dramatically reduces the absolute distance towards a constant $\mathcal{O}(1)$.

- **Rate:** Because the sequence maintains a constant rate, β remains 0.
- **Distance:** The relative distance δ decays much faster than $n^{-1/2}$. As d shrinks, α increases towards 1.
- **Metric:** Because the rate offers no counter-balancing improvement, the metric strictly worsens. $\gamma(1/2 - \epsilon) > 0.5$.

Theorem 2. (Uniqueness of the Optimal Operating Point). *For the family of modular codes over elementary abelian groups, the total decay metric $\gamma(\lambda)$ attains a strict global minimum at $\lambda = 1/2$: $\min_{\lambda \in [0,1]} \gamma(\lambda) = 0.5$. Consequently, the "Square-Root Barrier" represents the Pareto-optimal frontier for this algebraic structure.*

Proof. The proof follows from the convexity of the rate-distortion function for the group algebra. Let $H_r(x)$ be the entropy function governing the volume of the filtration slices. The rate exponent β is directly related to $1 - H_r(\lambda)$. The distance exponent α is related to the generalized Reed-Muller bound.

Near the center cut $\lambda = 1/2$, the entropy function is flat ($H_r'(1/2) = 0$), implying the rate decay exponent $\beta \approx 0$. However, the distance function has a non-zero derivative. Crucially, as λ increases, the term $1 - H_r(\lambda)$ grows quadratically due to the Gaussian tail, while the distance exponent improves only linearly. Because the quadratic penalty strictly overtakes the linear gain, the sum $\gamma(\lambda) = \alpha(\lambda) + \beta(\lambda)$ is strictly increasing for $\lambda > 1/2$. Symmetry arguments show identical worsening behavior for $\lambda < 1/2$. Therefore, the minimum is unique and located exactly at the peak of the density distribution, $\lambda = 1/2$. $\square$

3 Structural Alternatives for Asymptotic and Non-asymptotic Goodness

The asymptotic collapse of r-elementary group codes demonstrated in Sect. 2 is fundamentally driven by the geometry of the algebra's Jacobson radical. The multivariate quotient ring creates a "wide" combinatorial space where the degrees of the generators are highly concentrated, forcing the code spaces into the Square-Root Barrier. To seek codes with asymptotic goodness, we must structurally alter the underlying group to manipulate the radical filtration. We examine three distinct abelian alternatives.

3.1 The Cyclic Code: C_{r^m}

The most immediate structural pivot from the wide elementary abelian group C_r^m is the extremely narrow cyclic group of the same order, $G \cong C_{r^m}$. Over the field $\mathbb{F}_r$, the group algebra $\mathbb{F}_r[C_{r^m}]$ is isomorphic to the quotient ring:

$$\mathbb{F}_r[x]/\langle x^{r^m} - 1\rangle.$$

Because the characteristic of the field perfectly divides the group order, we have $x^{r^m} - 1 = (x-1)^{r^m}$. Let $y = x - 1$. The algebra is therefore a local ring generated by a single nilpotent variable, $\mathbb{F}_r[C_{r^m}] \cong \mathbb{F}_r[y]/\langle y^{r^m}\rangle$.

Unlike the elementary abelian case, this local ring is strictly uniserial. Its radical filtration is a single, linear chain of principal ideals:

$$\langle y^0\rangle \supset \langle y^1\rangle \supset \langle y^2\rangle \supset \cdots \supset \langle y^{r^m-1}\rangle \supset \{0\}.$$

The existence of exactly one cyclic code for every possible dimension k, completely eliminates the massive combinatorial explosion of invariant subspaces seen in the elementary algebra. However, this uniserial depth provides absolutely no asymptotic coding-theoretic advantage due to the following result.

Proposition 2 (The Bridge Concept). *Let $n = r^m$, where r is prime. Every cyclic code of length n over $\mathbb{F}_r$ is permutation-equivalent to an ideal in the elementary abelian group algebra $\mathbb{F}_r[H_{r,m}]$.*

Proof. Let $V = \mathbb{F}_r^n$ be the ambient vector space. We compare two distinct group structures acting on the index set $\{0, 1, \ldots, n-1\}$: the cyclic group C_{r^m} and the elementary abelian group C_r^m (see [1]).

We identify an integer index $j \in \{0, \ldots, r^m - 1\}$ with a vector $\mathbf{v} \in \mathbb{F}_r^m$ via the standard r-adic expansion:

$$j = \sum_{k=0}^{m-1} v_k r^k \qquad \xrightarrow{\ \pi\ } \qquad \mathbf{v} = (v_0, v_1, \ldots, v_{m-1}).$$

This bijection π induces a canonical linear isomorphism between the ambient vector spaces of the respective group algebras. We define two distinct types of linear operators acting on the standard basis vectors $\mathbf{e_v}$ of V:

- The Cyclic Shift (σ): The generator of the cyclic group action, representing arithmetic addition with carry, $\sigma(\mathbf{e_v}) = \mathbf{e}_{\pi(\pi^{-1}(\mathbf{v})+1 \ (\mathrm{mod}\ r^m))}$.
- The Elementary Translations (τ_k): The m independent generators of the elementary abelian action, representing component-wise addition without carry. For $0 \le k \le m - 1$, $\tau_k(\mathbf{e_v}) = \mathbf{e}_{\mathbf{v}+\mathbf{u}_k \ (\mathrm{mod}\ r)}$, where $\mathbf{u}_k$ is the k-th standard basis vector.

A linear subspace $\mathcal{C} \subseteq V$ is a cyclic code if and only if it is invariant under the cyclic shift, meaning $\sigma(\mathcal{C}) = \mathcal{C}$. To prove that $\mathcal{C}$ is also an elementary abelian group code (an ideal in $\mathbb{F}_r[H_{r,m}]$), we must show that σ-invariance inherently implies τ_k-invariance for all $k \in \{0, \ldots, m - 1\}$.

We analyze the action of σ^{r^k}, which adds r^k in integer arithmetic. This operator acts on the k-th coordinate v_k exactly like τ_k. However, if $v_k = r - 1$, it generates an arithmetic carry that cascades to the higher-order bits. Thus, the cyclic operator can be decomposed exactly into the elementary translation plus a correction term, $\sigma^{r^k} = \tau_k + N_k$. where N_k represents the "Noise" or carry effect.

In this ring of characteristic r, the linear operators $y_k = \tau_k - \mathrm{id}$ generate the Jacobson radical and are strictly nilpotent, satisfying $y_k^r = 0$. The operator $y_k^{r-1} = (\tau_k - \mathrm{id})^{r-1}$ acts as an exact conditional filter: when applied to a basis vector, it annihilates any component whose k-th coordinate has not reached the arithmetic threshold of $r - 1$. So $N_k \in \mathbb{F}_r[\tau_{k+1}, \ldots, \tau_{m-1}]$.

Finally, we prove that $\mathcal{C}$ is invariant under all τ_k by reverse induction on k.

Base Case ($k = m - 1$): At the highest coordinate bit, arithmetic addition modulo r^m inherently drops any overflow carry. Therefore, the operations are strictly identical: $\sigma^{r^{m-1}} = \tau_{m-1}$. Because $\mathcal{C}$ is a cyclic code, it is directly invariant under τ_{m-1}.

Inductive Step: Assume $\mathcal{C}$ is invariant under all higher translations τ_i for $i > k$. This implies $\mathcal{C}$ is invariant under any polynomial combination of these higher translations. Consequently, $\mathcal{C}$ is invariant under the noise operator N_k. Given the algebraic decomposition $\tau_k = \sigma^{r^k} - N_k$, and by the linearity of the subspace $\mathcal{C}$, the code must also be closed under τ_k. By induction, $\mathcal{C}$ is invariant under all elementary generators τ_k. Thus, $\mathcal{C}$ is formally an ideal in $\mathbb{F}_r[H_{r,m}]$. $\square$

Exact Parameters (The Dinh Bound). Unlike the probabilistic bounds required for the elementary abelian case, the minimum distance of repeated-root cyclic codes is completely deterministic. The exact distance profile was established by Dinh [3], providing a strict uniserial parallel to the Kasami-Lin-Peterson bound.

Lemma 2 (Dinh bound [3]). *Let $\mathcal{C}_k = \langle (x-1)^k \rangle$ be a cyclic code of length $n = r^m$ over $\mathbb{F}_r$. If the generator degree k falls within the interval:*

$$n - r^{m-t} + (\tau - 1)r^{m-t-1} < k \le n - r^{m-t} + \tau r^{m-t-1}$$

for some integers $0 \le t \le m-1$ and $1 \le \tau \le r-1$, then the minimum distance is precisely $d(\mathcal{C}_k) = (\tau + 1)r^t$.

Remark 1 (The Constant Rate Trap). To maintain a strictly positive asymptotic rate $R > 0$, the dimension $\dim(\mathcal{C}_k) = n - k$ must scale linearly with n. By the Dinh bound, if we desire a rate of $R = 1/r$, we must restrict $k \le n - n/r$. This forces the degree into the $t = 0$ regime, constraining the distance to evaluate to $d \le (\tau + 1)r^0 \le r$. Thus, any constant-rate modular cyclic code is trapped at a constant, absolute minimum distance.

3.2   The Hierarchical Abelian r-Group

To strike a structural balance between the uniserial depth of the cyclic group and the extreme combinatorial width of the elementary group, we consider hierarchical abelian r-groups. These groups take the mixed form $G \cong C_{r^{k_1}} \times C_{r^{k_2}} \times \cdots \times C_{r^{k_t}}$, where the invariants k_i form an arbitrary integer partition of the total exponent $m = \sum_{i=1}^{t} k_i$. The total order of this group is $n = r^m$.

Without loss of generality, we order the cyclic components such that $k_t = k_{max} = \max\{k_i\}$. If we define $y_i = g_i - 1$ for each of the t cyclic generators, the corresponding group algebra $\mathcal{R}_{hier} = \mathbb{F}_r[G]$ is isomorphic to a multivariate polynomial ring truncated by the specific orders of the cyclic components:

$$\mathcal{R}_{hier} \cong \mathbb{F}_r[y_1, y_2, \ldots, y_t]/\langle y_1^{r^{k_1}}, y_2^{r^{k_2}}, \ldots, y_t^{r^{k_t}} \rangle.$$

Combinatorially, the dimension distribution of the homogeneous components h_k corresponds to the convolution of discrete uniform distributions of widths r^{k_i}. By manipulating the partition $\{k_i\}$, we control the geometry of the filtration:

- **The Gaussian Regime:** If the invariants are kept small and balanced (e.g., $k_1 \approx k_2 \approx \cdots \approx k_t$), the Central Limit Theorem holds. The convolution converges into a sharp Gaussian peak, re-introducing the severe rate-concentration problem of the elementary abelian case.

- **The Plateau Regime:** If we deliberately unbalance the partition to break the Central Limit Theorem (forcing k_{max} to be overwhelmingly larger than the other invariants), the distribution flattens into a broad "plateau." This mathematically allows for the selection of ideal slices with strictly positive rates without being confined to a narrow density band.

However, any attempt to solve the rate concentration geometrically by entering the Plateau Regime fundamentally destroys the code's asymptotic minimum distance. Because y_t generates a cyclic code of length $r^{k_{max}}$, the weight of its powers y_t^k grows logarithmically, suffering from the exact uniserial topological failure outlined by the Dinh bound.

Proposition 3 (Distance Bound for Hierarchical Algebras). *Let $\{C_s\}$ be a family of codes constructed from the ideals of the hierarchical group algebra $\mathcal{R}_{hier}$ with a strictly positive asymptotic rate $R > 0$. For any integer partition of m, the minimum distance is strictly upper-bounded by the length of the largest cyclic component:*

$$d(C_s) \leq r^{k_{max}}.$$

Proof. Let C_n be an ideal in $\mathcal{R}_{hier}$ maintaining an asymptotic rate $R > 0$. We analyze the standard monomials of the ideal. Let $LT(C_n)$ be the set of leading monomials of C_n under a lexicographic term order where $y_1 \succ y_2 \succ \cdots \succ y_t$. The complement Δ of this set, containing all standard monomials not in $LT(C_n)$, forms a combinatorial "staircase" closed under division.

Asymptotically, the dimension sequence guarantees $\dim(C_n) \geq (R - \delta)n$ for some arbitrarily small $\delta > 0$. Therefore, the total volume of standard monomials in the staircase is bounded by the quotient ring's dimension:

$$|\Delta| \leq n - (R - \delta)n = (1 - R + \delta)n.$$

We determine the maximum possible height of this staircase along the pure y_t axis (the axis of the largest cyclic component). Let K be the smallest integer such that the pure power $y_t^K \in LT(C_n)$. Because Δ is closed under division, any monomial $y_1^{a_1} \ldots y_t^{a_t} \in \Delta$ must satisfy $a_t < K$.

The total number of available configurations for all the strictly smaller variables $(y_1, \ldots, y_{t-1})$ is exactly the order of their combined subgroup:

$$\prod_{i=1}^{t-1} r^{k_i} = \frac{n}{r^{k_{max}}}, \quad |\Delta| \leq K\left(\frac{n}{r^{k_{max}}}\right).$$

Equating our two bounds for the size of the staircase, we obtain:

$$(1 - R + \delta)n \leq K\left(\frac{n}{r^{k_{max}}}\right) \implies K \geq (1 - R + \delta)r^{k_{max}}.$$

Because $(1 - R + \delta)$ is a strictly positive constant, K scales linearly with $r^{k_{max}}$, guaranteeing that the pure power y_t^K appears as a leading term in the ideal: $y_t^K \in LT(C_n)$.

To evaluate the minimum distance, we invoke the Elimination Theorem for Gröbner bases [2]. Under our strict lexicographic ordering, any monomial containing the variables $y_1, \ldots, y_{t-1}$ is strictly larger than any pure power of y_t. Thus, the specific codeword $f \in \mathcal{C}_n$ whose leading term is y_t^K cannot contain any of those other variables.

Consequently, f must be a strictly univariate polynomial $f(y_t)$. This proves the ideal $\mathcal{C}_n$ has a non-trivial intersection with the isolated cyclic subalgebra:

$$f \in \mathcal{C}_n \cap \left(\mathbb{F}_r[y_t] / \langle y_t^{r^{k_{max}}} \rangle \right).$$

So $d(\mathcal{C}_n) \leq wt(f) \leq r^{k_{max}}$. $\qquad\qquad\square$

Corollary 2 (Distance Bound for the Canonical Plateau). *Let $G_{hier} \cong \prod_{i=1}^{t} C_{r^i}$ be the canonical hierarchical group designed to maximize the dimension plateau, with total order $n = r^{t(t+1)/2}$. Let $\{\mathcal{C}_n\}$ be a family of codes constructed from the ideals of $\mathbb{F}_r[G_{hier}]$ with a strictly positive asymptotic rate $R > 0$. The minimum distance is bounded by a sub-polynomial function:*

$$d(\mathcal{C}_n) \leq r^{\lceil \sqrt{2 \log_r n} \rceil} = n^{o(1)}.$$

Proof. By Proposition 3, the minimum distance is bounded by the largest cyclic component: $d(\mathcal{C}_n) \leq r^{k_{max}}$. In our canonical plateau construction, $k_i = i$, so $k_{max} = t$. The absolute maximum distance achievable is thus bounded by r^t.

We now express this upper bound strictly in terms of the total block length n. From the group order $n = r^{t(t+1)/2}$, we approximate the exponent:

$$\log_r n = \frac{t(t+1)}{2} \approx \frac{t^2}{2} \implies t \approx \sqrt{2 \log_r n}.$$

Substituting t back into our cyclic distance limit yields the exact bound $r^{\lceil \sqrt{2 \log_r n} \rceil}$.

To formally compare this against the elementary abelian square-root barrier ($d \sim n^{1/2}$), we rewrite this hierarchical distance bound strictly with n as the base:

$$r^{\sqrt{2 \log_r n}} = \left(r^{\log_r n} \right)^{\frac{\sqrt{2 \log_r n}}{\log_r n}} = n^{\sqrt{\frac{2}{\log_r n}}}.$$

As $n \to \infty$, the exponent $\sqrt{2/\log_r n}$ strictly approaches zero. Therefore, despite mathematically achieving a flat dimension plateau, the minimum distance collapses to $n^{o(1)}$. $\qquad\square$

3.3   The Dense Spectrum of Intermediate Topologies

Having established the extreme boundaries of modular abelian group codes—the Square-Root Barrier for the purely multilinear elementary group (C_r^m) and the absolute constant distance for the purely uniserial cyclic group (C_{r^m})—it is natural to ask whether intermediate hierarchical groups induce mathematical jumps or discontinuities in asymptotic performance.

By utilizing the Total Decay Metric (α, β) established in Sect. 2, we can mathematically prove that intermediate topologies smoothly and densely fill the entire spectrum of relative distance decay between the square-root exponent $(\alpha = 1/2)$ and the cyclic collapse $(\alpha = 1)$.

Proposition 4 (The Dense Continuum of Abelian Codes). *For any rational tuning parameter $\rho \in [0, 1]$, there exists a family of abelian r-group codes over $\mathbb{F}_r$ of length n that maintains a strictly positive asymptotic rate $(\beta = 0)$ while achieving a relative minimum distance decay exponent of:*

$$\alpha(\rho) = 1 - \frac{\rho}{2}.$$

Remark 2 (The Geometric Slider). This result proves that there are absolutely no gaps or discontinuous performance drops in the landscape of abelian group codes. By evaluating $\alpha(\rho)$ at the extremes, we recover the fundamental bounds:

- When $\rho = 1$ (the purely multilinear elementary group, $a = m$), the exponent is exactly $\alpha(1) = 0.5$.
- When $\rho = 0$ (the purely uniserial cyclic group, $a = 0$), the exponent collapses to $\alpha(0) = 1.0$.

General hierarchical groups act as a structural slider. By diluting the optimal $\mathcal{O}(n^{1/2})$ performance of the elementary group with the "dead weight" of a deep cyclic chain, one can mathematically dial in any rational exponent. As the size of the uniserial chain increases, the relative distance smoothly degrades from the Silver Standard straight down into the Logarithmic Trap.

3.4   Direct Product with a Semisimple Group Algebra

To break the structural limitations of the r-groups (which are entirely dominated by the nilpotent Jacobson radical), it is natural to investigate whether the asymptotic properties can be improved by tensoring with a "good" semisimple abelian algebra.

We consider the direct product $G \cong P \times H$, where P is an abelian r-group (the modular component, e.g., C_r^m) and H is an abelian group whose order is coprime to r (the semisimple component, meaning $\gcd(|H|, r) = 1$). The total block length of the code is $n = |P| \cdot |H|$.

The group algebra over G can be constructed as a tensor product of the constituent algebras, $\mathbb{F}_r[G] \cong \mathbb{F}_r[P] \otimes_{\mathbb{F}_r} \mathbb{F}_r[H]$.

By Maschke's Theorem the group algebra $\mathbb{F}_r[H]$ is strictly semisimple. It decomposes completely into a direct sum of finite field extensions determined by the cyclotomic cosets of r modulo $|H|$, $\mathbb{F}_r[H] \cong \bigoplus_{i=1}^{t} \mathbb{F}_{q_i}$, where each $q_i = r^{D_i}$. Note that $\sum_{i=1}^{t} D_i = |H|$.

Substituting this semisimple decomposition back into our tensor product, the full group algebra distributes over the direct sum:

$$\mathbb{F}_r[G] \cong \mathbb{F}_r[P] \otimes_{\mathbb{F}_r} \left(\bigoplus_{i=1}^{t} \mathbb{F}_{q_i} \right) \cong \bigoplus_{i=1}^{t} \left(\mathbb{F}_r[P] \otimes_{\mathbb{F}_r} \mathbb{F}_{q_i} \right) \cong \bigoplus_{i=1}^{t} \mathbb{F}_{q_i}[P].$$

This reveals that tensoring with a semisimple group does not fundamentally alter the topology of the modular radical; it simply shatters the algebra into a collection of disjoint copies of the original bad r-group algebra, scaled to larger alphabets $\mathbb{F}_{q_i}$. Any ideal $\mathcal{C}$ in $\mathbb{F}_r[G]$ decomposes as $\mathcal{C} = \bigoplus_{i=1}^{t} I_i$, where each component I_i is an ideal strictly within $\mathbb{F}_{q_i}[P]$.

Because these components are disjoint, the global minimum distance is determined by the weakest link, $d(\mathcal{C}) = \min_{1 \leq i \leq t} d(I_i)$.

Proposition 5 (The Dilution Bound). *Let $\mathcal{C} = \bigoplus_{i=1}^{t} I_i$ be a code in the mixed group algebra $\mathbb{F}_r[P \times H]$ with a strictly positive asymptotic rate $R_{global} > 0$. The global minimum distance is bottlenecked by the modular component P, and the relative distance δ decays multiplicatively with the size of H.*

Proof. The total dimension over the base field $\mathbb{F}_r$ is the sum of the dimensions of the components scaled by their field extension degrees:

$$K = \sum_{i=1}^{t} \dim_{\mathbb{F}_{q_i}} (I_i) \cdot D_i.$$

The total block length over $\mathbb{F}_r$ is $n = |P| \cdot |H| = |P| \sum_{i=1}^{t} D_i$. The global rate $R_{global} = K/n$ can be rewritten as:

$$R_{global} = \frac{\sum_{i=1}^{t} \dim_{\mathbb{F}_{q_i}} (I_i) \cdot D_i}{|P| \sum_{j=1}^{t} D_j} = \sum_{i=1}^{t} \left(\frac{D_i}{\sum_{j=1}^{t} D_j} \right) \left(\frac{\dim_{\mathbb{F}_{q_i}} (I_i)}{|P|} \right).$$

Let $R_i = \dim_{\mathbb{F}_{q_i}} (I_i)/|P|$ denote the "local rate" of the i-th component code. The equation above proves that the global rate is strictly a convex combination of the local rates R_i, where the weights $\frac{D_i}{|H|}$ sum exactly to 1.

By the averaging principle, if a weighted average equals R_{global}, at least one of the constituent values must be greater than or equal to that average. Therefore, there exists at least one "guilty" component index $k \in \{1, \ldots, t\}$ such that:

$$R_k \geq R_{global} > 0.$$

This component $I_k \subseteq \mathbb{F}_{q_k}[P]$ is forced to carry a high information density to sustain the global rate. Because I_k is an ideal in a purely modular group algebra, the structural bounds of the modular filtration remain intact.

As established in our comprehensive structural analysis of general r-group topologies, any ideal in $\mathbb{F}_{q_k}[P]$ maintaining a constant local rate $R_k > 0$ must inevitably intersect the dense, low-weight regions of its Jacobson radical— regardless of whether P is structured as multilinear, uniserial, or hierarchical. Consequently, the minimum distance of this specific modular component is trapped by the global maximum bound for abelian r-groups, mathematically capping it at the square-root barrier:

$$d(I_k) \leq \mathcal{O}\left(\sqrt{|P|} \right).$$

Since the global distance is the minimum over all components, we have $d(\mathcal{C}) = \min_i d(I_i) \leq d(I_k) \leq \mathcal{O}\left(\sqrt{|P|} \right).$ $\qquad\square$

Remark 3 (The Inescapable Decay). While the absolute distance is firmly capped at $\mathcal{O}(\sqrt{|P|})$ by the guilty modular component, the total block length n includes the semisimple factor $|H|$. Consequently, the relative minimum distance $\delta = d(\mathcal{C})/n$ suffers a severe dilution effect:

$$\delta = \frac{d(\mathcal{C})}{n} \leq \frac{\mathcal{O}(\sqrt{|P|})}{|P| \cdot |H|} = \mathcal{O}\left(\frac{1}{\sqrt{|P|} \cdot |H|}\right).$$

Comparing this to a purely modular code ($\delta \sim 1/\sqrt{|P|}$), the inclusion of the semisimple group actively accelerates the decay of the relative distance by a multiplicative factor of $1/|H|$. To bypass the modular radical and achieve the $\mathcal{O}(n/\log n)$ Logarithmic Barrier bounds afforded by semisimple algebras [7,8], the r-group must be abandoned entirely.

4 Beyond Asymptotic Goodness: The Universal Hierarchy

The binary classification of error-correcting codes as either asymptotically "good" (constant rate, constant relative distance) or "bad" is overly reductive for algebraic coding theory. To accurately evaluate the complex landscape of modular group algebras, we must abandon explicit polynomial representations in favor of a coordinate-free approach. By focusing entirely on how code parameters scale intrinsically through group homomorphisms, we can introduce a continuous spectrum of performance.

4.1 Generalized Asymptotic Goodness

We formalize this intrinsic plan by tracking the growth functions of the code parameters relative to the block length.

Definition 2 ($\Theta(f, g)$-Goodness). *A family of group codes $\{\mathcal{C}_n\}$ of length n is defined to be $\Theta(f, g)$-good if its dimension k_n and minimum distance d_n scale asymptotically as:*

$$k_n = \Theta(f(n)), \quad d_n = \Theta(g(n)).$$

Based on the results of the previous section and our structural bounds, this coordinate-free metric naturally stratifies all abelian group codes into a universal hierarchy of five canonical tiers:

1. **Tier I: The Linear Class** ($f = n, g = n$). Achieves the "Golden Standard" of true asymptotic goodness (constant rate, constant relative distance). This tier is exclusively populated by semisimple codes.
2. **Tier II: The Square-Root Class** ($f = n, g = \sqrt{n}$). Achieves constant rate but faces the Square-Root Barrier. This is the optimal "Silver Standard" for modular codes, seeded exclusively by purely multilinear elementary abelian p-groups (C_r^m).

3. **Tier III: The Logarithmic Class** ($f = n, g = \log n$). Maintains a constant rate but suffers extreme distance collapse. This tier acts as a "Coprime Trap." It is seeded by uniserial cyclic groups (C_{r^m}) and heavily imbalanced hierarchical groups.
4. **Tier IV: The Robustness Class** ($f = n^{1-\epsilon}, g = n$). Achieves linear distance by sacrificing constant rate (accepting polynomial dimension decay). This tier is typically populated by codes generated via subfield descent or by selecting ultra-deep slices of a radical filtration.
5. **Tier V: The Trivial Class** ($f = 1, g = n$). Achieves maximum absolute distance by completely collapsing the dimension to a constant (vanishing rate). The canonical examples are repetition codes (Fig. 1).

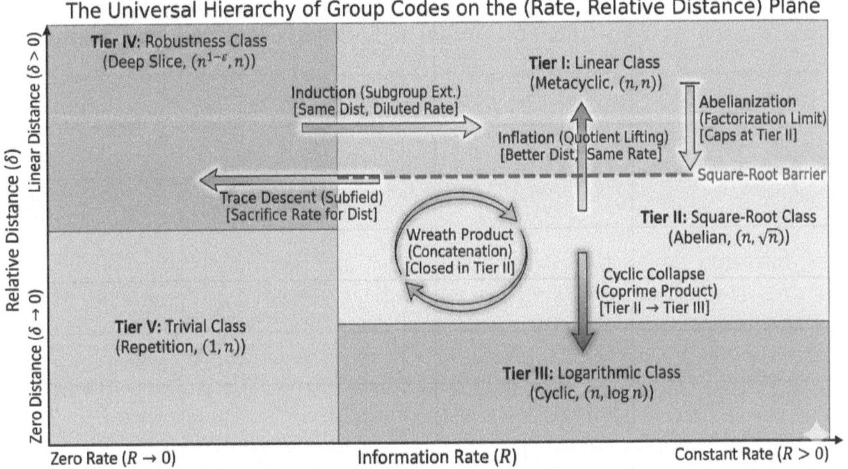

Fig. 1. The Universal Hierarchy: A (*Rate*, δ) classification of modular and semisimple group codes.

4.2   The Universal Reduction Machinery

To deterministically predict the asymptotic tier of complex, composite groups based on their simpler constituent subgroups, we formalize the intrinsic reduction of code parameters.

Definition 3 ((m, j, c)**-Reduction**). *A complex target family of codes C_{target} of length N is said to be reduced from a simpler seed family C_{seed} of length n if there exist scaling functions $m(n)$, $j(n)$, and $c(n)$ such that the target parameters map as:*

$$N = n \cdot m(n) \qquad \text{(Length Expansion / Index)}$$
$$K = k \cdot j(n) \qquad \text{(Dimension Scaling / Multiplicity)}$$
$$D = d \cdot c(n) \qquad \text{(Distance Scaling / Weight Gain)}$$

Theorem 3 (The Universal Transformation Theorem). *If the seed family is $\Theta(f,g)$-good, and it reduces to a target family via an (m,j,c)-reduction, then the target family is $\Theta(F,G)$-good, where the target growth functions are given by:*

$$F(N) = j(h(N)) \cdot f(h(N)), \quad G(N) = c(h(N)) \cdot g(h(N)).$$

where $h(N)$ is the inverse of the length scaling function (i.e., $n = h(N)$).

4.3   Vertical Propagation in the Group Lattice

We apply this machinery to the standard exact sequence of groups $1 \to K \to G \to Q \to 1$. Code properties propagate vertically through the lattice via two primary mechanisms.

Proposition 6 (Upward Propagation: Inflation/Quotient Lifting). *Let a code be defined on the quotient group Q and inflated to the total group G. The codewords are repeated $|K|$ times. The transformation is an $(|K|, 1, |K|)$-reduction, yielding target functions:*

$$F(N) = f(n), \quad G(N) = |K| \cdot g(n).$$

Effect: Inflation moves the code "North-West" in the rate-distance plane. It perfectly preserves the exact dimension while strictly boosting the absolute distance via redundancy.

Proposition 7 (Downward Propagation: Induction/Module Extension). *Let a code be induced from a subgroup H to the total group G. The code space becomes a direct sum of $t = [G : H]$ disjoint shifts. The transformation is a $(t, t, 1)$-reduction, yielding target functions:*

$$F(N) = t \cdot f(n), \quad G(N) = g(n).$$

Effect: Induction moves the code "East." It linearly scales the dimension, maintaining the relative rate, but the absolute distance remains rigidly fixed, severely diluting the relative minimum distance.

These propositions establish a fundamental rule of group codes: large kernels generate redundancy (improving distance), while large indices generate disjoint codeblocks (diluting distance).

4.4   Advanced Reductions and "No-Go" Theorems

By applying this intrinsic framework to specific group products, we can rigorously prove several "no-go" theorems regarding the modular barrier.

- **Wreath Products and Fractal Codes.** For a Wreath product $G \wr H$, the reduction geometrically combines the seeds such that the absolute distances multiply: $D \sim d_{\text{in}} \cdot d_{\text{out}}$. If both the inner and outer seeds are derived from Tier II Square-Root codes, the resulting distance scales as $\sqrt{n} \cdot \sqrt{m} = \sqrt{nm} = \sqrt{N}$. Consequently, **Tier II is mathematically closed under concatenation**. It is impossible to break the square-root barrier by fractally stacking square-root codes.

- **The Coprime Trap.** If a group is constructed as a direct product of coprime cyclic components, $G = C_{n_1} \times C_{n_2}$ with $\gcd(n_1, n_2) = 1$, the Chinese Remainder Theorem forces a total collapse into a single cyclic group $C_{n_1 n_2}$. The coprime factors destroy the multilinear "width" of the lattice, immediately reducing the topology to a uniserial chain. This explicitly forces an $(n, \sqrt{n})$ seed to plummet into Tier III $(n, \log n)$.

- **Subfield Descent (Trace Maps).** Mapping codes defined over an extension field $\mathbb{F}_{q^m}$ down to the base field $\mathbb{F}_q$ preserves the absolute distance but imposes a severe dimension penalty $(K = k/m)$. This reduction explicitly maps codes from Tier I or Tier II directly into Tier IV, exchanging rate for robust linear distance over small alphabets.

Table 1. The effect of standard group operations on the $\Theta(f, g)$-goodness tuple.

Operation	Transformation on (f, g)	Hierarchy Movement
Inflation (Lifting)	$(f(n),\ t \cdot g(n))$	$\searrow$ (Increases Dist, Same Dim)
Induction (Extension)	$(t \cdot f(n),\ g(n))$	$\rightarrow$ (Dilutes Rel. Dist, Same Rate)
Coprime Product	$(f, g) \rightarrow (n, \log n)$	$\downarrow$ (Collapses to Tier III)
Wreath Product	$(f_1 f_2,\ g_1 g_2)$	$\circlearrowright$ (Strictly closed in Tier II)
Trace Descent	$(\frac{1}{m} f(n),\ g(n))$	$\leftarrow$ (Drops to Tier IV)

5 Conclusion

The Universal Reduction Machinery acts as a calculus for algebraic topology, allowing us to compute the exact asymptotic trajectory of any group code operation (Table 1).

In this work, we have completely reframed the asymptotic evaluation of algebraic group codes. By moving beyond the binary and often reductive classification of codes as merely "good" or "bad," we have established a rigorous, quantitative hierarchy based on continuous asymptotic growth rates, denoted by the $\Theta(f(n), g(n))$-goodness metric. *Asymptotic goodness is an intrinsic property of the composition factors, not the group extensions. The structural limits of the algebra are inescapably inherited.*

Acknowledgments. This work was funded by the European Union through the Spanish National Cybersecurity Institute (INCIBE) within the Recovery, Transformation and Resilience Plan—NextGenerationEU, under the Spain Digital Agenda 2026 framework (project MRR-MAETD:24-INCIBE-001).

Disclosure of Interests. The authors have no competing interests to declare that are relevant to the content of this article.

References

1. Berman, S.D.: On the theory of group codes. Cybernetics **3**(1), 25–31 (1967). https://doi.org/10.1007/BF01072842
2. Cox, D.A., Little, J., O'Shea, D.: Ideals, Varieties, and Algorithms. UTM, Springer, Cham (2015). https://doi.org/10.1007/978-3-319-16721-3
3. Dinh, H.Q.: On the linear ordering of some classes of negacyclic and cyclic codes and their distance distributions. Finite Fields Appl. **14**(1), 22–40 (2008). https://doi.org/10.1016/j.ffa.2007.07.001
4. Feller, W.: An Introduction to Probability Theory, vol. II, Wiley (1968)
5. Hoeffding, W.: Probability inequalities for sums of bounded random variables. J. Am. Stat. Assoc. **58**(301), 13–30 (1963). https://doi.org/10.1080/01621459.1963.10500830
6. Kasami, T., Lin, S., Peterson, W.W.: Some results on generalized Reed-Muller codes. Inf. Control **11**(1–2), 38–49 (1968)
7. Lin, S., Weldon, E.J.: Long BCH codes are bad. Inf. Control **11**(4), 445–451 (1967). https://doi.org/10.1016/S0019-9958(67)90660-2
8. MacWilliams, F.J.: Binary codes which are ideals in the group algebra of an Abelian group. Bell Syst. Tech. J. **49**(6), 987–1011 (1970). https://doi.org/10.1002/j.1538-7305.1970.tb01812.x
9. MacWilliams, F.J., Sloane, N.J.A.: The Theory of Error-Correcting Codes. North-Holland (1977)

Algebraic Geometry Codes with Many Automorphisms Arising from Galois Points

Satoru Fukasawa$^{(\boxtimes)}$

Yamagata University, Yamagata 990-8560, Japan
`s.fukasawa@sci.kj.yamagata-u.ac.jp`

Abstract. This paper presents a method of constructing algebraic geometry codes with many automorphisms arising from Galois points for algebraic curves. This enables us to construct codes with automorphisms systematically, and gives a geometric interpretation of a condition in a well-known criterion by Stichtenoth for yielding automorphisms of AG codes. Compared with Stichtenoth's criterion, the main theorem does not require a condition on the genus. The proof relies on techniques from algebraic geometry, more precisely on investigating actions of automorphisms on a linear system coming from a morphism to a projective plane. This paper also provides a generalization of the main theorem, which recovers several known one-point AG codes with automorphisms.

Keywords: algebraic geometry code · Galois point · automorphism group · finite field

1 Introduction

A finite field of q elements is denoted by $\mathbb{F}_q$. The set of all $\mathbb{F}_q$-rational points of an algebraic variety V over $\mathbb{F}_q$ is denoted by $V(\mathbb{F}_q)$. V. D. Goppa discovered a construction of algebraic geometry codes, according to the following data: a smooth projective curve X defined over $\mathbb{F}_q$, a subset $S \subset X(\mathbb{F}_q)$, and a divisor D over $\mathbb{F}_q$ (see, for example, [10,11]). This paper presents a method of constructing algebraic geometry codes with many automorphisms, using the theory of *Galois points*.

Let $\varphi : X \to \mathbb{P}^2$ be a morphism defined over $\mathbb{F}_q$, which is birational onto $\varphi(X)$. A point $P \in \mathbb{P}^2 \setminus \varphi(X)$ is called an outer Galois point, if the field extension $\overline{\mathbb{F}}_q(X)/(\pi_P \circ \varphi)^* \overline{\mathbb{F}}_q(\mathbb{P}^1)$ of function fields induced by the projection $\pi_P : \varphi(X) \to \mathbb{P}^1$ from P is a Galois extension (see [2,8,12], see also [4, Remark 1] for the definition of a Galois point over $\mathbb{F}_q$). The main result is the following.

Theorem 1. *Let $G_1, G_2 \subset \mathrm{Aut}_{\mathbb{F}_q}(X)$ be different finite subgroups with $|G_1| = |G_2|$, and let $Q, Q' \in X(\overline{\mathbb{F}}_q)$. Assume that the following four conditions are satisfied:*

(a) $X/G_i \cong \mathbb{P}^1$ for $i = 1, 2$;
(b) $G_1 \cap G_2 = \{1\}$;

L. Batina and F. Özbudak (Eds.): WAIFI 2026, LNCS 16611, pp. 89–94, 2026.
https://doi.org/10.1007/978-3-032-27574-5_6

(c) $\sum_{\sigma \in G_1} \sigma(Q) = \sum_{\tau \in G_2} \tau(Q)$ *over* $\overline{\mathbb{F}}_q$, *and this divisor D is defined over* $\mathbb{F}_q$;
(d) $Q' \in X(\mathbb{F}_q)$, $Q' \notin \operatorname{supp}(D)$, *and the cardinality $\#S$ of the orbit $S :=$* $\langle G_1, G_2 \rangle \cdot Q'$ *is at least* $|G_1| + 1$.

Then the following hold.

(1) The evaluation map

$$\Phi : \mathcal{L}(D) \to \bigoplus_{R \in S} \mathbb{F}_q \cdot R; \quad f \mapsto (f(R))_{R \in S}$$

is injective, and the minimum distance of the code $\Phi(\mathcal{L}(D))$ is at least $\#S - |G_1|$.

(2) There exists an injective homomorphism

$$\langle G_1, G_2 \rangle \hookrightarrow \operatorname{Aut}(\Phi(\mathcal{L}(D))).$$

Remark 1. It follows from a criterion described by the present author in [3] that conditions (a), (b) and (c) are satisfied if and only if there exists a morphism $\varphi :$ $X \to \mathbb{P}^2$ of degree $|G_1|$ defined over $\mathbb{F}_q$, which is birational onto its image, and different outer Galois points $P_1, P_2 \in (\mathbb{P}^2 \setminus \varphi(X))(\mathbb{F}_q)$ exist for $\varphi(X)$ such that $G_{P_1} = G_1$, $G_{P_2} = G_2$, and points P_1, P_2, Q are collinear, where $G_{P_i} \subset \operatorname{Aut}_{\mathbb{F}_q}(X)$ is the Galois group at P_i for $i = 1, 2$.

Theorem 1 enables us to construct linear codes admitting automorphisms systematically. It is worth pointing out that an automorphism $\sigma \in G_P$ of an outer Galois point P is a typical example of an automorphism satisfying a condition $\sigma^*(D) = D$ for yielding an automorphism of an AG code $\Phi(\mathcal{L}(D))$ (see [9], [10, Section 8.2]); Theorem 1 gives a geometric interpretation of this condition. Furthermore, compared with Stichtenoth's criterion [10, Proposition 8.2.3 (b)], Theorem 1 (2) does not require a condition on the genus of X.

The proof relies on techniques from algebraic geometry, more pecisely on investigating actions of automorphisms on the linear system coming from the morphism $\varphi : X \to \mathbb{P}^2$. In the final section, this paper provides a generalization of Theorem 1, which recovers several known one-point AG codes with automorphisms.

2 Proof

Proof (Proof of Theorem 1). Let $f \in \mathcal{L}(D) \setminus \{0\}$. Since $\deg D = |G_1|$, it follows that the number of zeros of f is at most $|G_1|$. By condition (d), $\#S \geq |G_1| + 1$. Therefore, $f(R) \neq 0$ for some $R \in S$. This argument also implies that the minimum distance is at least $\#S - |G_1|$. Assertion (1) follows.

We prove assertion (2). By condition (c), it follows that $\gamma^* D = D$, namely, $\mathcal{L}(\gamma^* D) = \mathcal{L}(D)$, for any $\gamma \in \langle G_1, G_2 \rangle$. Furthermore, $\gamma(S) = S$ for any $\gamma \in \langle G_1, G_2 \rangle$, since S is an orbit. Therefore, we have a natural homomorphism

$$\langle G_1, G_2 \rangle \to \mathrm{Aut}(\Phi(\mathcal{L}(D))$$

(see also [9], [10, 8.2.3]). We prove that there does not exist $\gamma \in \langle G_1, G_2 \rangle \setminus \{1\}$ such that $\gamma|_S = 1$. Assume that $\gamma \in \langle G_1, G_2 \rangle$ and $\gamma|_S = 1$. Let $D_i := \sum_{\sigma \in G_i} \sigma(Q')$ for $i = 1, 2$. By conditions (a), (b) and (c), it follows from the criterion by the present author in [3] that the coverings $X \to X/G_1$, $X \to X/G_2$ are realized as projections from different outer Galois points P_1, P_2 with $G_{P_1} = G_1$, $G_{P_2} = G_2$. The birational embedding $X \to \mathbb{P}^2$ is denoted by φ. Then $D_1, D_2 \in \Lambda$ for $i = 1, 2$, where Λ is the sublinear system of $|D|$ corresponding to φ. Note that $D_1 \ne D_2$, $D_1 \ne D$, and $D_2 \ne D$. Since $\gamma|_S = 1$, it follows that $\gamma^* D_1 = D_1$ and $\gamma^* D_2 = D_2$. Since $\gamma^* D = D$, it follows that $\gamma^* \Lambda = \Lambda$, namely, there exists a linear transformation $\tilde{\gamma}$ of $\mathbb{P}^2$ such that $\tilde{\gamma} \circ \varphi = \varphi \circ \gamma$. Let ℓ, ℓ_1 and $\ell_2 \subset \mathbb{P}^2$ be the lines corresponding to D, D_1 and D_2, respectively. Then the linear transformation $\tilde{\gamma}$ fixes non-collinear points P_1, P_2 and $\varphi(Q') \in \mathbb{P}^2$, which are given by $\ell \cap \ell_1$, $\ell \cap \ell_2$ and $\ell_1 \cap \ell_2$, respectively. If there exists a point $R \in \varphi(S)$ with $R \notin \ell_1 \cup \ell_2$, then $\tilde{\gamma} = 1$, namely, $\gamma = 1$. Therefore, we can assume that $\varphi(S) \subset \ell_1 \cup \ell_2$. By condition (d), $\varphi(S) \not\subset \ell_1$. Therefore, there exists a point $R_2 \in \varphi(S) \cap \ell_2$ with $R_2 \notin \ell_1$. In particular, $R_2 \ne \varphi(Q')$. Similarly, there exists a point $R_1 \in \varphi(S) \cap \ell_1$ with $R_1 \ne \varphi(Q')$. Since $\tilde{\gamma}$ fixes three points $P_1, \varphi(Q'), R_1$ on the line ℓ_1, it follows that $\tilde{\gamma}|_{\ell_1} = 1$. Similarly, $\tilde{\gamma}|_{\ell_2} = 1$. Therefore, $\tilde{\gamma} = 1$, namely, $\gamma = 1$.

Remark 2. Theorem 1 can be generalized to the case where collinear outer Galois points $P_1, P_2, \dots, P_k$ exist, namely, there exists an $\mathbb{F}_q$-line $\ell \subset \mathbb{P}^2$ such that $P_1, P_2, \dots, P_k \in \ell$. Assume the following:

(d') $Q' \in X(\mathbb{F}_q)$, $Q' \notin \mathrm{supp}(\varphi^* \ell)$, and there exist i, j with $i \ne j$ such that $\#(\langle G_{P_i}, G_{P_j} \rangle \cdot Q') \ge |G_{P_i}| + 1$.

Let $S := \langle G_{P_1}, G_{P_2}, \dots, G_{P_k} \rangle \cdot Q'$. Then there exists an injective homomorphism

$$\langle G_{P_1}, G_{P_2}, \dots, G_{P_k} \rangle \hookrightarrow \mathrm{Aut}(\Phi(\mathcal{L}(\varphi^* \ell))).$$

3   Examples

Although the following three examples do not appear to yield new codes, they serve to illustrate Theorem 1 clearly. Example 3 arises from a recent result of Borges and the present author [1].

Example 1. Let $X \subset \mathbb{P}^2$ be the Fermat curve

$$X^{q+1} + Y^{q+1} + Z^{q+1} = 0.$$

It is confirmed that $P_1 = (1 : 0 : 0)$ and $P_2 = (0 : 1 : 0)$ are outer Galois points with $G_{P_i} \cong \mathbb{Z}/(q+1)\mathbb{Z}$ for $i = 1, 2$ (see also [7]). Conditions (a) and (b) in Theorem 1 are satisfied for G_{P_1}, G_{P_2}. Let ζ be a primitive $q+1$-th root of unity. Then

$$\langle G_{P_1}, G_{P_2} \rangle = \{(X : Y : Z) \mapsto (\zeta^i X : \zeta^j Y : Z) \mid 0 \leq i, j \leq q\}.$$

Let $Q \in X(\mathbb{F}_{q^2}) \cap \{Z = 0\}$. Then condition (c) in Theorem 1 is satisfied, and D is the divisor defined by $X \cap \{Z = 0\}$. We take a point $Q' = (a : b : 1) \in X(\mathbb{F}_{q^2})$ with $ab \neq 0$. Then

$$S = \{(\zeta^i a : \zeta^j b : 1) \mid 0 \leq i, j \leq q\}.$$

Therefore, condition (d) in Theorem 1 is satisfied for the 4-tuple $(G_{P_1}, G_{P_2}, Q, Q')$. It is well known that $\dim \mathcal{L}(D) = 3$ (see, for example, [6, Lemma 11.28]). Note that $q + 1$ points $(\zeta^i a : b : 1) \in S$ with $0 \leq i \leq q$ are zeros of the function $y - b \in \mathcal{L}(D)$. Then the code $\Phi(\mathcal{L}(D))$ has parameters

$$[(q + 1)^2, \ 3, \ (q + 1)q]_{q^2}$$

and admits an automorphism group

$$(\mathbb{Z}/(q + 1)\mathbb{Z})^{\oplus 2}.$$

Example 2. Let $X = \mathbb{P}^1$. Assume that q is odd and $q \geq 5$. Let ζ be a primitive $m := (q - 1)/2$-th root of unity, and let

$$G_1 := \{(s : t) \mapsto (\zeta^i s : t) \mid 0 \leq i \leq m - 1\},$$
$$G_2 := \{(s : t) \mapsto (\zeta^i s + (1 - \zeta^i)t : t) \mid 0 \leq i \leq m - 1\}.$$

By Lüroth's theorem, condition (a) in Theorem 1 is satisfied. Obviously, $G_1 \cap G_2 = \{1\}$, namely, condition (b) is satisfied. We take $Q = (1 : 0)$. Then condition (c) in Theorem 1 is satisfied, and $D = mQ$. Let $Q' = (0 : 1)$. It can be confirmed that

$$\langle G_1, G_2 \rangle \cong \mathbb{F}_q \rtimes (\mathbb{Z}/m\mathbb{Z}),$$

and that $\langle G_1, G_2 \rangle \cdot Q' = \mathbb{P}^1(\mathbb{F}_q) \setminus \{Q\}$. Therefore, condition (d) in Theorem 1 is satisfied for the 4-tuple (G_1, G_2, Q, Q'). It is well known that $\dim \mathcal{L}(D) = m + 1$ (see, for example, [10]). Then the (Reed–Solomon) code $\Phi(\mathcal{L}(D))$ has parameters

$$\left[q, \ \frac{q + 1}{2}, \ \frac{q + 1}{2}\right]_q$$

and admits an automorphism group of order $q(q - 1)/2$.

Example 3. Let $\varphi(X) \subset \mathbb{P}^2$ be defined by

$$(x^{q^2} + x)^{q+1} + (y^{q^2} + y)^{q+1} + 1 = 0,$$

which was studied in [1]. According to [1], the full automorphism group $G :=$ $\mathrm{Aut}(X)$ is generated by the Galois groups of all outer Galois points on the line $\{Z = 0\}$, and $|G| = q^5(q^2-1)(q+1)$. The divisor coming from $\varphi(X) \cap \{Z = 0\}$ is denoted by D. Let $a^{q+1} = -1$, let $x_0^{q^2} + x_0 = a$ and let $Q' \in X(\mathbb{F}_{q^4})$ with $\varphi(Q') = (x_0 : 0 : 1)$. It is easily verified that $\#(\langle G_{P_1}, G_{P_2}\rangle \cdot Q') > \deg \varphi(X) = |G_{P_1}|$ for Galois points $P_1 = (1 : 0 : 0)$ and $P_2 = (0 : 1 : 0)$. This implies that condition (d') in Remark 2 is satisfied. It can be confirmed that $\#(G \cdot Q') = q^5(q^2 - 1)$. It was proved that $\dim \mathcal{L}(D) = 3$ in [1, Proposition 1]. Note that $q^3 + q^2$ points $(x : 0 : 1) \in G \cdot Q'$ with $(x^{q^2} + x)^{q+1} + 1 = 0$ are zeros of the function $y \in \mathcal{L}(D)$. Then the code $\Phi(\mathcal{L}(D))$ has parameters

$$[q^7 - q^5, \ 3, \ q^7 - q^5 - q^3 - q^2]_{q^4}$$

and admits an automorphism group of order $q^5(q^2 - 1)(q + 1)$.

4   A Generalization

T. Hasegawa [5] suggested a generalization of our code in Example 1, according to replacing the divisor D by mD with $m < \#S/|G_1|$. Based on this, Theorem 1 was generalized by the author as follows.

Theorem 2. *Let (G_1, G_2, Q, Q') be a 4-tuple satisfying conditions (a), (b) and (c) in Theorem 1. We take a positive integer e so that $D = eD_0$, where $D_0 = \sum_{R \in \mathrm{supp}(D)} R$. Let m be a positive integer. Assume the following:*

(d") $Q' \in X(\mathbb{F}_q)$, $Q' \notin \mathrm{supp}(D)$, and $m < e \times \#S/|G_1|$, where $S := \langle G_1, G_2\rangle \cdot Q'$.

Then the following hold.

(1) The evaluation map

$$\Phi : \mathcal{L}(mD_0) \to \bigoplus_{R \in S} \mathbb{F}_q \cdot R; \ f \mapsto (f(R))_{R \in S}$$

is injective, and the minimum distance of the code $\Phi(\mathcal{L}(mD_0))$ is at least $\#S - m \times |G_1|/e$.
(2) There exists an injective homomorphism

$$\langle G_1, G_2\rangle \hookrightarrow \mathrm{Aut}(\Phi(\mathcal{L}(mD_0))).$$

Theorem 2 recovers several known one-point AG codes with automorphisms, as follows. If $G_1 \cdot Q = G_2 \cdot Q = \{Q\}$ for a 4-tuple (G_1, G_2, Q, Q') satisfying conditions in Theorem 1, then we obtain $D_0 = Q$, the injective evaluation map

$$\Phi : \mathcal{L}(mQ) \to \bigoplus_{R \in S} \mathbb{F}_q \cdot R; \ f \mapsto (f(R))_{R \in S},$$

and an injective homomorphism

$$\langle G_1, G_2 \rangle \hookrightarrow \mathrm{Aut}(\Phi(\mathcal{L}(mQ)))$$

for $m < \#S$. For example, if we take $m = q - 1$ in Example 2 and $D_0 = Q$, then for any integer m' with $1 \le m' \le q - 1$, an automorphism group $\mathbb{F}_q \rtimes \mathbb{F}_q^\times$ of the Reed–Solomon code with parameters $[q, m' + 1, q - m']$ is derived from our construction; this automorphism group coincides with the full automorphism group of the code, if $m' \le q - 3$ (see [9, Example 3.6]).

Acknowledgments. The author is grateful to Professor Masaaki Homma, Professor Takehiro Hasegawa, and Professor Kazuki Higashine for helpful comments.

This work was partially supported by JSPS KAKENHI Grant Numbers JP19K03438, JP22K03223, and JP26K06718.

References

1. Borges, H., Fukasawa, S.: An elementary abelian p-cover of the Hermitian curve with many automorphisms. Math. Z. **302**, 695–706 (2022)
2. Fukasawa, S.: Galois points for a plane curve in arbitrary characteristic. Geometriae Dedicata **139**, 211–218 (2009)
3. Fukasawa, S.: A birational embedding of an algebraic curve into a projective plane with two Galois points. J. Algebra **511**, 95–101 (2018)
4. Fukasawa, S.: Galois points and rational functions with small value sets. Hiroshima Math. J. **54**, 37–43 (2024)
5. Hasegawa, T.: Private communications (2022)
6. Hirschfeld, J.W.P., Korchmáros, G., Torres, F.: Algebraic Curves over a Finite Field. Princeton Univ. Press, Princeton (2008)
7. Homma, M.: Galois points for a Hermitian curve. Comm. Algebra **34**, 4503–4511 (2006)
8. Miura, K., Yoshihara, H.: Field theory for function fields of plane quartic curves. J. Algebra **226**, 283–294 (2000)
9. Stichtenoth, H.: On automorphisms of geometric Goppa codes. J. Algebra **130**, 113–121 (1990)
10. Stichtenoth, H.: Algebraic Function Fields and Codes, Graduate Texts in Mathematics, vol. 254. Springer-Verlag, Berlin Heidelberg (2009)
11. Tsfasman, M.A., Vlăduţ, S.G.: Algebraic-Geometric Codes, Mathematics and Its Applications, vol. 58. Kluwer Academic Publishers, Dordrecht (1991)
12. Yoshihara, H.: Function field theory of plane curves by dual curves. J. Algebra **239**, 340–355 (2001)

Classification of Hermitian-Relative Curves up to Projective Equivalence

Masaaki Homma[1] and Seon Jeong Kim[2]([✉])

[1] Department of Mathematics, Kanagawa University, Yokohama 221-8686, Japan
`homma@kanagawa-u.ac.jp`
[2] Department of Mathematics, and RINS, Gyeongsang National University,
Jinju 52828, Korea
`skim@gnu.ac.kr`

Abstract. Let q be an even power of a prime p, and let $\mathbb{F}_q$ be the finite field with q elements. For $A = (a_{ij}) \in GL(3, \mathbb{F}_q)$, C_A denotes the plane curve defined by the equation $(x^{\sqrt{q}}, y^{\sqrt{q}}, z^{\sqrt{q}})A\,(x, y, z)^t = 0$. We call C_A a Hermitian-relative curve. Let $A^* = (a_{ji}^{\sqrt{q}})$ be the conjugate transpose. Then C_A is a Hermitian curve if $A = A^*$, which is an important curve for various reasons. Two Hermitian-relative curves C_A and C_B are projectively equivalent if and only if there exists $T \in GL(3, \mathbb{F}_q)$ such that $B = T^*AT$. In previous works [4,5], we determined all possible pairs $(N_q(C_A), I_q(C_A))$, where $N_q(C)$ and $I_q(C)$ denote the number of rational points and rational inflexions on a curve C, respectively. The resulting pairs are: $(q\sqrt{q}+1, q\sqrt{q}+1)$, $(\sqrt{q}+1, \sqrt{q}+1)$, $(1, 1)$, $(q - \sqrt{q}+1, 0)$, $(q+1, 1)$, $(q+1, 2)$, $(q+\sqrt{q}+1, 1)$, and $(q+2\sqrt{q}+1, 0)$. Extending these results, we aim to categorize the curves into equivalence classes according to their projective equivalence. It is well known that the curves of type $(q\sqrt{q}+1, q\sqrt{q}+1)$ are Hermitian curves, which constitute a single equivalence class. We focus on partitioning curves of other types into their respective projective equivalence classes.

Keywords: Plane curve · Finite field · Rational point · Hermitian curve · Inflexion

1 Introduction

Throughout this paper, q is assumed to be an even power of a prime p, so that $\sqrt{q}$ is also a power of p. Let $\mathbb{F}_q$ denote a finite field of order q. Then $\mathbb{F}_{\sqrt{q}}$ and $\overline{\mathbb{F}}_q$ represent a subfield of order $\sqrt{q}$ and the algebraic closure of $\mathbb{F}_q$, respectively. For $A = (a_{ij}) \in GL(3, \mathbb{F}_q)$, let $A^* = (a_{ji}^{\sqrt{q}})$ be the matrix taking entry-wise $\sqrt{q}$-th power of A and being transposed. A matrix A that satisfies $A = A^*$ is called a Hermitian matrix. A Hermitian curve is a plane curve defined by

$$\left(x^{\sqrt{q}}\ y^{\sqrt{q}}\ z^{\sqrt{q}}\right) A \begin{pmatrix} x \\ y \\ z \end{pmatrix} = 0, \tag{1}$$

L. Batina and F. Özbudak (Eds.): WAIFI 2026, LNCS 16611, pp. 95–110, 2026.
https://doi.org/10.1007/978-3-032-27574-5_7

where A is a Hermitian matrix. If we replace the Hermitian matrix A in equation (1) with an arbitrary nonsingular matrix $A \in GL(3, \mathbb{F}_q)$, the curve defined by the given equation is still a nonsingular plane curve of degree $\sqrt{q}+1$. We denote this curve by C_A and refer to it as a relative of the Hermitian curve, or simply a Hermitian-relative curve, as introduced in [4].

All Hermitian-relative curves, including Hermitian curves themselves, are projectively equivalent to each other over the algebraic closure $\overline{\mathbb{F}}_q$ of $\mathbb{F}_q$, which is due to Pardini [6]. The Hermitian curve over $\mathbb{F}_q$ has $(q\sqrt{q}+1)$ rational points at which the tangent line contacts the curve with multiplicity $\sqrt{q}+1$. At a non-rational point of a Hermitian curve, the tangent line contacts with multiplicity $\sqrt{q}$. A point is called an inflexion point (or simply an inflexion) if the tangent line at that point intersects the curve with multiplicity $\sqrt{q}+1$. In [4], it is proved that the number $N_q(C)$ of rational points on any Hermitian-relative curve satisfies the following condition

$$N_q(C) \equiv 1 \pmod{\sqrt{q}},$$

which means, in particular, any Hermitian-relative curve C over $\mathbb{F}_q$ has at least one rational point. Unlike the Hermitian curve, a Hermitian-relative curve that is not Hermitian may possess $\mathbb{F}_q$-rational points that are not inflexions. Let $I_q(C)$ denote the number of inflexions among the $\mathbb{F}_q$-rational points of the curve. We classify the Hermitian-relative curves according to the pair $(N_q(C), I_q(C))$. A curve C is said to be of type (n, i) if $N_q(C) = n$ and $I_q(C) = i$.

The classification for the case $q = 4$ was completed by Hirschfeld in [1,2], and is summarized in [4, Table 2]. Specifically, Hermitian-relative curves over $\mathbb{F}_4$ are categorized into seven distinct types:

$$(9,9), (3,3), (1,1), (7,1), (5,1), (9,0), (3,0). \tag{2}$$

Among these, the types $(3,3), (7,1)$, and $(3,0)$ each split into two projective equivalence classes. Since each of the remaining four types consists of a unique projective equivalence class, the classification yields a total of ten projective equivalence classes.

For $A \in GL(3, \mathbb{F}_q)$, note that $C_A = C_{\alpha A}$ for any $\alpha \in \mathbb{F}_q^*$. Thus, a Hermitian-relative curve is uniquely determined by an element of $PGL(3, \mathbb{F}_q)$. Two matrices A and B are said to be congruent (denoted by $A \sim B$) if there exists a nonsingular matrix $T \in GL(3, \mathbb{F}_q)$ such that $B = T^*AT$. Let

$$\begin{pmatrix} x' \\ y' \\ z' \end{pmatrix} = T^{-1} \begin{pmatrix} x \\ y \\ z \end{pmatrix}.$$

Then we have

$$\begin{pmatrix} x^{\sqrt{q}} & y^{\sqrt{q}} & z^{\sqrt{q}} \end{pmatrix} A \begin{pmatrix} x \\ y \\ z \end{pmatrix} = \begin{pmatrix} x'^{\sqrt{q}} & y'^{\sqrt{q}} & z'^{\sqrt{q}} \end{pmatrix} T^*AT \begin{pmatrix} x' \\ y' \\ z' \end{pmatrix}.$$

This identity implies that the curves C_A and C_B are projectively equivalent. Consequently, two nonsingular matrices A and B are congruent if and only if the corresponding curves C_A and C_B are projectively equivalent. The remainder of this paper is organized as follows. In Sect. 2, we review the results obtained in the two previous papers [4,5]. Section 3 is devoted to the classification of all Hermitian-relative curves into projective equivalence classes by partitioning each type accordingly. Finally, in Sect. 4, we determine the automorphism groups for several specific curves. In this regard, the computational techniques employed for the classification in Sect. 3 are slightly modified and applied to determine these groups, providing a more complete characterization of the curves.

2   Previous Results

For a Hermitian-relative curve C_A, the curve C_{A^*} is called the mirror curve of C_A. Note that C_A is Hermitian if and only if $C_{A^*} = C_A$. The notation $\mathcal{L}_P(C)$ denotes the tangent line to the curve C at a point P.

Proposition 1 ([4]). *For a mirror pair (C_A, C_{A^*}), the following properties hold.*

(1) $C_A(\mathbb{F}_q) = C_{A^*}(\mathbb{F}_q)$.
(2) For $P \in C_A(\mathbb{F}_q) = C_{A^*}(\mathbb{F}_q)$, P is an inflexion of C_A if and only if P is an inflexion of C_{A^*}. In this case, $\mathcal{L}_P(C_A) = \mathcal{L}_P(C_{A^*})$.
(3) If $P \in C_A(\mathbb{F}_q)$ is a non-inflexion, then there is a point $P' \in C_A(\mathbb{F}_q) \setminus \{P\}$ so that $C_A.\mathcal{L}_P(C_A) = \sqrt{q}P + P'$. In this case, $C_{A^*}.\mathcal{L}_{P'}(C_{A^*}) = \sqrt{q}P' + P$. In particular P' is a non-inflexion of C_{A^*}.

Consider a map $\varphi_A : C_A \to C_A$ defined by $\varphi_A(P) = P'$ where $C_A.\mathcal{L}_P(C_A) = \sqrt{q}P + P'$. Note that P is an inflexion of C_A if and only if $\varphi_A(P) = P$.

Proposition 2 ([5]). *Let $A \in GL(3, \mathbb{F}_q)$ and let C_A be the corresponding Hermitian-relative curve. For any point $P = (x_0, y_0, z_0)^t \in C_A$, we have $\varphi_A(P) = A^{-1}A^* P^{(q)}$. In particular, if $P = (x_0, y_0, z_0)^t \in C_A(\mathbb{F}_q)$, then $\varphi_A(P) = A^{-1}A^* P$.*

In [4], all Hermitian-relative curves with at least 2 inflexions are classified. This classification yields three distinct types, as summarized in Table 1.

Table 1. Classification of curves with $I_q(C) \geq 2$

	$N_q(C_A)$	$I_q(C)$	Number of classes
(a)	$\sqrt{q}^3 + 1$	$\sqrt{q}^3 + 1$	1
(b)	$\sqrt{q} + 1$	$\sqrt{q} + 1$	$\sqrt{q}$
(c)	$q + 1$	2	$\frac{1}{2}(\sqrt{q} + 1)(\sqrt{q} - 2)$

Since we have covered all cases where $I_q(C) \geq 2$, the remaining candidate for the condition $N_q(C) = I_q(C)$ is the case where $N_q(C) = I_q(C) = 1$. Indeed, such curves exist.

Proposition 3 ([5]). *Any Hermitian-relative curve of type $(1,1)$ is projectively equivalent to a curve corresponding to* $\begin{pmatrix} 1 & 0 & 0 \\ 0 & \eta & 1 \\ 0 & 1 & 0 \end{pmatrix}$ *for some* $\eta \in \mathbb{F}_q \setminus \mathbb{F}_{\sqrt{q}}$.

From Table 1 and Proposition 3, we get the following:

Theorem 1 ([5]). *There exist Hermitian-relative curves of types $(q\sqrt{q}+1, q\sqrt{q}+1)$, $(\sqrt{q}+1, \sqrt{q}+1)$, and $(1,1)$. Moreover, if $N_q(C) = I_q(C)$ for a Hermitian-relative curve C, then its type must be one of these three cases.*

Next, we consider Hermitian-relative curves that possess at least one rational point that is not an inflexion (i.e., $N_q(C) \neq I_q(C)$)

Proposition 4 ([5]). *Any Hermitian-relative curve with $N_q(C) \neq I_q(C)$ is projectively equivalent to curves of the form*

$$C_{a,b} : \quad x^{\sqrt{q}}z + xy^{\sqrt{q}} + bxz^{\sqrt{q}} + ayz^{\sqrt{q}} = 0, \tag{3}$$

where $a, b \in \mathbb{F}_q$, $a \neq 0$.

We first classify the curves with $b = 0$ which corresponds to $\begin{pmatrix} 0 & 0 & 1 \\ 1 & 0 & 0 \\ 0 & a & 0 \end{pmatrix}$.

Theorem 2 ([5]). *The following hold;*

(1) If $\sqrt{q} \equiv 0 \pmod 3$, then $C_{a,0}$ is of type $(q+1, 1)$.
(2) If $\sqrt{q} \equiv 1 \pmod 3$, then $C_{a,0}$ is of type $(q+1, 2)$.
(3) If $\sqrt{q} \equiv 2 \pmod 3$, then $C_{a,0}$ is of type $(q + 2\sqrt{q} + 1, 0)$ for $a = \omega^{3j}$, $0 \leq j < \frac{q-1}{3}$ and $(q - \sqrt{q} + 1, 0)$, otherwise. Here ω is a primitive element of $\mathbb{F}_q$.

Finally, we provide a classification of the curves $C_{a,b}$ where $a, b \in \mathbb{F}_q^*$.

Proposition 5 ([5]). *When $b \neq 0$, any Hermitian-relative curve $C_{a,b}$ is projectively equivalent to a curve C_c for some $c \in \mathbb{F}_q^*$, where*

$$C_c : \quad x^{\sqrt{q}}z - cxy^{\sqrt{q}} + cxz^{\sqrt{q}} - yz^{\sqrt{q}} = 0 \tag{4}$$

which is corresponding to the matrix $M_c = \begin{pmatrix} 0 & 0 & 1 \\ -c & 0 & 0 \\ c & -1 & 0 \end{pmatrix}$. Conversely, any curve C_c with $c \neq 0$ is projectively equivalent to $C_{a,b}$ for some $a, b \in \mathbb{F}_q$, $a \neq 0$.

For $M_c = \begin{pmatrix} 0 & 0 & 1 \\ -c & 0 & 0 \\ c & -1 & 0 \end{pmatrix}$ with $c \neq 0$, let $U = M_c^{-1} M_c^*$. The characteristic polynomial $P_U(\lambda)$ of U is $P_U(\lambda) = \lambda^3 - c^{\sqrt{q}}\lambda^2 + c^{\sqrt{q}}\lambda - c^{\sqrt{q}-1}$.

Theorem 3 ([5]). *Any polynomial $f_c(\lambda) = \lambda^3 - c^{\sqrt{q}}\lambda^2 + c^{\sqrt{q}}\lambda - c^{\sqrt{q}-1}$ satisfies one of the following five conditions:*

(I) *f_c is an irreducible polynomial.*
(II) *f_c has a triple root.*
(III) *f_c has three simple roots, only one of them is a $(\sqrt{q}+1)$-th root of unity.*
(IV) *f_c has a double root and the other simple root, both of them are $(\sqrt{q}+1)$-th roots of unity.*
(V) *f_c has three simple roots which are all $(\sqrt{q}+1)$-th roots of unity.*

For each of the five cases of f_c given above, we obtain the corresponding type (n, i) as shown in Table 2. The table also provides the counts of such polynomials according to the congruence of $\sqrt{q}$ modulo 3.

Table 2. Classification of curves C_{A_c} with $c \in \mathbb{F}_q^*$

Case	type (n, i)	the number of polynomials f_c		
		$\sqrt{q} \equiv 0 \pmod 3$	$\sqrt{q} \equiv 1 \pmod 3$	$\sqrt{q} \equiv 2 \pmod 3$
(I)	$(q - \sqrt{q} + 1, 0)$	$\frac{q-\sqrt{q}}{3}$	$\frac{q-\sqrt{q}}{3}$	$\frac{q-\sqrt{q}-2}{3}$
(II)	$(q + 1, 1)$	0	1	1
(III)	$(q + 1, 2)$	$\frac{q-\sqrt{q}-2}{2}$	$\frac{q-\sqrt{q}-4}{2}$	$\frac{q-\sqrt{q}-2}{2}$
(IV)	$(q + \sqrt{q} + 1, 1)$	$\sqrt{q}$	$\sqrt{q}$	$\sqrt{q}$
(V)	$(q + 2\sqrt{q} + 1, 0)$	$\frac{q-\sqrt{q}}{6}$	$\frac{q-\sqrt{q}}{6}$	$\frac{q-\sqrt{q}-2}{6}$

Using Theorem 1, 2 and 3, we conclude the following:

Theorem 4 ([5]). *For any $q > 4$, there exist curves of exactly eight types:*

$$(q\sqrt{q} + 1, q\sqrt{q} + 1), (\sqrt{q} + 1, \sqrt{q} + 1), (1, 1),$$
$$(q - \sqrt{q} + 1, 0), (q + 1, 1), (q + 1, 2), (q + \sqrt{q} + 1, 1), (q + 2\sqrt{q} + 1, 0). \tag{5}$$

Remark 1. Comparing the list for $q = 4$ in (5) with the list for $q > 4$ in (5), we observe that the type $(q+1, 2) = (5, 2)$ is absent in the case where $q = 4$. In fact, Tables 1 and 2 show that the number of such polynomials is given by $\frac{q-\sqrt{q}-2}{2}$, which vanishes when $q = 4$.

3   The Number of Equivalence Classes for Each Type

In this section, we provide a complete classification of Hermitian-relative curves over $\mathbb{F}_q$. We first note that the classification for two specific types follows directly from existing literature, as summarized in Table 1. For curves of type $(q\sqrt{q} + 1, q\sqrt{q} + 1)$, there is exactly one equivalence class. It is a well-established result that any curve with exactly $(q\sqrt{q} + 1)$ $\mathbb{F}_q$-rational points is a Hermitian curve

[3]. Since all Hermitian curves are projectively equivalent to each other, they form a single equivalence class. Similarly, for curves of type $(\sqrt{q}+1, \sqrt{q}+1)$, the number of equivalence classes is exactly $\sqrt{q}$, as already presented in Table 1. In the following subsections, we focus on the other types.

As mentioned in Sect. 1, two Hermitian-relative curves C_A and C_B are projectively equivalent if and only if their associated matrices A and B are congruent; that is, there exists a nonsingular matrix T such that $B = T^*AT$. Since T acts as a projective equivalence between C_B and C_A, it must satisfy the relation $T\varphi_B = \varphi_A T$. Recall that two matrices M_1 and M_2 are said to be similar if there exists a nonsingular matrix T such that $M_2 = T^{-1}M_1T$. Indeed, we have the following lemma:

Lemma 1. *(1) If A and B are congruent, then $B^{-1}B^*$ and $A^{-1}A^*$ are similar, and hence they have the same characteristic polynomial.*
(2) If $P_U(\lambda) = \lambda^3 + u\lambda^2 + v\lambda + w$, then $P_{\alpha U}(\lambda) = \alpha^3 P_U(\frac{\lambda}{\alpha}) = \lambda^3 + \alpha u\lambda^2 + \alpha^2 v\lambda + \alpha^3 w$ for any $\alpha \in \mathbb{F}_q^*$.
(3) If two curves C_A and C_B are projectively equivalent, then the characteristic polynomials $P_U(\lambda)$ and $P_V(\lambda)$ of $V = B^{-1}B^*$ and $U = A^{-1}A^*$, respectively, satisfy $P_V(\lambda) = \alpha^3 P_U(\frac{\lambda}{\alpha})$ for some $\alpha \in \mu_{\sqrt{q}+1}$, where $\mu_{\sqrt{q}+1} = \{x \in \mathbb{F}_q^* \mid x^{\sqrt{q}+1} = 1\}$ is the set of $(\sqrt{q}+1)$-th root of unity in $\mathbb{F}_q$.

Proof. (1) Since $B = T^*AT$, we have $B^* = (T^*AT)^* = T^*A^*T$. It follows that

$$B^{-1}B^* = (T^*AT)^{-1}(T^*A^*T) = T^{-1}A^{-1}(T^*)^{-1}T^*A^*T = T^{-1}(A^{-1}A^*)T.$$

This implies that $B^{-1}B^*$ and $A^{-1}A^*$ are similar and thus share the same characteristic polynomial.

(2) We have $P_{\alpha U}(\lambda) = \det(\lambda I - \alpha U) = \det(\alpha(\frac{\lambda}{\alpha}I - U)) = \alpha^3 \det(\frac{\lambda}{\alpha}I - U) = \alpha^3 P_U(\frac{\lambda}{\alpha})$.

(3) Suppose that two curves C_A and C_B are projectively equivalent. Then there exist a nonsingular matrix $T \in GL(3, \mathbb{F}_q)$ and $\rho \in \mathbb{F}_q^*$ such that $B = \rho T^*AT$. Then $V = B^{-1}B^* = \rho^{\sqrt{q}-1}T^{-1}(A^{-1}A^*)T = \rho^{\sqrt{q}-1}T^{-1}UT$. Thus $P_V(\lambda) = \rho^{3(\sqrt{q}-1)}P_U(\frac{\lambda}{\rho^{\sqrt{q}-1}})$. □

3.1 Curves of Type $(1, 1)$

By Proposition 3, any Hermitian-relative curve of type $(1, 1)$ is projectively equivalent to a curve corresponding to $\begin{pmatrix} 1 & 0 & 0 \\ 0 & \eta & 1 \\ 0 & 1 & 0 \end{pmatrix}$ for some $\eta \in \mathbb{F}_q \setminus \mathbb{F}_{\sqrt{q}}$. By the theorem proved immediately below, the number of equivalence classes of curves of this type is exactly one.

Theorem 5. *For any $\eta_1, \eta_2 \in \mathbb{F}_q \setminus \mathbb{F}_{\sqrt{q}}$, two matrices $A_1 = \begin{pmatrix} 1 & 0 & 0 \\ 0 & \eta_1 & 1 \\ 0 & 1 & 0 \end{pmatrix}$ and*

$A_2 = \begin{pmatrix} 1 & 0 & 0 \\ 0 & \eta_2 & 1 \\ 0 & 1 & 0 \end{pmatrix}$ *are congruent, and hence two curves corresponding to these matrices are projectively equivalent.*

Proof. Since $\mathbb{F}_q$ is a two dimensional vector space over $\mathbb{F}_{\sqrt{q}}$, $\{1, \eta_1\}$ is a basis of $\mathbb{F}_q$ over $\mathbb{F}_{\sqrt{q}}$. Hence there exist α, β in $\mathbb{F}_{\sqrt{q}}$ with $\alpha \neq 0$ such that $\eta_2 = \alpha\eta_1 + \beta$. Since the norm and trace maps from $\mathbb{F}_q$ to $\mathbb{F}_{\sqrt{q}}$ are surjective, there exist $\zeta, \xi \in \mathbb{F}_q$ such that $\zeta^{\sqrt{q}+1} = \alpha$, and $\xi^{\sqrt{q}} + \xi = \beta$. Note that $\zeta \neq 0$. Let $T = \begin{pmatrix} 1 & 0 & 0 \\ 0 & \zeta & 0 \\ 0 & \zeta^{-\sqrt{q}}\xi & \zeta^{-\sqrt{q}} \end{pmatrix}$.

Then we have

$$T^*A_1T = \begin{pmatrix} 1 & 0 & 0 \\ 0 & \zeta^{\sqrt{q}} & \zeta^{-1}\xi^{\sqrt{q}} \\ 0 & 0 & \zeta^{-1} \end{pmatrix} \begin{pmatrix} 1 & 0 & 0 \\ 0 & \eta_1 & 1 \\ 0 & 1 & 0 \end{pmatrix} \begin{pmatrix} 1 & 0 & 0 \\ 0 & \zeta & 0 \\ 0 & \zeta^{-\sqrt{q}}\xi & \zeta^{-\sqrt{q}} \end{pmatrix}$$

$$= \begin{pmatrix} 1 & 0 & 0 \\ 0 & \zeta^{\sqrt{q}+1}\eta_1 + \xi^{\sqrt{q}} + \xi & 1 \\ 0 & 1 & 0 \end{pmatrix} = \begin{pmatrix} 1 & 0 & 0 \\ 0 & \eta_2 & 1 \\ 0 & 1 & 0 \end{pmatrix} = A_2.$$

Thus A_1 and A_2 are congruent, and hence the curves C_{A_1} and C_{A_2} are projectively equivalent. $\qquad\square$

3.2   Curves $C_{a,0}$ in (3)

Let $G = \{\alpha^3 k \mid \alpha \in \mathbb{F}_q^*,\ k \in \mathbb{F}_{\sqrt{q}}^*\}$. Then G is a subgroup of the multiplicative group $\mathbb{F}_q^*$. Specifically, $G = \langle \omega^3, \omega^{\sqrt{q}+1} \rangle$, where ω is a primitive element of $\mathbb{F}_q$. Thus

$$G = \begin{cases} \langle \omega \rangle = \mathbb{F}_q^* & \text{if } \gcd(3, \sqrt{q}+1) = 1 \ (\text{i.e., } \sqrt{q} \equiv 0, 1 \pmod{3}), \\ \langle \omega^3 \rangle \subsetneq \mathbb{F}_q^* & \text{if } \gcd(3, \sqrt{q}+1) = 3 \ (\text{i.e., } \sqrt{q} \equiv 2 \pmod{3}). \end{cases}$$

Lemma 2. *For $\eta, \xi \in \mathbb{F}_q^*$, two curves corresponding to matrices $A_\eta = \begin{pmatrix} 0 & 0 & 1 \\ 1 & 0 & 0 \\ 0 & \eta & 0 \end{pmatrix}$*

and $A_\xi = \begin{pmatrix} 0 & 0 & 1 \\ 1 & 0 & 0 \\ 0 & \xi & 0 \end{pmatrix}$ are projectively equivalent if and only if $\frac{\xi}{\eta} \in G$.

Proof. Suppose that two curves corresponding to A_η and A_ξ are projectively equivalent. Then there exists $T \in GL(3, \mathbb{F}_q)$ such that $A_\xi = \alpha T^*A_\eta T$. We have

$$\xi = \det A_\xi = \det(\alpha T^*A_\eta T) = \alpha^3 (\det T)^{\sqrt{q}+1} \det A_\eta = \alpha^3 (\det T)^{\sqrt{q}+1}\eta.$$

Thus $\frac{\xi}{\eta} = \alpha^3 (\det T)^{\sqrt{q}+1} \in G$.

Conversely, assume that $\frac{\xi}{\eta} \in G$. Then $\frac{\xi}{\eta} = \alpha^3 k$ for some $\alpha \in \mathbb{F}_q^*$, $k \in \mathbb{F}_{\sqrt{q}}^*$.

Let $T = \begin{pmatrix} (k\alpha^{\sqrt{q}+1})^{-1} & 0 & 0 \\ 0 & \alpha & 0 \\ 0 & 0 & \alpha^{\sqrt{q}} \end{pmatrix}$. Then

$$
\begin{aligned}
T^* A_\eta T &= \begin{pmatrix} (k\alpha^{\sqrt{q}+1})^{-1} & 0 & 0 \\ 0 & \alpha^{\sqrt{q}} & 0 \\ 0 & 0 & \alpha \end{pmatrix} \begin{pmatrix} 0 & 0 & 1 \\ 1 & 0 & 0 \\ 0 & \eta & 0 \end{pmatrix} \begin{pmatrix} (k\alpha^{\sqrt{q}+1})^{-1} & 0 & 0 \\ 0 & \alpha & 0 \\ 0 & 0 & \alpha^{\sqrt{q}} \end{pmatrix} \\
&= \begin{pmatrix} 0 & 0 & (k\alpha)^{-1} \\ (k\alpha)^{-1} & 0 & 0 \\ 0 & \alpha^2 \eta & 0 \end{pmatrix} = (k\alpha)^{-1} \begin{pmatrix} 0 & 0 & 1 \\ 1 & 0 & 0 \\ 0 & \alpha^3 k\eta & 0 \end{pmatrix} = (k\alpha)^{-1} A_\xi.
\end{aligned}
$$

Since A_η and A_ξ represent the same element in $PGL(3, \mathbb{F}_q)$ up to a scalar factor $(k\alpha)^{-1}$, the two curves corresponding A_η and A_ξ are projectively equivalent. $\square$

While the types of each curve were already specified in Theorem 2, this section focuses on classifying them into their respective equivalence classes.

Theorem 6. *Consider the curves $C_{a,0} : x^{\sqrt{q}}z + xy^{\sqrt{q}} + ayz^{\sqrt{q}} = 0$ with $a \in \mathbb{F}_q^*$.*

(1) *If $\sqrt{q} \equiv 0 \pmod 3$, then $C_{a,0}$, $a \in \mathbb{F}_q^*$ is of type $(q+1, 1)$ and all such curves are projectively equivalent.*

(2) *If $\sqrt{q} \equiv 1 \pmod 3$, then for any $a \in \mathbb{F}_q^*$, the curve $C_{a,0}$ is of type $(q+1, 2)$. All such curves are projectively equivalent.*

(3) *If $\sqrt{q} \equiv 2 \pmod 3$, the classification is as follows:*
  - *For $a \in \langle \omega^3 \rangle$, the curves $C_{a,0}$ are of type $(q + 2\sqrt{q} + 1, 0)$ and form a single equivalence class.*
  - *The remaining curves, where $a \in \omega\langle\omega^3\rangle \cup \omega^2\langle\omega^3\rangle$, are of type $(q - \sqrt{q} + 1, 0)$. These curves are partitioned into two distinct equivalence classes corresponding to the cosets $\omega\langle\omega^3\rangle$ and $\omega^2\langle\omega^3\rangle$.*

Proof. If $\sqrt{q} \equiv 0, 1 \pmod 3$, then $G = \mathbb{F}_q^*$. Hence, by Lemma 2, the number of equivalence classes is one. If $\sqrt{q} \equiv 2 \pmod 3$, then $G = \langle\omega^3\rangle \subsetneq \mathbb{F}_q^*$. Hence, by Lemma 2, the number of equivalence classes is three, since the order of the factor group $\mathbb{F}_q/G$ is three. $\square$

3.3   Curves C_c with $c \neq 0$ in (4)

By Proposition 4, any curve $C_{a,b}$ with $a, b \in \mathbb{F}_q^*$ is projectively equivalent to

$$
C_c : x^{\sqrt{q}}z - cxy^{\sqrt{q}} + cxz^{\sqrt{q}} - yz^{\sqrt{q}} = 0,
$$

which corresponds to the matrix $M_c := \begin{pmatrix} 0 & 0 & 1 \\ -c & 0 & 0 \\ c & -1 & 0 \end{pmatrix}$. To classify these curves into equivalence classes, we examine the characteristic polynomial of $M_c^{-1} M_c^*$, as discussed in Lemma 1.

Theorem 7. *Two curves corresponding to M_c and $M_{c'}$ are projectively equivalent if and only if $c = c'$. Consequently, the number of distinct equivalence classes among the curves C_c for $c \in \mathbb{F}_q$ is exactly equal to the number of corresponding polynomials f_c as listed in Table 2.*

Proof. Suppose that two curves corresponding to M_c and $M_{c'}$ are projectively equivalent. Then $M_{c'} = \rho T^* M_c T$ for some $T \in GL(3, \mathbb{F}_q)$ and $\rho \in \mathbb{F}_q^*$. If we let $U = M_c^{-1} M_c^*$ and $V = M_{c'}^{-1} M_{c'}^*$. Their characteristic polynomials are given by:

$$P_U(\lambda) = \lambda^3 - c^{\sqrt{q}}\lambda^2 + c^{\sqrt{q}}\lambda - c^{\sqrt{q}-1}$$
$$P_V(\lambda) = \lambda^3 - c'^{\sqrt{q}}\lambda^2 + c'^{\sqrt{q}}\lambda - c'^{\sqrt{q}-1}.$$

On the other hand, by Lemma 1,

$$P_V(\lambda) = \rho^{3(\sqrt{q}-1)} P_U\left(\frac{\lambda}{\rho^{\sqrt{q}-1}}\right) = \lambda^3 - \rho^{\sqrt{q}-1} c^{\sqrt{q}}\lambda^2 + (\rho^{\sqrt{q}-1})^2 c^{\sqrt{q}}\lambda - (\rho^{\sqrt{q}-1})^3 c^{\sqrt{q}-1}.$$

By comparing the coefficients of the quadratic and linear terms in the two expressions for $P_V(\lambda)$, we conclude that $\rho^{\sqrt{q}-1} = 1$. This implies $P_V(\lambda) = P_U(\lambda)$, and hence $c = c'$. $\qquad\square$

3.4 The Number of Equivalence Classes for Each Curve Type

Reviewing the classification process thus far, we first categorized curves with at least two inflexion points, leading to the results in Table 1 [4].

To classify curves satisfying $N_q(C) = I_q(C)$, we found curves C with $N_q(C) = I_q(C) = 1$, which are all projectively equivalent.

To classify curves with $N_q(C) > I_q(C)$, we found a simple form $C_{a,b}$ in (3) for curves possessing rational points that are not inflexion points. First, the case where $b = 0$ was analyzed, and the classification of curves $C_{a,0}$ yielded Theorem 6. Next, to classify curves $C_{a,b}$ with $b \neq 0$, we established the canonical form C_c in (4). The characteristic polynomial of $A_\eta^{-1} A_\eta^*$, where A_η is the matrix corresponding to $C_{a,0}$, is given by $\lambda^3 - \eta^{\sqrt{q}-1}$. According to Lemma 1, this implies that a curve $C_{a,0}$ and a curve C_c cannot be projectively equivalent. Consequently, there is no overlap between these two families. For curves of type $(q+1, 2)$ in Table 1, the number of equivalence classes was $\frac{1}{2}(\sqrt{q}+1)(\sqrt{q}-2)$. By summing the classes obtained from $C_{a,0}$ and C_c, we obtain $\frac{q-\sqrt{q}-2}{2}$, which is consistent with the result in Table 1. Especially, for type $(q+1, 1)$, when $\sqrt{q}$ is a multiple of 3, no corresponding curves were found in the C_c case, while exactly one class emerged from the $C_{a,0}$ case. By consolidating these classification results, we obtain the comprehensive summary presented in Table 3.

4 The Automorphism Groups of Some Curves

As mentioned in the Introduction, we now apply a slight modification of the classification techniques developed in Sect. 3 to determine the automorphism groups of the representative curves.

Table 3. Classification of projective equivalence classes for each curve type

Type of Curve (n, i)	The number of equivalence classes		
	$\sqrt{q} \equiv 0 \pmod 3$	$\sqrt{q} \equiv 1 \pmod 3$	$\sqrt{q} \equiv 2 \pmod 3$
$(q\sqrt{q}+1, q\sqrt{q}+1)$	1	1	1
$(\sqrt{q}+1, \sqrt{q}+1)$	$\sqrt{q}$	$\sqrt{q}$	$\sqrt{q}$
$(1, 1)$	1	1	1
$(q - \sqrt{q}+1, 0)$	$\frac{q-\sqrt{q}}{3}$	$\frac{q-\sqrt{q}}{3}$	$\frac{q-\sqrt{q}+4}{3}$
$(q+1, 1)$	1	1	1
$(q+1, 2)$	$\frac{q-\sqrt{q}-2}{2}$	$\frac{q-\sqrt{q}-2}{2}$	$\frac{q-\sqrt{q}-2}{2}$
$(q + \sqrt{q}+1, 1)$	$\sqrt{q}$	$\sqrt{q}$	$\sqrt{q}$
$(q + 2\sqrt{q}+1, 0)$	$\frac{q-\sqrt{q}}{6}$	$\frac{q-\sqrt{q}}{6}$	$\frac{q-\sqrt{q}+4}{6}$

4.1   Hermitian Curves

The automorphism group $\mathrm{Aut}(\mathcal{H})$ of the Hermitian curve $\mathcal{H}$ over $\mathbb{F}_q$ is isomorphic to the projective general unitary group $PGU(3, \sqrt{q})$, whose order is $q\sqrt{q}(q\sqrt{q}+1)(q-1)$ according to [7,8].

4.2   A Curve of Type $(\sqrt{q}+1, \sqrt{q}+1)$

In [4], any curve of type $(\sqrt{q}+1, \sqrt{q}+1)$ is projectively equivalent to the curve C_{A_η} corresponding to a matrix $A_\eta = \begin{pmatrix} 1 & 0 & 0 \\ 0 & 1 & 0 \\ 0 & 0 & \eta \end{pmatrix}$ with $\eta \in \mathbb{F}_q \setminus \mathbb{F}_{\sqrt{q}}$. Then C_{A_η} is

defined by the equation $x^{\sqrt{q}+1} + y^{\sqrt{q}+1} + \eta z^{\sqrt{q}+1} = 0$. Let $T = \begin{pmatrix} a_{11} & a_{12} & a_{13} \\ a_{21} & a_{22} & a_{23} \\ a_{31} & a_{32} & a_{33} \end{pmatrix}$ be

a nonsingular matrix in $GL(3, \mathbb{F}_q)$ which preserves the curve C_η, i.e., $T^* A_\eta T = \alpha A_\eta$ for some $\alpha \in \mathbb{F}_q^*$. Since C_η contains $\sqrt{q}+1$ rational points which are on the line $z = 0$, T preserves the line $z = 0$. So we have $a_{31} = a_{32} = 0$. Substituting A_η and T into $T^* A_\eta T = \alpha A_\eta$, we have

$$\alpha \begin{pmatrix} 1 & 0 & 0 \\ 0 & 1 & 0 \\ 0 & 0 & \eta \end{pmatrix} = \begin{pmatrix} a_{11}^{\sqrt{q}+1} + a_{21}^{\sqrt{q}+1} & a_{11}^{\sqrt{q}}a_{12} + a_{21}^{\sqrt{q}}a_{22} & a_{11}^{\sqrt{q}}a_{13} + a_{21}^{\sqrt{q}}a_{23} \\ a_{12}^{\sqrt{q}}a_{11} + a_{22}^{\sqrt{q}}a_{21} & a_{12}^{\sqrt{q}+1} + a_{22}^{\sqrt{q}+1} & a_{12}^{\sqrt{q}}a_{13} + a_{22}^{\sqrt{q}}a_{23} \\ a_{13}^{\sqrt{q}}a_{11} + a_{23}^{\sqrt{q}}a_{21} & a_{13}^{\sqrt{q}}a_{12} + a_{23}^{\sqrt{q}}a_{22} & a_{13}^{\sqrt{q}+1} + a_{23}^{\sqrt{q}+1} + a_{33}^{\sqrt{q}+1}\eta \end{pmatrix}.$$

Comparing each entries of both sides, we have

$$\begin{cases} a_{11}^{\sqrt{q}+1} + a_{21}^{\sqrt{q}+1} = a_{12}^{\sqrt{q}+1} + a_{22}^{\sqrt{q}+1} = \alpha \neq 0 \\ a_{11}^{\sqrt{q}}a_{12} + a_{21}^{\sqrt{q}}a_{22} = a_{11}^{\sqrt{q}}a_{13} + a_{21}^{\sqrt{q}}a_{23} = 0 \\ a_{12}^{\sqrt{q}}a_{11} + a_{22}^{\sqrt{q}}a_{21} = a_{12}^{\sqrt{q}}a_{13} + a_{22}^{\sqrt{q}}a_{23} = 0 \\ a_{13}^{\sqrt{q}}a_{11} + a_{23}^{\sqrt{q}}a_{21} = a_{13}^{\sqrt{q}}a_{12} + a_{23}^{\sqrt{q}}a_{22} = 0 \\ a_{13}^{\sqrt{q}+1} + a_{23}^{\sqrt{q}+1} + a_{33}^{\sqrt{q}+1}\eta = \alpha\eta \end{cases}$$

If we let $v_1 = (a_{11}, a_{21})$, $v_2 = (a_{12}, a_{22})$, $v_3 = (a_{13}, a_{23})$, and consider them in the Hermitian space, then above equations means that $\langle v_1, v_1 \rangle = \langle v_2, v_2 \rangle = \alpha \neq 0$, and $\langle v_1, v_2 \rangle = \langle v_1, v_3 \rangle = \langle v_2, v_3 \rangle = 0$. Thus $\{v_1, v_2\}$ forms a basis of $\mathbb{F}_q^2$, and we conclude that $v_3 = (0, 0)$ which means $a_{13} = a_{23} = 0$. Thus we have

$$\begin{cases} a_{11}^{\sqrt{q}+1} + a_{21}^{\sqrt{q}+1} = a_{12}^{\sqrt{q}+1} + a_{22}^{\sqrt{q}+1} = \alpha \neq 0 \\ a_{11}^{\sqrt{q}} a_{12} + a_{21}^{\sqrt{q}} a_{22} = 0 \\ a_{33}^{\sqrt{q}+1} = \alpha \end{cases}$$

Then we have $T = \begin{pmatrix} a_{11} & a_{12} & 0 \\ a_{21} & a_{22} & 0 \\ 0 & 0 & a_{33} \end{pmatrix}$. Letting $b_{ij} = \frac{a_{ij}}{a_{33}}$, $T = \begin{pmatrix} b_{11} & b_{12} & 0 \\ b_{21} & b_{22} & 0 \\ 0 & 0 & 1 \end{pmatrix}$ in $PGL(3, \mathbb{F}_q)$, where

$$\begin{cases} b_{11}^{\sqrt{q}+1} + b_{21}^{\sqrt{q}+1} = b_{12}^{\sqrt{q}+1} + b_{22}^{\sqrt{q}+1} = 1 \neq 0 \\ b_{11}^{\sqrt{q}} b_{12} + b_{21}^{\sqrt{q}} b_{22} = 0 \end{cases}$$

Thus we have

$$\mathrm{Aut}(C_\eta) = \left\{ \begin{pmatrix} a & b & 0 \\ c & d & 0 \\ 0 & 0 & 1 \end{pmatrix} \,\middle|\, a^{\sqrt{q}+1} + c^{\sqrt{q}+1} = b^{\sqrt{q}+1} + d^{\sqrt{q}+1} = 1,\ a^{\sqrt{q}} b + c^{\sqrt{q}} d = 0 \right\}$$

$$\cong \{ M \in GL(2, \mathbb{F}_q) \mid M^* M = I_2 \} = U(2, \sqrt{q}),$$

whose order is $\sqrt{q}(\sqrt{q} + 1)(q - 1)$.

4.3   A Curve of Type $(1, 1)$

Fix an element $\eta \in \mathbb{F}_q \setminus \mathbb{F}_{\sqrt{q}}$. Let $A_\eta = \begin{pmatrix} 1 & 0 & 0 \\ 0 & \eta & 1 \\ 0 & 1 & 0 \end{pmatrix}$ and corresponding curve

$C_\eta : \ x^{\sqrt{q}+1} + \eta y^{\sqrt{q}+1} + y^{\sqrt{q}} z + yz^{\sqrt{q}} = 0$. The unique rational point of C_η is $(0, 0, 1)^t$, at which the tangent line is given by $y = 0$. Let $T = \begin{pmatrix} a_{11} & a_{12} & a_{13} \\ a_{21} & a_{22} & a_{23} \\ a_{31} & a_{32} & a_{33} \end{pmatrix}$ be a nonsingular matrix in $GL(3, \mathbb{F}_q)$ which preserves the curve C_η, i.e., $T^* A_\eta T = \alpha A_\eta$ for some $\alpha \in \mathbb{F}_q^*$. Since T preserves the point $(0, 0, 1)^t$ and the tangent line $y = 0$, we have $a_{13} = a_{23} = a_{21} = 0$.

$$\alpha \begin{pmatrix} 1 & 0 & 0 \\ 0 & \eta & 1 \\ 0 & 1 & 0 \end{pmatrix} = \begin{pmatrix} a_{11}^{\sqrt{q}+1} & a_{11}^{\sqrt{q}} a_{12} + a_{31}^{\sqrt{q}} a_{22} & 0 \\ a_{12}^{\sqrt{q}} a_{11} + a_{22}^{\sqrt{q}} a_{31} & a_{12}^{\sqrt{q}+1} + a_{22}^{\sqrt{q}+1}\eta + a_{32}^{\sqrt{q}} a_{22} + a_{22}^{\sqrt{q}} a_{32} & a_{22}^{\sqrt{q}} a_{33} \\ 0 & a_{33}^{\sqrt{q}} a_{22} & 0 \end{pmatrix}.$$

Comparing each entries of both sides, we have

$$\begin{cases} \alpha = a_{11}^{\sqrt{q}+1} = a_{22}^{\sqrt{q}} a_{33} = a_{33}^{\sqrt{q}} a_{22} \in \mathbb{F}_{\sqrt{q}} \\ a_{11}^{\sqrt{q}} a_{12} + a_{31}^{\sqrt{q}} a_{22} = 0 = a_{12}^{\sqrt{q}} a_{11} + a_{22}^{\sqrt{q}} a_{31} \\ \alpha \eta = a_{22}^{\sqrt{q}+1} \eta + (a_{12}^{\sqrt{q}+1} + a_{32}^{\sqrt{q}} a_{22} + a_{22}^{\sqrt{q}} a_{32}) \end{cases}$$

Since $\alpha, a_{22}^{\sqrt{q}+1}, (a_{12}^{\sqrt{q}+1} + a_{32}^{\sqrt{q}}a_{22} + a_{22}^{\sqrt{q}}a_{32}) \in \mathbb{F}_{\sqrt{q}}$ and $\eta \in \mathbb{F}_q \setminus \mathbb{F}_{\sqrt{q}}$, we have

$$\begin{cases} \alpha = a_{22}^{\sqrt{q}+1} \\ a_{12}^{\sqrt{q}+1} + a_{32}^{\sqrt{q}}a_{22} + a_{22}^{\sqrt{q}}a_{32} = 0 \end{cases}$$

Also we get

$$\begin{cases} \alpha = a_{11}^{\sqrt{q}+1} = a_{22}^{\sqrt{q}}a_{33} = a_{33}^{\sqrt{q}}a_{22} \in \mathbb{F}_{\sqrt{q}} \\ a_{11}^{\sqrt{q}}a_{12} + a_{31}^{\sqrt{q}}a_{22} = 0 = a_{12}^{\sqrt{q}}a_{11} + a_{22}^{\sqrt{q}}a_{31} \\ 0 \neq \det T = a_{11}a_{22}a_{33} \end{cases}$$

Since we are working in $PGL(3, \mathbb{F}_q)$, we may normalize the matrix by setting $a_{33} = 1$ without loss of generality. Then $a_{22} \in \mathbb{F}_{\sqrt{q}}$ and $\alpha = a_{22} = a_{22}^{\sqrt{q}+1}$ and hence $a_{22} = 1$. Then, the above equations become

$$\begin{cases} a_{12}^{\sqrt{q}+1} + a_{32}^{\sqrt{q}} + a_{32} = 0 \\ \alpha = a_{11}^{\sqrt{q}+1} = 1 \\ a_{11}^{\sqrt{q}}a_{12} + a_{31}^{\sqrt{q}} = 0 = a_{12}^{\sqrt{q}}a_{11} + a_{31} \\ 0 \neq \det T = a_{11} \end{cases}$$

Thus

$$\mathrm{Aut}(C_\eta) = \left\{ \begin{pmatrix} a & b & 0 \\ 0 & 1 & 0 \\ -ab^{\sqrt{q}} & c & 1 \end{pmatrix} \mid a \in \mu_{\sqrt{q}+1},\ b \in \mathbb{F}_q, c^{\sqrt{q}} + c = -b^{\sqrt{q}+1} \right\}.$$

Indeed we can prove easily that this set is a subgroup of $GL(3, \mathbb{F}_q)$. For each $b \in \mathbb{F}_q$, there are exactly $\sqrt{q}$ values of c that satisfy the equation $c^{\sqrt{q}} + c = -b^{\sqrt{q}+1}$. Thus the order of $\mathrm{Aut}(C_\eta)$ is $(\sqrt{q}+1)q\sqrt{q}$.

4.4   Curves $C_{a,0}$ in (3)

Fix an element $\eta \in \mathbb{F}_q^*$. Let $A_\eta = \begin{pmatrix} 0 & 0 & 1 \\ 1 & 0 & 0 \\ 0 & \eta & 0 \end{pmatrix}$ and corresponding curve C_η :

$$x^{\sqrt{q}}z + xy^{\sqrt{q}} + \eta y z^{\sqrt{q}} = 0. \text{ Let } U = A_\eta^{-1}A_\eta^* = \begin{pmatrix} 0 & 0 & \eta^{\sqrt{q}} \\ \eta^{-1} & 0 & 0 \\ 0 & 1 & 0 \end{pmatrix} = \frac{1}{\eta}\begin{pmatrix} 0 & 0 & \eta^{\sqrt{q}+1} \\ 1 & 0 & 0 \\ 0 & \eta & 0 \end{pmatrix}.$$

Let $U_1 = \begin{pmatrix} 0 & 0 & \eta^{\sqrt{q}+1} \\ 1 & 0 & 0 \\ 0 & \eta & 0 \end{pmatrix}$. Note that U and U_1 induce the same projective trans-

formation φ on the projective plane $\mathbb{P}^2(\mathbb{F}_q)$. Recall that, for $P \in C_\eta$, $\varphi(P) = Q$ means that the divisor $\sqrt{q}P + Q$ of C_η is cut out by the tangent line $\mathcal{L}_P(C_\eta)$. If P is a rational point of C_η, then U_1 sends P to Q. P is an inflexion if and only if $\varphi(P) = P$. If P is not an inflexion, then $\varphi(P) = Q \neq P$. Note that $\varphi^2(P) = \varphi(Q) \neq P$. Indeed, if not, then the line PQ would cut out the divisors

$\sqrt{q}P + Q$ and $P + \sqrt{q}Q$ simultaneously, which is impossible. But $\varphi^3(P) = P$, since $U_1^3 = \mathrm{diag}(\eta^{\sqrt{q}+2}, \eta^{\sqrt{q}+2}, \eta^{\sqrt{q}+2}) = \eta^{\sqrt{q}+2}I_3$, which represents the identity in $PGL(3, \mathbb{F}_q)$. Thus, for any rational point P on C_η that is not an inflexion, the set $\{\varphi^n(P) \mid n \in \mathbb{N}\} = \{P, \varphi(P), \varphi^2(P)\}$ forms a 3-cycle under φ. Since any automorphism preserves tangent lines, it also preserves the 3-cycles. Fix a 3-cycle $\{(1,0,0)^t, (0,1,0)^t, (0,0,1)^t\}$. Indeed it is a 3-cycle under φ since $\varphi((1,0,0)^t) = (0,1,0)^t$, $\varphi((0,1,0)^t) = (0,0,1)^t$ and $\varphi((0,0,1)^t) = (1,0,0)^t$. Now for arbitrary rational point $P = (x_0, y_0, z_0)^t$ on C_η which is not an inflexion, the set

$$\{P, \varphi(P), \varphi^2(P)\} = \{(x_0, y_0, z_0)^t, (\eta^{\sqrt{q}+1}z_0, x_0, \eta y_0)^t, (\eta^{\sqrt{q}+1}y_0, \eta^{\sqrt{q}}z_0, x_0)^t\}$$

forms a 3-cycle under φ. Let T be the matrix whose columns are the three vectors in this 3-cycle multiplied by the scalars $1, \alpha$, and β, respectively. That is,

$$T = \left(P \ \alpha\varphi(P) \ \beta\varphi^2(P)\right) = \begin{pmatrix} x_0 & \alpha\eta^{\sqrt{q}+1}z_0 & \beta\eta^{\sqrt{q}+1}y_0 \\ y_0 & \alpha x_0 & \beta\eta^{\sqrt{q}}z_0 \\ z_0 & \alpha\eta y_0 & \beta x_0 \end{pmatrix}.$$

We want to prove that T is an automorphism of C_η. Compare the entries of matrices in both side of the equation $T^* A_\eta T = \delta A_\eta$.

(1,1)-entry: $x_0^{\sqrt{q}}z_0 + x_0 y_0^{\sqrt{q}} + \eta y_0 z_0^{\sqrt{q}} = 0$, since $P \in C_\eta$.

(1,2)-entry: $\alpha\eta(x_0 y_0^{\sqrt{q}} + x_0^{\sqrt{q}}z_0 + \eta y_0 z_0^{\sqrt{q}})^{\sqrt{q}} = 0$, since $P \in C_\eta$.

Similarly, it can be shown that the $(2,2), (2,3), (3,1)$, and $(3,3)$ entries also vanish. We now compute the remaining entries.

(1,3)-entry: $\beta(x_0^{\sqrt{q}+1} + \eta^{\sqrt{q}+1}y_0^{\sqrt{q}+1} + \eta^{\sqrt{q}+1}z_0^{\sqrt{q}+1}) = \delta.$

(2,1)-entry: $\alpha^{\sqrt{q}}(x_0^{\sqrt{q}+1} + \eta^{\sqrt{q}+1}y_0^{\sqrt{q}+1} + \eta^{\sqrt{q}+1}z_0^{\sqrt{q}+1}) = \delta.$

(3,2)-entry: $\alpha\beta^{\sqrt{q}}\eta(x_0^{\sqrt{q}+1} + \eta^{\sqrt{q}+1}y_0^{\sqrt{q}+1} + \eta^{\sqrt{q}+1}z_0^{\sqrt{q}+1}) = \delta\eta.$

Note that $x_0^{\sqrt{q}+1} + \eta^{\sqrt{q}+1}y_0^{\sqrt{q}+1} + \eta^{\sqrt{q}+1}z_0^{\sqrt{q}+1} \neq 0$ since $\det(T^* A_\eta T) \neq 0$. Thus, we obtain the following proportionality: $\alpha\beta^{\sqrt{q}}\eta : \alpha^{\sqrt{q}} : \beta = \eta : 1 : 1$ and hence $\alpha\beta^{\sqrt{q}} : \alpha^{\sqrt{q}} : \beta = 1 : 1 : 1$ which is equivalent to $\alpha\beta^{\sqrt{q}} = \alpha^{\sqrt{q}} = \beta$.

Lemma 3. *We have* $\alpha^{\sqrt{q}-2} = 1$, $\alpha^3 = 1$ *and* $\beta = \alpha^2$.

Proof. Since $\beta = \alpha^{\sqrt{q}}$, $\alpha^{\sqrt{q}} = \alpha\beta^{\sqrt{q}} = \alpha(\alpha^{\sqrt{q}})^{\sqrt{q}} = \alpha^2$. Hence $\alpha^{\sqrt{q}-2} = 1$. Since $q - 1 = (\sqrt{q} - 2)(\sqrt{q} + 2) + 3$, $1 = \alpha^{q-1} = (\alpha^{\sqrt{q}-2})^{\sqrt{q}+2}\alpha^3 = \alpha^3$. Finally, $\beta = \alpha^{\sqrt{q}} = \alpha^{\sqrt{q}-2}\alpha^2 = \alpha^2$. $\qquad\square$

The Case $3 \mid (\sqrt{q} - 1)$. Note that $\alpha^{\sqrt{q}-1} = (\alpha^3)^{\frac{\sqrt{q}-1}{3}} = 1$. Since $\alpha^{\sqrt{q}-2} = 1$, we have $\alpha = 1$ and hence $\beta = 1$. By Theorem 2, C_η is of type $(q + 1, 2)$. Hence C_η has $q - 1$ rational points, none of which are inflexion points. For any such point $P = (x_0, y_0, z_0)^t = \begin{pmatrix} x_0 \\ y_0 \\ z_0 \end{pmatrix}$, we construct an automorphism as $T_P = (P, \varphi(P), \varphi^2(P)) \in GL(3, \mathbb{F}_q)$. Thus,

$$\mathrm{Aut}(C_\eta) = \{T_P \mid P \text{ is a rational point which is not an inflexion}\}.$$

Therefore, the order of $\mathrm{Aut}(C_\eta)$ is $q - 1$.

The Case3 $\mid \sqrt{q}$. Since $\alpha^3 = 1$, $\alpha = 1$, and hence $\beta = 1$. By Theorem 2, C_η is of type $(q + 1, 1)$. Hence C_η has q rational points, none of which are inflexion points. For any such point $P = (x_0, y_0, z_0)^t = \begin{pmatrix} x_0 \\ y_0 \\ z_0 \end{pmatrix}$, we construct an automorphism as $T_P = (P, \varphi(P), \varphi^2(P)) \in GL(3, \mathbb{F}_q)$. Thus,

$$\mathrm{Aut}(C_\eta) = \{T_P \mid P \text{ is a rational point which is not an inflexion}\}.$$

Therefore, the order of $\mathrm{Aut}(C_\eta)$ is q.

The Case3 $\mid (\sqrt{q} + 1)$ Since $3 \mid (\sqrt{q} - 2)$, the condition $\alpha^3 = 1$ implies $\alpha^{\sqrt{q}-2} = 1$. The roots of $\alpha^3 = 1$ are $1, \rho, \rho^2$, where ρ is a third primitive root of unity. Then $(\alpha, \beta) = (1, 1), (\rho, \rho^2), (\rho^2, \rho)$. By Theorem 2, C_η is of type
$$\begin{cases} (q + 2\sqrt{q} + 1, 0) & \text{if } \eta \in \langle \omega^3 \rangle \\ (q - \sqrt{q} + 1, 0) & \text{otherwise.} \end{cases}$$
Note that these curves have no inflexions. For any such point $P = (x_0, y_0, z_0)^t = \begin{pmatrix} x_0 \\ y_0 \\ z_0 \end{pmatrix}$, we construct three automorphisms as $T_{P,1} = (P, \varphi(P), \varphi^2(P))$, $T_{P,2} = (P, \rho\varphi(P), \rho^2\varphi^2(P))$, and $T_{P,3} = (P, \rho^2\varphi(P), \rho\varphi^2(P)) \in GL(3, \mathbb{F}_q)$. Thus,

$$\mathrm{Aut}(C_\eta) = \{T_{P,1} \mid P \in C_\eta(\mathbb{F}_q)\} \cup \{T_{P,2} \mid P \in C_\eta(\mathbb{F}_q)\} \cup \{T_{P,3} \mid P \in C_\eta(\mathbb{F}_q)\}.$$

Therefore, the order of $\mathrm{Aut}(C_\eta)$ is $\begin{cases} 3(q + 2\sqrt{q} + 1) & \text{if } \eta \in \langle \omega^3 \rangle \\ 3(q - \sqrt{q} + 1) & \text{otherwise.} \end{cases}$

Theorem 8. *For the specific types of Hermitian-relative curves classified in Sect. 3, the orders of their automorphism groups are determined as summarized in Table 4.*

Table 4. Automorphism group orders of specific curves

representative matrix A	type of the curve C_A	order of $\mathrm{Aut}(C_A)$
$\begin{pmatrix} 1 & 0 & 0 \\ 0 & 1 & 0 \\ 0 & 0 & 1 \end{pmatrix}$	$(q\sqrt{q}+1, q\sqrt{q}+1)$	$q\sqrt{q}(q\sqrt{q}+1)(q-1)$
$\begin{pmatrix} 1 & 0 & 0 \\ 0 & 1 & 0 \\ 0 & 0 & \eta \end{pmatrix}, \eta \in \mathbb{F}_q - \mathbb{F}_{\sqrt{q}}$	$(\sqrt{q}+1, \sqrt{q}+1)$	$\sqrt{q}(\sqrt{q}+1)(q-1)$
$\begin{pmatrix} 1 & 0 & 0 \\ 0 & \eta & 1 \\ 0 & 1 & 0 \end{pmatrix}, \eta \in \mathbb{F}_q - \mathbb{F}_{\sqrt{q}}$	$(1,1)$	$q\sqrt{q}(\sqrt{q}+1)$
$\begin{pmatrix} 0 & 0 & 1 \\ 1 & 0 & 0 \\ 0 & \eta & 0 \end{pmatrix}, \eta \in \mathbb{F}_q^*, \sqrt{q} \equiv 0 \pmod 3$	$(q+1, 1)$	q
$\begin{pmatrix} 0 & 0 & 1 \\ 1 & 0 & 0 \\ 0 & \eta & 0 \end{pmatrix}, \eta \in \mathbb{F}_q^*, \sqrt{q} \equiv 1 \pmod 3$	$(q+1, 2)$	$q-1$
$\begin{pmatrix} 0 & 0 & 1 \\ 1 & 0 & 0 \\ 0 & \eta & 0 \end{pmatrix}, \eta \in \mathbb{F}_q^*, \sqrt{q} \equiv 2 \pmod 3,$ $\eta^{\frac{q-1}{3}} = 1$	$(q+2\sqrt{q}+1, 0)$	$3(q+2\sqrt{q}+1)$
$\begin{pmatrix} 0 & 0 & 1 \\ 1 & 0 & 0 \\ 0 & \eta & 0 \end{pmatrix}, \eta \in \mathbb{F}_q^*, \sqrt{q} \equiv 2 \pmod 3,$ $\eta^{\frac{q-1}{3}} \neq 1$	$(q-\sqrt{q}+1, 0)$	$3(q-\sqrt{q}+1)$

Acknowledgments. The second author was partially supported by the National Research Foundation of Korea(NRF) grant funded by the Korea government(MSIT) (2022R1A2C1012291).

References

1. Hirschfeld, J.W.P.: Projective geometries over finite fields. Oxford University Press, Oxford (1979)
2. Hirschfeld, J.W.P.: Projective geometries over finite fields, 2nd edn. Oxford University Press, Oxford (1998)
3. Hirschfeld, J.W.P., Storme, L., Thas, J.A., Voloch, J.F.A.: A characterization of Hermitian curves. J. Geom. **41**, 72–78 (1991)
4. Homma, M., Kim, S.J.: Relatives of the Hermitian curve. J. Geom. **117**, 11 (2026). https://doi.org/10.1007/s00022-026-00795-8

5. Homma, M., Kim, S.J.: Rational points and inflexions on Hermitian-relative curves, preprint
6. Pardini, R.: Some remarks on plane curves over fields of finite characteristic. Compositio Math. **60**, 3–17 (1986)
7. Rück, H.-G., Stichtenoth, H.: A characterization of Hermitian function fields over finite fields. J. Reine Angew. Math. **457**, 185–188 (1994)
8. Stichtenoth, H.: Algebraic function fields and codes (2 edn.), GTM 254, Springer-Verlag, Berlin and Heidelberg

Improvements to Jacobian Arithmetic in Global Function Fields

Vincent Macri$^{(\boxtimes)}$, Michael Jacobson Jr. , and Renate Scheidler

University of Calgary, Calgary, Canada
`{vincent.macri,jacobs,rscheidl}@ucalgary.ca`

Abstract. We present two improvements to arithmetic in the Jacobian of global function fields based on the approach of Hess. The first reduces the number of expensive reduction steps by optimizing for typical inputs rather than worst-case behavior, assuming the function field contains a degree-one place. This results in an expected asymptotic speed-up by a factor that is logarithmic in the genus. The second introduces a memory–time trade-off that speeds up computations by caching frequently used intermediate results. Our asymptotic analysis and empirical experiments show that our improved algorithms are significantly faster in practice than previously published methods. To the best of our knowledge, our publicly-available software implementation of Jacobian arithmetic is the first to support unique representatives of divisor classes.

Keywords: Global function field · Algebraic curve · Divisor · Jacobian

1 Introduction

1.1 Motivation

Elliptic curves are the subject of numerous notoriously challenging open problems, including the Koblitz–Zywina conjecture, the Lang–Trotter conjecture, the Sato–Tate conjecture, and perhaps most famously, the Birch and Swinnerton-Dyer conjecture [19,23]. While these conjectures are about elliptic curves over $\mathbb{Q}$, generating numerical evidence for (or against) them requires considering a given curve over finite fields $\mathbb{F}_q$ for many values of q and computing the number of points on the curve. Efficient point arithmetic is essential in this endeavour. This motivates the desire to compute the size of the group of points on an elliptic curve over a finite field. Although proofs of these conjectures are out of reach, there is much computational evidence supporting them.

A natural next step is to consider generalizations of these conjectures to curves that are not necessarily elliptic, and see whether computational evidence supports or refutes the generalized conjectures. In the more general setting, the points on the curve no longer form a group: instead, elements of the Jacobian are equivalence classes of degree zero divisors. Hence, rather than counting points over many finite fields, one instead needs to compute the order of the Jacobian of the curve over all these fields. In order to tackle this and similar problems,

L. Batina and F. Özbudak (Eds.): WAIFI 2026, LNCS 16611, pp. 111–128, 2026.
https://doi.org/10.1007/978-3-032-27574-5_8

it is essential to have fast algorithms for computing in the Jacobian group of a curve or its associated function field. Moreover, for algorithmic efficiency, it is beneficial if this arithmetic should be performed in a way that gives unique representatives of elements of the Jacobian group.

The most recent efficient algorithm for computing in the Jacobian with unique representatives of degree divisor classes is the approach of maximally reduced divisors from [7]. Other recent approaches to Jacobian arithmetic such as [10–12] do not give unique representatives of Jacobian elements.

1.2   Our Contributions

In [7], Hess showed how to compute in the divisor class group, which contains the Jacobian as a subgroup. By specializing to the Jacobian and requiring the existence of a degree one place (a restriction which is nearly always satisfied), we develop an improved algorithm that supports unique representation of divisor classes and is significantly faster than the approach recommended in [7]. We also demonstrate how effective use of caching can speed up Jacobian arithmetic even further, especially if the function field contains a degree one infinite place. Our improved Jacobian algorithm is both faster asymptotically for typical inputs, and faster in practice when implemented. Complexity analysis indicates that our algorithm should give a speedup of approximately $\log_2(g)/2$, where g is the genus of the function field. A performance comparison confirms this: for example, for degree 3 function fields of genus 100, our implementation of our improved algorithm is approximately three times faster than our implementation of the unique Jacobian arithmetic algorithm suggested in [7]. An implementation of our algorithm is available in SageMath [22], since the 10.9.beta7 release.

1.3   Outline of the Paper

In Sect. 2 we establish notation and summarize the necessary background material. Section 3 describes our modifications to Hess' maximally reduced divisors from [7] that facilitate our improved algorithms for Jacobian arithmetic presented in Sect. 4. We conduct a theoretical complexity analysis in Sect. 5 before analyzing the actual performance of our implementations in Sect. 6 and proposing directions for future research on Sect. 7.

2   Background

For details, the reader is referred to [18]. Throughout, let K be a finite field and F/K a geometric global function field of genus g. We write $F = K(x, y)$ with $x, y \in F$, where $K(x)$ is a rational function field and y has minimal polynomial $f(t) = t^n + \sum_{i=0}^{n-1} t^i a_i(x) \in K[x, t]$, with $a_i(x) \in K[x]$ for $0 \leq i \leq n-1$. Then $F/K(x)$ has degree n in t and $f(y) = 0$. The quantity

$$C_f := \max \left\{ \left\lceil \frac{\deg(a_i)}{n-i} \right\rceil : 0 \leq i \leq n-1 \right\} \tag{1}$$

determines the size of F and will appear in our complexity statements [7, p. 427].

We denote the set of places of F by $\mathbb{P}(F)$, partitioned into the infinite places (the poles of x) and the finite places of F. The finite and infinite maximal orders of $F/K(x)$ are denoted by $\mathfrak{O}_{F,0}$ and $\mathfrak{O}_{F,\infty}$ respectively. Divisors of F are formal finite sums $D = \sum_{P \in \mathbb{P}(F)} v_P(D)P$, where $v_P(D) \in \mathbb{Z}$ is the valuation of D at P. We write $\operatorname{supp}(D) = \{P \in \mathbb{P}(F) : v_P(D) \neq 0\}$ (the support of D) and let $\deg(D)$ denote the degree of D. The respective sub-sums of D over the finite places and the infinite places in $\operatorname{supp}(D)$ are denoted by D^0 and D^∞, so $D = D^0 + D^\infty$.

The principal divisor of a non-zero function $a \in F$ is denoted $\operatorname{div}(a)$. Two divisors $D, D' \in \operatorname{Div}(F)$ are (linearly) equivalent, denoted $D \equiv D'$, if they differ by a principal divisor. Let $\operatorname{Div}(F)$ and $\operatorname{Div}^0(F)$ denote the groups of divisors and degree zero divisors of F, respectively, and let $\operatorname{Cl}(F)$ and $\operatorname{Cl}^0(F)$ denote the groups of divisor classes and degree zero divisor classes of F, respectively, under linear equivalence. The group $\operatorname{Cl}^0(F)$ is also referred to as the *Jacobian* of F and is the main protagonist of our work herein.

The Riemann-Roch space of a divisor $D \in \operatorname{Div}(F)$ is the finite-dimensional K-vector space $\mathcal{L}(D) = \{a \in F : D + \operatorname{div}(a) \geq 0\} \cup \{0\}$; its K-dimension is denoted by $\ell(D)$. A partial order on $\operatorname{Div}(F)$ is given by $D \leq D'$ if $v_P(D) \leq v_P(D')$ for all $P \in \mathbb{P}(F)$. Riemann-Roch spaces and their dimensions are monotonic with respect to this partial order; this monotonicity will play a key role in our main algorithm (Algorithm 1).

Lemma 1. *[18, Lemma 1.4.8] Let $D, D' \in \operatorname{Div}(F)$ with $D \leq D'$. Then $\mathcal{L}(D) \subseteq \mathcal{L}(D')$ and $\ell(D') - \ell(D) = \dim(\mathcal{L}(D')/\mathcal{L}(D)) \leq \deg(D') - \deg(D)$.*

We conclude this section with the notion of a maximally reduced divisor as presented in [7]. For convenience, we restate the definition here.

Definition 1. *[7, Definition 8.1] Let $A \in \operatorname{Div}(F)$ with $\deg(A) \geq 1$. A divisor $\tilde{D}$ is called* maximally reduced along A *if $\tilde{D} \geq 0$ and $\ell(\tilde{D} - sA) = 0$ for all integers $s \geq 1$. The representation of a divisor D as $D = \tilde{D} - rA - \operatorname{div}(a)$, with $\tilde{D}$ maximally reduced along A, $r \in \mathbb{Z}$, and $a \in F^*$ is called a* maximal reduction *of D along A.*

Hess [7, Proposition 8.2] showed that if $\deg(A) = 1$ then the maximal reduction of a divisor D along A is unique.

3   Unique Hess Representation

The notion of maximally reduced divisors (Definition 1) was originally defined for elements of $\operatorname{Cl}(F)$, but restricting to $\operatorname{Cl}^0(F)$ and imposing the restriction that A is a degree one place allows for us to achieve improved algorithmic performance. We note that the following notion is restricted to degree zero divisors.

Definition 2. *Let $D \in \operatorname{Div}^0(F)$ and $A \in \mathbb{P}(F)$ with $\deg(A) = 1$. The* Hess-reduction *of D (along A) is the maximal reduction of D along A. We call D* Hess-reduced *(along A) if D is equal to its Hess-reduction (along A) and refer to it as the* Unique Hess representation *of its divisor class.*

By [7, Proposition 8.2], the Hess-reduction of a degree zero divisor is unique. Here, A can be finite or infinite, but the latter case will allow for more efficient implementations later on as it creates additional opportunities to exploit caching of intermediate results. In practice, requiring the existence of a degree one place is a very mild restriction. By the Hasse-Weil Bound [18, Theorem 5.2.3], every function field $F/\mathbb{F}_q$ of genus g has a degree one place if q is sufficiently large relative to g. Hence, the existence of a degree one place can be guaranteed by extending scalars to a sufficiently large extension of $\mathbb{F}_q$. Even for prime fields $\mathbb{F}_q$, the probability that a function field has no degree one place is quite low; see [8] for a heuristic argument.

We require two key results for Hess-reduced divisors.

Proposition 1. *Let $A \in \mathbb{P}(F)$ with $\deg(A) = 1$, and let $\tilde{D} \in \mathrm{Div}^0(F)$ be Hess-reduced along A. Then $A \notin \mathrm{supp}(\tilde{D})$.*

Proof. Since $\tilde{D}$ is maximally reduced along A, we have $\ell(\tilde{D} - A) = 0$. It follows that $\tilde{D} - A \not\geq 0$, so $v_A(\tilde{D}) \leq 0$. On the other hand $\tilde{D} \geq 0$, so $v_A(\tilde{D}) = 0$. □

In [7], Hess noted that for a divisor $\tilde{D}$ maximally reduced along A, we have $\ell(\tilde{D}) \leq \deg(A)$ and $\deg(\tilde{D}) < g + \deg(A)$. We strengthen these statements for the setting of Hess-reduced divisors.

Proposition 2. *Let $A \in \mathbb{P}(F)$ with $\deg(A) = 1$, and let $D = \tilde{D} - rA$ be Hess-reduced along A. Then $\deg(\tilde{D}) = r$ and $0 \leq r \leq g$.*

Proof. We have $\deg(\tilde{D}) = \deg(rA) = r$. Since $\tilde{D}$ is maximally reduced along A, we have $\tilde{D} \geq 0$ and $\ell(\tilde{D} - A) = 0$, so Lemma 1 yields

$$\ell(\tilde{D}) = \ell(\tilde{D}) - \ell(\tilde{D} - A) \leq \deg(\tilde{D}) - \deg(\tilde{D} - A) = \deg(A) = 1.$$

Riemann's Theorem [18, Theorem 1.4.17] implies $\deg(\tilde{D}) \leq \ell(\tilde{D}) + g - 1 \leq g$. □

4   Improvements to the Hess Reduction Algorithm

In this section, we introduce two new modifications to the algorithm of [7] for computing maximal reductions that effect substantial efficiency improvements when restricting to degree zero divisors. The first employs a search strategy for finding maximal reductions that is different from Hess's and performs better in practice. The second improvement introduces caching of intermediate computational results which effects a further speed-up.

4.1   Improvements to Reduction Algorithm

In [7], Hess provided an efficient algorithm for computing a basis of a Riemann-Roch space and described how to use it for finding maximal reductions of divisors. After specializing to degree zero divisors and the framework of Unique Hess representations in the Jacobian of F, this task translates to the following problem (see [14, Section 4.1] for details).

Problem 1 (HR-Min). Let $A \in \mathbb{P}(F)$ with $\deg(A) = 1$. Given $D \in \mathrm{Div}^0(F)$, find the minimal integer $0 \leq r \leq g$ such that $\ell(D + rA) = 1$.

By the monotonicity of ℓ, an equivalent problem is to find the maximal integer r' such that $\ell(D + r'A) = 0$. Then $r = r' + 1$ solves HR-Min.

Once a solution to HR-Min for an input divisor $D \in \mathrm{Div}^0(F)$ is found, it is fairly straightforward to compute the Hess-reduction of D, assuming we can compute Riemann-Roch bases. Suppose that r solves HR-Min for D and that we can compute a non-zero element $a \in \mathcal{L}(D + rA)$, which is then a basis of this space. Then $\mathrm{div}(a)$ is unique because $\ell(D + rA) = 1$. Moreover, $D + rA + \mathrm{div}(a) \geq 0$, so setting $\tilde{D} := D + rA + \mathrm{div}(a)$, we see that $\tilde{D} \geq 0$ and $\ell(\tilde{D} - sA) = 0$ for all integers $s \geq 1$. So $\tilde{D}$ is maximally reduced along A, and hence $D \equiv \tilde{D} - rA$ is the unique Hess-reduction of D along A. See Algorithm 2 for an algorithmic description of this process.

Finding the solution r to HR-Min can be accomplished by trying each possible candidate value for r between 0 and g, but each trial requires computing a basis of a Riemann-Roch space, which is by far the most expensive part of the computation. In [7], Hess suggested to find r via binary search, which takes $O(\log g)$ iterations, but Hess was working in $\mathrm{Cl}(F)$ rather than $\mathrm{Cl}^0(F)$. When working over $\mathrm{Cl}^0(F)$, a heuristic argument [14, Subsection 4.2.1] shows that with high likelihood, for a randomly chosen $D \in \mathrm{Div}^0(F)$, the solution to HR-Min is $r = g$. Empirically, we find that for $K = \mathbb{F}_q$, the value $r = g$ solves HR-Min with probability approaching $1 - q^{-1}$ as q grows. This stems from the observation that a random divisor of degree at most g is expected to have degree exactly g with the same likelihood that a random polynomial of degree at most g in $\mathbb{F}_q[x]$ is expected to have degree exactly g, which is just the probability that its coefficient at x^g is non-zero. Furthermore, the same heuristic suggests that when $r = g$ is not the solution to HR-Min, the next most likely candidate is $r = g - 1$, then $g - 2$, and so on, with the solution $r = 0$ corresponding to the case when D is principal.

This observation suggests that a linear search in decreasing order to solving HR-Min should outperform the binary search approach suggested in [7], by significantly decreasing the number of Riemann-Roch basis computations. Our implementation bears this out, and this idea forms the basis of Algorithm 1, our improved algorithm for solving HR-Min. The key idea of this algorithm is to try the most likely values for r first. A formal proof of correctness is given in Theorem 1.

The approach for computing the Hess-reduction of a divisor outlined above relies on the ability to compute the Riemann-Roch dimension of a divisor E. In practice, this is done by computing a basis of $\mathcal{L}(E)$ and returning the number of basis elements. In our setting, by the monotonicity of the Riemann-Roch dimension (Lemma 1), it suffices to only compute one non-zero element of $\mathcal{L}(E)$ rather than a full basis, which may be faster. An easy modification of the Riemann-Roch basis algorithm from [7] effects this. This variant is called a *short-circuited Riemann-Roch* operation, denoted by SCRR [14, Definition 3.6.2]. This operation takes as input a divisor $E \in \mathrm{Div}(F)$ and outputs a non-zero element of

$\mathcal{L}(E)$ if $\ell(E) > 0$, or the string `None` if $\mathcal{L}(E) = \{0\}$. Note that if $\ell(E) = 1$, then $\mathrm{SCRR}(E)$ returns a basis of $\mathcal{L}(E)$, and $\mathrm{div}(\mathrm{SCRR}(E))$ is unique.

Algorithm 1. Optimized Hess reduction strategy via linear search

---

Input: $D \in \mathrm{Div}^0(F)$, $A \in \mathbb{P}(F)$ with $\deg(A) = 1$
Output: (r, a) where r solves `HR-Min` and $0 \neq a \in \mathcal{L}(D + rA)$
1: $\tilde{D} \leftarrow D + (g - 1)A$
2: $a \leftarrow \mathrm{SCRR}(\tilde{D})$
3: **if** $a = $ `None` **then** $\triangleright$ If $\ell(D + (g-1)A) = 0$ then $\ell(D + gA) = 1$.
4:     $\tilde{D} \leftarrow \tilde{D} + A$ $\triangleright$ $\tilde{D} = D + gA$
5:     $a \leftarrow \mathrm{SCRR}(\tilde{D})$ $\triangleright$ $a \in \mathcal{L}(D + gA)$
6:     **return** g, a
7: **end if**
8: **for** $m \leftarrow g - 2, \ldots, 0$ **do**
9:     $\tilde{D} \leftarrow \tilde{D} - A$ $\triangleright$ $\tilde{D} = D + mA$
10:     $a' \leftarrow \mathrm{SCRR}(\tilde{D})$
11:     **if** $a' = $ `None` **then** $\triangleright$ $\ell(D + (m+1)A) = 1$
12:         **return** $m + 1, a$ $\triangleright$ $a \in \mathcal{L}(D + (m+1)A)$
13:     **end if**
14:     $a \leftarrow a'$
15: **end for**
16: **return** $0, a$

Theorem 1. *Let F/K be a function field of genus $g \geq 2$. Given $D \in \mathrm{Div}^0(F)$, Algorithm 1 returns (r, a) where r solves* `HR-Min` *for D and $0 \neq a \in \mathcal{L}(D + rA)$.*

Proof. Let $D \in \mathrm{Div}^0(F)$. Suppose first that Algorithm 1 returns from Line 6. Then $\ell(D + (g-1)A) = 0$. By [7, Proposition 8.2], `HR-Min` has a solution, and by the monotonicity of Riemann-Roch spaces (Lemma 1), we have $\ell(D + gA) = 1$ in order for this solution to exist. Therefore $r = g$ is the solution to `HR-Min`, and since $a \in \mathcal{L}(D + gA)$, the output (r, a) of Algorithm 1 is correct.

Next, assume that Algorithm 1 returns from Line 12. In order for this line to be reached, we must have $\ell(D + (g-1)A) \geq 1$. By monotonicity $\ell(D + gA) \geq 1$, so $r = g$ does not solve `HR-Min`. Let m be the value of the loop variable at the point where Algorithm 1 reaches Line 12. Then $\ell(D + mA) = 0$, so by monotonicity, $\ell(D + m'A) = 0$ for all $m' < m$. Since the condition in Line 11 is not satisfied for $m + 1$, we have $\ell(D + (m + 1)A) \geq 1$. By monotonicity and the existence of a minimal solution, $r = m + 1$ is the minimal integer that solves `HR-Min`. As $a \in \mathcal{L}(D + (m+1)A)$, Algorithm 1 returns the correct value on Line 12.

Finally suppose that Algorithm 1 returns from Line 16. For this point to be reached we must have $\ell(D + mA) \geq 0$ for all $0 \leq m \leq g$. Since a solution to `HR-Min` with $0 \leq r \leq g$ exists, it follows that $r = 0$ must be the solution. Accordingly, our algorithm returns $r = 0$ as the solution to `HR-Min` in this case. Furthermore, since $g \geq 2$ by assumption, the for loop will run at least once, so we have $a \in \mathcal{L}(D)$. $\qquad\square$

In the most likely case where $r = g$ solves HR-Min, Algorithm 1 completes with only two invocations of SCRR, in contrast to binary search which requires $O(\log g)$ such calls. We also note that the algorithm handles the second most likely case where $r = g - 1$ is the solution with the same number of steps as the $r = g$ case, so the optimal scenario occurs with expected probability $1 - q^{-2}$.

Our main strategy for designing Algorithm 1 was to minimize the number of costly calls to SCRR. The remainder of the work is given by the divisor additions in Lines 1, 4 and 9. In practice, software implementations such as SageMath [22] represent divisors as pairs of fractional ideals of the finite and infinite maximal orders, as suggested in [7]. With this representation, we are able to effect an additional performance gain at the cost of a reasonable increase to memory usage by exploiting caching, which we will describe next.

4.2 Caching

About half of the work of traditional generic Jacobian arithmetic entails arithmetic on the at most n infinite places of F. When A is chosen to be a degree one infinite place, caching intermediate results involving these places leads to effective speed-ups in Algorithm 1, at a modest increase in memory cost. For instance, cached values can be used in Lines 1, 4 and 9. The approach still works when A is a finite place of degree one, but offers fewer opportunities for caching.

For function fields with a unique infinite place of degree one, arithmetic on degree zero divisors is entirely governed by arithmetic on their finite part. Therefore, we assume henceforth that F/K is a function field of genus g with exactly $t \geq 2$ infinite places, denoted $\infty_1, \ldots, \infty_t$, with $\deg(\infty_t) = 1$. Let $D_1, D_2 \in \mathrm{Div}^0(F)$ be Hess-reduced along ∞_t, so $D_1 = \tilde{D}_1 - r_1 \infty_t$ and $D_2 = \tilde{D}_2 - r_2 \infty_t$ with $0 \leq r_1, r_2 \leq g$. Let $D_3 = D_1 + D_2$; we need to compute $D_3^0 = D_1^0 + D_2^0$ and $D_3^\infty = D_1^\infty + D_2^\infty$. We call the process of computing either D_3^0 or D_1^∞ a *partial divisor addition*. Since D_1 and D_2 are Hess-reduced, the number of possibilities for D_3^∞ is effectively bounded, suggesting that caching is worthwhile. A lengthy but elementary combinatorial derivation (see [14, Subsection 4.5.3]) yields a loose upper bound of $O(g^{t-1})$ ideals of $\mathfrak{O}_{F,\infty}$ on the cache for sums of infinite parts of Hess-reduced divisors. A similar technique can be applied to one of the steps in SCRR, with a memory cost of $O(g^{2t+1})$. Here, we cache a map from $\mathfrak{O}_{F,\infty}$-ideals to $n \times n$ matrices whose entries are rational functions over K of numerator and denominator degree at most g.

Empirically, the actual number of cached elements is substantially smaller than these bounds. In our experiments, the size of the caches was never an issue, and was inconsequential over larger finite fields. The typical case where $r = g$ solves HR-Min happens less frequently for smaller finite fields, causing more distinct intermediate values to be cached. For $g = 25$, $t = 2$, and $\mathbb{F}_{32\,771}$, after approximately 5000 Jacobian additions, the observed cache size was 60 for the addition of the infinite parts of divisors and 33 for the SCRR cache. See [14, Table 4.2] for a table of observed cache sizes.

Beyond caching, the remaining memory cost of Algorithm 2 is simply the memory cost of the input and output divisors, along with some intermediate expressions of similar size, so we do not include a full memory cost analysis.

5   Time Complexity of Computing Hess-Reductions

We now present a time complexity analysis of Algorithm 1, with and without caching, and compare it to the time complexity for the binary search approach suggested in [7]. To ensure a fair comparison, we also give the time complexity for the binary search with and without using our caching technique. We do not consider the aforementioned impact of caching on the subroutine SCRR, as it only speeds up one small part of the SCRR computation.

An appropriate measure of the size of a divisor is given by its height, which will be used in our complexity statements.

Definition 3. *[7, p. 434] Let $D \in \mathrm{Div}(F)$. We define the* (divisor) height *of D to be*

$$\mathrm{h}(D) := \sum_{P \in \mathbb{P}(F)} |v_P(D)| \cdot \deg(P).$$

Note that $\mathrm{h}(D) = \deg(D)$ when $D \geq 0$. We express our running times in terms of the costs of Riemann-Roch basis computations and divisor additions, using the following notation:

Definition 4. *[14, Notation 4.3.3] Denote by $\mathbf{RR}(h)$ the time complexity, expressed in field operations in K, of a short-circuited Riemann-Roch computation on input a divisor $D \in \mathrm{Div}(F)$ with $h = \mathrm{h}(D)$.*

Let $D_1, D_2 \in \mathrm{Div}^0(F)$ be Hess-reduced. Denote by $\mathbf{I}$ the time complexity, expressed in field operations in K, of computing either D_3^0 or D_3^∞, where $D_3 = D_1 + D_2$.

5.1   Time Complexity in High-Level Operations

Theorem 2. *Let $D = \bar{D} - \bar{r}A$ be the unreduced sum of two Hess-reduced divisors of F. Assume that $(g - 1)A$ and $-A$ have both been precomputed. Then the following hold:*

1. In the worst case, the complexity of Algorithm 1 on input D is

$$g\mathbf{I} + \sum_{m=0}^{g-1} \mathbf{RR}(\mathrm{h}(D + mA)). \tag{2}$$

2. In the heuristically average case, the complexity of Algorithm 1 on input D is

$$2\mathbf{I} + \mathbf{RR}(3g + 1) + \mathbf{RR}(3g). \tag{3}$$

Proof. The worst case complexity follows from the design of the algorithm. For the heuristically average case, let D_1 and D_2 be the two Hess-reduced divisors such that $D = D_1 + D_2$. Assume that $r = g$ solves HR-Min, and that $D_1 = \tilde{D}_1 - r_1 A$ and $D_2 = \tilde{D}_2 - r_2 A$ are also both typical divisors, hence $r_1 = r_2 = g$. We may write $\bar{D} = \tilde{D}_1 + \tilde{D}_2$ and $\bar{r} = 2g$, so $\deg(\bar{D}) = 2g$. Algorithm 1 computes the divisor sums $D + (g-1)A$ and $D + (g-1)A + A = D + gA$, giving $2\mathbf{I}$. Algorithm 1 also computes $\mathrm{SCRR}(D + (g-1)A)$ and $\mathrm{SCRR}(D + gA)$. Noting that $\bar{D} \geq 0$ and that $A \notin \mathrm{supp}(\bar{D})$ by Proposition 1, the heights of these divisors are

$$\mathrm{h}(D + (g-1)A) = \mathrm{h}(\bar{D}) + \mathrm{h}(-2gA + (g-1)A) = 3g + 1,$$
$$\mathrm{h}(D + gA) = \mathrm{h}(\bar{D}) + \mathrm{h}(-2gA + gA) = 3g,$$

whence the heuristically average case complexity follows. $\square$

For completeness, we give the analogous result for computing Hess-reductions using binary search for solving HR-Min. In the typical case where $r_1 = r_2 = r = g$, this complexity is worse than that of Theorem 2 for $g \geq 4$.

Theorem 3. *Let $D = \bar{D} - \bar{r}A$ be the unreduced sum of two Hess-reduced divisors. Assume that mA has been precomputed for all $0 \leq m \leq g$. Then both the worst case and heuristically average case complexity of finding the solution r to HR-Min on input D with binary search and computing $a \in \mathcal{L}(D + rA)$ is*

$$\lceil \log_2(g+1)\rceil \mathbf{I} + \sum_{i=1}^{\lceil \log_2(g+1)\rceil} \mathbf{RR}\left(\mathrm{h}\left(D + \left\lceil \frac{g\left(2^i - 1\right)}{2^i}\right\rceil A\right)\right).$$

Proof. In the worst case for binary search, the sought value lies at an endpoint of the search range, which holds in the heuristically average case where $r = g$ solves HR-Min. Here binary search needs to compute $D + mA$ and $\mathrm{SCRR}(D + mA)$ for all $m = \left\lceil g \cdot \frac{2^i - 1}{2^i}\right\rceil$ for all $1 \leq i \leq \lceil \log_2(g + 1)\rceil$. $\square$

5.2   Time Complexity in Base Field Operations

To give the time complexity in base field operations, we must first fix a representation of divisors, then substitute the best known values for the two subroutines $\mathbf{I}$ and $\mathbf{RR}(h)$ into Theorems 2 and 3.

In practice, divisors are represented by a pair (I, J) where I and J are ideals of $\mathfrak{O}_{F,0}$ and $\mathfrak{O}_{F,\infty}$, respectively. They are represented as $n \times n$ matrices in Hermite Normal Form whose entries are polynomials in $K[x]$ of degree at most g when the divisor is Hess-reduced.

The cost of $\mathbf{I}$ is simply the complexity of multiplying two ideals and converting the resulting matrix to Hermite Normal Form. The best asymptotic complexity for $\mathbf{I}$ that we are aware of arises from replacing naive polynomial multiplication with FFT in [20, Theorem 5.2.2], yielding

$$\mathbf{I} = \tilde{O}(n^6(g\log(gn)\log\log(gn) + g^2)). \tag{4}$$

To the best of our knowledge, the fastest asymptotic complexity for $\mathbf{RR}(h)$ can be found in [1, Corollary 4.1.5] and is given by

$$\mathbf{RR}(h) = O(n^5(h + n^2 C_f)^2). \tag{5}$$

In practice, the performance is better than predicted by Eqs. (4) and (5), as seen in Figs. 1 and 2.

Jacobian arithmetic with Hess-reduced divisors is performed using Algorithm 2. So the complexity of Jacobian arithmetic is simply the complexity of Algorithm 2 with whichever algorithm is used to compute (r, a) in Line 2.

Algorithm 2. Jacobian arithmetic with Hess-reduced divisors

Input: Hess-reduced divisors $D_1 = \tilde{D}_1 - r_1 A$ and $D_2 = \tilde{D}_2 - r_2 A$
Output: $D_3 = \tilde{D}_3 - r_3 A$ with D_3 Hess-reduced and $D_3 \equiv D_1 + D_2$
1: $E \leftarrow D_1 + D_2$
2: Compute (r, a) where r is a solution to $\mathtt{HR\text{-}Min}$ for input E and $a \in \mathcal{L}(D + rA) \setminus \{0\}$
3: $D_3 \leftarrow E + \mathrm{div}(a)$

We can now give the time complexity of Jacobian arithmetic with Algorithm 1 or with binary search, both with and without caching. We account for the impact of caching by assuming that adding the infinite parts of divisors is amortized as $O(1)$.

Theorem 4. *Computing the Hess-reduction of two Hess-reduced divisors of F has the following asymptotic time complexity, expressed in operations in K.*

1. *Using Algorithm 1 in Line 2 of Algorithm 2 and no caching, the worst case time complexity is*
$$O(C_f^2 g n^9 + C_f g^2 n^7 + g^3 n^6)$$
and the heuristically average case complexity is
$$O(C_f^2 n^9 + C_f g n^7 + g^2 n^6 + g n^6 \log(gn) \log\log(gn))$$

2. *Using Algorithm 1 in Line 2 of Algorithm 2 and caching, the worst case time complexity is*
$$O(C_f^2 g n^9 + C_f g^2 n^7 + g^3 n^5)$$
and the heuristically average case complexity is
$$O(C_f^2 n^9 + C_f g n^7 + g^2 n^6 + g n^6 \log(gn) \log\log(gn))$$

3. *Using binary search in Line 2 of Algorithm 2 and no caching, the time complexity is*
$$O(\log(g)(C_f^2 n^9 + C_f g n^7 + g^2 n^6 + g n^6 \log(gn) \log\log(gn)))$$

4. *Using binary search in Line 2 of Algorithm 2 and caching, the time complexity is*

$$O(\log(g)(C_f^2 n^9 + C_f g n^7 + g^2 n^5) + g^2 n^6 + g n^6 \log(gn) \log \log(gn))$$

Proof. In Algorithm 2, we add the finite and infinite parts of D_1 and D_2 to compute E, for a cost of $2\mathbf{I}$. One of these $\mathbf{I}$ operations is on the infinite parts of D_1 and D_2, hence amortized as $O(1)$ if caching is used. We then use either Algorithm 1 or binary search to find (r, a), with the corresponding costs given in Theorems 2 and 3. Finally, we compute $D_3 = E + \mathrm{div}(a)$ for a cost of $2\mathbf{I}$, and again one of these is $O(1)$ if caching is used. The computation of $\mathrm{div}(a)$ from a is dominated by the other costs in the algorithm. After accounting for caching in Algorithm 1 and in the binary search approach and substituting Eqs. (4) and (5) into Theorems 2 and 3, we obtain the asserted complexity estimates. $\qquad\square$

We determine the overall asymptotic complexities when one of g, n is fixed and the other varies. Although caching makes a difference in practical performance, after fixing either g or n and allowing the other to vary (see Figs. 1 and 2), it only impacts lower order terms in the overall complexities.

Corollary 1. *Computing the Hess-reduction of two Hess-reduced divisors of F has the following asymptotic time complexity, expressed in operations in K.*
When g is fixed and $n \to \infty$: $O(C_f^2 n^9)$
When n is fixed and $g \to \infty$:

(a) $O(C_f^2 g)$ in the worst case when using Algorithm 1 in Line 2 of Algorithm 2;
(b) $O(C_f^2)$ heuristiclly on average when using Algorithm 1 in Line 2 of Algorithm 2;
(c) $O(C_f^2 \log(g))$ when using binary search in Line 2 of Algorithm 2.

Proof. For fixed g and varying n, the term $C_f^2 n^9$ dominates in all the expressions of Theorem 4. For fixed n and varying g, we use the identity

$$g \le \frac{1}{2}(C_f n - 2)(n - 1); \tag{6}$$

established in [7, Cor. 5.6] and identify the dominant term in each of the complexity estimates of Theorem 4. $\qquad\square$

Asymptotically, for fixed n and varying g, the heuristic average complexities of using linear versus binary search in Algorithm 2 differ by a factor of $O(\log(g))$. Indeed, for typical Hess-reduced input divisors, Algorithm 1 will complete after 2 reductions, whereas binary search is expected to perform $\lceil \log_2(g+1) \rceil$ reductions, yielding a relative factor of $\lceil \log_2(g + 1) \rceil / 2$.

6   Timing Experiments

We implemented Jacobian arithmetic on Unique Hess representations in Sage using both Algorithm 1 and binary search for reduction, both with and without caching, and profiled the results. Our implementation of Unique Hess arithmetic using Algorithm 1 was accepted into SageMath, available since the SageMath 10.9.beta7 release [22]. Our implementation of the binary search approach as well as our code used to profile the implementations is available at [15]. Our implementations of Algorithm 1 and the binary search variant were mostly written in Python, with some performance-critical subroutines written in Cython.

We carried out our timing experiments on a server running Red Hat Enterprise Linux 9.7 with an Intel Xeon CPU E7-8891 v4 with 80 64-bit CPU cores at 2.80 GHz, and 256 GiB of memory. Our experiments were run on a custom fork of Sage 10.7.beta7, using Python 3.12.5. SageMath has over 100 software dependencies, so we will only list major relevant dependencies: Singular 4.4.1 [24], FLINT 3.4.0 [21], GMP 6.3.0 [5], and numpy 2.3.2 [6]. We also used Magma 2.27–6 [2] to perform some genus computations that took too long in SageMath. Our use of Magma was restricted to computing genera when initializing some function fields, and so it has no impact on our reported timing results.

6.1   Experimental Setup

For testing, we generated global function fields both via an ad-hoc method and via the method described in [20]. The latter provides an easy way to generate global function fields with two infinite places of degree one. These function fields are given by a defining polynomial $f(t) \in K[x, t]$ of degree $n = [F : K(x)]$, irreducible over $K(x)$, and of the form:

$$f(t) = t^n + \sum_{i=0}^{n-1} a_i t^i,$$

with $a_i \in K[x]$ for $0 \leq i \leq n - 1$, and

$$\deg(a_{n-1}) = C_f,$$
$$\deg(a_i) < (n - i)C_f \quad \text{for } 1 \leq i \leq n - 1,$$
$$\deg(a_0) = nm - 1.$$

We conjecture that for these function fields, equality always holds in (6). We verified this for all the fields we tested for producing Fig. 1, so our timing results do not depend on this assumption. Note that equality in (6) implies that any two of g, n and C_f uniquely determine the third.

For Fig. 2, we instead generated function fields via an ad-hoc method, by trying random polynomials $f(x, t)$, monic and of degree 3 in t, with appropriate bounds on C_f, verifying that they were irreducible over $K(x)$ and that the corresponding function field contained a degree one infinite place. Equality did not hold in Eq. (6) for the function fields generated by this ad-hoc method.

Due to space constraints, we only report timing tests over the finite field $\mathbb{F}_{32771}$, a 16 bit prime field. In [14, Chapter 5] the first author conducted extensive timing experiments on a wide range of parameter sets. The data presented here is representative of the findings in that source, with the possible exception of very small primes like $p = 2, 3$ where the generic case for which Algorithm 1 is optimized does not happen sufficiently frequently.

For each triple (g, n, p), we performed tests in 5 function fields. In each function field we performed Unique Hess arithmetic on 5 Fibonacci-style addition chains of length 1000. Reported results are the average over all function fields and addition chains for each (g, n, p).

To test the correctness of our implementations, we relied on SageMath's robust testing framework that is able to automatically create a test suite and check correctness by verifying that the abelian group axioms hold on a set of examples.

6.2   Timing Results

Overall, Algorithm 1 significantly outperformed the binary search approach. In our figures, we plotted the average CPU time in milliseconds to perform a single Jacobian addition for various parameters and implementations, alongside the corresponding asymptotic complexities. For Algorithm 1 we used the heuristically average case complexity. For the asymptotic complexity plots in Fig. 1, we assumed equality in Eq. (6) and replaced C_f accordingly. In Fig. 2 we treated C_f as a constant because C_f^2 is negligible compared to n^9 asymptotically, and the largest value of C_f in our data plotted in Fig. 2 was $C_f = 6$. To account for constant factors in the plots of our asymptotic estimates, we computed a constant a such that if the complexity is $O(f(x))$, then $af(x)$ matches the data at a single measured value. For Fig. 1, since our improvement is by a factor of $O(\log(g))$, we matched the constant a to the largest data point $g = 55$ in order to more accurately capture the dependence on g. In Fig. 2, we matched the constant a to the first data point as our improvements do not affect the asymptotic complexity in n.

Figure 1 shows that caching effects a noticeable speed-up, but the most significant performance gains come from using Algorithm 1. In particular, for $g \geq 13$, the non-caching implementation using Algorithm 1 outperforms the caching implementation using binary search. As predicted by our complexity analysis, the difference in performance between Algorithm 1 and the binary search approach grows with g. Our data show an increase in speed that is slightly less than the predicted factor of $\lceil \log(g + 1) \rceil / 2$ for linear versus binary search, likely because this speed-up only applies in the most expensive part of Jacobian arithmetic (reduction to a unique representative), and the lower degree terms that were excluded from the analysis in Corollary 1 are not entirely negligible for our parameter range. For genus 100 (not depicted in Fig. 1), the linear search variant was about 3 times faster both with and without caching. Here, we would expect the performance to increase by a factor of about $\lceil \log_2(101) \rceil / 2 = 3.5$, which is reasonably close to our measured speed-up of a factor of 3. The relative impact of caching is smaller as the genus grows, since the computations that are cached take up a smaller proportion of the total time with increasing genus. For the linear search variant, caching gave a speedup of a factor of 1.63 for genus 4, was still noticeable at 1.15 for genus 25, but was negligible at 1.01 for genus 100.

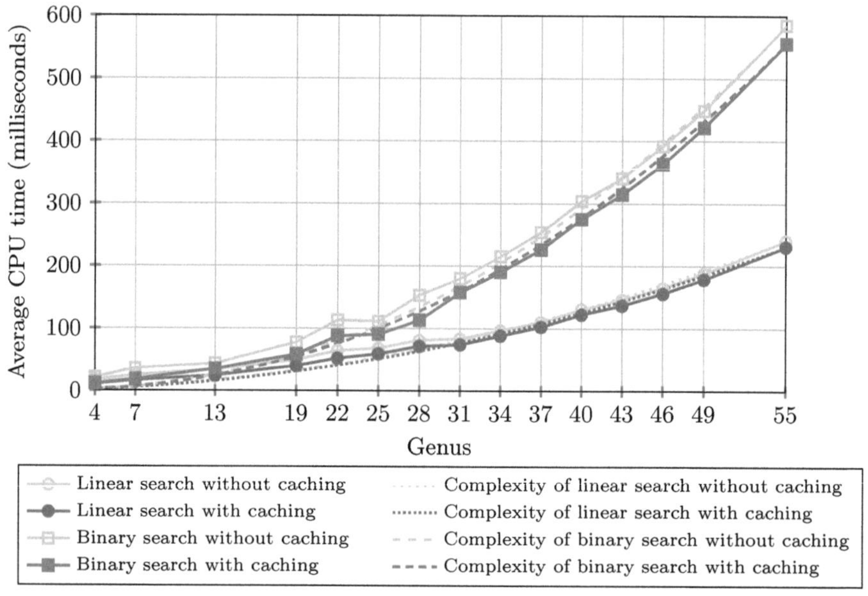

Fig. 1. Addition performance for $4 \leq g \leq 55$, $n = 3$, $p = 32771$.

For fixed g and varying n, all our algorithms have same the asymptotic complexity by Corollary 1. With g fixed, caching effects a speed-up by a constant factor in the number of partial divisor additions, with a more pronounced performance gain for the computationally more intense binary search compared to Algorithm 1. However, this constant factor speed-up is dominated by the cost of the short-circuited Riemann-Roch computations. Linear search is expected to be faster on average than binary search by a constant factor of approximately $\lceil \log_2(g+1) \rceil / 2$ as explained earlier; for $g = 15$, this factor is 2. The data depicted in Fig. 2 bear this out and also suggest that the asymptotic upper bound of $O(C_f^2 n^9)$ is an over-estimate of the actual complexity of Jacobian arithmetic. For $g = 15$, caching produced a speedup of a factor ranging from 1.20 to 1.30, without any strong trends dependent on n.

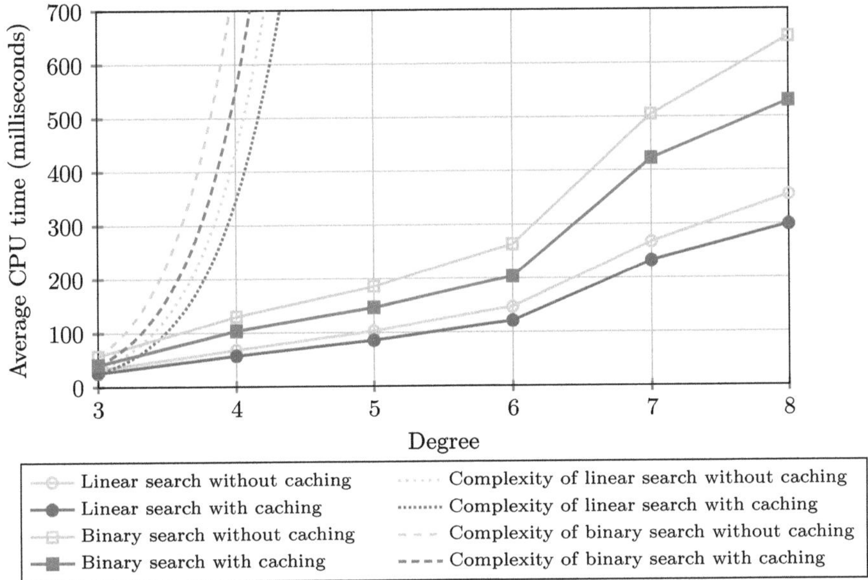

Fig. 2. Addition performance for $3 \leq n \leq 8$, $g = 15$, $p = 32771$.

7   Conclusions and Further Research

Our reduction algorithm (Algorithm 1) was designed specifically for optimal performance in the most frequent cases. Further speed-ups will likely require improvements to the underlying subroutines, namely Riemann-Roch space computation and ideal multiplication. Additional information about the input divisor might potentially narrow down the possible candidate solutions for HR-Min, but given the overwhelming likelihood of the generic case, it is difficult to see how that approach would lead to a significant performance gain in this situation. Even *a priori* ruling out the case where $r = g - 1$ solves HR-Min without performing a short-circuited Riemann-Roch computation would only speed up the algorithm by a constant factor.

The timing experiments of Sect. 6 demonstrate that Algorithm 1 is the fastest unique divisor reduction algorithm in practice, at least for the function fields we tested. As the non-caching version of the linear search implementation outperformed the caching version of the binary search implementation, we do not need to assume the existence of a degree one infinite place for the linear search method to outperform the binary search approach. Our timing experiments also demonstrate that caching computations involving the infinite parts of divisors gives a noticeable speedup when the function field has a degree one infinite place.

7.1   Future Work

We list several possible directions for future work.

For short-circuited Riemann-Roch computations, we used SageMath's modified version of the algorithm for computing K-bases of Riemann-Roch spaces from [7]. Algorithm 1 demonstrates that short-circuited Riemann-Roch computations are sufficient for some applications. It would be beneficial to have a potential SCRR approach that is faster than simply exiting early from an algorithm designed to perform a full Riemann-Roch basis computation.

For hyperelliptic curves, the fastest known algorithm for Jacobian arithmetic is NUCOMP [9,13]. The traditional approach to divisor arithmetic first computes the sum of two input divisors and subsequently reduces the result. NUCOMP in essence interleaves divisor addition and reduction, thereby operating on smaller inputs and avoiding the costly computation of bases for non-reduced intermediate divisors. It is worthwhile to explore whether a similar strategy can be realized for Jacobian arithmetic for non-hyperelliptic curves.

The first framework for arithmetic on unique representatives of elements in the Jacobian of a split hyperelliptic curve (i.e. a hyperelliptic curve with two infinite places) is due to Paulus and Rück [17]. In addition to reduction, the Paulus-Rück arithmetic requires additional adjustment steps to obtain such unique representatives. These adjustment steps were eliminated in the balanced divisor arithmetic introduced by Galbraith, Harrison, and Mireles Morales [3,4,16]. In [14], it was shown that unique reduced divisor class representatives of [17] are a special case of the Hess-reductions introduced in Definition 2 when the curve is split hyperelliptic. This relationship could potentially be leveraged to explore whether the notion of a balanced divisor could be extended to non-hyperelliptic curves with two infinite places, or even to curves with more infinite places, with the aim to simplify Jacobian arithmetic and/or improve its performance.

The notion of a semi-reduced divisor plays a crucial role in Jacobian arithmetic for hyperelliptic curves. Divisor addition produces a semi-reduced divisor that is equivalent to the sum of two input divisors, which is subsequently reduced. The concept of semi-reduced divisors was generalized to arbitrary function fields in [14]; in essence, a divisor D is semi-reduced if no sub-sum of D is the conorm of a divisor of $K(x)$. The author of [14] also proved that Hess-reductions along an infinite place of degree one are semi-reduced. Future work could explore whether the notion of semi-reducedness can be generalized further in such a way that Hess-reductions along a finite degree one place, or even along a degree one divisor as considered in [7], are semi-reduced. An appropriate notion of semi-reducedness may be a key ingredient for an efficient reduction algorithm that produces unique representatives of divisor classes.

Acknowledgments. All three authors were supported by the Natural Sciences and Engineering Research Council of Canada. The first author also received funding from the Province of Alberta through an AGES Research scholarship. We are grateful to three anonymous reviewers for their valuable comments. We also thank Kwankyu Lee for writing much of the SageMath function field machinery that we built on, and for reviewing the first author's pull request https://github.com/sagemath/sage/pull/41453 adding Unique Hess Jacobian arithmetic to SageMath.

Disclosure of Interests. The authors have no competing interests.

References

1. Bauch, J.D.: Lattices over polynomial rings and applications to function fields. Ph.D. thesis, Universitat Autònoma de Barcelona, Bellaterra (2014)
2. Bosma, W., Cannon, J., Playoust, C.: The Magma algebra system I: the user language. J. Symb. Comput. **24**(3–4), 235–265 (1997). https://doi.org/10.1006/jsco.1996.0125
3. Galbraith, S.D.: Mathematics of Public Key Cryptography. 2 edn. (2018)
4. Galbraith, S.D., Harrison, M., Mireles Morales, D.J.: Efficient hyperelliptic arithmetic using balanced representation for divisors. In: van der Poorten, A.J., Stein, A. (eds.) ANTS 2008. LNCS, vol. 5011, pp. 342–356. Springer, Heidelberg (2008). https://doi.org/10.1007/978-3-540-79456-1_23
5. Granlund, T., et al.: GNU Multiple Precision Arithmetic Library. Free Software Foundation, Inc (2023)
6. Harris, C.R., et al.: Array programming with NumPy. Nature **585**(7825), 357–362 (2020). https://doi.org/10.1038/s41586-020-2649-2
7. Hess, F.: Computing Riemann-Roch spaces in algebraic function fields and related topics. J. Symb. Comput. **33**(4), 425–445 (2002). https://doi.org/10.1006/jsco.2001.0513
8. Howe, E., Lauter, K.E., Top, J.: Pointless curves of genus three and four. In: Arithmetic, Geometry and Coding Theory (AGCT 2003). Séminaires et Congrès, vol. 11, pp. 125–141. Société Mathématique de France, Paris (2005)
9. Jacobson, Jr., M.J., van der Poorten, A.J.: Computational aspects of NUCOMP. In: Algorithmic number theory (Sydney, 2002), Lecture Notes in Comput. Sci., vol. 2369, pp. 120–133. Springer, Berlin (2002). https://doi.org/10.1007/3-540-45455-1_10
10. Junge, M.: Asymptotically Fast Arithmetic in the Picard Group of Algebraic Curves & Related Topics. Ph.D. thesis, Universität Oldenburg (2022)
11. Khuri-Makdisi, K.: Linear algebra algorithms for divisors on an algebraic curve. Math. Comput. **73**(245), 333–357 (2004). https://doi.org/10.1090/S0025-5718-03-01567-9
12. Khuri-Makdisi, K.: Asymptotically fast group operations on Jacobians of general curves. Math. Comput. **76**(260), 2213–2239 (2007). https://doi.org/10.1090/S0025-5718-07-01989-8
13. Lindner, S., Imbert, L., Jacobson, M.J.: Balanced NUCOMP. In: Boulier, F., England, M., Sadykov, T.M., Vorozhtsov, E.V. (eds.) CASC 2020. LNCS, vol. 12291, pp. 402–420. Springer, Cham (2020). https://doi.org/10.1007/978-3-030-60026-6_23
14. Macri, V.: Comparison of and improvements to degree zero divisor class group arithmetic in algebraic function fields. Master's thesis, University of Calgary (2025). https://doi.org/10.11575/PRISM/50422
15. Macri, V.: Improvements to Jacobian arithmetic in global function fields (2026). https://github.com/vincentmacri/Improvements-to-Jacobian-Arithmetic-in-Global-Function-Fields
16. Mireles Morales, D.J.: Efficient Arithmetic on Hyperelliptic Curves with Real Representation. Ph.D. thesis, The University of London (2008)

17. Paulus, S., Rück, H.G.: Real and imaginary quadratic representations of hyperelliptic function fields. Math. Comput. **68**(227), 1233–1241 (1999). https://doi.org/10.1090/S0025-5718-99-01066-2
18. Stichtenoth, H.: Algebraic Function Fields and Codes, Graduate Texts in Mathematics, 2 edn., vol. 254. Springer, Berlin, Heidelberg (2009). https://doi.org/10.1007/978-3-540-76878-4
19. Sutherland, A.V.: Fast Jacobian arithmetic for hyperelliptic curves of genus 3. Open Book Ser. **2**(1), 425–442 (2019). https://doi.org/10.2140/obs.2019.2.425
20. Tang, A.: Infrastructure of function fields of unit rank one. Ph.D. thesis, University of Calgary (2011)
21. The FLINT team: FLINT: Fast Library for Number Theory (2025)
22. The Sage Developers: SageMath, the Sage Mathematics Software System (2026)
23. Wiles, A.: The Birch and Swinnerton-Dyer conjecture
24. Wolfram, D., Greuel, G.M., Pfister, G., Schönemann, H.: SINGULAR 4-4-1 — a computer algebra system for polynomial computations (2025)

On Secret Sharing from Extended Norm-Trace Curves

Olav Geil$^{(\boxtimes)}$ [ID]

Department of Mathematical Sciences, Aalborg University, Aalborg, Denmark
`olav@math.aau.dk`

Abstract. In [4] Camps-Moreno et al. treated (relative) generalized Hamming weights of codes from extended norm-trace curves and they gave examples of resulting good asymmetric quantum error-correcting codes employing information on the relative distances. In the present paper we study ramp secret sharing schemes which are objects that require an analysis of higher relative weights and we show that not only do schemes defined from one-point algebraic geometric codes from extended norm-trace curves have good parameters, they also possess a second layer of security along the lines of [11]. It is left undecided in [4, page 2889] if the "footprint-like approach" as employed by Camps-Moreno herein is strictly better for codes related to extended norm-trace codes than the general approach for treating one-point algebraic geometric codes and their likes as presented in [12]. We demonstrate that the method used in [4] to estimate (relative) generalized Hamming weights of codes from extended norm-trace curves can be viewed as a clever application of the enhanced Goppa bound in [12] rather than a competing approach.

Keywords: Extended norm-trace curves · One-point algebraic geometric codes · Ramp secret sharing · Relative generalized Hamming weights

1 Introduction

Secret sharing is a branch of cryptography in which a central dealer shares a secret among a set of participants by giving them each a share. This is done in such a way that if few of them pool their shares then they obtain no information on the secret, but if many do then they can recover it in full. One often requires the schemes to be linear meaning that (partial) recovery, whenever possible, can be done by simple and fast linear algebra algorithms. In most schemes studied in the literature the shares and the secrets are elements from the same set, often a finite field $\mathbb{F}_q$. This is in contrast to linear ramp secret sharing where the secret belongs to $\mathbb{F}_q^\ell$, and the shares to $\mathbb{F}_q$, with $\ell \geq 1$. Ramp schemes were originally introduced by Blakley and Meadows in [2] and by Yamamoto in [21], and have gained a lot of interest due to their rich structures and their application in connection with for instance storage of bulk data and secure multiparty computation [7,8]. A linear ramp scheme is synonymous to a pair of nested linear

L. Batina and F. Özbudak (Eds.): WAIFI 2026, LNCS 16611, pp. 129–144, 2026.
https://doi.org/10.1007/978-3-032-27574-5_9

codes and worst case (partial) information leakage as well as worst case (partial) information recovery can be described by relative parameters of the pair of codes and their duals.

For large number of participants compared to the field size knowledge on the worst case scenarios, however, by far gives the complete picture, but a deeper analysis is typically very difficult. In [11] the author coined the concept of maximum non-i-qualifying sets to address this difficulty. These are sets of maximum size being not able to recover $i \log_2(q)$ bits of information. It was demonstrated that two specific classes of monomial-Cartesian codes possess families of maximum non-i-qualifying sets with a very rich structure and based on this information one can then impose a second layer of security. By elaborating on results in [4] in the present paper we demonstrate that also ramp secret sharing schemes constructed from nested one-point algebraic geometric codes from the extended norm-trace curves posses families of well-structured maximum non-i-qualifying sets giving again rise to a second layer of security.

The exposition in [4] uses the language of Gröbner basis theory, but as we demonstrate their strategy for deriving estimates on code parameters related to the extended norm-trace curve is not a competing method to the enhanced Goppa kind of bounds in [12, Eq. 13, Eq. 17] for general algebraic function fields of transcendence degree 1, but can be viewed as a clever application thereof. By proving this we answer an open question raised by Camps-Moreno et al. in [4, page 2889].

The paper is organized as follows. In Sect. 2 we give the necessary background on linear ramp secret sharing. We next treat general theory of relative generalized Hamming weights of nested one-point codes and their relatives in Sect. 3. Then in Sect. 4 we elaborate on material in [4] regarding (relative) generalized Hamming weights of nested decreasing norm-trace codes and their duals, and demonstrate the relationship between the method used in [4] and that of [12]. Finally, in Sect. 5 we apply such results to produce the alluded linear ramp schemes with desirable structure and consequently a second layer of security. Section 6 contains concluding remarks.

2   Linear Ramp Secret Sharing

In linear ramp secret sharing the shares given to the individual participants belong to a finite field $\mathbb{F}_q$, but the secret is a tuple in $\mathbb{F}_q^\ell$, where ℓ is allowed to be strictly larger than 1. The linearity means that if $c_1^{(1)}, \ldots, c_n^{(1)}$ are shares for a secret $\boldsymbol{s}^{(1)}$ and $c_1^{(2)}, \ldots, c_n^{(2)}$ are shares for a secret $\boldsymbol{s}^{(2)}$ then also $ac_1^{(1)} + bc_1^{(2)}, \ldots, ac_n^{(1)} + bc_n^{(2)}$ work as shares for the secret $a\boldsymbol{s}^{(1)} + b\boldsymbol{s}^{(2)}$ for any $a, b \in \mathbb{F}_q$. Such schemes by [7, Sec. 4.2] can be put into the form of a nested code construction as follows. Let $C_2 \subseteq C_1 \subseteq \mathbb{F}_q^n$, $\dim C_2 = k_2 < \dim C_1 = k_1$ and $\ell = k_1 - k_2$. Consider bases $\{\boldsymbol{b}_1, \ldots, \boldsymbol{b}_{k_2}\}$ and $\{\boldsymbol{b}_1, \ldots, \boldsymbol{b}_{k_2}, \boldsymbol{b}_{k_2+1}, \ldots, \boldsymbol{b}_{k_1}\}$, respectively, for C_2 and C_1, respectively, as vector spaces over $\mathbb{F}_q$. A secret $\boldsymbol{s} = (s_1, \ldots, s_\ell) \in \mathbb{F}_q^\ell$ is encoded to

$$(c_1, \ldots, c_n) = r_1 \boldsymbol{b}_1 + \cdots + r_{k_2} \boldsymbol{b}_{k_2} + s_1 \boldsymbol{b}_{k_2+1} + \cdots + s_\ell \boldsymbol{b}_{k_1} \tag{1}$$

and participant i, then receives c_i as a share $i = 1, \ldots, n$.

The crucial parameters of a linear ramp secret sharing scheme are the number of participants, which is n, the dimension of space of secrets, which is ℓ, the privacy numbers $t_1, \ldots, t_\ell$, and the reconstruction numbers $r_1, \ldots, r_\ell$. Here, for $i = 1, \ldots, \ell$, t_i is largest possible and r_i is smallest possible such that

- no set of t_i participants is able to recover i times $\log_2(q)$ bits of information about $\boldsymbol{s}$
- any set of r_i participants can recover i times $\log_2(q)$ bits of information.

Of special interest are full privacy and full recovery corresponding to $t = t_1$ and $r = r_\ell$. As is well-known [12, 16]

$$t_m = M_m(C_2^\perp, C_1^\perp) - 1 \tag{2}$$
$$r_m = n - M_{\ell-m+1}(C_1, C_2) + 1 \tag{3}$$

where

$$M_t(C_1, C_2) = \min\{\#\mathrm{Supp}(D) \mid D \subseteq C_1, D \cap C_2 = \{\boldsymbol{0}\}, \dim D = t\}$$

is called the tth relative generalized Hamming weight.

As is always the case in coding theory, one cannot choose parameters freely. If, say, q, n, ℓ and $1 \leq r_1 < \cdots < r_\ell \leq n$ are fixed numbers then among the set of all pairs of nested codes $C_2 \subseteq C_1 \subseteq \mathbb{F}_q^n$ with such parameters there are limitations as to how close t_i can be to r_i, $i = 1, \ldots, \ell$, see [6]. However, by definition for any i there exist groups of size $r_i - 1$ who cannot recover i times $\log_2(q)$ bits of information and when we have knowledge about the structure of such groups we may employ it as a second layer of security as demonstrated in the recent paper [11]. The importance of the mentioned groups justify that we give them a particular name [11, Def. 8].

Definition 1. *Given a linear ramp secret sharing scheme defined from $C_2 \subseteq C_1 \subseteq \mathbb{F}_q^n$, a set $A \subseteq \{1, \ldots, n\}$ is called maximum non-i-qualifying if $\#A = n - M_{\ell-i+1}(C_1, C_2)$, but from the corresponding shares one cannot recover $i \log_2(q)$ bits of information.*

To establish information on the access-structure (who can recover how much) and in particular information on the maximum non-i-qualifying sets we may employ the following result [11, Thm. 3] which is a slight reformulation of [7, Thm. 10] (see also [18]).

Theorem 1. *Let $A = \{i_1 < \cdots < i_m\} \subseteq \{1, \ldots, n\}$. Assume $c'_{i_1}, \ldots, c'_{i_m}$ are simultaneously realizable shares in those positions, see (1). The amount of possible secrets $\boldsymbol{s}$ corresponding to such shares equals q^s with*

$$s = \max\{\dim D \mid D \subseteq C_1, D \cap C_2 = \{\boldsymbol{0}\}, \mathrm{Supp}(D) \subseteq \bar{A}\}$$

where $\bar{A} = \{1, \ldots, n\} \setminus A$.

In [11] Theorem 1 was employed to show that ramp schemes coming from the particular nested monomial-Cartesian codes as described in [9, Sec. 4] does not only have good parameters $n, \ell, r_1, \ldots, r_\ell, r_1, \ldots, t_\ell$, but also support a second layer of security by possessing maximum non-i-qualifying sets of desirable well-structured form. This insight was then used in [11, Sec. 4] to construct monomial-Cartesian code based schemes with larger ℓ maintaining desirable second layer of security by possessing well-structured maximum non-i-qualifying sets whilst paying less interest in the worst case security $t_1, \ldots, t_\ell$. The aim of the present contribution is to investigate what can be said regarding a second layer of security for ramp schemes based on decreasing norm-trace codes.

3 Relative Parameters of One-Point Algebraic Geometric Codes and Their Relatives

Before treating codes from extended norm-trace curves we revisit the general theory of relative generalized Hamming weights of any pair of nested one-point algebraic geometric codes and their relatives [12]. Consider an arbitrary algebraic function field of transcendence degree 1 defined over some finite field $\mathbb{F}_Q$. Let $P_1, \ldots, P_n, Q$ be pairwise different rational places. Let $H(Q)$ be the Weierstrass semigroup of Q, i.e.

$$H(Q) = -\nu_Q(R = \cup_{m=0}^\infty \mathcal{L}(mQ))$$

where ν_Q is the discrete valuation corresponding to Q. Define

$$H^*(Q) = \{\lambda \in H(Q) \mid C_{\mathcal{L}}(G = P_1 + \cdots + P_n, \lambda Q) \neq C_{\mathcal{L}}(G, (\lambda - 1)Q)\}.$$

Clearly, $\#H^*(Q) = n$. We now recall material from [12] on how to estimate (relative) generalized Hamming weights. Let $D \subseteq \mathbb{F}_Q^n$ be any subspace, of dimension say m. There exist functions $f_1, \ldots, f_m \in R$ such that

$$\{(f_i(P_1), \ldots, f_i(P_n)) \mid i = 1, \ldots, m\}$$

is a basis for D and such that

$$- \nu_Q(f_1) < \cdots < -\nu_Q(f_m) \tag{4}$$

holds, where without loss of generality we may assume that all numbers in (4) belong to $H^*(Q)$. These values do not depend on the choice of the f_i's, but are invariants of D and we may therefore define $\rho(D) = \{-\nu_Q(f_1), \ldots, -\nu_Q(f_m)\} \subseteq H^*(Q)$. From [12, Prop. 17] we have the bound

$$\#\mathrm{Supp}(D) \geq \#\left(H^*(Q) \cap \left(\cup_{s=1}^m (\gamma_s + H(Q))\right)\right) \tag{5}$$

where $\rho(D) = \{\gamma_1, \ldots, \gamma_m\}$.

With proper care one can apply (5) to any set of nested codes defined from R, but for nested one-point codes the situation is immediate [12, Thm. 19]. Let $\lambda_2 <

λ_1 be elements in $H^*(Q)$ and consider $m \leq \dim C_{\mathcal{L}}(G, \lambda_1 Q) - \dim C_{\mathcal{L}}(G, \lambda_2 Q)$.
We have

$$M_m(C_{\mathcal{L}}(G, \lambda_1 Q), C_{\mathcal{L}}(G, \lambda_2 Q))$$
$$\geq \min \left\{ \# \left(H^*(Q) \cap \left(\cup_{s=1}^m (\gamma_s + H(Q)) \right) \right) \mid \gamma_s \in H^*(Q) \text{ for } s = 1, \ldots, m, \right.$$
$$\left. \lambda_2 < \gamma_1 < \cdots < \gamma_m \leq \lambda_1 \right\}. \qquad (6)$$

That this bound (and in larger generality (5)) can be seen as an enhancement
of the Goppa bound for a one-point Goppa code i.e. the bound

$$d(C_{\mathcal{L}}(G, \lambda Q)) \geq n - \lambda$$

follows from [14, Lem. 5.15] which says that for any numerical semigroup Λ
and any element λ herein we have $\lambda = \#(\Lambda \backslash (\lambda + \Lambda))$. In fact (6) can be a
strict improvement even when applied to estimate the minimum distance of a
one-point code. This happens when some of the elements in $\Lambda \backslash (\lambda + \Lambda)$ are not
contained in $H^*(Q)$.

As noted in [12] a translation of [12, Thm 14] into the particular case of
one-point algebraic geometric codes and their relatives produces a similar result
as (5), but for dual spaces. We now fill in the missing details by stating such
result. Consider a subspace $D \subseteq \mathbb{F}_Q^n$ of dimension m and let $\eta_1, \ldots, \eta_m$ be the
unique numbers such that for $i = 1, \ldots, m$ there exists some $c \in D$ satisfying $c \in$
$C_{\mathcal{L}}^{\perp}(G, (\eta_i - 1)Q)$, but $c \notin C_{\mathcal{L}}^{\perp}(G, \eta_i Q)$. We shall write $\kappa(D) = \{\eta_1, \ldots, \eta_m\} \subseteq$
$H^*(Q)$. The counterpart to (5) is

$$\#\mathrm{Supp}(D) \geq \#\left(H(Q) \cap \left(\cup_{s=1}^m (\eta_s - H(Q)) \right) \right) \qquad (7)$$

From this one immediately obtains [12, Thm. 20] which we now state

$$M_m(C_{\mathcal{L}}^{\perp}(G, \lambda_2 Q), C_{\mathcal{L}}^{\perp}(G, \lambda_1 Q))$$
$$\geq \min \left\{ \# \left(H(Q) \cap \left(\cup_{s=1}^m (\gamma_s - H(Q)) \right) \right) \mid \gamma_s \in H^*(Q) \text{ for } s = 1, \ldots, m, \right.$$
$$\left. \lambda_2 < \gamma_1 < \cdots < \gamma_m \leq \lambda_1 \right\}. \qquad (8)$$

(Note that in the above formula $\gamma_1, \ldots, \gamma_m$ play the role of $\eta_1, \ldots, \eta_m$).

4   Bounds on Relative Parameters of Codes from the Extended Norm-Trace Curve

The extended norm-trace curve is the curve over $\mathbb{F}_{q^s}$ given by the equation

$$x^u = y^{q^{s-1}} + y^{q^{s-2}} + \cdots + y, \qquad (9)$$

where u is a positive divisor of $\frac{q^s-1}{q-1}$. Recall, that the right hand side of (9) corresponds to the trace map from $\mathbb{F}_{q^s}$ to $\mathbb{F}_q$, and similarly that $x^{\frac{q^s-1}{q-1}}$ corresponds to the norm map. Related codes are therefore generalizations of norm-trace codes [10,17] which again are generalizations of Hermitian codes [1,20,22]. Codes from the extended norm-trace curve have been extensively studied in [3–5,15] and one of the crucial observations employed here is the systematic structure of the affine variety, i.e. the set of affine roots of (9). As is easily seen from the property of the norm-map and the trace-map the affine point set is the disjoint union of the following sets [5, Lem. 3.1],

$$A_0 = \{(0,b) \mid b^{q^{s-1}} + \cdots + b = 0\} = \Gamma_0^{(1)} \times \Gamma_0^{(2)}$$

and for $i = 1, \ldots, q-1$

$$A_i = \{(a,b) \mid a^u = \alpha^i, b^{q^{s-1}} + \cdots + b = \alpha^i\} = \Gamma_i^{(1)} \times \Gamma_i^{(2)},$$

where α is a primitive element of $\mathbb{F}_q$. Clearly, $\#A_0 = \#\Gamma_0^{(2)} = q^{s-1}$ and for $i = 1, \ldots, q-1$ we have $\#A_i = uq^{s-1}$ where $\#\Gamma_i^{(1)} = u$ and $\#\Gamma_i^{(2)} = q^{s-1}$, [5, Lem. 3.1]. The algebraic function field of transcendence degree 1 over $\mathbb{F}_{q^s}$ defined from (9) has exactly $1 + \sum_{i=0}^{q-1} \#A_i$ rational places [5, Sec. 4.2], each of them, but one, being related to an affine point, and the last being the unique place at infinity. Following [5, Sec. 4.2] and [19] the Weierstrass semigroup corresponding to the unique place Q at infinity equals

$$H(Q) = \langle u, q^{s-1} \rangle$$

and we have

$$R = \cup_{m=0}^\infty \mathcal{L}(mQ) = \mathbb{F}_{q^s}[X,Y]/I$$

where $I = \langle X^u - Y^{q^{s-1}} - \cdots - Y \rangle$. Writing

$$x^i y^j = X^i Y^j + \langle X^u - Y^{q^{s-1}} - \cdots - Y \rangle$$

and following [5, Sec. 4.2] one sees that

$$\{x^i y^j \mid 0 \le i, 0 \le j < q^{s-1}\}$$

is a basis for R as a vector space over $\mathbb{F}_{q^s}$ and clearly $-\nu_Q(x^i y^j) = iq^{s-1} + ju$, where ν_Q indicates the discrete valuation corresponding to Q. No two elements of this basis have the same ν_Q-value, i.e. there is a one-to-one correspondence between the basis and $H(Q)$.

We shall denote by $P_1, \ldots, P_n$ the rational places different from Q, where of course $n = (u(q-1)+1)q^{s-1}$. Writing $G = P_1 + \cdots + P_n$ and as before

$$H^*(Q) = \{\lambda \in H(Q) \mid C_{\mathcal{L}}(G, \lambda Q) \ne C_{\mathcal{L}}(G, (\lambda-1)Q)\},$$

it is not difficult to see that

$$H^*(Q) = \{iq^{s-1} + ju \mid 0 \le i < u(q-1)+1, 0 \le j < q^{s-1}\}$$

and that

$$\{(f(P_1), \ldots, f(P_n)) \mid f = x^i y^j, \text{ where } 0 \leq i < u(q-1)+1, 0 \leq j < q^{s-1}\}$$

is a basis for $\mathbb{F}_{q^s}^n$ as a vector space over $\mathbb{F}_{q^s}$, and in particular that

$$\{(f(P_1), \ldots, f(P_n)) \mid f = x^i y^j, \text{ where } 0 \leq i < u(q-1)+1, 0 \leq j < q^{s-1},$$
$$iq^{s-1} + ju \leq \lambda\}$$

is a basis for $C_{\mathcal{L}}(G, \lambda Q)$. In [5] similar results were proved using Gröbner basis theoretical arguments.

We next turn our attention to the problem of estimating relative generalized Hamming weights. Let $D \subseteq \mathbb{F}_{q^s}^n$ be any subspace, of dimension say m and consider $\rho(D) = \{\gamma_1, \ldots, \gamma_m\} \subseteq H^*(Q)$ as in Sect. 3. We introduce the function $\iota : H^*(Q) \to \{(i,j) \mid 0 \leq i < u(q-1)+1 \text{ and } 0 \leq j \leq q^{s-1} - 1\}$ given by $\iota(\lambda = iq^{s-1} + ju) = (i,j)$. In [4] $\#\mathrm{Supp}(D)$ was estimated for the considered curve using a Gröbner basis approach. We explain their result and demonstrate the relationship with (5). Let $(a,b) \in \iota(\{\gamma_1, \ldots, \gamma_m\}) = \{(i_1, j_1), \ldots, (i_m, j_m)\}$ be chosen such that $a = \min\{i_1, \ldots, i_m\}$, then the bound in [4, Eq. (3)+ Eq. (4)] reads

$$\#\mathrm{Supp}(D) \geq \#\{(i,j) \mid 0 \leq i < u(q-1)+1, 0 \leq j < q^{s-1}, (i_s, j_s) \leq_p (i,j)$$
$$\text{for some } s \in \{1, \ldots, m\} \text{ or } (a+u, 0) \leq_p (i,j)\}. \quad (10)$$

Here, we used the partial ordering $\leq_p$ given by $(\alpha, \beta) \leq_p (\epsilon, \delta)$ if and only if $\alpha \leq \epsilon$ and $\beta \leq \delta$. Note, that the condition $(a+u, 0) \leq_p (i,j)$ of course only comes into action if $a + u < u(q-1)+1$. To see that (10) is a consequence of (5) we only need to show that $(a+u)q^{s-1} + 0u$ belongs to $(aq^{s-1}, bu) + H(Q)$ whenever $0 < b$. But, $(aq^{s-1} + bu) + (0q^{s-1} + (q^{s-1} - b)u) = (a+u)q^{s-1} + 0u$ and we are through.

Combining (10) and (6) one obtains

$$M_m(C_{\mathcal{L}}(G, \lambda_1 Q), C_{\mathcal{L}}(G, \lambda_2 Q))$$
$$\geq \min\{\#\{(i,j) \mid 0 \leq i < u(q-1)+1, 0 \leq j < q^{s-1}, \iota(\gamma_s) \leq_p (i,j)$$
$$\text{for some } s \in \{1, \ldots, m\} \text{ or } (a+u, 0) \leq_p (i,j)\},$$
$$\text{for } s = 1, \ldots, m, \quad \lambda_2 < \gamma_1 < \cdots < \gamma_m \leq \lambda_1\}. \quad (11)$$

We continue the study of relative generalized Hamming weights but now turn to dual codes. Given a subspace $D \subseteq \mathbb{F}_{q^s}^n$ of dimension m let $\kappa(D) = \{\eta_1, \ldots, \eta_m\}$ be as in Sect. 3. Choose

$$(a,b) \in \iota(\{\eta_1, \ldots, \eta_m\}) = \{(i_1, j_1), \ldots, (i_m, j_m)\}$$

with a maximal. We have

$$\#\mathrm{Supp}(D) \geq \#\{(i,j) \mid 0 \leq i < u(q-1)+1, 0 \leq j < q^{s-1}, (i,j) \leq_p (i_s, j_s)$$
$$\text{for some } s \in \{1, \ldots, m\} \text{ or } (i,j) \leq_p (a-u, q^{s-1} - 1)\}. \quad (12)$$

We prove this by applying (7) and by using similar arguments as above. We only need to demonstrate that $(a - u)q^{s-1} + (q^s - 1)u \in (aq^{s-1} + bu) - H(Q)$ for $a \geq u$. We have $(aq^{s-1} + bu) - (b+1)u = (a-u)q^{s-1} + (q^{s-1} - 1)u$ and we are through.

Combining (12) and (8) one obtains

$$M_m(C_{\mathcal{L}}^{\perp}(G, \lambda_2 Q), C_{\mathcal{L}}^{\perp}(G, \lambda_1 Q))$$
$$\geq \min\{\#\{(i,j) \mid 0 \leq i < u(q-1)+1, 0 \leq j < q^{s-1}, (i,j) \leq_p \iota(\gamma_s)$$
$$\text{for some } s \in \{1, \ldots, m\} \text{ or } (i,j) \leq_p (a-u, q^s - 1)\},$$
$$\text{for } s = 1, \ldots, m, \lambda_2 < \gamma_1 < \cdots < \gamma_m \leq \lambda_1\}. \quad (13)$$

One of the important insights from [4] is that for so-called decreasing norm-trace codes $C_2 \subseteq C_1$ the estimate on $M_m(C_1, C_2)$ inferred from (10) becomes sharp. A code of dimension k is called a decreasing norm-trace code if it equals the span of functions g_i, $i = 1, \ldots, k$ evaluated at $P_1, \ldots, P_n$ where we have $g_i = x^{\alpha_i} y^{\beta_i}$ with $0 \leq \alpha_i < (q-1)u + 1, 0 \leq \beta_i < q^{s-1}$, and where for any $0 \leq \alpha \leq \alpha_i$ and $0 \leq \beta \leq \beta_i$ there exists a $j \in \{1, \ldots, k\}$ such that $g_j = x^\alpha y^\beta$. Hence, the one-point algebraic geometric codes related to the extended norm-trace curve are of this type. As a consequence of the sharpness in case of decreasing codes we can conclude that in the case of the extended norm-trace curve the right hand side of (5) and the right hand side (10) are identical (this could alternatively have been proved directly by using [14, Lem. 5.15]).

In [4, Thm. 5.3] the dual of decreasing norm-trace code is shown to be equivalent to another decreasing norm-trace code. We leave it for the reader to inspect that the estimate one obtains by applying such correspondence to the relative generalized Hamming weights of a pair of duals of nested decreasing norm-trace codes is in fact the same as one gets by applying (12) directly. From the above we conclude that both (11) and (13) are sharp.

5   Schemes with a Second Layer of Security

As mentioned in the previous section it is proved in [4] that all their bounds are sharp in that for any $\gamma_1 < \cdots < \gamma_m$ in $H^*(Q)$ there exists a corresponding space D of dimension m satisfying $\rho(D) = \{\gamma_1, \ldots, \gamma_m\}$ and with equality in (10). Moreover, among such spaces there exist some which can be written

$$D = \text{Span}_{\mathbb{F}_{q^s}}\{(f_1(P_1), \ldots, f_1(P_n)), \ldots, (f_m(P_1), \ldots, f_m(P_n))\}$$

with $-\nu_Q(f_i) = \gamma_i$, $i = 1, \ldots, m$ and each element f_i being a product of linear factors. Our treatment of ramp secret sharing shall rely heavily on such observations.

With the aim of establishing ramp secret sharing schemes with maximum non-i-qualifying sets allowing for a second layer of security we now revisit some of the results in [4]. Here, by the first level of security we refer to the parameters $t_1, \ldots, t_\ell$ and $r_1, \ldots, r_\ell$, and the second layer means that we have systematic families of sets of maximal size of participants that cannot all be ignored when $i \log_2(q)$ bits of information $i \in \{1, \ldots, \ell\}$ are to be retrieved. The following theorem is a combination of [4, Lem. 3.1] and Case 1.1 in the proof of [4, Thm. 3.2] adapted to our language. Well-structured maximum non-i-qualifying sets derived from the theorem are discussed in the subsequent remark, corollary and examples.

Theorem 2. *Consider pairwise different* $\gamma_1, \ldots, \gamma_w \in H^*(Q)$ *and write*

$$\iota(\{\gamma_1, \ldots, \gamma_w\}) = \{(a_1, b_1), \ldots, (a_w, b_w)\}.$$

Assume $a_1 < \cdots < a_w$ *and* $b_w < \cdots < b_1$ *and that* $a_w - a_1 < u$. *Assume* $a_1 \leq u(q-2) + 1$ *Then the right hand side of (10) reads*

$$n - \left(a_1 q^{s-1} + b_w u + \sum_{i=1}^{w-1} (a_{i+1} - a_i)(b_i - b_w) \right). \tag{14}$$

The following functions $f_1, \ldots, f_w \in R$ *satisfy* $-\nu_Q(f_j) = a_j q^{s-1} + b_j u$, $j = 1, \ldots, w$ *and for*

$$D = Span_{\mathbb{F}_{q^s}}\{(f_1(P_1), \ldots, f_1(P_n)), \ldots, (f_w(P_1), \ldots, f_w(P_n))\}$$

equality holds in (10) meaning that $\#Supp(D)$ *is equal to the right hand side of (14). Let* $i' \in \{1, \ldots, q-1\}$ *and choose*

$$\alpha_1, \ldots, \alpha_{a_1} \in \Gamma_0^{(1)} \times \Gamma_1^{(1)} \times \cdots \times \Gamma_{i'-1}^{(1)} \times \Gamma_{i'+1}^{(1)} \times \cdots \times \Gamma_{q-1}^{(1)}. \tag{15}$$

Enumerate $\Gamma_{i'}^{(1)} = \{\alpha'_1, \ldots, \alpha'_u\}$ *and* $\Gamma_{i'}^{(2)} = \{\beta_1, \ldots, \beta_{q^{s-1}}\}$. *Finally for* $j = 1, \ldots, w$ *define*

$$f_j = \prod_{i=1}^{a_1}(x - \alpha_i) \prod_{i=1}^{a_j - a_1} (x - \alpha'_i) \prod_{i=1}^{b_j}(y - \beta_i). \tag{16}$$

Proof. See [4, Lem. 3.1] and the first part of the proof of [4, Thm. 3.2]. $\quad\square$

To make Theorem 2 operational in connection with deriving second layer of security we shall need the following lemma which is new.

Lemma 1. *Consider* $\gamma_1 < \cdots < \gamma_w$ *in* $H^*(Q)$. *Assume* $\gamma_w - \gamma_1 < \min\{u, q^{s-1}\}$ *and write*

$$\{\iota(\gamma_1), \ldots, \iota(\gamma_w)\} = \{(a_1, b_1), \ldots, (a_w, b_w)\}. \tag{17}$$

The enumeration on the right hand side of (17) can be done in such a way that

$$a_1 < \cdots < a_w \text{ and } b_w < \cdots < b_1. \tag{18}$$

It holds that $a_w - a_1 < u$. *Under the additional condition* $\gamma_1 \leq (u(q-2)+w)q^{s-1}$ *it holds that* $a_1 \leq u(q-2) + 1$.

Proof. The assumption $\gamma_j - \gamma_i < \min\{u, q^{s-1}\}$ for $1 \le i < j \le w$ implies (18). Aiming for a contradiction assume $a_w - a_1 \ge u$, and recall that by the very definition of the function ι we have $b_1 - b_w \le q^{s-1} - 1$. We obtain

$$(a_w q^{s-1} + b_w u) - (a_1 q^{s-1} + b_1 u) \ge u q^{s-1} - (q^{s-1} - 1)u = u$$

which by assumption is impossible. Finally, the additional condition in combination with $\gamma_w - \gamma_1 < \min\{u, q^{s-1}\}$ ensures that $\iota(\gamma_i) \not\succ_p (u(q - 2) + w, 0)$, $i = 1, \ldots, w$. But, the a_is constitute a strictly increasing sequence and we are through (to avoid confusion be aware that $\iota(\gamma_1)$ needs not be equal to (a_1, b_1)).

Corollary 1. *Consider $\lambda_2 < \lambda_1$ with $\lambda_2 + 1, \lambda_1 \in H^*(Q)$ and $\lambda_1 - (\lambda_2 + 1) < \min\{u, q^{s-1}\}$ and $\lambda_2 < (u(q - 2) + w)q^{s-1}$ and consider the nested codes $C_2 = C_{\mathcal{L}}(G, \lambda_2 Q) \subseteq C_{\mathcal{L}}(G, \lambda_1 Q) = C_1$ the co-dimension ℓ of which equals $\#(H^*(Q) \cap \{\lambda_2 + 1, \ldots, \lambda_1\})$. For $w \in \{1, \ldots, \ell\}$ let $\{\gamma_1, \ldots, \gamma_w\} \subseteq H^*(Q)$ with $\lambda_2 < \gamma_1 < \cdots < \gamma_w \le \lambda_1$ be such that the minimum value is attained in (10) among all possible choices of $\{\gamma_1, \ldots \gamma_w\}$ of this form. I.e. the right hand side of (10) for the given $\gamma_1, \ldots, \gamma_w$ achieves the value of $M_w(C_1, C_2)$. The set of positions corresponding to common roots of related functions $f_1, \ldots, f_w$ as in Theorem 2 constitutes a maximum non-$(\ell - w + 1)$-qualifying set. In other words, by leaving out all participants corresponding to non-common roots of these functions the remaining participants cannot detect $(\ell - w + 1) \log_2(q)$ bits of information.*

Corollary 1 in combination with the particular structure of $\{P_1, \ldots, P_n\}$ as well as the particular structure of each of $f_1, \ldots, f_w$ imposes a second layer of security in a large family of ramp secret sharing schemes defined from nested one-point algebraic geometric codes over the extended norm-trace curves. Such schemes have families of well-structured sets of participants who cannot all be left out if $(\ell - w + 1) \log_2(q)$ bits of information are to be retrieved. This is discussed in the following remark.

Remark 1. Recall, that the affine variety of the extended norm-trace curve equals the disjoint union $\cup_{v=0}^{q-1} A_v$ where $A_i = \Gamma_i^{(1)} \times \Gamma_i^{(2)}$, and where $\#\Gamma_i^{(2)} = q^{s-1}$ for $i = 0, \ldots, q - 1$, where $\#\Gamma_0^{(1)} = 1$ and where for $i = 1, \ldots, q - 1$ $\#\Gamma_i^{(1)} = u$. To illustrate the second layer of security assume in the following that we have an organization with $q - 1$ large departments each having uq^{s-1} members. Say, A_i corresponds to a large department i, $i = 1, \ldots, q - 1$. Assume further that we have a single small department of size q^{s-1}. This department corresponds to A_0. If $a_1 < u(q - 2) + 1$ then one may choose $\alpha_1, \ldots, \alpha_{a_1}$ in Theorem 2 in such a way that the common roots of the $f_1, \ldots, f_w$ in (16) correspond to the disjoint union of the following sets: $\lfloor \frac{a_1}{u} \rfloor$ entire large departments $A_{i_1} \cup \cdots \cup A_{i_{\lfloor a_1/u \rfloor}}$, and for some $i'' \ne i'$ both belonging to $\{1, \ldots, q - 1\} \setminus \{i_1, \ldots, i_{\lfloor a_1/u \rfloor}\}$ a subset $S \times \Gamma_{i''}^{(2)} \subseteq A_{i''}$, where $S \subseteq \Gamma_{i''}^{(1)}$, $\#S = a_1 - \lfloor \frac{a_1}{u} \rfloor$ and finally a subset of $A_{i'}$ of size equal to $b_w u + \sum_{i=1}^{w-1}(a_{i+1} - a_i)(b_i - b_w)$ with the form of some (possibly irregular) staircase. We call the latter subset "the set locally induced by $\gamma_1, \ldots, \gamma_w$" (or "the locally induced set" for short). Observe, that i' plays the exact same role as in Theorem 2, whereas i'' is introduced to reflect a particular systematic way for choosing the elements in the left hand side of (15). The subset $S \times \Gamma_{i''}^{(2)}$ as well a the locally induced set can be given their own meaning. This is done by dividing $A_{i''}$ and $A_{i'}$, respectively, into u horizontal levels each containing q^{s-1} elements. Here, we enumerate the levels according to $\Gamma_{i''}^{(1)}$ and $\Gamma_{i'}^{(1)}$, respectively. Similarly, we can in an obvious way divide $A_{i''}$ and $A_{i'}$, respectively, into q^{s-1} vertical levels each containing u members. Hence, for instance $S \times \Gamma_{i''}^{(2)}$ consist of $a_1 - \lfloor \frac{a_1}{u} \rfloor$ horizontal levels. Note, that for fixed $A_{i''}$ there are $\binom{u}{a_1 - \lfloor a_1/u \rfloor}$ possibilities for that. If $a_1 = u(q-2)+1$ then there is in the set of common roots no set $S \times \Gamma_{i''}^{(2)}$ and we need to include the entire set A_0. Returning to the case $a_1 < u(q-2)+1$ we may of course also include A_0 in the zero-set, which then causes a minor change in how we may include the other departments. The many different ways one can define $f_1, \ldots, f_w$ given fixed $\gamma_1, \ldots, \gamma_w$ provides us with many different very systematic patterns of common roots, and similarly, of course, of the same number of different very systematic patterns of non-common roots, the latter being those participants if all left out, the remaining participants cannot detect $\ell - w + 1 \log_2(q)$ bits of information. This concludes the remark.

We illustrate the idea with a couple of examples where the first describes a situation where the set of common roots (and therefore also the set of non-common roots) are as simple as can possibly be.

Example 1. Let $q \geq 3$ be a prime-power, $s \geq 2$, and $u \geq 2$. Consider codes $C_2 = C_{\mathcal{L}}(G, \lambda_2 Q) \subseteq C_{\mathcal{L}}(G, \lambda_1 Q) = C_1 \subseteq \mathbb{F}_{q^s}^n$ of co-dimension $\ell = 1$ defined by $\iota(\lambda_1) = (\tau u, 0)$ with $\tau \in \{1, \ldots, q-2\}$. The conditions of Corollary 1 are clearly satisfied and by applying (2), (3), (14) and (13) we obtain

$$t = M_1(C_2^\perp, C_1^\perp) - 1 = (\tau - 1)uq^{s-1} + q^{s-1} + u - 1$$
$$r = n - M_1(C_1, C_2) + 1 = \tau u q^{s-1} + 1$$
$$r - t = (q^{s-1} - 1)(u - 1) + 1$$

In the following we use the language of Remark 1. Now for all $1 \leq i_1 < \cdots < i_\tau \leq q - 1$ the set $A_{i_1} \cup \cdots \cup A_{i_\tau}$ constitutes a maximum non-1-qualifying set. Hence, by leaving out all members of $(q - 1) - \tau$ large departments and the small department one does not obtain any information. In particular if $\tau = q - 2$ one cannot leave out an entire large department in combination with the small department if one wants to generate information. Next let $\iota(\lambda_1) = (\tau u + 1, 0)$ with $\tau \in \{1, \ldots, q-2\}$. We obtain

$$r = (\tau u + 1)q^{s-1} + 1$$

with the same value of $r - t$ as before. By leaving out all members of any set of $(q - 1) - \tau$ large departments one does not obtain any information. And by leaving out any $(q - 1) - \tau - 1$ large departments, the small department as well as $S \times \Gamma_i^{(2)}$ where A_i $i \in \{1, \ldots, q - 1\}$ is not one of the already left out large department and where $S \subseteq \Gamma_i^{(1)}$ is of size equal to $u - 1$ one neither obtains any information. The complementary sets are maximum non-1-qualifying. In particular if $\tau = q - 2$ one cannot leave out an entire large department and obtain any information. Similarly, one cannot leave out the small department and $S \times \Gamma_i^{(2)}$, $i \in \{1, \ldots, q - 1\}$ where S is given as above.

Example 2. Consider the Hermitian curve $x^{q+1} - y^q - y$ over $\mathbb{F}_{q^2}$. We have $u = q + 1$, $s = 2$, and $q^{s-1} = q$. Let $\lambda_2 + 1 = a_\ell q$ where $q - 1 \le a_\ell < u(q - 2) + 1 = (q+1)(q-2)+1$. Let $\ell = q$ and $\lambda_1 = \lambda_2 + \ell$. Then $\lambda_2 + 1, \ldots, \lambda_2 + \ell$ all belong to $H^*(Q)$. Moreover, we have $\iota(\lambda_2 + 1 + i) = (a_\ell - i, i)$ for $i = 0, \ldots, \ell - 1$. Hence, with $C_2 \subseteq C_1$ as in Corollary 1 all conditions therein are satisfied. Consider the related ramp secret sharing scheme. We have

$$t_1 = M_1(C_2^\perp, C_1^\perp) - 1 = (a_\ell + 1) + (q - 1)(a_\ell - q) - 1$$

$$t_\ell = M_\ell(C_2^\perp, C_1^\perp) - 1 = \left((a_\ell - (q - 1))q + \sum_{i=0}^{\ell-1} i\right) - 1 = (a_\ell - q + 1)q + \frac{(q - 1)q}{2} - 1$$

$$r_1 = n - M_\ell(C_1, C_2) + 1 = a_\ell q - \left(\sum_{i=1}^{\ell}(i - 1)\right) + 1 = a_\ell q - \frac{(q - 1)q}{2} + 1$$

$$r_\ell = n - M_1(C_1, C_2) + 1 = n - (n - \lambda_1) + 1 = (a_\ell + 1)q.$$

Using the approach described in Remark 1 we have maximum non-1-qualifying sets of the following form. Namely, the disjoint union of $\lfloor \frac{a_\ell - (q-1)}{q+1} \rfloor$ large departments, the following subset of a large department $S \times \Gamma_{i''} \subseteq A_{i''}$, with $\#S = a_\ell - (q - 1) - \lfloor (a_\ell - (q - 1))/(q + 1) \rfloor$ and finally a locally induced set (subset of $A_{i'}$) which has the form of a staircase each step having height equal to 1. We obtain maximum non-ℓ-qualifying sets as above, but with the latter locally induced set being replaced by $q - 1$ vertical levels each containing $q + 1$ elements. One, of course can also investigate non-i-qualifying sets with $1 < i < \ell = q$, but we shall refrain from that in the present exposition.

Example 3. In this example we consider $q = 4$, $s = 3$ and $u = \frac{(q^s - 1)}{(q - 1)3} = 7$. We obtain codes over $\mathbb{F}_{64}$ of length $n = (u(q - 1) + 1)q^{s-1} = 352$. We have $88, 89 \notin H^*(Q) \subseteq \langle 7, 16 \rangle$, but $87, 90, 91, 92 \in H^*(Q)$. We shall consider a pair of nested codes of co-dimension $\ell = 3$, namely, $C_2 \subseteq C_1$ where $C_2 = C_\mathcal{L}(G, 87Q)$ and $C_1 = C_\mathcal{L}(G, 92Q)$ for which we note that all conditions of Corollary 1 are satisfied. We have $\iota(90) = (3, 6)$, $\iota(91) = (0, 13)$, and $\iota(92) = (4, 4)$ Applying (14) we calculate

$$M_1(C_1, C_2) = \min\{262, 261, 260\} = 260 \tag{19}$$

$$M_2(C_1, C_2) = \min\{289, 276, 274\} = 274 \tag{20}$$

$$M_3(C_1, C_2) = 295, \tag{21}$$

where the three values on the right hand side of (19) each corresponds to a calculation concerning one of the values $90, 91, 92$ the minimum being attained for 92. The three values on the right hand sides of (20) each corresponds to a calculation concerning a pair of such values, the minimum being attained for $\{90, 92\}$. Finally, (21) corresponds to a calculation concerning all three numbers $90, 91, 92$ at the same time. To calculate the relative weights of the dual codes we apply (13) directly. We obtain

$$M_1(C_2^{\perp}, C_1^{\perp}) = \min\{14, 28, 25\} = 14 \tag{22}$$
$$M_2(C_2^{\perp}, C_1^{\perp}) = \min\{35, 34, 33\} = 33 \tag{23}$$
$$M_3(C_2^{\perp}, C_1^{\perp}) = 40. \tag{24}$$

From the above in combination with (2) and (3) we obtain $t_1 = 259$, $t_2 = 273$, $t_3 = 294$, $r_1 = 313$, $r_2 = 320$, and $r_3 = 339$. To detect maximum non-1-qualifying sets we should apply Remark 1 to $\{\iota(90)\iota(91), \iota(92)\}$. Here, $a_1 = 0$ and therefore all the sets we obtain consist of a locally induced set and nothing else. So leaving out the entire set of members from all departments but one large department, there are limits as to which patterns of members from the remaining department that could be left out. To detect maximum non-2-qualifying sets we should consider $\{\iota(90), \iota(92)\}$. Here, $a_1 = 3$. Hence, we obtain sets consisting of a Cartesian product $S \times \Gamma_{i''}^{(2)}$, $\#S = 3$ in combination with a locally induced set. Finally, to detect maximum non-1-qualifying sets we should consider $\{\iota(92)\}$. Here, $a_1 = 4$, and we obtain sets that are the union of a Cartesian product $S \times \Gamma_{i'''}^{(2)}$, $\#S = 4$ and a locally induced set.

Remark 2. Revisiting Theorem 2 and the conditions therein keep the assumption $\gamma_1 < \cdots < \gamma_w$ in $H^*(Q)$ again with $\iota(\{\gamma_1, \ldots, \gamma_w\}) = \{(a_1, b_1), \ldots, (a_w, b_w)\}$ and with the condition that $a_1 < \cdots < a_w$ and $b_w < \cdots < b_1$. But instead of requiring $a_1 \le u(q - 2) + 1$ assume now $u(q - 2) + 1 < a_1$. Inspecting (10) we see that the last option does not come into action as $u(q - 1) + 1 < a_1 + u$ which does not correspond to the first coordinate of any $\iota(\delta)$ where $\delta \in H^*(Q)$. Hence, the right hand side of (10) becomes

$$\#\{(i, j) \mid 0 \le i < u(q - 1) + 1, 0 \le j \le q^{s-1} - 1, (a_s, b_s) <_p (i, j)$$
$$\text{for some } s \in \{1, \ldots, w\}\}.$$

But then the situation is similar to that of monomial-Cartesian codes regarding parameters of primary codes (see [11, Eq. (5)]) and the maximum non-i-qualifying sets have a similar structure, however with a little less freedom of choice (see [11, Thm. 6]). The advantage of using codes from extended norm-trace curves for such high value of a_1 does not lie in the reconstruction numbers, nor the structure of maximum non-i-qualifying sets. Rather, it is the privacy numbers that are improved.

We illustrate Remark 2 with a final example.

Example 4. We first consider nested one-point algebraic geometric codes over $\mathbb{F}_{16}$ using the Hermitian curve $X^5 - Y^4 - Y$. We have $\iota(66) = (14, 2)$ and $\iota(67) = (13, 3)$ both 66 and 67 belonging to $H^*(Q)$ as

$$\iota(H^*(Q)) = \{(i, j) \mid 0 \le i < 16, 0 \le j < 4\}.$$

Hence, the co-dimension of $C_2 = C_{\mathcal{L}}(G, 65Q) \subseteq C_{\mathcal{L}}(G, 67Q) = C_1$ is $\ell = 2$. We have

$$M_2(C_1, C_2) = \#\{(14, 2), (13, 3), (14, 3), (15, 2), (15, 3)\} = 5$$
$$M_1(C_1, C_2) = \min\{\#\{((13, 3), (14, 3), (15, 3)\}, \#\{(14, 2), 14, 3), (15, 2), (15, 3)\}\}$$
$$= 3$$

We have maximum non-1-qualifying sets of the form: all 64 points in the entire affine variety except

$$\{(\alpha_1', \beta_1), (\alpha_2', \beta_1), (\alpha_1', \beta_2), (\alpha_2', \beta_2), (\alpha_3', \beta_2)\}$$

where we use the notation from Theorem 2 with the arbitrary enumeration from there. We have maximum non-2-qualifying sets of the form: all 64 points except

$$\{(\alpha_1', \beta_1), (\alpha_2', \beta_1), (\alpha_3', \beta_1)\}.$$

Turning to comparable monomial-Cartesian codes we consider as point set (affine variety) now $\mathbb{F}_{16} \times S_2 = \{P_1', \ldots, P_{64}'\}$ i.e. $\#S_2 = 4$. We write $\mathbb{F}_{16} = \{\delta_1, \ldots, \delta_{16}\}$ and $S_2 = \{\epsilon_1, \ldots, \epsilon_4\}$ and consider an evaluation map ev $: \mathbb{F}_{16}[X, Y] \to \mathbb{F}_{16}^{64}$ given by ev$(F) = (F(P_1'), \ldots, F(P_{64}'))$. Define

$$C_2' = \mathrm{Span}_{\mathbb{F}_{16}}\{\mathrm{ev}(X^i Y^j) \mid 0 \leq i < 16, 0 \leq j < 4, 4i + 5j \leq 65\} \subseteq$$
$$C_1' = \mathrm{Span}_{\mathbb{F}_{16}}\{\mathrm{ev}(X^i Y^j) \mid 0 \leq i < 16, 0 \leq j < 4, 4i + 5j \leq 67\}.$$

The relative parameters are the same as before (see [11, Eq. (5)]), but now we have that the entire affine variety (still of size 64) minus

$$\{(\delta_1, \epsilon_1), (\delta_2, \epsilon_1), (\delta_1, \epsilon_2), (\delta_2, \epsilon_2), (\delta_3, \epsilon_2)\}$$

is a maximum non-1-qualifying set (see [11, Thm. 6]). Similarly, the entire affine variety minus

$$\{(\delta_1, \epsilon_1), (\delta_2, \epsilon_1), (\delta_3, \epsilon_1)\}$$

is a maximum non-2-qualifying set. Due to the many ways one can enumerate the elements of $\mathbb{F}_{16}$ and similarly enumerate the elements of S_2 the second layer of security becomes much stronger for the secret sharing scheme defined from the latter pair of codes.

In this paper we only considered the case of not too large co-dimension of two one-point algebraic geometric codes $C_2 \subseteq C_1$ imposing a situation where $a_1 < \cdots < a_w$ and $b_w < \cdots < b_1$. It is also possible to consider higher co-dimension for which we refer the reader to employ the last part of [4, Prf. of Thm. 3.2].

6   Concluding Remarks

The second layer of security resulting from families of maximum non-i-qualifying sets of systematic form was first considered in [11] in the case of monomial-Cartesian codes. In the present work it was analyzed for schemes defined from nested one-point algebraic geometric codes defined over extended norm-trace curves. There should be other algebraic codes for which the resulting schemes have a second layer of security. In this paper we demonstrated that the "footprint-like approach" applied in [4] is merely a clever application of the enhanced Goppa bound in [12] than a competing method. A similar remark could be made concerning [10] which, however, predates [12]. In the opinion of the author of this contribution, indeed the "footprint-like approach'n' may be beneficial, but merely in connection with affine variety codes not defined from $\cup_{m=0}^{\infty}\mathcal{L}(mQ)$ where Q is a rational place in an algebraic function field of transcendence degree 1. The literature contains several examples of such studies including codes defined from other structures related to function fields of transcendence degree 1 than the above union of $\mathcal{L}$-spaces, e.g. [13].

The author would like to express his gratitude to the reviewers for their careful reading of the paper and for their thoughtful feedback.

Acknowledgement. The author would like to express his gratitude to the reviewers for their careful reading of the paper and for their thoughtful feedback.

References

1. Barbero, A.I., Munuera, C.: The weight hierarchy of Hermitian codes. SIAM J. Discrete Math, **13**(1), 79–104 (electronic) (2000)
2. Blakley, G.R., Meadows, C.: Security of ramp schemes. In: Advances in cryptology (Santa Barbara, Calif., 1984), Lecture Notes in Computer Science, vol. 196, pp. 242–268. Springer, Berlin (1985)
3. Bras-Amorós, M., O'Sullivan, M.E.: Duality for some families of correction capability optimized evaluation codes. Adv. Math. Commun. **2**(1), 15–33 (2008)
4. Camps-Moreno, E., López, H.H., Matthews, G.L., San-José, R.: The weight hierarchy of decreasing norm-trace codes. Des. Codes Cryptogr. **93**(7), 2873–2894 (2025)
5. Carvalho, C., López, H.H., Matthews, G.L.: Decreasing norm-trace codes. Des. Codes Cryptogr. **92**(5), 1143–1161 (2024)
6. Cascudo, I., Cramer, R., Xing, C.: Bounds on the threshold gap in secret sharing and its applications. IEEE Trans. Inform. Theory **59**(9), 5600–5612 (2013)
7. Chen, H., Cramer, R., Goldwasser, S., de Haan, R., Vaikuntanathan, V.: Secure Computation from Random Error Correcting Codes. In: Naor, M. (ed.) EUROCRYPT 2007. LNCS, vol. 4515, pp. 291–310. Springer, Heidelberg (2007). https://doi.org/10.1007/978-3-540-72540-4_17
8. Csirmaz, L.: Ramp secret sharing and secure information storage, Preprint (2009).
9. Galindo, C., Geil, O., Hernando, F., Ruano, D.: Improved constructions of nested code pairs. IEEE Trans. Inform. Theory **64**(4), 2444–2459 (2018)
10. Geil, O.: On codes from norm-trace curves. Finite Fields Appl. **9**(3), 351–371 (2003)
11. Geil, O.: Considerate ramp secret sharing. Des. Codes Cryptogr. **94**(3), 49 (2026)
12. Geil, O., Martin, S., Matsumoto, R., Ruano, D., Luo, Y.: Relative generalized Hamming weights of one-point algebraic geometric codes. IEEE Trans. Inform. Theory **60**(10), 5938–5949 (2014)
13. Geil, O., Özbudak, F.: On affine variety codes from the Klein quartic. Cryptogr. Commun. **11**(2), 237–257 (2019)

14. Høholdt, T., van Lint, J.H., Pellikaan, R.: Algebraic geometry codes. In: Pless, V.S., Huffman, W.C. (eds.) Handbook of Coding Theory, vol. 1, pp. 871–961. Elsevier, Amsterdam (1998)
15. Janwa, H., Piñero, F.L.: On parameters of subfield subcodes of extended norm-trace codes (2016). arXiv preprint arXiv:1604.05777
16. Kurihara, J., Uyematsu, T., Matsumoto, R.: Secret sharing schemes based on linear codes can be precisely characterized by the relative generalized Hamming weight. IEICE Trans. Fundam. **E95-A** (11), 2067–2075 (2012)
17. Matthews, G.L., Murphy, A.W.: Norm-trace-lifted codes over binary fields. In: Proceedings of 2022 IEEE International Symposium on Information Theory (ISIT), pp. 3079–3084. IEEE (2022)
18. Patel, S., Persiano, G., Seo, J.Y., Yeo, K.: Efficient secret sharing for large-scale applications. In: Proceedings of the 2024 ACM SIGSAC Conference on Computer and Communications Security, pp. 3065–3079 (2024)
19. Pellikaan, R.: On the existence of order functions. J. Statist. Plann. Inference **94**(2), 287–301 (2001)
20. Stichtenoth, H.: A note on hermitian codes over $GF(q^2)$. IEEE Trans. Inform. Theory **34**(5), 1345–1348 (1988)
21. Yamamoto, H.: Secret sharing system using (k, L, n) threshold scheme. Electron. Comm. Japan Part I Comm. **69**(9), 46–54 (1986)
22. Yang, K., Kumar, P.V.: On the true minimum distance of Hermitian codes. In: Coding theory and algebraic geometry, pp. 99–107. Springer (1992)

Self-orthogonal Codes from p-ary Quadratic Forms via Character Sums with Applications to LCD and Arbitrary Hull Dimensional Codes

Virginio Fratianni[1(✉)] and Sihem Mesnager[1,2]

[1] Université Paris 8, Laboratoire Analyse, Géométrie et Applications, LAGA, Université Sorbonne Paris Nord, CNRS, UMR 7539, 93430 Villetaneuse, France
virginio.fratianni@etud.univ-paris8.fr, smesnager@univ-paris8.fr
[2] Telecom Paris, Polytechnic Institute of Paris, Palaiseau 91120, France

Abstract. Self-orthogonal codes hold significant importance in coding theory and cryptography, particularly due to their diverse applications in lattices and quantum error correction. Recently, researchers have constructed new families of self-orthogonal codes from specific classes of functions, largely driven by a novel self-orthogonality criterion in odd characteristic based on p-divisibility, introduced by Li and Heng in 2024 [18]. In this paper, following the latter approach, we explore linear codes derived from a quadratic function in the second generic construction framework. By explicitly evaluating the weight enumerators via exponential sums, we show that the resulting codes have four weights, are p-divisible, and, consequently, are self-orthogonal. Furthermore, we apply these results to construct classes of Linear Complementary Dual (LCD) codes, as well as linear codes with a fixed Hull dimension. We focus on low-dimensional Hull codes, as they play a pivotal role in determining the automorphism group of a linear code, efficiently checking permutational equivalence between codes, and constructing entanglement-assisted quantum error-correcting codes (EAQECCs).

Keywords: Linear codes · q-ary functions · LCD codes · Hull codes

1 Introduction

Coding theory is a highly interdisciplinary field at the intersection of information theory, computer science, and mathematics, dedicated to analyzing the properties of codes for applications such as data compression, secure cryptography, and wireless transmission. Despite its broad connections to various applied domains, the discipline is fundamentally anchored in rigorous algebraic frameworks. This is particularly true for linear codes, where code construction and analysis rely heavily on advanced mathematical tools, including Galois fields, cyclotomic field

L. Batina and F. Özbudak (Eds.): WAIFI 2026, LNCS 16611, pp. 145–162, 2026.
https://doi.org/10.1007/978-3-032-27574-5_10

theory, Gauss and exponential sums, and Fourier transforms. Today, as modern infrastructures—from cloud storage networks to wireless communications—depend entirely on robust coding schemes, the continuous mathematical exploration of codes remains of paramount importance.

A linear code is said to be *self-orthogonal* if it is contained in its dual code under a specific inner product. The construction and study of self-orthogonal codes have recently become an attractive research topic in coding theory, particularly due to the diverse applications in lattices and quantum error correction. To better classify the relationship between a code and its dual, we consider the Hull of a linear code, defined as the intersection of the code and its dual. A linear code with a trivial (zero-dimensional) Hull is called a *Linear Complementary Dual* (LCD) code. They were initially introduced by Massey in 1992 [20]. Later, Carlet and Guilley [4] demonstrated the critical importance of these codes in cryptographic applications, particularly in mitigating side-channel attacks (SCA) and fault injection attacks (FIA). Consequently, the literature on LCD codes has expanded significantly since 2015.

Hulls of linear codes have been introduced to classify finite projective planes in [1]. Then, it was shown that the Hull plays a pivotal role in determining the complexity of certain algorithms for checking permutational equivalence between two linear codes and for computing the automorphism group of a linear code [26,27]. Furthermore, in quantum coding theory, codes with an explicitly known, low-dimensional Hull are fundamental for constructing Entanglement-Assisted Quantum Error-Correcting Codes (EAQECCs), as the Hull dimension strictly dictates the number of required maximally entangled states [12].

The construction of linear codes from functions over finite fields constitutes a profoundly rich and rapidly expanding branch of coding theory [22]. The core philosophy of this approach relies on evaluating highly non-linear cryptographic functions such as bent, semi-bent, and weakly regular plateaued functions over specific defining sets to generate the codewords (for instance, see [6,11,23,30] and the references therein). The resulting linear codes often exhibit highly favorable algebraic characteristics. This methodology, particularly through the frameworks of the first and second generic constructions [9], has yielded a vast body of literature. Over the last two decades, researchers have successfully utilised this intersection of cryptography and coding theory to produce infinite families of optimal or near-optimal codes, including few-weight and minimal codes, which are well-suited for modern applications such as secret sharing schemes and secure multi-party computation. In 2024, Li and Heng introduced a novel self-orthogonality criterion in odd characteristic for linear codes, based on p-divisibility [18]. This development has sparked renewed interest in constructing self-orthogonal codes from cryptographic functions, largely driven by their significant applications in LCD and quantum codes (see, for instance, [3,14,24,32]).

This paper aligns with the active line of research described above. In particular, we propose an alternative approach to that of Chen et al. [5]. Within the framework of the second generic construction, we focus on the specific function $\mathrm{Tr}_{q/p}(x^2 + x)$, where $q = p^m$ and p is an odd prime. Instead of relying on its

Walsh spectrum, we compute the code parameters, most notably the complete weight enumerator, directly using character theory. This computation is facilitated by two novel technical lemmas on exponential sums. Consequently, we obtain a family of four self-orthogonal codes with four weights. We also apply these codes to construct Linear Complementary Dual (LCD) codes and we leverage these self-orthogonal codes to systematically build a family of linear codes with a fixed Hull dimension, giving special attention to the one-dimensional case.

The remainder of this paper is organized as follows. Section 2 introduces the preliminary concepts regarding coding theory, functions over finite fields, and character theory. In particular, it presents the two results on exponential sums. Section 3 details the construction of the self-orthogonal code, explicitly calculating its parameters and weight distribution. Finally, Sect. 3 also outlines the derivations of the associated LCD codes and ℓ-dimensional Hull codes.

2   Preliminaries

Throughout this paper, unless otherwise stated, we will assume $\mathbb{F}_q$ to be the finite field of q elements, where $q = p^m$ for a prime p.

This section presents some preliminary results in coding theory, functions over finite fields and character theory.

2.1   Coding Theory

An $[n, k, d]$ linear code $\mathcal{C}$ over $\mathbb{F}_q$ is a linear subspace of $\mathbb{F}_q^n$ with dimension k and minimum (Hamming) distance d. The value $n - k$ is the codimension of $\mathcal{C}$. Given a linear code $\mathcal{C}$ of length n over $\mathbb{F}_q$, its Euclidean dual code is denoted by $\mathcal{C}^\perp$. The code $\mathcal{C}^\perp$ is defined by

$$\mathcal{C}^\perp = \left\{ (b_0, b_1, \ldots, b_{n-1}) \in \mathbb{F}_q^n : \sum_{i=0}^{n-1} b_i c_i = 0 \text{ for every } (c_0, c_1, \ldots, c_{n-1}) \in \mathcal{C} \right\}.$$

Definition 1 (Hull of a code). *The* Hull *of a linear code $\mathcal{C}$ is defined as the intersection of the code and its dual, that is:*

$$Hull(\mathcal{C}) = \mathcal{C} \cap \mathcal{C}^\perp.$$

A linear code $\mathcal{C}$ is said to be *self-orthogonal* if $\mathcal{C} \subseteq \mathcal{C}^\perp$, which is equivalent to stating that $Hull(\mathcal{C}) = \mathcal{C}$. On the opposite end of the spectrum, we define Linear Complementary Dual codes.

Definition 2 (LCD code). *A linear code $\mathcal{C}$ is said to be a* Linear Complementary Dual *(LCD) code if its Hull is trivial, meaning:*

$$Hull(\mathcal{C}) = \{0\}.$$

The following Theorem provides a fundamental algebraic characterization of LCD codes.

Theorem 1 ([20]). *Let C be an $[n, k, d]$ linear code over $\mathbb{F}_q$ and let G be its generator matrix. Then C is an LCD code if and only if the matrix GG^T is nonsingular. Moreover, if C is an LCD code, then $\pi_C = G^T(GG^T)^{-1}G$ is the orthogonal projector from $\mathbb{F}_q^n$ onto C.*

Similarly, we have an algebraic characterization for self-orthogonal codes based on the same Gram matrix.

Theorem 2 ([15]). *Let C be an $[n, k, d]$ linear code over $\mathbb{F}_q$ and let G be its generator matrix. Then C is self-orthogonal if and only if $GG^T = 0$.*

In the binary case, there is a well-known sufficient condition for a code to be self-orthogonal that depends on the weights of its codewords.

Theorem 3 ([15]). *Let C be an $[n, k, d]$ linear code over $\mathbb{F}_2$. Then C is self-orthogonal if every codeword of C has a weight divisible by 4.*

In 2024, Li and Heng introduced the following self-orthogonality criterion in odd characteristic based on p-divisibility.

Theorem 4 ([18]). *Let $q = p^m$, where p is an odd prime. Let C be an $[n, k, d]$ linear code over $\mathbb{F}_q$ with $\mathbf{1} \in C$, where $\mathbf{1}$ is the all-1 vector of length n. If C is p-divisible, then C is self-orthogonal.*

In our computations, we will need the *Pless Power Moments* (see [15]), which provide a set of identities relating the moments of the weight distribution of a linear code to those of its dual. Let $(A_0, A_1, \ldots, A_n)$ be the weight distribution of the code C, where A_i is the number of codewords of weight i. Similarly, let $(B_0, B_1, \ldots, B_n)$ be the weight distribution of the dual code $C^\perp$. The first four identities are explicitly given by:

$$\sum_{i=0}^{n} A_i = q^k,$$

$$\sum_{i=0}^{n} i A_i = q^{k-1}[n(q-1) - B_1],$$

$$\sum_{i=0}^{n} i^2 A_i = q^{k-2}\left[n(q-1)(n(q-1)+1) - (2(n-1)(q-1)+q)B_1 + 2B_2\right],$$

$$\sum_{i=0}^{n} i^3 A_i = q^{k-3}\Big[n(q-1)\big(n^2(q-1)^2 + 3n(q-1) + 1\big)$$
$$- \Big(3(n-1)(q-1)\big(n(q-1)+1\big) + q(3n(q-1)+1)\Big)B_1$$
$$+ \big(6(n-2)(q-1) + 6q\big)B_2 - 6B_3\Big].$$

Finally, we recall the standard definition of code equivalence.

Definition 3 (Monomially Equivalent Codes). *Two linear codes C and C' of length n over $\mathbb{F}_q$ are said to be* monomially equivalent *if there exists a monomial matrix M (a matrix with exactly one non-zero entry in each row and column) such that $C' = \{cM \mid c \in C\}$. This is equivalent to saying that C' is obtained from C by a permutation of coordinates followed by scaling each coordinate by a non-zero element of $\mathbb{F}_q$.*

2.2   Functions over Finite Fields

Again, let $q = p^m$ for a prime p. The absolute trace function $\mathrm{Tr}_{q/p} : \mathbb{F}_q \to \mathbb{F}_p$ is defined as

$$\mathrm{Tr}_{q/p}(x) := \sum_{i=0}^{m-1} x^{p^i} = x + x^p + x^{p^2} + \cdots + x^{p^{m-1}}.$$

Definition 4 (Functions over finite fields). *Let $q = p^m$ where p is a prime. A vectorial function $\mathbb{F}_{q^n} \to \mathbb{F}_{q^r}$ is called an (n,r)-q-ary function. When $q = 2$, an (n,r)-2-ary function will be simply denoted an (n,r)-function. A* Boolean function *is an $(n,1)$-function, that is a function $\mathbb{F}_{2^n} \to \mathbb{F}_2$.*

Let $f : \mathbb{F}_q \to \mathbb{F}_p$ be a p-ary function, then the Walsh transform of f is given by:

$$\hat{\chi}_f(\lambda) = \sum_{x \in \mathbb{F}_q} \zeta_p^{f(x) - \mathrm{Tr}_{q/p}(\lambda x)},$$

for $\lambda \in \mathbb{F}_q$, where $\zeta = e^{\frac{2\pi i}{p}}$ is a complex primitive p-th root of the unity.

A p-ary function $f : \mathbb{F}_q \longrightarrow \mathbb{F}_p$ is called bent if all its Walsh-Hadamard coefficients satisfy $|\hat{\chi}_f(b)|^2 = p^m$. A bent function f is called regular bent if, for every $b \in \mathbb{F}_{p^m}$,

$$p^{-\frac{m}{2}} \hat{\chi}_f(b) = \zeta_p^{f^*(b)}$$

for some p-ary function $f^* : \mathbb{F}_q \to \mathbb{F}_p$. The bent function f is called weakly regular bent if there exists a complex number u with $|u| = 1$ and a p-ary function f^* such that

$$u p^{-\frac{m}{p}} \hat{\chi}_f(b) = \zeta_p^{f^*(b)}$$

for all $b \in \mathbb{F}_{p^m}$. Such a function $f^*(x)$ is called the *dual* of $f(x)$.

2.3   Character Theory

To determine the parameters of the code, and in particular the weight enumerator, we need the following technical lemmas on exponential sums.

Let $\overline{\chi}$ and χ be the canonical additive characters of $\mathbb{F}_p$ and $\mathbb{F}_q$ respectively, and let $\overline{\eta}$ and η be the canonical quadratic multiplicative characters of $\mathbb{F}_p$ and $\mathbb{F}_q$ respectively. The *Gauss sums* $G(\eta, \chi)$ and $G(\overline{\eta}, \overline{\chi})$ are defined by

$$G(\eta, \chi) = \sum_{c \in \mathbb{F}_q^*} \eta(c)\chi(c) \quad \text{and} \quad G(\overline{\eta}, \overline{\chi}) = \sum_{c \in \mathbb{F}_p^*} \overline{\eta}(c)\overline{\chi}(c).$$

Lemma 1 ([19], Theorem 5.15). *Let the symbols be the same as before.*
Then

$$G(\eta, \chi) = (-1)^{(m-1)} \sqrt{-1}^{\frac{(p-1)^2 m}{4}} \sqrt{q}, \quad G(\bar{\eta}, \bar{\chi}) = \sqrt{-1}^{\frac{(p-1)^2}{4}} \sqrt{p}.$$

Lemma 2 ([10], Lemma 7). *Let the symbols be the same as before. Then*

1. *if $m \geq 2$ is even, then $\eta(y) = 1$ for each $y \in \mathbb{F}_p^*$;*
2. *if m is odd, then $\eta(y) = \bar{\eta}(y)$ for each $y \in \mathbb{F}_p^*$.*

Lemma 3 ([19], Theorem 5.33). *Let χ be a nontrivial additive character of $\mathbb{F}_q$, and let $f(x) = a_2 x^2 + a_1 x + a_0 \in \mathbb{F}_q[x]$ with $a_2 \neq 0$. Then,*

$$\sum_{x \in \mathbb{P}_4} \chi(f(x)) = \chi\left(a_0 - a_1^2 (4a_2)^{-1}\right) \eta(a_2) G(\eta, \chi)$$

We will also need to evaluate the following exponential sums.

Lemma 4 ([5]). *Let $q = p^m$, $b \in \mathbb{F}_q$, $c \in \mathbb{F}_p$ and $\zeta_p = e^{\frac{2\pi i}{p}}$. Then:*

$$\sum_{x \in \mathbb{F}_q} \sum_{y \in \mathbb{F}_p^*} \zeta_p^{y(\mathrm{Tr}_{q/p}(bx)+c)} = \begin{cases} 0 & \text{if } b \neq 0; \\ q(p-1) & \text{if } b = 0 \text{ and } c = 0; \\ -q & \text{if } b = 0 \text{ and } c \neq 0. \end{cases}$$

Lemma 5 ([17]). *Let the symbols be the same as before and $G = G(\eta, \chi)$. We have*

$$\sum_{y \in \mathbb{F}_p^*} \sum_{x \in \mathbb{F}_q} \zeta_p^{y \, \mathrm{Tr}_{q/p}(x^2+x)} = \begin{cases} (p-1)G, & \text{if } 2 \mid m \text{ and } p \mid m, \\ -G, & \text{if } 2 \mid m \text{ and } p \nmid m, \\ 0, & \text{if } 2 \nmid m \text{ and } p \mid m, \\ \eta(-m)GG, & \text{if } 2 \nmid m \text{ and } p \nmid m. \end{cases}$$

In addition, we prove the following technical lemma that will be used in the computation of the weight enumerator.

Lemma 6. *Let $q = p^m$ where $m \geq 2$ is even and p is an odd prime such that $p \mid m$. Let χ be the canonical additive character of $\mathbb{F}_q$, η the canonical quadratic multiplicative character of $\mathbb{F}_q$, and $G = G(\eta, \chi)$. Then, for any $b \in \mathbb{F}_q$ and $c \in \mathbb{F}_p$:*

$$\sum_{x \in \mathbb{F}_q} \sum_{y \in \mathbb{F}_p^*} \sum_{z \in \mathbb{F}_p^*} \zeta_p^{\mathrm{Tr}_{q/p}(yx^2+yx+zbx)+zc} =$$

$$= \begin{cases} (p-1)^2 G & \text{if } \mathrm{Tr}_{q/p}(b^2) = 0 \text{ and } 2c = \mathrm{Tr}_{q/p}(b); \\ (1-p)G & \text{if } \mathrm{Tr}_{q/p}(b^2) = 0 \text{ and } 2c \neq \mathrm{Tr}_{q/p}(b); \\ (1-p)G & \text{if } \mathrm{Tr}_{q/p}(b^2) \neq 0 \text{ and } 2c = \mathrm{Tr}_{q/p}(b); \\ G & \text{if } \mathrm{Tr}_{q/p}(b^2) \neq 0 \text{ and } 2c \neq \mathrm{Tr}_{q/p}(b). \end{cases}$$

Proof. Let S denote the sum to be evaluated. By expanding the trace map and grouping the terms involving x, we have:

$$S = \sum_{y \in \mathbb{F}_p^*} \sum_{z \in \mathbb{F}_p^*} \zeta_p^{zc} \sum_{x \in \mathbb{F}_q} \chi(yx^2 + (y + zb)x)$$

The innermost sum over $x \in \mathbb{F}_q$ is a standard quadratic Gauss sum. For $A, B \in \mathbb{F}_q$ with $A \neq 0$, we recall the identity given in Lemma 3:

$$\sum_{x \in \mathbb{F}_q} \chi(Ax^2 + Bx) = \chi(-\frac{B^2}{4A})\eta(A)G(\eta, \chi).$$

Here, $A = y \in \mathbb{F}_p^*$ and $B = y + zb$. Since m is even, every element of the subfield $\mathbb{F}_p^*$ is a quadratic residue in $\mathbb{F}_q$, which implies $\eta(y) = 1$ for all $y \in \mathbb{F}_p^*$. Evaluating the argument of the additive character:

$$-\frac{B^2}{4A} = -\frac{(y + zb)^2}{4y} = -\frac{y}{4} - \frac{zb}{2} - \frac{z^2b^2}{4y}$$

Applying the trace map $\mathrm{Tr}_{q/p}$ to this expression, we note that $\mathrm{Tr}_{q/p}(-\frac{y}{4}) = -\frac{y}{4}\mathrm{Tr}_{q/p}(1) = -\frac{y}{4}m$. Since $p \mid m$, it follows that $m \equiv 0 \pmod{p}$, and thus $\mathrm{Tr}_{q/p}(-\frac{y}{4}) = 0$. The exponent simplifies to:

$$\mathrm{Tr}_{q/p}\left(-\frac{B^2}{4A}\right) = -\frac{z}{2}\,\mathrm{Tr}_{q/p}(b) - \frac{z^2}{4y}\,\mathrm{Tr}_{q/p}(b^2)$$

Substituting this back into S and letting $G = G(\eta, \chi)$ (noting that the assumption $p \mid m$ implies $\chi(-4^{-1}) = 1$), we separate the variables y and z:

$$S = G \sum_{z \in \mathbb{F}_p^*} \zeta_p^{z\left(c - \frac{\mathrm{Tr}_{q/p}(b)}{2}\right)} \left(\sum_{y \in \mathbb{F}_p^*} \zeta_p^{-\frac{z^2\,\mathrm{Tr}_{q/p}(b^2)}{4y}} \right)$$

We now evaluate the inner sum over y. As y ranges over $\mathbb{F}_p^*$, its inverse y^{-1} also ranges over all non-zero elements of $\mathbb{F}_p$. This reduces to the sum of a linear additive character over $\mathbb{F}_p^*$. If $\mathrm{Tr}_{q/p}(b^2) = 0$, the exponent is 0 for all y, and the sum evaluates to $p - 1$. If $\mathrm{Tr}_{q/p}(b^2) \neq 0$, the exponent covers all non-zero values of $\mathbb{F}_p$ (since $z \neq 0$), and the sum of the non-trivial character over $\mathbb{F}_p^*$ evaluates to -1.

Finally, we evaluate the outer sum over z, which depends on the coefficient $c - \frac{1}{2}\mathrm{Tr}_{q/p}(b)$. If $2c = \mathrm{Tr}_{q/p}(b)$, the exponent is 0, yielding a factor of $p - 1$. If $2c \neq \mathrm{Tr}_{q/p}(b)$, the sum evaluates to -1. Combining these conditions yields the four discrete cases stated in the lemma.

We also present a modified version that could be useful for other code constructions.

Lemma 7. *Let $m \geq 2$ be even, $\overline{\chi}$ the canonical additive character of $\mathbb{F}_p$, χ the canonical additive character of $\mathbb{F}_q$, η the canonical quadratic multiplicative character of $\mathbb{F}_q$, $\overline{\eta}$ the canonical quadratic multiplicative character of $\mathbb{F}_p$, and let $K = \chi(-4^{-1})G(\eta,\chi)$. Then, for $a,b,c \in \mathbb{F}_q$,*

$$\sum_{x\in\mathbb{F}_q} \sum_{y\in\mathbb{F}_p^*} \sum_{z\in\mathbb{F}_p^*} \zeta_p^{\mathrm{Tr}_{q/p}(y^2x^2+yx+zbx)+zc} =$$

$$= \begin{cases} (1-p)K & \text{if } \mathrm{Tr}_{q/p}(b^2)=0,\ \mathrm{Tr}_{q/p}(b)\neq 0 \text{ and } c=0; \\ K & \text{if } \mathrm{Tr}_{q/p}(b^2)=0,\ \mathrm{Tr}_{q/p}(b)\neq 0 \text{ and } c\neq 0; \\ (p-1)^2K & \text{if } \mathrm{Tr}_{q/p}(b^2)=0,\ \mathrm{Tr}_{q/p}(b)=0 \text{ and } c=0; \\ (1-p)K & \text{if } \mathrm{Tr}_{q/p}(b^2)=0,\ \mathrm{Tr}_{q/p}(b)=0 \text{ and } c\neq 0; \\ (p-1)K(W-1) & \text{if } \mathrm{Tr}_{q/p}(b^2)\neq 0 \text{ and } c=0; \\ -K(W-1) & \text{if } \mathrm{Tr}_{q/p}(b^2)\neq 0 \text{ and } c\neq 0, \end{cases}$$

where $W = \overline{\chi}\left(\frac{\mathrm{Tr}_{q/p}(b)^2}{4\,\mathrm{Tr}_{q/p}(b^2)}\right)\overline{\eta}(-\,\mathrm{Tr}_{q/p}(b^2))G(\overline{\eta},\overline{\chi})$.

Proof. From the direct calculation we obtain

$$\sum_{x\in\mathbb{F}_q} \sum_{y\in\mathbb{F}_p^*} \sum_{z\in\mathbb{F}_p^*} \zeta_p^{\mathrm{Tr}_{q/p}(y^2x^2+yx+zbx)+zc}$$

$$= \sum_{y\in\mathbb{F}_p^*} \sum_{z\in\mathbb{F}_p^*} \zeta_p^{zc} \sum_{x\in\mathbb{F}_q} \zeta_p^{\mathrm{Tr}_{q/p}(y^2x^2+x(y+zb))}$$

Thanks to the well known technical Lemma 3, we obtain

$$\sum_{y\in\mathbb{F}_p^*} \sum_{z\in\mathbb{F}_p^*} \zeta_p^{zc}\chi(-(y+zb)^2(4y^2)^{-1})\eta(y^2)G(\eta,\chi)$$

Since m is even, from Lemma 2 we get that $\eta(y^2)=1$, so we derive

$$G(\eta,\chi) \sum_{y\in\mathbb{F}_p^*} \sum_{z\in\mathbb{F}_p^*} \zeta_p^{zc}\chi(-(y+zb)^2(4y^2)^{-1})$$

Since $y \mapsto \frac{1}{2y}$ defines a bijection of $\mathbb{F}_p^*$, we deduce, by this change of variable, that

$$G(\eta,\chi)\chi(-4^{-1}) \sum_{z\in\mathbb{F}_p^*} \zeta_p^{zc} \sum_{y\in\mathbb{F}_p^*} \chi(-zby-(zby)^2)$$

$$= G(\eta,\chi)\chi(-4^{-1}) \sum_{z\in\mathbb{F}_p^*} \zeta_p^{zc} \sum_{y\in\mathbb{F}_p^*} \chi(-zy(b+zyb^2))$$

$$= G(\eta,\chi)\chi(-4^{-1}) \sum_{z\in\mathbb{F}_p^*} \zeta_p^{zc} \sum_{y\in\mathbb{F}_p^*} \chi(-zy\,\mathrm{Tr}_{q/p}(b)-z^2y^2\,\mathrm{Tr}_{q/p}(b^2)).$$

Let $K = G(\eta,\chi)\chi(-4^{-1})$. Since $\mathrm{Tr}_{q/p}(b)$ and $\mathrm{Tr}_{q/p}(b^2)$ belong to $\mathbb{F}_p$, we can rewrite the inner character using the canonical additive character $\overline{\chi}$ of $\mathbb{F}_p$. By applying the substitution $w = zy$, as y runs through $\mathbb{F}_p^*$, w also runs through $\mathbb{F}_p^*$ for any fixed $z \in \mathbb{F}_p^*$. Thus, the sums over z and w separate:

$$K \left(\sum_{z \in \mathbb{F}_p^*} \zeta_p^{zc} \right) \left(\sum_{w \in \mathbb{F}_p^*} \overline{\chi}(-w \operatorname{Tr}_{q/p}(b) - w^2 \operatorname{Tr}_{q/p}(b^2)) \right)$$

The first sum evaluates to $p - 1$ if $c = 0$, and to -1 if $c \neq 0$. For the second sum, let $I = \sum_{w \in \mathbb{F}_p^*} \overline{\chi}(-w \operatorname{Tr}_{q/p}(b) - w^2 \operatorname{Tr}_{q/p}(b^2))$. We can evaluate it over the whole $\mathbb{F}_p$ by adding and subtracting the term for $w = 0$:

$$I = \left(\sum_{w \in \mathbb{F}_p} \overline{\chi}(-w \operatorname{Tr}_{q/p}(b) - w^2 \operatorname{Tr}_{q/p}(b^2)) \right) - 1$$

We can now evaluate I in three distinct cases:

- If $\operatorname{Tr}_{q/p}(b^2) = 0$ and $\operatorname{Tr}_{q/p}(b) \neq 0$, the sum is over a non-trivial additive character, yielding 0. Thus $I = -1$.
- If $\operatorname{Tr}_{q/p}(b^2) = 0$ and $\operatorname{Tr}_{q/p}(b) = 0$, we are summing $\overline{\chi}(0) = 1$ over p elements. Thus $I = p - 1$.
- If $\operatorname{Tr}_{q/p}(b^2) \neq 0$, completing the square in the exponent yields:

$$\sum_{w \in \mathbb{F}_p} \overline{\chi}\left(-\operatorname{Tr}_{q/p}(b^2)\left(w + \frac{\operatorname{Tr}_{q/p}(b)}{2\operatorname{Tr}_{q/p}(b^2)} \right)^2 + \frac{\operatorname{Tr}_{q/p}(b)^2}{4\operatorname{Tr}_{q/p}(b^2)} \right)$$

$$= \overline{\chi}\left(\frac{\operatorname{Tr}_{q/p}(b)^2}{4\operatorname{Tr}_{q/p}(b^2)} \right) \overline{\eta}(-\operatorname{Tr}_{q/p}(b^2)) G(\overline{\eta}, \overline{\chi})$$

Which means $I = \overline{\chi}\left(\frac{\operatorname{Tr}_{q/p}(b)^2}{4\operatorname{Tr}_{q/p}(b^2)} \right) \overline{\eta}(-\operatorname{Tr}_{q/p}(b^2)) G(\overline{\eta}, \overline{\chi}) - 1$.

Multiplying the constant K, the result of the first sum (depending on c), and the result of the second sum I (depending on $\operatorname{Tr}_{q/p}(b)$ and $\operatorname{Tr}_{q/p}(b^2)$), we obtain exactly the six cases of our thesis.

2.4   The Second Generic Construction of Linear Codes

This section briefs the state of the art of the second generic construction of linear codes. A complete exposition is given in [22].

The second generic construction of linear codes from functions is obtained by fixing a set $D = \{d_1, d_2, \cdots, d_n\}$ in $\mathbb{F}_q$, where $q = p^m$, and by defining a linear code involving D as follows:

$$\mathcal{C}_D = \{(\operatorname{Tr}_{q/p}(xd_1), \operatorname{Tr}_{q/p}(xd_2), \cdots, \operatorname{Tr}_{q/p}(xd_n)) \mid x \in \mathbb{F}_q\}.$$

The set D is usually called the *defining-set* of the code $\mathcal{C}_D$. The resulting code $\mathcal{C}_D$ is linear over $\mathbb{F}_p$ of length n with dimension at most m. Indeed, the function

$d \in \mathbb{F}_q \to \mathrm{Tr}_{q/p}(xd)$ matches once with each linear form over $\mathbb{F}_q$ when x ranges over $\mathbb{F}_q$.

The idea of constructing a linear code from a defining set goes back to Wolfmann [29] in 1975, who was inspired by the trace construction of irreducible cyclic codes given by Baumert and McEliece in 1972 in [2]. However, the first apparition in the literature of the proper second generic construction goes back to 2007 by Ding and Niederreiter in [9].

In [16], it is possible to find the complete weight enumerators of this kind of linear code in three specific cases:

1. When D is a skew Hadamard difference set or Paley type partial difference set in $\mathbb{F}_q$;
2. When $D = \{f(x) : x \in \mathbb{F}_q\} \setminus \{0\}$, where $f(x)$ is a quadratic form over $\mathbb{F}_q$;
3. When $D = \{x \in \mathbb{F}_q : \mathrm{Tr}_{q/p}(x^{p^s+1}) = 0\}$, where $m = 2s$ is an even integer.

In [7], Ding studied some classes of two generic construction codes from specific defining sets. In particular, we would like to consider the preimage $f^{-1}(b)$ for some functions f from $\mathbb{F}_q$ to $\mathbb{F}_p$.

In the boolean case, for a function f from $\mathbb{F}_{2^m}$ to $\mathbb{F}_2$, Ding was able to determine the complete weight enumerator by the use of the Walsh transform. In particular, let $D_f := f^{-1}(1)$ be the support of f and let $n_f = |D_f|$; then we have the following theorem, with specific corollaries when the function f is bent or semi-bent. We recall that $n_f = 2^{m-1} \pm 2^{(m-2)/2}$.

Theorem 5 ([7], Theorem 9). *If f is not an affine function, then C_{D_f} is a binary linear code of length n_f and dimension m, and its weight distribution is given by the following multiset*

$$\left\{\left\{\frac{2n_f + \hat{\chi}_f(w)}{4} : w \in \mathbb{F}_{2^m}^*\right\}\right\} \cup \{\{0\}\}.$$

We now consider the specific case in which f is a bent function.

Corollary 1 ([7], Corollary 10). *Let f be a bent function from* GF (2^m) *to* GF(2) *with $f(0) = 0$, where $m \geq 4$ and is even. Then C_{D_f} is an $\left[n_f, m, \left(n_f - 2^{(m-2)/2}\right)/2\right]$ two-weight binary code with the following weight distribution*

Weight	Frequency
0	1
$\dfrac{n_f}{2} - 2^{\frac{m-4}{2}}$	$\dfrac{2^m - 1 - n_f 2^{-\frac{m-2}{2}}}{2}$
$\dfrac{n_f}{2} + 2^{\frac{m-4}{2}}$	$\dfrac{2^m - 1 + n_f 2^{-\frac{m-2}{2}}}{2}$

Interestingly, these codes C_{D_f} happen to be self-orthogonal and minimal.

Proposition 1. *Let f be a bent function from* GF (2^m) *to* GF(2) *with* $f(0) = 0$, *where $m \geq 8$ and is even. Then C_{D_f} is a minimal self-orthogonal*
$$\left[n_f, m, \left(n_f - 2^{(m-2)/2}\right)/2\right] \text{ two-weight binary code.}$$

Proof. In Corollary 1, we have the complete enumerator of the code C_{D_f}. Thanks to Theorem 3 for binary codes, we can conclude that the codes are self-orthogonal.

We define w_1 as the smaller weight of the codewords and w_2 as the bigger one. Following Theorem 11 in [8], to conclude that the code is minimal, it suffices to prove that $2w_1 \neq w_2$.

3   Families of Self-orthogonal, LCD and Low-Dimensional Hull Codes from a Quadratic Form

In this section, we focus on the function $f(x) = \mathrm{Tr}_{q/p}(x^2 + x)$ to build the associated second generic construction code using the zero set

$$D = \{x \in \mathbb{F}_q : \mathrm{Tr}_{q/p}(x^2 + x) = 0\},$$

where $q = p^m$, with p an odd prime. This code has been studied by Li et al. [17] to obtain a family of few-weight codes.

We now consider the augmented code

$$\overline{C}_D = \{(\mathrm{Tr}_{q/p}(xd_1) + c, \ldots, \mathrm{Tr}_{q/p}(xd_n) + c) : x \in \mathbb{F}_q, c \in \mathbb{F}_p\},$$

where $D = \{d_1, \ldots, d_n\}$, assuming $d_n = 0$. For a primitive element α of $\mathbb{F}_q$, we obtain the generator matrix

$$M = \begin{pmatrix} \mathrm{Tr}_{q/p}(\alpha^0 d_1) & \cdots & \mathrm{Tr}_{q/p}(\alpha^0 d_{n-1}) & 0 \\ \cdots & \cdots & \cdots & 0 \\ \mathrm{Tr}_{q/p}(\alpha^{m-1} d_1) & \cdots & \mathrm{Tr}_{q/p}(\alpha^{m-1} d_{n-1}) & 0 \\ 1 & \cdots & 1 & 1 \end{pmatrix}. \tag{1}$$

In the following Theorem, we determine all parameters of the code using a direct character-sum approach.

Theorem 6. *Let $q = p^m$ where $m \geq 2$ is an even integer and p is an odd prime such that $p \mid m$. Let C be the linear code over $\mathbb{F}_p$ defined by the codewords*

$$\mathbf{c}_{bc} = (\mathrm{Tr}_{q/p}(bd) + c)_{d \in D}$$

for $b \in \mathbb{F}_q$ and $c \in \mathbb{F}_p$, where $D = \{x \in \mathbb{F}_q \mid \mathrm{Tr}_{q/p}(x^2 + x) = 0\}$.
Let $G = -(-1)^{\frac{m(p-1)}{4}} p^{\frac{m}{2}}$. Then C is a self-orthogonal $[p^{m-1} + p^{-1}(p-1)G, m+1]$ linear code and its weight distribution is given by the following table:

Weight	Frequency
0	1
$p^{m-1} + p^{-1}(p-1)G$	$p - 1$
$p^{m-2}(p-1)$	$p^{m-1} + p^{-1}(p-1)G - 1$
$p^{m-2}(p-1) + p^{-1}(p-1)G$	$2p^{m-1}(p-1) + p^{-1}(p-1)(p-2)G - (p-1)$
$p^{m-2}(p-1) + p^{-1}(p-2)G$	$p^{m-1}(p-1)^2 - p^{-1}(p-1)^2G$

The minimum distance d is given by:

$$d = \begin{cases} p^{m-2}(p-1) & \text{if } 8 \nmid m(p-1); \\ p^{m-2}(p-1) + p^{-1}(p-1)G & \text{otherwise.} \end{cases}$$

The dual code has parameters $[p^{m-1} + p^{-1}(p-1)G, p^{m-1} + p^{-1}(p-1)G - m - 1, 3]$.

Proof. For the length of the code, we can refer to Li et al. [17]; notice that in this case $0 \in D$. For $b \in \mathbb{F}_q$ and $c \in \mathbb{F}_p$, define the codeword

$$\mathbf{c}_{bc} = (\mathrm{Tr}_{q/p}(bd_1) + c, \ldots, \mathrm{Tr}_{q/p}(bd_n) + c).$$

Now we compute the weights:

$$\begin{aligned}
\mathrm{w}(\mathbf{c}_{bc}) &= n - |\{d \in D \mid \mathrm{Tr}_{q/p}(bd) + c = 0\}| \\
&= n - |\{x \in \mathbb{F}_q \mid \mathrm{Tr}_{q/p}(x^2 + x) = 0, \mathrm{Tr}_{q/p}(bx) = -c\}| \\
&= n - \frac{1}{p^2} \sum_{x \in \mathbb{F}_q} \sum_{y,z \in \mathbb{F}_p} \zeta_p^{y\,\mathrm{Tr}_{q/p}(x^2+x)+z(\mathrm{Tr}_{q/p}(bx)+c)} \\
&= n - \frac{1}{p^2} \sum_{x \in \mathbb{F}_q} \left(\sum_{y \in \mathbb{F}_p^*} \zeta_p^{y\,\mathrm{Tr}_{q/p}(x^2+x)} + 1 \right) \left(\sum_{z \in \mathbb{F}_p^*} \zeta_p^{z(\mathrm{Tr}_{q/p}(bx)+c)} + 1 \right) \\
&= n - \frac{1}{p^2} \left(p^m + \sum_{x \in \mathbb{F}_q} \sum_{y \in \mathbb{F}_p^*} \zeta_p^{y\,\mathrm{Tr}_{q/p}(x^2+x)} + \sum_{x \in \mathbb{F}_q} \sum_{z \in \mathbb{F}_p^*} \zeta_p^{z(\mathrm{Tr}_{q/p}(bx)+c)} \right. \\
&\qquad\qquad \left. + \sum_{x \in \mathbb{F}_q} \sum_{y \in \mathbb{F}_p^*} \sum_{z \in \mathbb{F}_p^*} \zeta_p^{\mathrm{Tr}_{q/p}(yx^2+yx+zbx)+zc} \right).
\end{aligned}$$

Hence, we have an expression of the form:

$$\mathrm{w}(\mathbf{c}_{bc}) = n - \frac{1}{p^2}\left(p^m + S_0 + S_1 + S_2\right),$$

where

$$S_0 = \sum_{x\in\mathbb{F}_q}\sum_{y\in\mathbb{F}_p^*} \zeta_p^{y\,\mathrm{Tr}_{q/p}(x^2+x)}, \quad S_1 = \sum_{x\in\mathbb{F}_q}\sum_{z\in\mathbb{F}_p^*} \zeta_p^{z(\mathrm{Tr}_{q/p}(bx)+c)}$$

and

$$S_2 = \sum_{x\in\mathbb{F}_q}\sum_{y\in\mathbb{F}_p^*}\sum_{z\in\mathbb{F}_p^*} \zeta_p^{\mathrm{Tr}_{q/p}(yx^2+yx+zbx)+zc}.$$

From Lemma 5 we have that $S_0 = (p-1)G$ and applying Lemmas 4 and 6, we evaluate the weight $\mathrm{w}(\mathbf{c}_{bc})$ according to the values of b and c:

1. If $b = 0$ and $c = 0$, $\mathrm{w}(\mathbf{c}_{00}) = 0$.
2. If $b = 0$ and $c \neq 0$, $S_1 = -p^m$ and $S_2 = (1-p)G$, yielding $\mathrm{w}(\mathbf{c}_{0c}) = p^{m-1} + p^{-1}(p-1)G$.
3. If $b \neq 0$, $\mathrm{Tr}_{q/p}(b^2) = 0$, and $2c = \mathrm{Tr}_{q/p}(b)$, $S_1 = 0$ and $S_2 = (p-1)^2 G$, yielding $\mathrm{w}(\mathbf{c}_{bc}) = p^{m-2}(p-1)$.
4. If $b \neq 0$, $\mathrm{Tr}_{q/p}(b^2) = 0$, and $2c \neq \mathrm{Tr}_{q/p}(b)$, $S_1 = 0$ and $S_2 = (1-p)G$, yielding $\mathrm{w}(\mathbf{c}_{bc}) = p^{m-2}(p-1) + p^{-1}(p-1)G$.
5. If $b \neq 0$, $\mathrm{Tr}_{q/p}(b^2) \neq 0$, and $2c = \mathrm{Tr}_{q/p}(b)$, $S_1 = 0$ and $S_2 = (1-p)G$, yielding the same weight as in case 4.
6. If $b \neq 0$, $\mathrm{Tr}_{q/p}(b^2) \neq 0$, and $2c \neq \mathrm{Tr}_{q/p}(b)$, $S_1 = 0$ and $S_2 = G$, yielding $\mathrm{w}(\mathbf{c}_{bc}) = p^{m-2}(p-1) + p^{-1}(p-2)G$.

To determine the frequencies of these weights, we count the number of elements $b \in \mathbb{F}_q$ satisfying the condition $\mathrm{Tr}_{q/p}(b^2) = 0$. Let N_0 denote this number. Evaluating the associated quadratic form, we obtain:

$$N_0 = p^{m-1} + p^{-1}(p-1)G.$$

The number of elements for which $\mathrm{Tr}_{q/p}(b^2) \neq 0$ is $N_1 = p^m - N_0$. Since p is an odd prime, the linear equation $2c = \mathrm{Tr}_{q/p}(b)$ has exactly one solution for $c \in \mathbb{F}_p$ given b, leaving $p - 1$ choices where $2c \neq \mathrm{Tr}_{q/p}(b)$. Combining the occurrence of these roots with our six cases gives the exact frequencies A_w listed in the table. Because $\mathrm{w}(\mathbf{c}_{bc}) = 0 \iff (b, c) = (0, 0)$, the map $(b, c) \mapsto \mathbf{c}_{bc}$ is injective, giving the code dimension $k = m + 1$.

The expression of the minimum distance d derives from the fact that, if $8 \nmid m(p-1)$, then G is positive; otherwise it is negative.

The self-orthogonality condition is derived from the p-divisibility criterion in Theorem 4. Regarding the minimum distance of the dual code, it can be determined by employing the first three Pless power moments.

3.1 Construction of a General Class of LCD Codes

In the following Proposition, we derive a class of LCD codes constructed from self-orthogonal codes.

Proposition 2. *Let $\mathcal{C}$ be the self-orthogonal linear code over $\mathbb{F}_p$ with parameters $[n, m + 1, d]$ as defined in Theorem 6, where $n = p^{m-1} + p^{-1}(p - 1)G$ and d is its minimum distance. Let M be a generator matrix of $\mathcal{C}$. The linear code $\overline{\mathcal{C}}$ generated by the block matrix*

$$\overline{M} = [\mathbb{I}_{m+1} \mid M]$$

is a $[p^{m-1} + p^{-1}(p - 1)G + m + 1, m + 1, \geq d + 1]$ LCD code over $\mathbb{F}_p$.

Furthermore, the dual code $\overline{\mathcal{C}}^{\perp}$ is a $[p^{m-1} + p^{-1}(p - 1)G + m + 1, p^{m-1} + p^{-1}(p - 1)G, 2]$ LCD code over $\mathbb{F}_p$.

Proof. By Theorem 6, the code $\mathcal{C}$ is self-orthogonal, which implies that $MM^T = 0$. The Gram matrix of $\overline{\mathcal{C}}$ is given by:

$$\overline{M}\,\overline{M}^T = (\mathbb{I}_{m+1} \mid M)\binom{\mathbb{I}_{m+1}}{M^T} = \mathbb{I}_{m+1} + MM^T = \mathbb{I}_{m+1}$$

Since the identity matrix is non-singular, $\overline{\mathcal{C}}$ is an LCD code. The dimension $m+1$ and the length $n + m + 1$ follow directly from the construction of the matrix $\overline{M}$. The inclusion of the identity matrix guarantees that the minimum distance increases by at least one compared to $\mathcal{C}$, hence it is $\geq d + 1$.

The dual code $\overline{\mathcal{C}}^{\perp}$ is LCD because the dual of an LCD code is always LCD. Its dimension is simply the length minus the dimension of $\overline{\mathcal{C}}$. Finally, the minimum distance of the dual code is exactly 2. This occurs because the generator matrix M of the original code contains two identical columns (specifically, columns consisting of all zeros except for a 1 in the last position). These identical columns remain present in the extended matrix $\overline{M}$, implying a linear dependence between two columns. The difference of these two columns directly yields a codeword of weight 2 in the dual code $\overline{\mathcal{C}}^{\perp}$.

Remark 1. Choosing a different generator matrix M for the code $\mathcal{C}$ would produce another LCD code, potentially with a different minimum distance.

Remark 2. In 2023, Youcef [31] introduced the definition of a *pure* LCD code. We recall that a linear code is said to be pure LCD if it is LCD and every monomially equivalent code is also LCD.

Notice that the linear codes constructed in Proposition 2 are LCD, but are generally not pure; that is, there may exist a monomially equivalent code with a nonzero Hull.

3.2 Construction of the Associated ℓ-Dimensional Hull Codes

In this section, we would like to build an ℓ-dimensional Hull code from the defining set

$$D = \{x \in \mathbb{F}_q : \mathrm{Tr}_{q/p}(x^2 + x) = 0\} = \{d_1, \ldots, d_n\}.$$

First, we recall the following construction due to Sok in 2022.

Proposition 3 ([28], Codes with an ℓ-dimensional Hull). *Let $\mathbb{F}_q$ be a finite field with $q > 3$, and let $\mathcal{C}$ be a self-orthogonal linear code over $\mathbb{F}_q$ of length n and dimension k, with a generator matrix in standard form $[\mathbb{I}_k \mid P]$. Let $0 \leq \ell \leq k$ be an integer, and let $\lambda \in \mathbb{F}_q^*$ such that $\lambda^2 \neq 1$. We construct the diagonal matrix of order k:*

$$\Lambda = diag(\underbrace{\lambda, \ldots, \lambda}_{k-\ell \; times}, \underbrace{1, \ldots, 1}_{\ell \; times}).$$

Then, the linear code $\mathcal{L}$ generated by the matrix $[\Lambda \mid P]$ is a linear code of length n, dimension k, and has a Hull of dimension exactly ℓ.

We now apply this construction to the specific family of codes defined by the trace function.

Proposition 4. *Let $q = p^m$ where $m \geq 2$ is an even integer and $p > 3$ is a prime such that $p \mid m$. Let $\mathcal{C}$ be the self-orthogonal $[n, m + 1, d]$ linear code over $\mathbb{F}_p$ described in Theorem 6, with length $n = p^{m-1} + p^{-1}(p - 1)G$ and generator matrix in systematic form $G_{sys} = [\mathbb{I}_{m+1} \mid P]$.*

Let $0 \leq \ell \leq m+1$ and let $\mathcal{L}$ be the code with an ℓ-dimensional Hull generated by the matrix $G' = [\Lambda \mid P]$ as constructed in Proposition 3.

Then, the code $\mathcal{L}$ is monomially equivalent to $\mathcal{C}$. Consequently, $\mathcal{L}$ has the exact same weight distribution as $\mathcal{C}$, and its minimum distance d is:

$$d = \begin{cases} p^{m-2}(p - 1) & \text{if } 8 \nmid m(p - 1); \\ p^{m-2}(p - 1) + p^{-1}(p - 1)G & \text{otherwise.} \end{cases}$$

Furthermore, the dual code $\mathcal{L}^\perp$ has parameters $[n, n - m - 1, 3]$ and maintains an ℓ-dimensional Hull.

Proof. By construction, the generator matrix of the new code $\mathcal{L}$ is $[\Lambda \mid P]$, where Λ is a diagonal matrix with diagonal entries $\lambda_i \in \{1, \lambda\}$ for some $\lambda \in \mathbb{F}_p^*$ such that $\lambda^2 \neq 1$. Since 0 is not an eigenvalue of Λ, all its diagonal entries are non-zero elements of $\mathbb{F}_p$.

The code $\mathcal{L}$ can be viewed as the image of the original code $\mathcal{C}$ under a linear transformation that multiplies the j-th coordinate of every codeword by λ_j for $1 \leq j \leq m + 1$, and by 1 for the remaining coordinates. This transformation corresponds to right-multiplying the codewords of $\mathcal{C}$ by the $(n \times n)$ diagonal matrix:

$$D = \begin{pmatrix} \Lambda & \mathbf{0} \\ \mathbf{0} & \mathbb{I}_{n-m-1} \end{pmatrix}.$$

Since D is a full-rank diagonal matrix, the code $\mathcal{L} = \mathcal{C}D$ is monomially equivalent to $\mathcal{C}$.

Monomial equivalence preserves the Hamming weight of every vector, because scaling a coordinate by a non-zero scalar does not change whether that coordinate is zero or non-zero. Thus, the weight distribution of $\mathcal{L}$ is strictly identical to that of $\mathcal{C}$, yielding the exact same minimum distance d.

By standard coding theory, if two codes are monomially equivalent via a diagonal matrix D, their dual codes are also monomially equivalent via the matrix D^{-1}. Therefore, $\mathcal{L}^{\perp}$ shares the exact same minimum distance as $\mathcal{C}^{\perp}$, which is exactly 3.

Remark 3. By deliberately choosing small values of the parameter ℓ, this construction provides a systematic method for generating linear codes with a low-dimensional Hull. In particular, setting $\ell = 1$ immediately yields a code with a one-dimensional Hull, while perfectly preserving the weight distribution and the minimum distance of the original self-orthogonal code.

Example 1. Choosing $p = 5$, $m = 10$, and $\ell = 1$, we obtain a linear code over $\mathbb{F}_5$ with Hull-dimension one.

Remark 4. We emphasize that, although the function $\mathrm{Tr}_{q/p}(x^2 + x)$ is known to be weakly regular bent over finite fields of odd characteristic (see Chap. 13 in [21]), our approach differs from several previous works in the literature. Some authors investigated codes derived from such functions via the Walsh transform and their spectral properties, for instance [3] and [13]. In contrast, our method relies on the explicit evaluation of certain exponential sums via character theory, thereby enabling us to determine the weight distribution and other structural properties of the resulting codes directly.

The defining set $D = \{x \in \mathbb{F}_q : \mathrm{Tr}_{q/p}(x^2 + x) = 0\}$ may also serve as a useful tool for constructing combinatorial objects such as strongly regular graphs, t-designs, or difference sets (see, for instance, [7]). Indeed, the exponential sums evaluated in this paper determine the intersection structure of the corresponding sets, which is a fundamental ingredient in the study of such combinatorial configurations. Consequently, the techniques developed in this article provide useful tools for analyzing these structures from a combinatorial perspective, in addition to their applications in coding theory, including the construction of few-weight codes, self-orthogonal codes, codes with arbitrary Hull dimension, and their potential use.

4   Conclusion

In this paper, we proposed a character sum approach to explore the construction of linear codes derived from the quadratic function $\mathrm{Tr}_{q/p}(x^2 + x)$ within the framework of the second generic construction. By leveraging character theory and introducing two novel technical lemmas on exponential sums, we explicitly determined the weight enumerators of the proposed codes, which have four weights and are self-orthogonal.

Building upon these self-orthogonal codes, we constructed a class of Linear Complementary Dual (LCD) and we derived linear codes with a fixed ℓ-dimensional Hull while strictly preserving the weight distribution and minimum distance of the original self-orthogonal code. This methodology is particularly valuable for low-dimensional cases (for instance $\ell = 1$), as such codes serve as fundamental building blocks for the construction of Entanglement-Assisted Quantum Error-Correcting Codes (EAQECCs), as well as for the efficient computation of automorphism groups of linear codes and for testing permutational equivalence between codes.

Future research directions include extending the direct exponential-sum approach developed in this paper to other highly nonlinear cryptographic functions, as well as exploring analogous constructions over fields of even characteristic, where different self-orthogonality criteria apply. Such investigations may also lead to new results on exponential sums. Another promising direction is to derive explicit parameters of the quantum error-correcting codes arising from the ℓ-dimensional Hull codes constructed in this work. Finally, it would also be worthwhile to investigate similar constructions based on weakly regular plateaued functions rather than weakly regular bent functions, as considered for instance in the articles of Mesnager, Özbudak and Sınak [24, 25].

Acknowledgments. The authors sincerely thank the anonymous reviewers for their insightful comments and suggestions, which improved the quality and presentation of this work.

Disclosure of Interests. The authors declare that they have no competing interests.

References

1. Assmus, E.F., Key, J.D.: Affine and projective planes. Discret. Math. **83**(2–3), 161–187 (1990)
2. Baumert, L.D., McEliece, R.J.: Weights of irreducible cyclic codes. Inf. Control **20**(2), 158–175 (1972)
3. Çakmak, M., Sinak, A., Yayla, O.: Constructions of self-orthogonal codes and LCD codes over finite fields. In: 17th International Conference on Information Security and Cryptology (ISCTürkiye), p. 1 (2024)
4. Carlet, C., Guilley, S.: Complementary dual codes for counter-measures to side-channel attacks. Adv. Math. Commun. **10**(1), 131–150 (2016)
5. Chen, W., Zhang, J., Shi, M.: A new family of p-ary self-orthogonal code from weakly regular plateaued functions. J. Appl. Math. Comput. **71**(2), 2327–2348 (2025)
6. Ding, C.: A construction of binary linear codes from Boolean functions. Discret. Math. **339**(9), 2288–2303 (2016)
7. Ding, C.: Linear codes from some 2-designs. IEEE Trans. Inf. Theory **60**, 3265–3275 (2015)
8. Ding, C., Heng, Z., Zhou, Z.: Minimal binary linear codes. IEEE Trans. Inf. Theory (2018)
9. Ding, C., Niederreiter, H.: Cyclotomic linear codes of order 3. IEEE Trans. Inf. Theory **53**(6) (2007)
10. Ding, K., Ding, C.: Binary linear codes with three weights. IEEE Commun. Lett. **18**(11), 1879–1882 (2014)
11. Fratianni, V., Mesnager, S.: Constructions of linear codes from vectorial plateaued functions and their subfield codes with applications to quantum CSS codes. arXiv preprint: arXiv:2602.14832 (2026)

12. Guenda, K., Jitman, S., Gulliver, T.A.: Constructions of entanglement-assisted quantum error-correcting codes from classical linear codes. Des. Codes Cryptogr. **86**(1), 121–136 (2018)
13. Heng, Z., Li, D., Liu, F.: Ternary self-orthogonal codes from weakly regular bent functions and their application in LCD Codes. Des. Codes Cryptogr. **91**, 3953–3976 (2023)
14. Heng, Z., Yang, M., Ming, Y.: A family of self-orthogonal divisible codes with locality 2. Discret. Math. **348**(10), 114529 (2025)
15. Huffman, W.C., Pless, V.: Fundamentals of Error-Correcting Codes. Cambridge University Press (2003)
16. Li, C., Bae, S., Ahn, J., Yang, S., Yao, Z.-A.: Complete weight enumerators of some linear codes and their applications. Des. Codes Cryptogr. (2016)
17. Li, F., Wang, Q., Lin, D.: A class of three-weight and five-weight linear codes. Discret. Appl. Math. **241**, 25–38 (2018)
18. Li, X., Heng, Z.: Self-orthogonal codes from p-divisible codes. IEEE Trans. Inf. Theory (2024)
19. Lidl, R., Niederreiter, H.: Introduction to Finite Fields and Their Applications. Cambridge University Press (1994)
20. Massey, J.M.: Linear codes with complementary duals. Discret. Math. (1992)
21. Mesnager, S.: Bent Functions-Fundamentals and Results, pp. 1–544. Springer (2016). ISBN 978-3-319-32593-4
22. Mesnager, S.: Linear codes from functions. In: A Concise Encyclopedia of Coding Theory, ch. 20. CRC Press, London (2021)
23. Mesnager, S.: Linear codes with few weights from weakly regular bent functions based on a generic construction. Cryptogr. Commun. **9**(1), 71–84 (2017)
24. Mesnager, S., Sınak, A.: Further designs for self-orthogonal and LCD codes developed from functions over finite fields. Comput. Appl. Math. **44**(7), 341 (2025)
25. Mesnager, S., Özbudak, F., Sınak, A.: Linear codes from weakly regular plateaued functions and their secret sharing schemes. Des. Codes Cryptogr. **87**(2–3), 463–480 (2019)
26. Sendrier, N.: Finding the permutation between equivalent codes: the support splitting algorithm. IEEE Trans. Inf. Theory **46**, 1193–1203 (2000)
27. Sendrier, N., Skersys, G.: On the computation of the automorphism group of a linear code. In: Proceedings of IEEE ISIT 2001, Washington, DC, p. 13 (2001)
28. Sok, L.: A new construction of linear codes with one-dimensional Hull. Des. Codes Cryptogr. (2022)
29. Wolfmann, J.: Codes projectifs à deux ou trois poids associés aux hyperquadriques d'une géométrie finie. Discret. Math. **13**(2), 185–211 (1975)
30. Xiang, C.: Linear codes from a generic construction. Cryptogr. Commun. **8** (2016)
31. Youcef, M.: Linear codes with arbitrary dimensional Hull and pure LCD code. arXiv preprint: arXiv:2306.00285 (2023)
32. Zhou, Z., Li, X., Tang, C., Ding, C.: Binary LCD codes and self-orthogonal codes from a generic construction. IEEE Trans. Inf. Theory **65** (2018)

Cryptography and Boolean Functions

On the (Non-)Existence of Efficient Class Group Orbits for Collision Search in CSIDH

Ryo Negishi[1], Kazuki Komine[1], Akira Katayama[2], and Masaya Yasuda[1]([✉])

[1] Department of Mathematics, Rikkyo University, Tokyo, Japan
{5072908,241c005w,myasuda}@rikkyo.ac.jp
[2] NTT Social Informatics Laboratories, NTT Corporation, Tokyo, Japan
akira0404.katayama@ntt.com

Abstract. Let $p = 4\ell_1 \cdots \ell_n - 1$ be a CSIDH-type prime for distinct odd primes $\ell_1, \ldots, \ell_n$. Let $\mathcal{O} = \mathbb{Z}[\sqrt{-p}]$, and let $Ell_{\mathbb{F}_p}(\mathcal{O})$ denote the set of $\mathbb{F}_p$-isomorphism classes of supersingular elliptic curves defined over $\mathbb{F}_p$ whose $\mathbb{F}_p$-endomorphism ring is isomorphic to $\mathcal{O}$. The action of the ideal class group of $\mathcal{O}$ on $Ell_{\mathbb{F}_p}(\mathcal{O})$ underlies the CSIDH scheme and its variants, whose security of key recovery relies on the computational hardness of the group action inverse problem (GAIP). In this work, we introduce a general framework for constructing class group orbits that enables an efficient meet-in-the-middle approach to the GAIP. Let L be the lattice in $\mathbb{Z}^n$ spanned by vectors $e = (e_1, \ldots, e_n) \in \mathbb{Z}^n$ such that the ideal class of $\mathfrak{l}_1^{e_1} \cdots \mathfrak{l}_n^{e_n}$ acts trivially on $Ell_{\mathbb{F}_p}(\mathcal{O})$, where each $\mathfrak{l}_i$ is a prime ideal of $\mathcal{O}$ lying above ℓ_i. If there exists a sub-lattice M of $\mathbb{Z}^n$ such that $L \subseteq M$, we can construct class group orbits of size $[M : L]$. The typical lattice $M = L + \mathbb{Z}e_0$ with $e_0 = (1, \ldots, 1) \in \mathbb{Z}^n$ yields efficiently computable orbits of size 3. We investigate the (non-)existence of other efficiently computable orbits by analyzing the structure of torsion sections on an elliptic surface whose fibers are Montgomery curves representing $Ell_{\mathbb{F}_p}(\mathcal{O})$.

Keywords: Supersingular elliptic curves · Ideal class groups · Group action inverse problem

1 Introduction

In [10], Castryck, Lange, Panny, Martindale, and Renes proposed the CSIDH scheme, a post-quantum, efficient, non-interactive key exchange scheme with compact public keys. For a prime $p \geq 5$, let $\mathcal{O} = \mathbb{Z}[\sqrt{-p}]$ be the order of the imaginary quadratic field $\mathbb{Q}(\sqrt{-p})$. CSIDH uses the action of the ideal class group $Cl(\mathcal{O})$ of $\mathcal{O}$ on the set $Ell_{\mathbb{F}_p}(\mathcal{O})$ of $\mathbb{F}_p$-isomorphism classes of supersingular elliptic curves E defined over $\mathbb{F}_p$, whose $\mathbb{F}_p$-endomorphism ring $\mathrm{End}_{\mathbb{F}_p}(E)$ is isomorphic to $\mathcal{O}$. After that, Beullens, Kleinjung, and Vercauteren [6] proposed the CSI-FiSh scheme, an efficient isogeny-based signature scheme using class group computations. Following the proposals of CSIDH and CSI-FiSh, several techniques and strategies have been developed in [4,9,27,29,33] to accelerate the computation of the class group action, including computations based on Vélu's formulas. In contrast, many constant-time implementations and variants of CSIDH have been proposed in [3,5,11,13,21,26,32] to prevent side-channel attacks (see also [24] for a mathematical survey on these developments).

The security of CSIDH relies on the hardness of the computational and decisional Diffie-Hellman (DH) problems for the class group action [10, Section 7]. The underlying problem of these DH problems is the group action inverse problem (GAIP) (see [6], and also [28] for the quantum equivalence of the computational DH problem). Fix $[E_0] \in Ell_{\mathbb{F}_p}(\mathcal{O})$. Given $[E]$ sampled uniformly at random from $Ell_{\mathbb{F}_p}(\mathcal{O})$, the GAIP asks for the ideal class $[\mathfrak{a}] \in Cl(\mathcal{O})$ such that $[\mathfrak{a}] \star [E_0] = [E]$. The best known classical algorithm for solving the GAIP is a meet-in-the-middle algorithm with complexity $O(\sqrt{h(\mathcal{O})})$, where $h(\mathcal{O}) = \#Cl(\mathcal{O})$ denotes the class number of $\mathcal{O}$ and satisfies $h(\mathcal{O}) \sim \sqrt{p}$. In contrast, the best known quantum algorithm is Kuperberg's algorithm [22] for the hidden shift problem, which has subexponential complexity in the size of $h(\mathcal{O})$, namely $2^{O(\sqrt{\log h(\mathcal{O})})}$. (See [7] for a trade-off between classical and quantum circuits in cryptanalysis of CSIDH.) Chi-Domínguez, Esser, Kunzweiler, and May [12] generalized a low-memory Pollard-style algorithm to the restricted effective group action (REGA) setting, introduced in [1], in order to solve the GAIP, which is referred to as the REGA-DLOG problem in a general setting in [12]. In 2025, May and Ostuzzi [25] studied multiple group action discrete logarithms (including CSIDH and its variants) and showed that solving them yields computational speedups.

A practical security analysis of CSIDH was discussed in the original CSIDH paper [10]. In particular, it was noted in [10, Subsection 7.1] that there seems to be no straightforward way to apply Pohlig-Hellman-style algorithms via proper subgroups, since the set $Ell_{\mathbb{F}_p}(\mathcal{O})$ does not form a group with efficiently computable operations that are compatible with the class group action. In this work, we introduce a general framework for constructing class group orbits that enables solving the GAIP via a meet-in-the-middle approach. For a positive integer n, let $p = 4\ell_1 \ldots \ell_n - 1$ be a prime of the CSIDH form for distinct odd primes $\ell_1, \ldots, \ell_n$. We then consider the group homomorphism

$$\Phi : \mathbb{Z}^n \longrightarrow Cl(\mathcal{O}), \quad e = (e_1, \ldots, e_n) \longmapsto \left[\mathfrak{l}_1^{e_1} \cdots \mathfrak{l}_n^{e_n}\right], \tag{1}$$

where each $\mathfrak{l}_i = \left(\ell_i, \sqrt{-p}-1\right)$ is a prime ideal of $\mathcal{O}$ lying above ℓ_i. Set $L = \ker \Phi$. If there exists a sub-lattice M of $\mathbb{Z}^n$ such that $L \subseteq M$, we can construct orbits of size $[M : L]$ under the action of $\Phi(M)$ on $Ell_{\mathbb{F}_p}(\mathcal{O})$. In addition, if the action of $\Phi(M)$ is efficiently computable, these orbits can be leveraged to solve the GAIP efficiently. A typical choice is the lattice $M = L + \mathbb{Z}e_0$ with $e_0 = (1, \ldots, 1) \in \mathbb{Z}^n$. It was shown in [33] that the index $[M : L]$ is equal to 3. An explicit formula is also provided in [33, Theorem 7] for constructing efficiently computable orbits of size 3 via the action of $\pm\Phi(e_0) \in Cl(\mathcal{O})$, without computing isogenies. We aim to investigate the (non-)existence of formulas analogous to [33, Theorem 7] for constructing other efficiently computable orbits. Specifically, elements of $Ell_{\mathbb{F}_p}(\mathcal{O})$

can be represented by $\mathbb{F}_p$-isomorphism classes of Montgomery curves whose $\mathbb{F}_p$-endomorphism ring is isomorphic to $\mathcal{O}$. By considering an elliptic surface whose fibers are Montgomery curves, we study the structure of generic torsion points on these curves (i.e., torsion sections on the elliptic surface) to determine whether analogous formulas can be derived using the methodology in [33].

2 Mathematical Preliminaries

In this section, we recall definitions and notations mainly on supersingular elliptic curves defined over finite fields (e.g., see [35,39] for elliptic curves, and [24] for mathematical background on CSIDH). Throughout this section, we fix a prime $p \geq 5$, and let $\overline{\mathbb{F}}_p$ denote an algebraic closure of the finite field $\mathbb{F}_p$ with p elements.

2.1 Basics on Elliptic Curves over Finite Fields

An elliptic curve defined over $\mathbb{F}_p$ is given by the short Weierstrass form $y^2 = x^3 + ax + b$ $(a, b \in \mathbb{F}_p)$ with $4a^3 + 27b^2 \neq 0$. For an algebraic extension field K of $\mathbb{F}_p$ (e.g., $K = \overline{\mathbb{F}}_p$), two elliptic curves are said to be *isomorphic over K* if their short Weierstrass equations are connected by a change of variables of the form $x' = u^2 x$ and $y' = u^3 y$ for some $u \in K^\times$. Let $E(K)$ denote the abelian group of K-rational points on E, with the point at infinity ∞_E serving as the zero point. For an integer $n \geq 2$, let $E[n]$ denote the n-torsion subgroup of $E(\overline{\mathbb{F}}_p)$.

A non-constant rational map $\phi : E \to E'$ between two elliptic curves is called a (non-zero) *isogeny* if it induces a group homomorphism from $E(\overline{\mathbb{F}}_p)$ to $E'(\overline{\mathbb{F}}_p)$. The induced homomorphism is surjective and has a finite kernel, denoted by $\ker(\phi)$. In particular, we say that ϕ is *defined over K* if the coefficients of the rational map defining ϕ belong to K. An isogeny ϕ also induces an embedding $\phi^* : \overline{\mathbb{F}}_p(E') \longrightarrow \overline{\mathbb{F}}_p(E)$ between two function fields of E and E', whose extension degree is defined as the degree of ϕ. In particular, we say that ϕ is *separable* if the induced field extension is separable. Every isogeny $\phi : E \to E'$ of degree d has a unique *dual isogeny* $\widehat{\phi} : E' \to E$ satisfying $\widehat{\phi} \circ \phi = [d]$, where $[d]$ denotes the multiplication-by-d map on E. Given a finite subgroup C of $E(\overline{\mathbb{F}}_p)$, there exists a separable isogeny from E with kernel C. The codomain curve, denoted by E/C, is unique up to isomorphism and can be computed using Vélu's formulas [37].

An isogeny from E to itself is called an *endomorphism* of E. The set $\mathrm{End}(E)$ of endomorphisms of E, together with the zero endomorphism $[0]$, forms a ring under addition and composition. The *trace* of an endomorphism $\phi \in \mathrm{End}(E)$ is defined as $\mathrm{tr}(\phi) := 1 + \deg(\phi) - \deg(1 - \phi) \in \mathbb{Z}$. We have $|\mathrm{tr}(\phi)| \leq 2\sqrt{\deg(\phi)}$ and $\phi^2 - [\mathrm{tr}(\phi)]\phi + [\deg(\phi)] = 0$ in $\mathrm{End}(E)$. Let $\mathrm{Aut}(E)$ denote the group of *automorphisms* of E that are the invertible elements of $\mathrm{End}(E)$.

2.2 Supersingular Elliptic Curves

Let $\pi : (x, y) \mapsto (x^p, y^p)$ denote the (*p*-th power) *Frobenius endomorphism* of E. The order of the group of $\mathbb{F}_p$-rational points on E is given by $\#E(\mathbb{F}_p) = p + 1 - t$, where $t = \mathrm{tr}(\pi)$ is the trace of π. We say that E is *supersingular* if $p \mid t$, equivalently, $t = 0$ since $|t| \leq 2\sqrt{p}$ and $p \geq 5$. (Otherwise, E is called *ordinary*.)

Let E be a supersingular elliptic curve E defined over $\mathbb{F}_p$. Then E has no $\overline{\mathbb{F}}_p$-rational point of order p, that is, $E[p] = \{\infty_E\}$. Moreover, the endomorphism ring $\mathrm{End}(E)$ is isomorphic to an order in a quaternion algebra. In particular, the Frobenius endomorphism of E satisfies $\pi^2 + [p] = 0$ in $\mathrm{End}(E)$ since $\deg(\pi) = p$. Let $\mathrm{End}_{\mathbb{F}_p}(E)$ denote the subring of $\mathrm{End}(E)$ of endomorphisms defined over $\mathbb{F}_p$. Then $\mathrm{End}_{\mathbb{F}_p}(E)$ is isomorphic to an order $\mathcal{O}$ in the imaginary quadratic field $K = \mathbb{Q}(\sqrt{-p})$. Since $\mathrm{End}_{\mathbb{F}_p}(E)$ contains $\mathbb{Z}[\pi] \cong \mathbb{Z}[\sqrt{-p}]$, the order $\mathcal{O}$ is equal to either $\mathbb{Z}[\sqrt{-p}]$ or the ring $\mathcal{O}_K$ of integers of K. We say that E lies on the *floor* (resp. *surface*) if $\mathrm{End}_{\mathbb{F}_p}(E)$ is isomorphic to $\mathbb{Z}[\sqrt{-p}]$ (resp. $\mathcal{O}_K$). We note that when $p \equiv 1 \pmod 4$, the floor and the surface coincide since $\mathcal{O}_K = \mathbb{Z}[\sqrt{-p}]$. In contrast, CSIDH uses a prime $p \geq 5$ of special form satisfying $p \equiv 3 \pmod 4$ and deals with supersingular elliptic curves on the floor.

2.3   Class Group and Its Action on Supersingular Curves

Let $\mathcal{O}$ be an order in a number field K. A *fractional ideal* of $\mathcal{O}$ is a nonzero $\mathcal{O}$-submodule $\mathfrak{a}$ of K for which there exists a nonzero element $r \in \mathcal{O}$ satisfying $r\mathfrak{a} \subseteq \mathcal{O}$ (that is, r can clear the denominators of $\mathfrak{a}$). We say that a fractional ideal $\mathfrak{a}$ is *invertible* if there exists a fractional ideal $\mathfrak{b}$ of $\mathcal{O}$ such that $\mathfrak{a}\mathfrak{b} = \mathcal{O}$. We denote such $\mathfrak{b}$ by $\mathfrak{a}^{-1}$. (When $\mathcal{O}$ is a Dedekind domain, such as the ring of integers of K, every fractional ideal of $\mathcal{O}$ is invertible.) A fractional ideal of $\mathcal{O}$ is called *principal* if it is generated by a single element of K. Clearly, every principal fractional ideal is invertible. Let $I(\mathcal{O})$ denote the set of invertible fractional ideals of $\mathcal{O}$, which forms a group under the product of fractional ideals (every $\mathfrak{a} \in I(\mathcal{O})$ has the inverse $\mathfrak{a}^{-1}$). Let $P(\mathcal{O})$ denote the subgroup of $I(\mathcal{O})$ of principal fractional ideals of $\mathcal{O}$. The *ideal class group* (or simply *class group*) of $\mathcal{O}$ is the quotient group $Cl(\mathcal{O}) := I(\mathcal{O})/P(\mathcal{O})$. It is finite and its cardinality is called the *class number* of $\mathcal{O}$, denoted by $h(\mathcal{O})$.

We take a prime with $p \equiv 3 \pmod 4$, as used in the CSIDH setting [10]. Set $K = \mathbb{Q}(\sqrt{-p})$, whose ring of integers is $\mathcal{O}_K = \mathbb{Z}[\theta]$ with $\theta = \frac{-1+\sqrt{-p}}{2}$. Consider the order $\mathcal{O} = \mathbb{Z}[\sqrt{-p}]$ in K. Let $Ell_{\mathbb{F}_p}(\mathcal{O})$ denote the set of $\mathbb{F}_p$-isomorphism classes $[E]$ of supersingular elliptic curves E defined over $\mathbb{F}_p$ lying on the floor, that is, satisfying $\mathcal{O} \cong \mathrm{End}_{\mathbb{F}_p}(E)$, where $\sqrt{-p}$ corresponds to the Frobenius endomorphism π of E. (Note that if E and E' are isomorphic over $\mathbb{F}_p$, then $\mathrm{End}_{\mathbb{F}_p}(E) \cong \mathrm{End}_{\mathbb{F}_p}(E')$.) For a fractional ideal $\mathfrak{a}$ of $\mathcal{O}$, we may view $r\mathfrak{a}$ as a subset of $\mathrm{End}_{\mathbb{F}_p}(E)$ under the isomorphism $\mathcal{O} \cong \mathrm{End}_{\mathbb{F}_p}(E)$, where r is a nonzero element in $\mathcal{O}$ clearing the denominators of $\mathfrak{a}$. We define an action of $Cl(\mathcal{O})$ on $Ell_{\mathbb{F}_p}(\mathcal{O})$ as follows. Let $[\mathfrak{a}] \in Cl(\mathcal{O})$ and $[E] \in Ell_{\mathbb{F}_p}(\mathcal{O})$. We may choose $\mathfrak{a}$ as an integral representative of $[\mathfrak{a}]$, that is, $\mathfrak{a} \subseteq \mathcal{O}$. We consider the isogeny $\varphi_\mathfrak{a} : E \longrightarrow E_\mathfrak{a}$ with kernel $\cap_{\alpha \in \mathfrak{a}} \ker(\alpha)$ under the isomorphism $\mathcal{O} \cong \mathrm{End}_{\mathbb{F}_p}(E)$. The degree of $\varphi_\mathfrak{a}$ is equal to the norm of $\mathfrak{a}$ (see [38, Proposition 42.2.16]). Since the isogeny $\varphi_\mathfrak{a}$ is defined over $\mathbb{F}_p$ by construction, its codomain curve $E_\mathfrak{a}$ is also defined over $\mathbb{F}_p$, and is unique up to $\mathbb{F}_p$-isomorphism. Moreover, we have $\mathrm{End}_{\mathbb{F}_p}(E_\mathfrak{a}) \cong \mathcal{O}$ via the injective ring homomorphism

$$\iota : \mathrm{End}(E/\mathfrak{a}) \longrightarrow \mathrm{End}(E) \otimes_{\mathbb{Z}} \mathbb{Q} \cong \mathcal{O} \otimes_{\mathbb{Z}} \mathbb{Q} = K, \quad \phi \longmapsto \frac{1}{\deg \varphi_{\mathfrak{a}}} \widehat{\varphi}_{\mathfrak{a}} \phi \varphi_{\mathfrak{a}}.$$

(See [38, Lemma 42.2.9], or [36, Chapter II] for the analogue over $\mathbb{C}$.) We then define $[\mathfrak{a}] \star [E]$ to be the $\mathbb{F}_p$-isomorphism class of $E_{\mathfrak{a}}$, that is, $[\mathfrak{a}] \star [E] := [E_{\mathfrak{a}}]$. It follows from [40, Theorem 4.5] that the map

$$Cl(\mathcal{O}) \times Ell_{\mathbb{F}_p}(\mathcal{O}) \longrightarrow Ell_{\mathbb{F}_p}(\mathcal{O}), \quad ([\mathfrak{a}], [E]) \longmapsto [\mathfrak{a}] \star [E] \tag{2}$$

defines a free and transitive group action.

2.4 Computing the Class Group Action

For an integer $n \geq 1$, we consider a prime of the form $p = 4\ell_1 \cdots \ell_n - 1$ with n distinct odd small primes $\ell_1, \ldots, \ell_n$, as used in the CSIDH setting [10]. Let E be a supersingular elliptic curve defined over $\mathbb{F}_p$. Since $\#E(\mathbb{F}_p) = p+1 = 4\ell_1 \cdots \ell_n$, the group $E(\mathbb{F}_p)$ is isomorphic to either

$$\mathbb{Z}/4\mathbb{Z} \times \mathbb{Z}/\ell_1\mathbb{Z} \times \cdots \times \mathbb{Z}/\ell_n\mathbb{Z} \quad \text{or} \quad \mathbb{Z}/2\mathbb{Z} \times \mathbb{Z}/2\mathbb{Z} \times \mathbb{Z}/\ell_1\mathbb{Z} \times \cdots \times \mathbb{Z}/\ell_n\mathbb{Z}. \tag{3}$$

By [16, Theorem 2.7], the former (resp. latter) case occurs if and only if E lies on the floor (resp. surface), that is, $\mathrm{End}_{\mathbb{F}_p}(E)$ is isomorphic to $\mathcal{O} = \mathbb{Z}[\sqrt{-p}]$ (resp. the ring of integers of $\mathbb{Q}(\sqrt{-p})$). Therefore, one can determine whether E lies on the floor or on the surface by counting the number of 2-torsion points in $E(\mathbb{F}_p)$.

Action of Special Ideal Classes. Since each small prime ℓ_i divides $p+1$, the ideal $\ell_i \mathcal{O}$ splits into two distinct prime ideals in $\mathcal{O}$ as $\ell_i \mathcal{O} = \mathfrak{l}_i \bar{\mathfrak{l}}_i$, where

$$\mathfrak{l}_i = (\ell_i, \sqrt{-p} - 1) \quad \text{and} \quad \bar{\mathfrak{l}}_i = (\ell_i, \sqrt{-p} + 1). \tag{4}$$

We note that $\bar{\mathfrak{l}}_i$ is the conjugate of $\mathfrak{l}_i$, and $[\bar{\mathfrak{l}}_i] = [\mathfrak{l}_i]^{-1}$ in the class group $Cl(\mathcal{O})$. Given $[E] \in Ell_{\mathbb{F}_p}(\mathcal{O})$, let $\varphi_{\mathfrak{l}_i} : E \to E_{\mathfrak{l}_i}$ be the separable isogeny corresponding to $\mathfrak{l}_i$, whose codomain curve is a representative of $[\mathfrak{l}_i] \star [E] = [E_{\mathfrak{l}_i}]$. Its kernel is $E[\ell_i] \cap \{P \in E(\overline{\mathbb{F}}_p) \mid \pi(P) = P\} = E[\ell_i] \cap E(\mathbb{F}_p)$, since $\sqrt{-p}$ corresponds to the Frobenius endomorphism π of E. Therefore, the codomain curve $E_{\mathfrak{l}_i}$ can be computed from the kernel using Vélu's formulas. Similarly, the separable isogeny corresponding to $\bar{\mathfrak{l}}_i$ has the kernel $E[\ell_i] \cap \{P \in E(\overline{\mathbb{F}}_p) \mid \pi(P) = -P\}$, which lies in $E(\mathbb{F}_{p^2})$. Therefore, a representative of $[\bar{\mathfrak{l}}_i] \star [E] = [\mathfrak{l}_i]^{-1} \star [E]$ can be computed from its kernel. In particular, both codomain curves $E_{\mathfrak{l}_i}$ and $E_{\bar{\mathfrak{l}}_i}$ can be computed explicitly from an ℓ_i-torsion point P lying in $E(\mathbb{F}_{p^2})$, using Vélu's formulas. More generally, we can evaluate the action of any ideal class of the form $[\mathfrak{l}_1]^{e_1} \cdots [\mathfrak{l}_n]^{e_n} \in Cl(\mathcal{O})$ for $(e_1, \ldots, e_n) \in \mathbb{Z}^n$.

Montgomery Curves. A *Montgomery curve* defined over $\mathbb{F}_p$ is an elliptic curve given by an algebraic equation of the form $E_{A,B} : By^2 = x^3 + Ax^2 + x$ $(A, B \in \mathbb{F}_p)$ with $B(A^2 - 4) \neq 0$. When $B = 1$, we simply write E_A for $E_{A,1}$. Any Montgomery curve can be transformed into a short Weierstrass form. However, the converse does not hold, since every Montgomery curve $E_{A,B}$ has the $\mathbb{F}_p$-rational 2-torsion point $(0,0)$, whereas there exist elliptic curves without any $\mathbb{F}_p$-rational 2-torsion points. Scalar multiplication on Montgomery curves can be efficiently computed using only the x-coordinate, the so-called Montgomery ladder (e.g. see [8]). For example, the doubling formula for the x-coordinate of a point P of E_A (or, more generally, of $E_{A,B}$) is given by

$$x_{[2]P} = \frac{(x_P^2 - 1)^2}{4x_P\left(x_P^2 + Ax_P + 1\right)}, \tag{5}$$

where x_R denotes the x-coordinate of a point R of the curve.

When considering the class group action in the CSIDH setting, we often use Montgomery curves of the form E_A, as justified by the following theorem [10, Proposition 8]. (By Eq. (3), every supersingular elliptic curve defined over $\mathbb{F}_p$ has at least one $\mathbb{F}_p$-rational 2-torsion point.)

Theorem 1 ([10]). *Let $p \geq 5$ be a prime with $p \equiv 3 \pmod 8$, and E a supersingular elliptic curve defined over $\mathbb{F}_p$. Then E lies on the floor if and only if there exists $A \in \mathbb{F}_p$ such that E is isomorphic over $\mathbb{F}_p$ to the Montgomery curve E_A. Moreover, if such an A exists, it is unique. In other words, if another element $B \in \mathbb{F}_p$ satisfies $E_A \cong E_B$ over $\mathbb{F}_p$, then $A = B$.*

This theorem implies that each $\mathbb{F}_p$-isomorphism class $[E] \in Ell_{\mathbb{F}_p}(\mathcal{O})$ corresponds uniquely to an element $A \in \mathbb{F}_p$ such that $E \cong E_A$ over $\mathbb{F}_p$. Moreover, Vélu's formulas on Montgomery curves E_A are provided in [14, Theorem 1].

3   Solving CSIDH Using Class Group Orbits

In this section, we focus on the following problem, which underlies the security of key recovery in CSIDH (see [10, Problem 10 in Sect. 7] or [6, Definition 1]). As in the CSIDH setting, we take a prime of the form $p = 4\ell_1 \ldots \ell_n - 1$ for distinct odd primes $\ell_1, \ldots, \ell_n$, and set the order $\mathcal{O} = \mathbb{Z}[\sqrt{-p}]$ of $K = \mathbb{Q}(\sqrt{-p})$.

Definition 1 (Group Action Inverse Problem, GAIP). *We fix an element $[E_0]$ of $Ell_{\mathbb{F}_p}(\mathcal{O})$. Given $[E]$ uniformly sampled at random from $Ell_{\mathbb{F}_p}(\mathcal{O})$, the GAIP asks for an ideal class $[\mathfrak{a}] \in Cl(\mathcal{O})$ satisfying $[\mathfrak{a}] \star [E_0] = [E]$.*

3.1   General Framework Using Lattices

We present a concrete framework for solving the GAIP using class group orbits. It relies on a meet-in-the-middle algorithm, as discussed in [10, Subsection 7.1].

Construction of Class Group Orbits. We consider the group homomorphism $\Phi : \mathbb{Z}^n \to Cl(\mathcal{O})$, defined in Eq. (1). Heuristically, it is surjective for most primes p, as discussed in [10, Subsection 7.1]. Specifically, any element of $Cl(\mathcal{O})$ can be written as $\Phi(e)$ with $e = (e_1, \ldots, e_n) \in [-B, B]^n$, for any integer $B > 0$ such that $(2B+1)^n > h(\mathcal{O})$. Set $L = \ker \Phi$. Then L is a full-rank sub-lattice of

$\mathbb{Z}^n$, where the image of any element acts trivially on the set $Ell_{\mathbb{F}_p}(\mathcal{O})$. In other words, each element of L defines an endomorphism of a supersingular elliptic curve defined over $\mathbb{F}_p$ lying on the floor. A basis of L can be obtained by collecting many relations among prime ideals of small norm, as described in [6, Section 3]. (The complexity of obtaining such a basis is subexponential in the size of $h(\mathcal{O}) \sim \sqrt{p}$.)

We consider another sub-lattice M of $\mathbb{Z}^n$ such that $L \subseteq M \subseteq \mathbb{Z}^n$. We then define an equivalent relation on the set $Ell_{\mathbb{F}_p}(\mathcal{O})$ by

$$[E] \sim_M [E'] \iff [E'] = \Phi(\boldsymbol{m}) \star [E] \text{ for some } \boldsymbol{m} \in M$$

for two $\mathbb{F}_p$-isomorphism classes $[E], [E'] \in Ell_{\mathbb{F}_p}(\mathcal{O})$. We call it the M-equivalent relation. In other words, the M-equivalence class of $[E] \in Ell_{\mathbb{F}_p}(\mathcal{O})$ is given by $[E]_M := \{ \Phi(\boldsymbol{m}) \star [E] \in Ell_{\mathbb{F}_p}(\mathcal{O}) \mid \boldsymbol{m} \in M \}$, which can be regarded as the orbit of $[E]$ under the action of the subgroup $\Phi(M)$ of $Cl(\mathcal{O})$. Moreover, we denote by the M-equivalence classes of $Ell_{\mathbb{F}_p}(\mathcal{O})$ by $\left[Ell_{\mathbb{F}_p}(\mathcal{O}) \right]_M := Ell_{\mathbb{F}_p}(\mathcal{O}) / \sim_M$.

Meet-in-the-Middle Approach for Solving the GAIP. We fix the smallest integer $B > 0$ such that $(2B + 1)^n > h(\mathcal{O})$. Then there exists a vector $\boldsymbol{e} = (e_1, \ldots, e_n) \in [-B, B]^n$ with $[E] = [\mathfrak{l}_1^{e_1} \cdots \mathfrak{l}_n^{e_n}] \star [E_0]$. We set $m = \left\lceil \frac{n}{2} \right\rceil$ and consider the relation

$$[\mathfrak{l}_1^{e_1} \cdots \mathfrak{l}_m^{e_m}] \star [E_0] = [\mathfrak{l}_{m+1}^{e_{m+1}} \cdots \mathfrak{l}_n^{e_n}]^{-1} \star [E] = \left[\mathfrak{l}_{m+1}^{-e_{m+1}} \cdots \mathfrak{l}_n^{-e_n} \right] \star [E]. \tag{6}$$

To recover the solution vector $\boldsymbol{e}$ of the GAIP between $[E_0]$ and $[E]$, we adopt a meet-in-the-middle algorithm based on a baby-step giant-step strategy, as mentioned in [10, Subsection 7.1]. Specifically, we carry out the following two steps:

1. We compute the set

$$S := \left\{ \left[\mathfrak{l}_1^{b_1} \cdots \mathfrak{l}_m^{b_m} \right] \star [E_0] \in Ell_{\mathbb{F}_p}(\mathcal{O}) \;\middle|\; b_1, \ldots, b_m \in [-B, B] \right\},$$

   which contains the element on the left-hand side of Eq. (6). The cardinality of S is $(2B + 1)^m = O\left(\sqrt{h(\mathcal{O})} \right) \sim O\left(\sqrt[4]{p} \right)$.

2. We perform random walks of the form $\left[\mathfrak{l}_{m+1}^{g_{m+1}} \cdots \mathfrak{l}_n^{g_n} \right] \star [E]$, as in the element on the right-hand side of Eq. (6), with $g_{m+1}, \ldots, g_n \in [-B, B]$, until they fall into S. For such a walk, i.e., when $\left[\mathfrak{l}_{m+1}^{g_{m+1}} \cdots \mathfrak{l}_n^{g_n} \right] \star [E] = \left[\mathfrak{l}_1^{b_1} \cdots \mathfrak{l}_m^{b_m} \right] \star [E_0]$, the vector

$$(b_1, \ldots, b_m, -g_{m+1}, \ldots, -g_n) \in [-B, B]^n \tag{7}$$

   provides a solution vector of the GAIP between $[E_0]$ and $[E]$.

Collision Search over Equivalence Classes. For a collision search over the set of M-equivalence classes $\left[Ell_{\mathbb{F}_p}(\mathcal{O}) \right]_M$, it suffices to consider a representative of the M-equivalence class of each element of $Ell_{\mathbb{F}_p}(\mathcal{O})$. Specifically, instead of S, we may use the set of M-equivalence classes of S as

$$S_M := \{[s]_M \mid s \in S\} \subseteq \left[Ell_{\mathbb{F}_p}(\mathcal{O})\right]_M. \tag{8}$$

In particular, we have $S_L = S$. Let $d = [M : L]$ denote the index of L in M (note that d divides $h(\mathcal{O})$). Then $\#\left[Ell_{\mathbb{F}_p}(\mathcal{O})\right]_M = h(\mathcal{O})/d$ and $\#S_M = \#S/d$. In practice, we may reduce the region $[-B, B]^n$ to a smaller one, such as $[-B, B]^{n-k} \times [-C, D]^k$ for some integers k, C, D with $0 < k < n$ and $0 < C, D < B$, in order to represent the elements of $\left[Ell_{\mathbb{F}_p}(\mathcal{O})\right]_M$. The number of elements computed in the meet-in-the-middle algorithm over $\left[Ell_{\mathbb{F}_p}(\mathcal{O})\right]_M$ is estimated as

$$O\left(\sqrt{\#\left[Ell_{\mathbb{F}_p}(\mathcal{O})\right]_M}\right) = O\left(\sqrt{h(\mathcal{O})/d}\right).$$

Therefore, the time complexity is reduced by a factor of $O\left(\sqrt{d}\right)$ compared to that over the set $Ell_{\mathbb{F}_p}(\mathcal{O})$, provided that the overhead of computing a representative of each M-equivalence class is negligible. Moreover, the space complexity is reduced by a factor of $O(d)$, since $\#S_M = \#S/d$.

3.2   Typical Orbits in CSIDH

The order $\mathcal{O} = \mathbb{Z}[\sqrt{-p}]$ has conductor $\mathfrak{f} = 2$ in the ring of integers $\mathcal{O}_K = \mathbb{Z}[\theta]$ of $K = \mathbb{Q}(\sqrt{-p})$ with $\theta = \frac{-1+\sqrt{-p}}{2}$, that is, $2\mathcal{O}_K \subseteq \mathcal{O}$. Then it follows from the proof of [30, Theorem 12.12, Chap. 1] that we have the exact sequence on abelian groups

$$1 \longrightarrow R^\times/S^\times \xrightarrow{\ \iota\ } Cl(\mathcal{O}) \xrightarrow{\ \kappa\ } Cl(\mathcal{O}_K) \longrightarrow 1, \tag{9}$$

where R denotes the quotient ring $\mathcal{O}_K/2\mathcal{O}_K$ and $S = \mathcal{O}/2\mathcal{O}_K$ its subring. Since the integer $\frac{p+1}{4}$ is odd by the form of p, we have $R \cong \mathbb{F}_2[\theta] \cong \mathbb{F}_4$ and $S \cong \mathbb{F}_2$ (note that θ satisfies $\theta^2 - \theta + \frac{p+1}{4} = 0$). Therefore, the quotient group $R^\times/S^\times \cong \mathbb{F}_4^\times$ is generated by the class of $\theta \in \mathcal{O}_K$, whose order is equal to 3 in the quotient group. Moreover, the image of the class of θ under ι is the ideal class

$$[\theta\mathcal{O}] = [\mathfrak{l}_1 \cdots \mathfrak{l}_n] \in Cl(\mathcal{O}), \tag{10}$$

which also has order 3. (Note that since $\theta\mathcal{O}$ is not principal in the order $\mathcal{O}$ since $\theta \notin \mathcal{O}$, whereas $\theta\mathcal{O}_K$ is principal in $\mathcal{O}_K$. See [33, Theorem 3] for details.)

Lattice Construction. We consider the composition of surjective group homomorphisms $\mathbb{Z}^n \xrightarrow{\ \Phi\ } Cl(\mathcal{O}) \xrightarrow{\ \kappa\ } Cl(\mathcal{O}_K)$, and denote its kernel by M. Then M is a full-rank lattice in $\mathbb{Z}^n$, since $L \subseteq M$ by construction. It follows from the exact sequence (9) that the index $[M : L]$ is equal to 3, since $Cl(\mathcal{O}) \cong \mathbb{Z}^n/L$ and $Cl(\mathcal{O}_K) \cong \mathbb{Z}^n/M$. We see from Eq. (10) that the quotient M/L is generated by the class of $e_0 := (1, \ldots, 1) \in M$ (note that $\kappa \circ \Phi(e_0) = [\theta\mathcal{O}_K] = [\mathcal{O}_K] \in Cl(\mathcal{O}_K)$

and thus $e_0 \in M$). Moreover, an explicit formula [33, Theorem 7] is available for computing the action of $\Phi(e_0)$ and $\Phi(-e_0) \in Cl(\mathcal{O})$ on the $\mathbb{F}_p$-isomorphism classes $[E_A] \in Ell_{\mathbb{F}_p}(\mathcal{O})$ of Montgomery curves E_A on the floor with $A \in \mathbb{F}_p$. Specifically, we have

$$[E_{A'}] = \Phi(e_0) \star [E_A] \quad \text{and} \quad [E_{A''}] = \Phi(-e_0) \star [E_A],$$

where

$$A' = -2\frac{A+6}{A-2} \quad \text{and} \quad A'' = 2\frac{A-6}{A+2}. \tag{11}$$

This implies that the A-coefficients of the Montgomery representatives of the two classes $\Phi(e_0) \star [E_A]$ and $\Phi(-e_0) \star [E_A] \in Ell_{\mathbb{F}_p}(\mathcal{O})$ can be computed directly over $\mathbb{F}_p$ without performing any isogeny computation.

Interpretation on Supersingular Isogeny Graphs. In *supersingular isogeny graphs* over $\mathbb{F}_p$, the vertices are $\mathbb{F}_p$-isomorphism classes of supersingular elliptic curves defined over $\mathbb{F}_p$, and the edges are equivalence classes of separable isogenies over $\mathbb{F}_p$ between these curves. (We say that two separable isogenies are *equivalent* if they have the same kernel.) Since $p \equiv 3 \pmod 4$, the graph exhibits a so-called volcano structure of level 2, consisting of surface and floor components [16, Theorem 2.7]. In particular, the surface and floor components are connected solely by 2-isogenies, where each surface vertex is adjacent to exactly three floor vertices. In the following lemma, we describe the relationship among these three floor vertices.

Lemma 1. *Assume that $p \equiv 3 \pmod 4$. Let E_A be a Montgomery curve lying on the floor with $A \in \mathbb{F}_p$. Let $\widetilde{E}$ be a 2-isogenous curve over $\mathbb{F}_p$ to E_A lying on the surface, unique up to $\mathbb{F}_p$-isomorphism. Then the two Montgomery curves $E_{A'}, E_{A''}$ lying on the floor are 2-isogenous over $\mathbb{F}_p$ to $\widetilde{E}$, where A', A'' are given by Eq. (11). In other words, the floor vertices adjacent to the surface vertex $[\widetilde{E}]$ are the three $\mathbb{F}_p$-isomorphism classes $[E_A]$, $[E_{A'}]$, and $[E_{A''}] \in Ell_{\mathbb{F}_p}(\mathcal{O})$.*

Proof. Let $P = (0,0)$ be the unique $\mathbb{F}_p$-rational point on E_A of exact order 2. Then $\widetilde{E}$ is isomorphic over $\mathbb{F}_p$ to the quotient $E_A/\langle P \rangle$. By Vélu's formulas [37] for a general Weierstrass equation (or see [39, Theorem 12.16]), a defining equation for $\widetilde{E}$ is given by

$$\widetilde{E} : Y^2 = X^3 + AX^2 - 4X - 4A = (X+A)(X+2)(X-2). \tag{12}$$

We transform the defining equation by $X \leftarrow X + 2$ to obtain

$$\widetilde{E} : Y^2 = X^3 + (A+6)X^2 + 4(A+2)X.$$

For the 2-torsion point $\widetilde{P} = (0,0)$ on $\widetilde{E}$, we apply Vélu's formulas again to obtain

$$\widetilde{E}/\langle \widetilde{P} \rangle : y^2 = x^3 + (A+6)x^2 - 16(A+2)x - 16(A+6)(A+2).$$

We also transform the defining equation by $x \leftarrow x - (A+6)$ to obtain

$$\widetilde{E}/\langle \widetilde{P} \rangle : y^2 = x^3 - 2(A+6)x^2 + (A-2)^2 x.$$

Moreover, we apply the change of variables $x \leftarrow u^{-2}x$ and $y \leftarrow u^{-3}y$, where $u \in \mathbb{F}_p^{\times}$ is a square root of $A - 2 \neq 0$ (the element $A - 2$ is a square in $\mathbb{F}_p$, as mentioned in the proof of [33, Theorem 7]), to obtain the Montgomery form

$$y^2 = x^3 - 2\frac{A+6}{A-2}x^2 + x.$$

Therefore $\widetilde{E}/\langle \widetilde{P} \rangle$ is isomorphic over $\mathbb{F}_p$ to $E_{A'}$. Similarly, the Montgomery form of $E_{A''}$ is obtained in the same way by applying the transformation $X \leftarrow X - 2$ to Eq. (12) defining $\widetilde{E}$. $\qquad\square$

The above lemma implies that using the typical orbits of size 3 is equivalent to ascending each vertex from the floor to the surface of the volcano structure and solving the GAIP (i.e., vectorisation) on the surface (cf. cryptanalysis of SCALLOP and OSIDH in [15,19] using specific structures). This yields a theoretical speedup by a factor of $\sqrt{3}$. In practice, however, the speedup is somewhat limited by the overhead of computing orbit representatives (see Sect. 5 for experimental results).

4   (Non-)Existence of Other Efficient Orbits

In this section, we investigate the (non-)existence of formulas analogous to [33, Theorem 7] for constructing efficiently computable orbits beyond the typical orbits of size 3, described in Subsect. 3.2. We also present a direct method to construct general orbits using lattice algorithms.

4.1   Non-Existence of Formulas for Construction of Orbits

To discuss the possibility of obtaining similar formulas to [33, Theorem 7], we recall the proof of the formula. For $\mathcal{O} = \mathbb{Z}[\pi]$ with $\pi = \sqrt{-p}$, let $\mathfrak{c} = 4\mathcal{O} + (\pi - 1)\mathcal{O}$ and P_+ (resp. P_-) be a point of E_A whose x-coordinate is 1 (resp. -1). They prove that

$$[\mathfrak{l}_1 \cdots \mathfrak{l}_n] \star E_A = [\bar{\mathfrak{c}}] \star E_A, \quad E_A[\bar{\mathfrak{c}}] = \langle P_+ \rangle,$$
$$[\mathfrak{l}_1^{-1} \cdots \mathfrak{l}_n^{-1}] \star E_A = [\mathfrak{c}] \star E_A, \quad E_A[\mathfrak{c}] = \langle P_- \rangle$$

and therefore

$$[\mathfrak{l}_1 \cdots \mathfrak{l}_n] \star E_A \cong E_A/\langle P_+ \rangle, \quad [\mathfrak{l}_1^{-1} \cdots \mathfrak{l}_n^{-1}] \star E_A \cong E_A/\langle P_- \rangle.$$

Then we obtain the desired formulas by computing $E_A/\langle P_+ \rangle$ and $E_A/\langle P_- \rangle$ with Vélu's formula. This formula is derived from the fact that there exist 4-torsion points P_+, P_- universally defined for any $A \in \mathbb{F}_p \setminus \{\pm 2\}$. Therefore, if we attempt to obtain a similar formula using the same method, it is necessary to find an n-torsion point universally defined for each A.

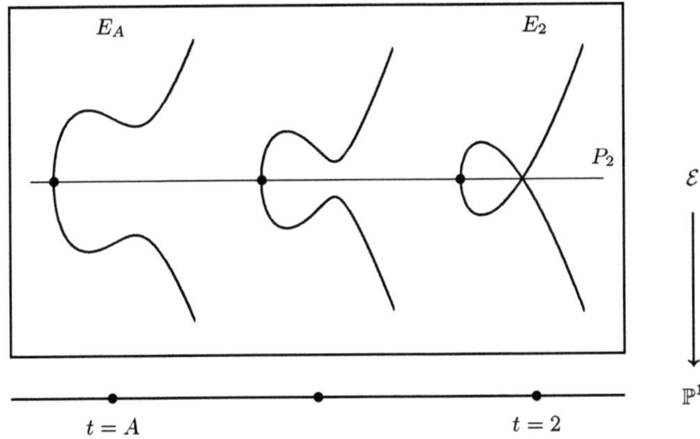

Fig. 1. Elliptic surface whose fibers are Montgomery curves.

Elliptic Surface with Montgomery Curve Fibers. Let us clarify the mathematical concept of "an n-torsion point universally defined for each A". It is precisely an n-torsion point parameterized by algebraic functions of A, e.g., $P_+ = (1, \sqrt{A+2})$, and these correspond to torsion points of the generic fiber of the elliptic surface of Montgomery curves as follows. First, we focus on the points that are parameterized by rational functions. The elliptic surface of Montgomery curves $\mathcal{E}/\mathbb{P}^1_{\mathbb{F}_p}$ is a bundle on $\mathbb{P}^1_{\mathbb{F}_p}$ such that the fiber over each $A \in \mathbb{P}^1_{\mathbb{F}_p}$ is E_A (E_A has a singularity when $A = \pm 2, \infty$, see Fig. 1). A section of the elliptic surface $\mathcal{E}/\mathbb{P}^1_{\mathbb{F}_p}$ is a morphism $s : \mathbb{P}^1_{\mathbb{F}_p} \to \mathcal{E}$ such that $s(A)$ lies in the fiber E_A for every A. More generally, a rational map from $\mathbb{P}^1_{\mathbb{F}_p}$ to $\mathcal{E}$ that is a section on some open subset gives a morphism since $\mathcal{E}$ is proper, and such a morphism is actually a section. Let η be the generic point of $\mathbb{P}^1_{\mathbb{F}_p}$ and we denote the function field of $\mathbb{P}^1_{\mathbb{F}_p}$ as $k(\mathbb{P}^1_{\mathbb{F}_p})$. Then a section $s : \mathbb{P}^1_{\mathbb{F}_p} \to \mathcal{E}$ gives a rational point $\operatorname{Spec} k(\mathbb{P}^1_{\mathbb{F}_p}) \to \mathcal{E}_\eta$ and vice versa.

Putting the above discussion together, the points that are parameterized by rational functions correspond to rational points of the generic fiber of $\mathcal{E}$ (Fig. 1).

Proposition 1. *Let F be the function field $\mathbb{F}_p(t) \cong k(\mathbb{P}^1_{\mathbb{F}_p})$. Let E be the generic fiber of the elliptic surface $\mathcal{E}$ namely, E is the elliptic curve defined by the equation $y^2 = x^3 + tx^2 + x$ over F. The set $E(F)$ contains the 2-torsion point $P_2 = (0,0)$ and no other nonzero F-rational points. In other words, P_2 is the only nonzero point parameterized by rational functions.*

Proof. By Tate's algorithm, we can determine the types of singular fibers. In our case, the fibers at $t = \pm 2$ are of type I_1 and the fiber at $t = \infty$ is of type I_4^*.

The number of components of the fibers at $t = \pm 2$ (resp. $t = \infty$) is 1 (resp. 9) over $\overline{\mathbb{F}}_p$[35, Appendix C. Table 15.1]. Moreover, the Picard number $\rho(\mathcal{E}_{\overline{\mathbb{F}}_p})$ is 10 since $\mathcal{E}_{\overline{\mathbb{F}}_p}$ is a rational elliptic surface [34, Lemma 10.2]. By applying Shioda-Tate formula [34, Theorem 10.3], we have $\operatorname{rank} E(\overline{\mathbb{F}}_p(t)) = 0$, hence $\operatorname{rank} E(F) = 0$.

Let $P \neq \infty_E$ be a 2-torsion point of E. The y-coordinate of P is 0 since $P = -P = (x_P, -y_P)$. The equation $x^3 + tx^2 + x = 0$ has no solution in F other than 0, $P = P_2 = (0,0)$. If $P \in E(F)$ is a point of exact order 4, then $[2]P = (0,0)$. By the doubling formula (5), the x-coordinate of P is ± 1; however, this is impossible because the equation $y^2 = t \pm 2$ has no solution in F.

Let v be the $(t-2)$-adic valuation of F. We denote F_v as the completion of F, and we will denote an algebraic closure of F_v by $\overline{F}_v$. Since the reduction of E at $t = 2$ is split multiplicative, E_v/F_v is isomorphic to the Tate curve [35, Appendix C. Theorem 14.1. (b)], in particular, there exists an element $q \in F_v$ such that $v(q) = -j(E_v) = 1$ and a group isomorphism $E_v(F_v) \cong F_v^\times/q^{\mathbb{Z}}$. Then

$$E_v(F_v)[p] \cong \{[u] \in F_v^\times/q^{\mathbb{Z}} \mid \exists m \in \mathbb{Z}, u^p = q^m\},$$

where $[u]$ denotes the congruent class of $u \in F_v^\times$ modulo $q^{\mathbb{Z}}$. Let $P \in E(F) \subset E(F_v)$ be a p-torsion point, which corresponds to an element $[u] \in F_v^\times/q^{\mathbb{Z}}$ such that $u^p = q^m$ for some $m \in \mathbb{Z}$. Then $\frac{m}{p} = \frac{m}{p}v(q) = v(u) \in \mathbb{Z}$ and $\frac{u}{q^{m/p}} \in \mu_p(F_v) = \{1\}$. This means that $[u] = 1$ and thus $P = \infty_E$.

Let ℓ be an odd prime with $\ell \neq p$. Let v be the valuation of F at $t = \infty$ and let F_v be the completion. We denote $\tilde{E}_v/k$ as the reduction of E_v and $\tilde{E}_v^{\mathrm{ns}}(k)$ as the group of non-singular points of $\tilde{E}_v(k)$ where $k = \mathbb{F}_p$ is the residue field of F_v. For the reduction map $E_v(F_v) \to \tilde{E}_v(k)$, we define two subsets of $E_v(F_v)$ as

$$E_v^0(F_v) = \{P \in E_v(F_v) \mid \tilde{P} \in \tilde{E}_v^{\mathrm{ns}}(k)\}, \quad E_v^1(F_v) = \{P \in E_v(F_v) \mid \tilde{P} = \infty_{\tilde{E}_v}\}.$$

Then there exists an exact sequence [35, VII. Proposition 2.1]

$$0 \longrightarrow E_v^1(F_v) \longrightarrow E_v^0(F_v) \longrightarrow \tilde{E}_v^{\mathrm{ns}}(k) \longrightarrow 0.$$

Since we have $\tilde{E}_v^{\mathrm{ns}}(k) \cong k$ as additive groups [35, Appendix C. Table 15.1] and $E_v^1(F_v)[\ell] = 0$ [35, IV. Proposition 2.3, VII. Proposition 2.2], we obtain $E_v^0(F_v)[\ell] = 0$. Hence, the morphism $E_v(F_v)[\ell] \to (E_v(F_v)/E_v^0(F_v))[\ell]$ is injective, and we conclude that $E_v(F_v)[\ell] = 0$ because the order of $E_v(F_v)/E_0(F_v)$ divides 4 [35, Appendix C. Table 15.1]. $\qquad\square$

Generic Torsion Points on Montgomery Curves. Next, we consider the points that are parameterized by algebraic functions that are not necessarily rational functions. However, such points are no longer necessarily defined over $\mathbb{F}_p$ at each fiber E_A. To ensure that $E_A/\langle P \rangle$ is defined over $\mathbb{F}_p$, it is reasonable to consider only points whose x-coordinate is parameterized by a rational function, such as P_+ and P_-.

Theorem 2. *Let $p \neq 3$ be a prime such that $p \equiv 3 \bmod 4$. Then*

$$\{P \in E_{\mathrm{tors}} \mid P = \infty_E \text{ or } x\text{-coordinate of } P \text{ lies in } F\} = \{\infty_E, P_2, \pm P_+, \pm P_-\}.$$

Proof. Through this proof, $P \neq \infty_E$ denotes a point of E whose x-coordinate belongs to F. Let $\ell \neq p$ be an odd prime and assume that $P \in E[\ell]$. For any $A \in \mathbb{F}_p \setminus \{\pm 2\}$, the reduction homomorphism $\phi_A : E[\ell] \to E_A[\ell]$ is injective

[35, VII. Proposition 3.1]. Since $\phi_A(P)$ is mapped to $\pm\phi_A(P)$ by the Frobenius action, it follows that

$$|E_A(\mathbb{F}_p)| \equiv p + 1 \pm (p + 1) \bmod \ell. \tag{13}$$

When $\ell \nmid p+1$, the supersingular curve E_0 does not satisfy Eq. (13). Consider the case of $\ell \mid p + 1$. We set $a = 4$ (resp. $a = 8$) when $p \equiv 3 \bmod 8$ (resp. $p \equiv 7 \bmod 8$). From [40, Theorem 4.1], there exists an elliptic curve E' over $\mathbb{F}_p$ whose Frobenius trace is a; in particular, we have $|E'(\mathbb{F}_p)| = 1 - a + p \equiv 0 \bmod 8$. Then E' has a Montgomery model $E_{A,B}$ because $|E'(\mathbb{F}_p)|$ is divisible by 8 [31]. If B is a square in $\mathbb{F}_p$, then $|E_{A,B}(\mathbb{F}_p)| = |E_A(\mathbb{F}_p)|$; otherwise, $|E_{A,B}(\mathbb{F}_p)| = |E_{-A}(\mathbb{F}_p)|$ since E_{-A} is a quadratic twist of E_A when $p \equiv 3 \bmod 4$. Hence E_A or E_{-A} does not satisfy Eq. (13) since $|E'(\mathbb{F}_p)| \equiv -a \not\equiv 0 \bmod \ell$.

Next, we consider p-torsions. Let v be the $(t - 2)$-adic valuation; we use the notation in the proof of the previous proposition. In this part, we regard $E_v[p]$ as an algebraic group over F_v. Assume that $P \in E_v[p](\overline{F}_v)$. Let F_v^{sep} be the separable closure of F_v in the algebraic closure $\overline{F}_v$ and put $G_{F_v} := Gal(F_v^{\mathrm{sep}}/F_v)$. P lies in $E_v[p](F_v^{\mathrm{sep}})$ and the cyclic group $\langle P \rangle$ is a G_{F_v}-stable subgroup of $E_v[p](F_v^{\mathrm{sep}})$. For each p-torsion point $[u] \in \overline{F}_v^{\times}/q^{\mathbb{Z}} \cong E_v[p](\overline{F}_v)$, there exists an integer m_u such that $u^p = q^{m_u}$. If $u' = uq^m$ is another representative of $[u]$, then $m_{u'} \equiv m_u \bmod p$. Consider the map

$$E_v[p](\overline{F}_v) \cong \overline{F}_v^{\times}/q^{\mathbb{Z}} \longrightarrow \mathbb{Z}/p\mathbb{Z}, \quad [u] \longmapsto m_u.$$

This is a group homomorphism whose kernel is $\mu_p(\overline{F}_v) = \{1\}$. The homomorphism induces an isomorphism between $\langle P \rangle$ and $\mathbb{Z}/p\mathbb{Z}$ that is compatible with the action of G_{F_v} where G_{F_v} acts on $\mathbb{Z}/p\mathbb{Z}$ trivially. Then there exists a finite étale subgroup H of $E_v[p]$ defined over F_v such that $H(F_v^{\mathrm{sep}}) = \langle P \rangle$. We find that H is isomorphic to the constant algebraic group $\mathbb{Z}/p\mathbb{Z}$ from the basic fact about finite étale groups. Thus, P lies in $H(F_v) \subset E_v[p](F_v)$. However, since $E_v[p](F_v) = \{\infty\}$ by the previous proposition, no such point P exists.

Finally, we consider points whose orders are powers of 2. If P is a 2-torsion point of E, then $P = P_2$ as in the proof of the previous proposition. When P is a point of order 4, the x-coordinate of P is ± 1 by the doubling formula (5); hence, $P = \pm P_+$ or $\pm P_-$. We assume that P is a point of exact order 8. The x-coordinate of $[2]P$ is again contained in F, so $[2]P = \pm P_+$ or $\pm P_-$. Then, we obtain an equation $x_P^4 \pm 4x_P^3 - (2 \pm 4t)x_P^2 \pm 4x_P = -1$ on the x-coordinate x_P of P by the doubling formula. Let v be the valuation of F defined by $v(f/g) = -\deg f + \deg g$ for any polynomials $f, g \in k[t] \setminus \{0\}$. If $v(x_P) > 0$ (resp. $v(x_P) < 0$), then $v(x_P^4 \pm 4x_P^3 - (2 \pm 4t)x_P^2 \pm 4x_P)$ is $v(x_P)$ (resp. $4v(x_P)$). This is impossible because this valuation must be $v(-1) = 0$. When $v(x_P) = 0$, $v(x_P^4 \pm 4x_P^3 - (2 \pm 4t)x_P^2 \pm 4x_P) = v(4t) = -1$, it is also impossible. $\qquad\square$

To summarize, it appears infeasible to derive a formula analogous to [33, Theorem 7] in the same manner. Instead, we present below a brute-force method based on the lattice structure of the class group.

4.2   Direct Construction of General Orbits via Lattice Algorithms

Let $M = \ker(\kappa \circ \Phi)$ be the sub-lattice of $\mathbb{Z}^n$ such that $\mathbb{Z}^n/M \cong Cl(\mathcal{O}_K)$, as described in Subsect. 3.2. (Recall that $M = L + \mathbb{Z}e_0$ with $L = \ker \Phi$ and $e_0 = (1,\ldots,1) \in \mathbb{Z}^n$.) Assume that the class number $h(\mathcal{O}_K)$ has a small cofactor $d > 1$. Fix a vector $v \in \mathbb{Z}^n$ such that the element $\Phi(v)$ has order d in $Cl(\mathcal{O})$ with $3 \nmid d$. We consider the lattice $N := M + \mathbb{Z}v \subseteq \mathbb{Z}^n$ to construct orbits of size $3d$ by combining two actions of $\Phi(v), \Phi(e_0) \in Cl(\mathcal{O})$. When v is long, it is inefficient to compute the action of $\Phi(v)$. It follows from [14, Theorem 1] that computing an isogeny of odd prime degree ℓ using Vélu's formulas requires $O(\ell)$ operations over $\mathbb{F}_p$ (cf. the square-root Vélu algorithm [4] with $\tilde{O}(\sqrt{\ell})$ time complexity, which is available for large ℓ such as $\ell \geq 113$). Thus, we introduce the *ℓ-weighted (Euclidean) norm* of $v = (v_1,\ldots,v_n) \in \mathbb{Z}^n$ as

$$\|v\|_\ell := \sqrt{(\ell_1 v_1)^2 + \cdots + (\ell_n v_n)^2}$$

with distinct prime factors $\ell_1,\ldots,\ell_n$ of $\frac{p+1}{4}$. The cost of computing the action of $\Phi(v)$ is approximately proportional to the ℓ-weighted norm $\|v\|_\ell$. We then need to find a representative of v modulo M whose ℓ-weighted norm is as small as possible. We adopt the approach in [18] to find such a representative. Let C be a basis matrix of M whose rows span M. Let $T := \mathrm{diag}(\ell_1,\ldots,\ell_n)$ denote the diagonal matrix whose i-th diagonal entry is ℓ_i for $i = 1,\ldots,n$. We transform C and v into CT and vT, respectively. We reduce CT using a lattice reduction algorithm, such as the LLL algorithm [23], and use a closest vector problem algorithm over a reduced basis, such as Babai's nearest plane algorithm [2], to find a vector u in the lattice spanned by the rows of CT such that u is close to vT in the standard Euclidean norm. Finally, we take $w = v - uT^{-1}$ as a short representative of v modulo M with respect to the ℓ-weighted norm (note that $uT^{-1} \in M$ since u is included in the lattice spanned by the rows of CT). If necessary, we replace w with $w + ke_0$ for $k \in \{\pm 1\}$, so that the element $\Phi(w)$ has order d in $Cl(\mathcal{O})$.

Example 1. We take $n = 5$ and $p = 4 \cdot 3 \cdot 5 \cdot 11 \cdot 17 \cdot 61 - 1 = 684419$, that is, $(\ell_1,\ldots,\ell_5) = (3, 5, 11, 17, 61)$. By computing a basis of $L = \ker \Phi$, we obtain

$$Cl(\mathcal{O}) \cong \mathbb{Z}/705\mathbb{Z} \cong \mathbb{Z}/3\mathbb{Z} \times \mathbb{Z}/5\mathbb{Z} \times \mathbb{Z}/47\mathbb{Z},$$

$$\langle [\mathfrak{l}_1 \mathfrak{l}_2 \mathfrak{l}_3 \mathfrak{l}_4 \mathfrak{l}_5] \rangle \times \langle [\mathfrak{l}_1 \mathfrak{l}_3 \mathfrak{l}_4^{-1} \mathfrak{l}_5^{-1}] \rangle \times \langle [\mathfrak{l}_1^2 \mathfrak{l}_3 \mathfrak{l}_5^{-1}] \rangle.$$

Furthermore, an LLL-reduced basis matrix of $M = \ker(\kappa \circ \Phi)$ is computed as

$$C = \begin{pmatrix} 1 & 1 & 1 & 1 & 1 \\ 1 & -1 & 0 & 2 & -1 \\ 1 & -1 & 2 & -2 & 0 \\ 1 & -1 & -1 & 0 & 3 \\ 2 & 3 & -2 & -1 & -1 \end{pmatrix},$$

whose rows span M. To construct orbits of size 5, we consider the ideal class $[\mathfrak{l}_1 \mathfrak{l}_3 \mathfrak{l}_4^{-1} \mathfrak{l}_5^{-1}] = \Phi(v)$ with $v = (1, 0, 1, -1, -1)$, which is a generator of the subgroup of $Cl(\mathcal{O})$ of order 5. We apply the above procedure to obtain a short representative of $\boldsymbol{v}$ modulo M with respect to the ℓ-weighted norm as $\boldsymbol{w} = (-4, -2, 0, -1, 0)$. The ℓ-weighted norm of $\boldsymbol{w}$ is $\sqrt{533}$, whereas that of $\boldsymbol{v}$ is $\sqrt{4140}$. Despite this reduction, the overhead of computing the action of $\Phi(\boldsymbol{w}) \in Cl(\mathcal{O})$ remains significant in practice (see Sect. 5 for experimental results).

5 Implementation and Experiments

In this section, we implement a meet-in-the-middle algorithm using several orbits in SageMath [17]. We also report experimental results for solving the GAIP.

5.1 Implementation in SageMath

To use class group orbits, we precompute the structure of the class group $Cl(\mathcal{O})$ of $\mathcal{O} = \mathbb{Z}[\sqrt{-p}]$ for a prime of the CSIDH form $p = 4\ell_1 \cdots \ell_n - 1$. Specifically, we implemented algorithms in [7,20] to compute the class group and its generators. Unlike [20], we adopted the heuristic in [10] that $\mathfrak{l}_1, \ldots, \mathfrak{l}_n$ generate the class group. For these computations, we used the SageMath functions `hnf()` and `smith_form()` to compute the Hermite and Smith normal forms of a matrix. We also implemented [10, Algorithm 2] (or [24, Algorithm 4]) for evaluating the class group action. In the algorithm, we used Vélu's formulas on Montgomery curves (see [14, Theorem 1]) to compute isogenies corresponding to class group actions.

5.2 Experiments

In Table 1, we show three CSIDH primes $p = 4\ell_1 \ldots \ell_n - 1$ and the structure of the corresponding class groups of $\mathcal{O} = \mathbb{Z}[\sqrt{-p}]$ for our experiments. For each p, we present a vector $\boldsymbol{w} \in \mathbb{Z}^n$ such that the element $\Phi(\boldsymbol{w})$ has order 5 in $Cl(\mathcal{O})$. In particular, the vector $\boldsymbol{w}$ is reduced modulo the lattice $M = \ker(\kappa \circ \Phi)$ with respect to the ℓ-weighted norm, as in Example 1, in order to construct orbits of size 5 as efficiently as possible. We fix $E_0 : y^2 = x^3 + x$, and take a random Montgomery curve E_A defined over $\mathbb{F}_p$ lying on the floor in our experiments. We execute the meet-in-the-middle algorithm using orbits of three sizes for solving the GAIP between $[E_0]$ and $[E_A] \in Ell_{\mathbb{F}_p}(\mathcal{O})$. Specifically, we consider three lattices $L = \ker \Phi$, M, and $N = M + \mathbb{Z}\boldsymbol{w}$, to construct orbits of sizes 1, 3, and 15, respectively. In our implementation, we choose the smallest Montgomery coefficient within each orbit as its representative.

In Table 2, we show experimental results for the three parameter sets listed in Table 1. For each parameter set, we performed 100 times with randomly chosen curves E_A, while we computed the three sets S_L, S_M, and S_N only once. All experiments were conducted on a single core of Intel Xeon Gold 5222 CPU@3.80 GHz with 32.0 GByte memory. As shown in Table 2, when using orbits of size 3 (i.e., the lattice M), we can reduce the search space (e.g., S_M), making the meet-in-the-middle algorithm more efficient than when orbits are not used

Table 1. Three CSIDH primes $p = 4\ell_1 \cdots \ell_n - 1$ and the class groups of $\mathcal{O} = \mathbb{Z}[\sqrt{-p}]$ for experiments (The vector $\boldsymbol{w}$ is reduced modulo the lattice $M = \ker(\kappa \circ \Phi)$ with respect to the ℓ-weighted norm)

(1)	$p = 1050510250358939$ (50-bit)
	$n = 10, (\ell_1, \ldots, \ell_n) = (3, 5, 7, 37, 43, 53, 59, 71, 73, 97)$
	$Cl(\mathcal{O}) \cong \mathbb{Z}/53926395\mathbb{Z} \cong \mathbb{Z}/3\mathbb{Z} \times \mathbb{Z}/5\mathbb{Z} \times \mathbb{Z}/7\mathbb{Z} \times \mathbb{Z}/107\mathbb{Z} \times \mathbb{Z}/33599\mathbb{Z}$
	$\left[\mathfrak{l}_1^{-2}\mathfrak{l}_2^{2}\mathfrak{l}_3^{-1}\mathfrak{l}_4^{-1}\mathfrak{l}_5^{-3}\mathfrak{l}_6^{2}\mathfrak{l}_8^{-3}\mathfrak{l}_{10}^{1}\right]$ is a generator of $Cl(\mathcal{O})$
	$\boldsymbol{w} = (6, 2, 0, -1, 2, 0, 0, 0, -1, -1)$ such that $\Phi(\boldsymbol{w})$ has order 5 in $Cl(\mathcal{O})$
(2)	$p = 9062879235581005 79$ (60-bit)
	$n = 11, (\ell_1, \ldots, \ell_n) = (3, 5, 7, 29, 59, 79, 97, 103, 107, 109, 137)$
	$Cl(\mathcal{O}) \cong \mathbb{Z}/1136478765\mathbb{Z} \cong \mathbb{Z}/3\mathbb{Z} \times \mathbb{Z}/5\mathbb{Z} \times \mathbb{Z}/101\mathbb{Z} \times \mathbb{Z}/750151\mathbb{Z}$
	$\left[\mathfrak{l}_1^{-2}\mathfrak{l}_3^{-3}\mathfrak{l}_4^{4}\mathfrak{l}_5^{-1}\mathfrak{l}_6^{1}\mathfrak{l}_8^{1}\mathfrak{l}_{11}^{1}\right]$ is a generator of $Cl(\mathcal{O})$
	$\boldsymbol{w} = (2, -2, 8, 4, 2, -1, 0, 0, -1, 1)$ such that $\Phi(\boldsymbol{w})$ has order 5
(3)	$p = 868133722131234381371$ (70-bit)
	$n = 12, (\ell_1, \ldots, \ell_n) = (3, 11, 23, 41, 53, 71, 83, 97, 101, 127, 131, 137)$
	$Cl(\mathcal{O}) \cong \mathbb{Z}/56368685655\mathbb{Z} \cong \mathbb{Z}/9\mathbb{Z} \times \mathbb{Z}/5\mathbb{Z} \times \mathbb{Z}/1252637459\mathbb{Z}$
	$\left[\mathfrak{l}_1^{2}\mathfrak{l}_2^{-3}\mathfrak{l}_3^{-2}\mathfrak{l}_4^{1}\mathfrak{l}_5^{4}\mathfrak{l}_6^{-2}\mathfrak{l}_7^{-1}\mathfrak{l}_9^{2}\mathfrak{l}_{10}^{-3}\mathfrak{l}_{11}^{2}\right]$ is a generator of $Cl(\mathcal{O})$
	$\boldsymbol{w} = (0, -6, -3, -2, 3, -4, -2, -1, 0, 0, -2, 0)$ such that $\Phi(\boldsymbol{w})$ has order 5

Table 2. Experimental results for solving the GAIP using the meet-in-the-middle algorithm with several orbits ($L = \ker \Phi$, $M = \ker(\kappa \circ \Phi)$, and $N = M + \mathbb{Z}\boldsymbol{w}$ are lattices for constructing orbits of sizes 1, 3, and 15, respectively, where $\boldsymbol{w}$ is given in Table 1 such that $\Phi(\boldsymbol{w})$ has order 5 in $Cl(\mathcal{O})$)

Parameter in Table 1	Lattice to be used	Size of orbits	Average running time (seconds)	Number of stored elements
(1)	L	1	19.6	$\#S_L = 6,480$
(p: 50-bit)	M	3	16.7	$\#S_M = 2,592$
	N	15	95.6	$\#S_N = 1,250$
(2)	L	1	111.8	$\#S_L = 67,228$
(p: 60-bit)	M	3	83.2	$\#S_M = 33,614$
	N	15	804.5	$\#S_N = 15,552$
(3)	L	1	1380.1	$\#S_L = 196,608$
(p: 70-bit)	M	3	1085.2	$\#S_M = 98,304$
	N	15	6244.1	$\#S_N = 33,614$

(i.e., the case of using the lattice L). While a theoretical speedup of $\sqrt{3} \approx 1.73$ can be expected, the actual speedup observed in our experiments was limited to about 1.3 times on average, since the search space is not reduced by a factor of 3 in practice. On the other hand, when using orbits of size 15, the search space can be reduced even further, but the running time increases considerably due to the computational overhead of evaluating the action of $\Phi(\boldsymbol{w}) \in Cl(\mathcal{O})$.

6   Conclusion

We presented a general framework for constructing class group orbits to solve the GAIP via a meet-in-the-middle algorithm. In the CSIDH setting, efficiently computable orbits of size 3 can be obtained by leveraging the formula provided in [33, Theorem 7]. However, our investigation in Subsect. 4.1 suggests that an analogous formula for constructing other efficiently computable orbits does not exist. Consequently, we proposed a brute-force method using lattice algorithms to construct general orbits that remain as computationally efficient as possible. Nevertheless, even with this method, the overhead of computing a representative of each general orbit remains a significant bottleneck. Indeed, our experiments demonstrated that using typical orbits of size 3 makes the meet-in-the-middle algorithm consistently effective. In contrast, using general orbits can reduce the required memory storage, but it remains impractical. In conclusion, using the typical orbits of size 3 is effective for cryptanalysis of CSIDH and its variants, and a theoretical speedup of $\sqrt{3}$ can be expected. In practice, however, this speedup is somewhat limited due to the overhead of computing orbit representatives. Furthermore, our framework is also applicable to low-memory Pollard-style random walk algorithms, such as algorithms of [12].

Acknowledgements. This work was supported by JST K Program Grant Number JPMJKP24U2, and JSPS KAKENHI Grant Numbers JP23K18469 and JP25H00399, Japan.

References

1. Alamati, N., De Feo, L., Montgomery, H., Patranabis, S.: Cryptographic group actions and applications. In: Moriai, S., Wang, H. (eds.) ASIACRYPT 2020. LNCS, vol. 12492, pp. 411–439. Springer, Cham (2020). https://doi.org/10.1007/978-3-030-64834-3_14
2. Babai, L.: On Lovász' lattice reduction and the nearest lattice point problem. Combinatorica **6**, 1–13 (1986)
3. Banegas, G., et al.: CTIDH: faster constant-time CSIDH. IACR Trans. Cryptogr. Hardware Embed. Syst. **2021**(4), 351–387 (2021)
4. Bernstein, D.J., De Feo, L., Leroux, A., Smith, B.: Faster computation of isogenies of large prime degree. Open Book Ser. **4**(1), 39–55 (2020)
5. Bernstein, D.J., Lange, T., Martindale, C., Panny, L.: Quantum circuits for the CSIDH: optimizing quantum evaluation of isogenies. In: Ishai, Y., Rijmen, V. (eds.) EUROCRYPT 2019. LNCS, vol. 11477, pp. 409–441. Springer, Cham (2019). https://doi.org/10.1007/978-3-030-17656-3_15
6. Beullens, W., Kleinjung, T., Vercauteren, F.: CSI-FiSh: efficient isogeny based signatures through class group computations. In: Galbraith, S.D., Moriai, S. (eds.) ASIACRYPT 2019. LNCS, vol. 11921, pp. 227–247. Springer, Cham (2019). https://doi.org/10.1007/978-3-030-34578-5_9

7. Biasse, J.-F., Bonnetain, X., Pring, B., Schrottenloher, A., Youmans, W.: A trade-off between classical and quantum circuit size for an attack against CSIDH. J. Math. Cryptol. **15**(1), 4–17 (2020)

8. Bos, J.W., Lenstra, A.K.: Topics in computational number theory inspired by. Presented at the (2017)

9. Castryck, W., Decru, T.: CSIDH on the Surface. In: Ding, J., Tillich, J.-P. (eds.) PQCrypto 2020. LNCS, vol. 12100, pp. 111–129. Springer, Cham (2020). https://doi.org/10.1007/978-3-030-44223-1_7

10. Castryck, W., Lange, T., Martindale, C., Panny, L., Renes, J.: CSIDH: an efficient post-quantum commutative group action. In: Peyrin, T., Galbraith, S. (eds.) ASIACRYPT 2018. LNCS, vol. 11274, pp. 395–427. Springer, Cham (2018). https://doi.org/10.1007/978-3-030-03332-3_15

11. Cervantes-Vázquez, D., Chenu, M., Chi-Domínguez, J.-J., De Feo, L., Rodríguez-Henríquez, F., Smith, B.: Stronger and faster side-channel protections for CSIDH. In: Schwabe, P., Thériault, N. (eds.) LATINCRYPT 2019. LNCS, vol. 11774, pp. 173–193. Springer, Cham (2019). https://doi.org/10.1007/978-3-030-30530-7_9

12. Chi-Domínguez, J.-J., Esser, A., Kunzweiler, S., May, A.: Low memory attacks on small key CSIDH. In: Applied Cryptography and Network Security (ACNS 2023), volume 13906 of Lecture Notes in Computer Science, pp. 276–304. Springer (2023)

13. Chi-Domínguez, J.-J., Rodríguez-Henríquez, F.: Optimal strategies for CSIDH. Adv. Math. Commun. **16**(2), 383–411 (2022)

14. Costello, C., Hisil, H.: A simple and compact algorithm for SIDH with arbitrary degree isogenies. In: Takagi, T., Peyrin, T. (eds.) ASIACRYPT 2017. LNCS, vol. 10625, pp. 303–329. Springer, Cham (2017). https://doi.org/10.1007/978-3-319-70697-9_11

15. Dartois, P., De Feo, L.: On the security of OSIDH. In: Public-Key Cryptography-PKC 2022, volume 13177 of Lecture Notes in Computer Science, pp. 52–81. Springer (2022)

16. Delfs, C., Galbraith, S.D.: Computing isogenies between supersingular elliptic curves over $\mathbb{F}_p$. Des., Codes, Cryptogr. **78**, 425–440 (2016)

17. The Sage Developers. SageMath, the Sage Mathematics Software System (Version 10.7) (2020). http://www.sagemath.org

18. Ding, J., Kudo, M., Okumura, S., Takagi, T., Tao, C.: Cryptanalysis of a public key cryptosystem based on Diophantine equations via weighted LLL reduction. Jpn. J. Ind. Appl. Math. **35**(3), 1123–1152 (2018). https://doi.org/10.1007/s13160-018-0316-x

19. De Feo, L., et al.: SCALLOP: scaling the CSI-FiSh. In: Public-Key Cryptography-PKC 2023, volume 13940 of Lecture Notes in Computer Science, pp. 345–375. Springer (2023)

20. Hafner, J.L., McCurley, K.S.: A rigorous subexponential algorithm for computation of class groups. J. Am. Math. Soc. **2**(4), 837–850 (1989)

21. Hutchinson, A., LeGrow, J., Koziel, B., Azarderakhsh, R.: Further optimizations of CSIDH: a systematic approach to efficient strategies, permutations, and bound vectors. In: Conti, M., Zhou, J., Casalicchio, E., Spognardi, A. (eds.) ACNS 2020. LNCS, vol. 12146, pp. 481–501. Springer, Cham (2020). https://doi.org/10.1007/978-3-030-57808-4_24

22. Kuperberg, G.: A subexponential-time quantum algorithm for the dihedral hidden subgroup problem. SIAM J. Comput. **35**(1), 170–188 (2005)

23. Lenstra, A.K., Lenstra, H.W., Lovász, L.: Factoring polynomials with rational coefficients. Math. Ann. **261**(4), 515–534 (1982)

24. Maino, L., Mula, M., Pintore, F.: A review of mathematical and computational aspects of CSIDH algorithms. J. Algebra Appl. **23**(7), 2530002 (2024)
25. May, A., Ostuzzi, M.: Multiple group action DLogs with(out) precomputation. In: Public-Key Cryptography-PKC 2025, volume 15676 of Lecture Notes in Computer Science, pp. 364–387. Springer (2025)
26. Meyer, M., Campos, F., Reith, S.: On lions and Elligators: an efficient constant-time implementation of CSIDH. In: Ding, J., Steinwandt, R. (eds.) PQCrypto 2019. LNCS, vol. 11505, pp. 307–325. Springer, Cham (2019). https://doi.org/10.1007/978-3-030-25510-7_17
27. Meyer, M., Reith, S.: A faster way to the CSIDH. In: Progress in Cryptology-INDOCRYPTO 2018, volume 11356 of Lecture Notes in Computer Science, pp. 137–152. Springer (2018)
28. Montgomery, H., Zhandry, M.: Full quantum equivalence of group action DLog and CDH, and more. J. Cryptol. **37**(4), 39 (2024)
29. Nakagawa, K., Onuki, H., Takayasu, A., Takagi, T.: L^1-norm ball for CSIDH: optimal strategy for choosing the secret key space. Discret. Appl. Math. **328**, 70–88 (2023)
30. Neukirch, J.: Algebraic number theory, volume 322 of Graduate Texts in Mathematics. Springer Science & Business Media (2013)
31. Okeya, K., Miyazaki, K., Sakurai, K.: A note on the number of Montgomery-form elliptic curves for cryptosystem. Jpn. Soc. Ind. Appl. Math. **12**(4), 255–268 (2002)
32. Onuki, H., Aikawa, Y., Yamazaki, T., Takagi, T.: A constant-time algorithm of CSIDH keeping two points. IEICE Trans. Fundam. Electron. Commun. Comput. Sci. **103**(10), 1174–1182 (2020)
33. Onuki, H., Takagi, T.: On collisions related to an ideal class of order 3 in CSIDH. In: Aoki, K., Kanaoka, A. (eds.) IWSEC 2020. LNCS, vol. 12231, pp. 131–148. Springer, Cham (2020). https://doi.org/10.1007/978-3-030-58208-1_8
34. Shioda, T.: On the Mordell-Weil lattices. Commentarii Mathematici Universitatis Sancti Pauli **39**(2), 211–240 (1990)
35. Silverman, J.H.: The Arithmetic of Elliptic Curves, volume 106 of Graduate Texts in Mathematics, second edition . Springer, New York(2009)
36. Silverman, J.H.: Advanced topics in the arithmetic of elliptic curves, volume 151 of Graduate Texts in Mathematics. Springer Science & Business Media (2013)
37. Vélu, J.: Isogénies entre courbes elliptiques. Comptes-Rendus de l'Académie des Sci. **273**, 238–241 (1971)
38. Voight, J.: Quaternion Algebras, volume 288 of Graduate Texts in Mathematics. Springer, Cham (2021)
39. Washington, L.C.: Elliptic Curves: Number Theory and Cryptography, 2nd edn. CRC Press, Boca Raton (2008)
40. Waterhouse, W.C.: Abelian varieties over finite fields. Annales Scientifiques de L'É.N.S **2**(4), 521–560 (1969)

Christian Kaspers$^{(\boxtimes)}$ and Alexander Pott

Otto von Guericke University Magdeburg, Faculty of Mathematics, Institute for
Algebra and Geometry, 39106 Magdeburg, Germany
{christian.kaspers,alexander.pott}@ovgu.de

Abstract. Locally-APN functions are a generalization of APN power
functions which satisfy the APN condition only locally. In this paper, we
naturally extend the original definition of locally-APN functions from
power functions to arbitrary functions on $\mathbb{F}_2^n$. We then present a new
construction of locally-APN functions which uses a spread and a Sidon
set. This combinatorial approach leads to a large family of locally-APN
functions containing several known locally-APN power functions. We precisely determine the differential spectrum and the algebraic degree of the
functions in this family.

Keywords: vectorial function · almost perfect nonlinear · locally-APN
function · differential spectrum · Sidon set · spread · Kloosterman sum

1 Introduction

A function $f\colon \mathbb{F}_2^n \to \mathbb{F}_2^n$ is called almost perfect nonlinear (APN) if the equation
$f(x + a) + f(x) = b$ has 0 or 2 solutions for all nonzero $a \in \mathbb{F}_2^n$ and all $b \in$
$\mathbb{F}_2^n$. Nyberg [19] introduced APN functions in 1994 as functions with optimal
differential properties which makes them intriguing for cryptographers, but they
have applications in coding theory and design theory, as well [7,20].

Despite extensive research in the past 30 years, the number of known APN
functions is still limited. While there are, by now, many sporadic examples in
small dimensions that were found by computer search—for example, Beierle et
al. [2] recently discovered millions of APN functions on $\mathbb{F}_2^8$—only few infinite
families are known. According to a recent list by Li and Kaleyski [16], we know
six infinite families of APN power functions and 23 infinite families of APN
non-power functions.

Note that for a power function $f(x) = x^d$ the APN property is satisfied if it
holds for $a = 1$, so if $f(x + 1) + f(x) = b$ has 0 or 2 solutions for all $b \in \mathbb{F}_2^n$. The
small number of such functions motivated Blondeau, Canteaut, and Charpin [5]
to introduce the following relaxation of almost perfect nonlinearity for power
functions: they call a power function $f(x) = x^d$ *locally-APN* if the equation
$f(x + 1) + f(x) = b$ has at most two solutions for all $b \in \mathbb{F}_2^n \backslash \mathbb{F}_2$.

L. Batina and F. Özbudak (Eds.): WAIFI 2026, LNCS 16611, pp. 184–198, 2026.
https://doi.org/10.1007/978-3-032-27574-5_12

Clearly, every APN power function is locally-APN, but there are also locally-APN functions which are not APN: all the known ones exist on $\mathbb{F}_{2^{2m}}$, and the most prominent example is the inverse function $x^{2^{2m}-2}$. To our knowledge, only the following other locally-APN power functions are known: Blondeau, Canteaut, and Charpin [5] showed that for $m \geq 3$ the functions x^{2^m-1} and $x^{2^{m+1}-1}$ are locally-APN. More recently, Hu et al. [14] proved that $x^{k(2^m-1)}$ is locally-APN if $\gcd(k, 2^m+1) = 1$, and Xie et al. [23] showed that $x^{s(2^m-1)+1}$ is locally-APN whenever there is a positive integer k such that $\gcd(k,m) = \gcd(2^k+1, 2^m+1) = 1$ and $s(2^k+1) \equiv 1 \pmod{2^m+1}$. According to Xie et al. [23], those families cover all locally-APN power functions that are not APN for $m \leq 10$.

Studying the original definition of a locally-APN power function f, note that it implies that the equation $f(x+a) + f(x) = b$ has at most two solutions for all nonzero $a \in \mathbb{F}_2^n$ and all $b \in \mathbb{F}_2^n \setminus \{0, a^d\}$. This observation motivates the following natural extension of locally-APNness to arbitrary functions:

Definition 1. *A function* $f \colon \mathbb{F}_2^n \to \mathbb{F}_2^n$ *is called* locally-APN *if for any nonzero* $a \in \mathbb{F}_2^n$, *there exists a set* B_a *with* $|B_a| = 2$ *such that for all* $b \in \mathbb{F}_2^n \setminus B_a$, *the equation*

$$f(x+a) + f(x) = b$$

has at most two solutions.

Clearly, the locally-APN power functions from the original definition in [5] are covered by Definition 1. With our more general definition, however, there exist many more functions satisfying this property.

In this paper, we present a new construction of locally-APN functions on $\mathbb{F}_2^n$ with n even that uses a spread, i. e. a subspace partition of $\mathbb{F}_2^n$, and a Sidon set. Our construction leads to a large number of locally-APN functions including the families introduced in [5] and in [14], and our computations hint that many of them are CCZ-inequivalent. While the locally-APNness of a function itself may be of cryptographic relevance [6], we not only prove that our new functions are locally-APN, but we moreover determine their full differential spectrum and their algebraic degree, which are important parameters of a function for cryptography and coding theory [4,5,7,10].

We only use Sidon sets as a tool to construct locally-APN functions, but Sidon sets are intriguing objects of research on their own. For more information and some recent results on them, e. g. their strong relation with APN functions, we refer the reader to [8,9,11,12,18,21,22]. For more information on spreads, which can be constructed from semifields and translation planes, and their applications, e. g. in the construction of bent functions, we refer to [1,13,15].

2   Preliminaries

In this section, we present all the definitions and basic results needed in the course of the paper. We start with some preliminaries on vectorial Boolean functions and, afterwards, introduce spreads and Sidon sets.

A vectorial Boolean function is a function from $\mathbb{F}_2^n$ to $\mathbb{F}_2^n$. We identify the n-dimensional vector space $\mathbb{F}_2^n$ with the finite field $\mathbb{F}_{2^n}$, and we denote the set of nonzero elements, i. e. the multiplicative group, of $\mathbb{F}_{2^n}$ by $\mathbb{F}_{2^n}^*$. We often focus on the case that $n = 2m$ is even.

We represent a function f on $\mathbb{F}_2^n$ either by n Boolean coordinate functions $f_1, \ldots, f_n \colon \mathbb{F}_2^n \to \mathbb{F}_2$ or by a univariate polynomial function on $\mathbb{F}_{2^n}$. One important property of f is its *algebraic degree* $\deg_{alg}(f)$, which we define as the maximum degree of its coordinate functions when given in algebraic normal form (see [7] for more background). Furthermore, we introduce the following differential properties of f.

Definition 2. *Let* $f \colon \mathbb{F}_2^n \to \mathbb{F}_2^n$. *For any* $a, b \in \mathbb{F}_2^n$ *define the* differential multiplicity *of* f *with respect to* a *and* b *by*

$$\delta_f(a, b) = |\{x \in \mathbb{F}_2^n : f(x + a) + f(x) = b\}|.$$

The differential uniformity $\delta(f)$ *of* f *is defined as*

$$\delta(f) = \max_{a,b \in \mathbb{F}_2^n, a \neq 0} \delta_f(a, b).$$

The differential spectrum $\mathbb{S}_f$ *of* f *is the multiset*

$$\mathbb{S}_f = \{\delta_f(a, b) : a, b \in \mathbb{F}_2^n, a \neq 0\},$$

and we denote the multiplicity of an element $\delta \in \mathbb{S}_f$ *by* ω_δ.

Using the notation from Definition 2, a function f on $\mathbb{F}_2^n$ is APN if and only if $\delta(f) = 2$, and it is locally-APN if and only if $\delta_f(a, b) \leq 2$ for all nonzero $a \in \mathbb{F}_2^n$ and all $b \in \mathbb{F}_2^n \setminus B_a$.

We next consider spreads, which are subspace partitions of $\mathbb{F}_2^{2m}$.

Definition 3. *We define a* spread *of* $\mathbb{F}_2^{2m}$ *as a set* $P = \{U_1, \ldots, U_{2^m+1}\}$ *of* m-*dimensional subspaces* $U_1, \ldots, U_{2^m+1}$ *of* $\mathbb{F}_2^{2m}$ *such that* $U_i \cap U_j = \{0\}$ *for all* $i \neq j$.

Clearly, a spread has the following properties: $|U_i| = 2^m$ for all i, $U_1 \cup \cdots \cup U_{2^m+1} = \mathbb{F}_2^{2m}$, and $U_i \oplus U_j = \mathbb{F}_2^{2m}$ for all $i \neq j$. A classical construction of a spread of $\mathbb{F}_2^{2m}$ uses the finite field $\mathbb{F}_{2^{2m}}$: the subfield $\mathbb{F}_{2^m}$ and its multiplicative cosets $\alpha \mathbb{F}_{2^m}$, $\alpha \in \mathbb{F}_{2^{2m}}$, form a spread of $\mathbb{F}_{2^{2m}}$. We call this spread the *classical spread* and denote it by P_c.

We eventually consider Sidon sets.

Definition 4. *A set* $S \subseteq \mathbb{F}_2^n$ *is called* Sidon *if* $x + y \neq z + u$ *for all distinct* $x, y, z, u \in S$.

Clearly, every subset of a Sidon set is Sidon, too. An interesting question about Sidon sets is whether we can extend them without violating the Sidon condition.

Definition 5. *A Sidon set $S \subseteq \mathbb{F}_2^n$ is called* maximal *if for all $x \in \mathbb{F}_2^n \backslash S$ the set $S \cup \{x\}$ is not Sidon. It is said to be* sum-free *if $x + y \notin S$ for all $x, y \in S$.*

If S is a sum-free Sidon set, then $0 \notin S$ and $S \cup \{0\}$ is a Sidon set of size $|S| + 1$ that is not sum-free. Consequently, a sum-free Sidon set is not maximal. In the course of the paper, we also need k-sums of the elements of a Sidon set.

Definition 6. *For a subset $S \subseteq \mathbb{F}_2^n$ and a positive integer $k \in \mathbb{N}$, define the multiset $S^{(k)}$ of k-sums of S by*

$$S^{(k)} = \left\{ \sum_{x \in K} x : K \subseteq S \text{ and } |K| = k \right\},$$

and denote the multiplicity of $a \in S^{(k)}$ by $\mathrm{mult}_{S,k}(a)$.

In other words, $S^{(k)}$ contains precisely those elements $a \in \mathbb{F}_2^n$ for which there exist distinct $x_1, \ldots, x_k \in S$ such that $x_1 + \cdots + x_k = a$, and $\mathrm{mult}_{S,k}(a)$ denotes the number of such unordered representations of a. For simplicity, we set $\mathrm{mult}_{S,k}(a) = 0$ for all $a \notin S^{(k)}$.

We summarize some easy results on the Sidon property which follow immediately from Definition 4 and have been pointed out by Czerwinski and Pott [11].

Proposition 1. *Let $S \subseteq \mathbb{F}_2^n$. The following statements are equivalent:*

1. *S is Sidon.*
2. *$\mathrm{mult}_{S,2}(x) \leq 1$ for all $x \in \mathbb{F}_2^n$, which means $|S^{(2)}| = \binom{|S|}{2}$.*
3. *$\mathrm{mult}_{S,3}(x) = 0$ for all $x \in S$, which means S and $S^{(3)}$ are disjoint.*
4. *$\mathrm{mult}_{S,4}(0) = 0$, which means $0 \notin S^{(4)}$.*

Additionally, we remark that for a Sidon set S, we have $\mathrm{mult}_{S,3}(0) = \frac{1}{3}|S \cap S^{(2)}|$. This can be seen as follows: if $s_1, s_2, s_3 \in S$ are distinct and satisfy $s_1 + s_2 + s_3 = 0$, then this implies that $s_1, s_2, s_3 \in S^{(2)}$ since $s_1 = s_2 + s_3$, $s_2 = s_1 + s_3$ and $s_3 = s_1 + s_2$. As a consequence, we obtain the following helpful result on $\mathrm{mult}_{S,3}(0)$:

Proposition 2. *Let $S \subseteq \mathbb{F}_2^n$ be Sidon. Then $\mathrm{mult}_{S,3}(0) \leq \frac{|S|}{3}$, and equality holds if and only if $S \subseteq S^{(2)}$. Moreover, $\mathrm{mult}_{S,3}(0) = 0$ if and only if S is sum-free.*

The k-sum multiplicity $\mathrm{mult}_{S,k}$ is closely related to the Fourier transform of the set S. For its definition, recall the *absolute trace function* $\mathrm{tr} \colon \mathbb{F}_{2^n} \to \mathbb{F}_2$ defined by $\mathrm{tr}(x) = \sum_{i=0}^{n-1} x^{2^i}$.

Definition 7. *Let $f \colon \mathbb{F}_{2^n} \to \mathbb{Z}$, let $S \subseteq \mathbb{F}_{2^n}$, and define the indicator function $1_S \colon \mathbb{F}_{2^n} \to \mathbb{Z}$ of S by $1_S(x) = 1$ if $x \in S$ and $1_S(x) = 0$ if $x \notin S$. The* Fourier-Hadamard transform *of f is the function $\hat{f} \colon \mathbb{F}_{2^n} \to \mathbb{Z}$ defined by*

$$\hat{f}(a) = \sum_{x \in \mathbb{F}_{2^n}} (-1)^{\mathrm{tr}(ax)} f(x),$$

and the Fourier transform *of S is the the Fourier-Hadamard transform $\widehat{1_S}$ of 1_S.*

It follows from the definition of 1_S that

$$\widehat{1_S}(a) = \sum_{x \in S} (-1)^{\mathrm{tr}(ax)}.$$

The following two results may be well-known. We refer to Thornburgh [22] for a proof and more details.

Lemma 1. *Let $S \subseteq \mathbb{F}_2^n$, and let $a \in \mathbb{F}_2^n$ and $k \in \mathbb{N}$. We have*

$$\widehat{(1_S)^k}(a) = 2^n \left| \left\{ (x_1, \ldots, x_k) \in S^k : x_1 + \cdots + x_k = a \right\} \right|.$$

So $\frac{1}{2^n}\widehat{(1_S)^k}(a)$ equals the number of ordered representations of $a \in \mathbb{F}_2^n$ as the sum $a = x_1 + \cdots + x_k$ where, unlike in the definition of $\mathrm{mult}_{S,k}(a)$, now $x_1, \ldots, x_k \in S$ are not necessarily distinct. For $k = 3$, which will be the most relevant case in our paper, we can say even more:

Proposition 3. *Let S be a Sidon set in $\mathbb{F}_{2^n}$, and let $a \in \mathbb{F}_{2^n} \backslash S$. Then*

$$\mathrm{mult}_{S,3}(a) = \frac{1}{3 \cdot 2^{n+1}} \widehat{(1_S)^3}(a).$$

In our work, the following well-known [9] Sidon set of size $2^m + 1$ in $\mathbb{F}_{2^{2m}}$, which we will denote by S_m, plays an important role.

Proposition 4. *In $\mathbb{F}_{2^{2m}}$, the set $S_m = \left\{ x \in \mathbb{F}_{2^{2m}} : x^{2^m+1} = 1 \right\}$ is Sidon. It is sum-free if and only if m is even.*

Note that S_m is the unique subgroup of order $2^m + 1$ of $\mathbb{F}_{2^{2m}}^*$. For a proof of Proposition 4, we refer to Carlet and Mesnager [9, Propositions 4.4 and 5.2], who studied subgroups of $\mathbb{F}_{2^n}^*$ that are (sum-free) Sidon sets in $\mathbb{F}_{2^n}$. If m is even, then S_m being sum-free implies that $S_m \cup \{0\}$ is also Sidon, and every (2^m+1)-subset of $S_m \cup \{0\}$ is Sidon as well. Nagy [18, Theorem 2 and Proposition 14] showed that S_m is maximal if m is odd and $S_m \cup \{0\}$ is maximal if m is even. We add some additional results about $\mathrm{mult}_{S_m,3}$.

Proposition 5. *For the Sidon set $S_m \subseteq \mathbb{F}_{2^{2m}}$, we have $\mathrm{mult}_{S_m,3}(0) = 0$ if m is even and $\mathrm{mult}_{S_m,3}(0) = \frac{2^m+1}{3}$ if m is odd.*

Proof. If m is even, the result follows from S_m being sum-free. For odd m, we use Proposition 2 and show that $S_m \subseteq S_m^{(2)}$. We begin by proving $\mathbb{F}_4^* \subseteq S_m$: First, note that S_m is the subgroup of $(2^m - 1)$th powers in $\mathbb{F}_{2^{2m}}^*$. Let α be a generator of $\mathbb{F}_{2^{2m}}^*$. Then the subgroup $\mathbb{F}_4^*$ of $\mathbb{F}_{2^{2m}}^*$ is generated by $\beta := \alpha^{(2^{2m}-1)/3} = \alpha^{(2^m+1)(2^m-1)/3}$. Since 3 divides $2^m + 1$ if m is odd, β is a $(2^m - 1)$th power in $\mathbb{F}_{2^{2m}}^*$, and, thus, $\beta \in S$ and $\mathbb{F}_4^*$ is a subgroup of S_m. Consequently, we have $1 = \beta + \beta^2$, and multiplying this equation by $s \in S_m$ proves $S_m \subseteq S_m^{(2)}$. $\square$

We conclude this section by pointing out a connection between the Fourier transform $\widehat{1_{S_m}}$ of S_m and Kloosterman sums.

Definition 8. *For $a \in \mathbb{F}_{2^n}$, define the* Kloosterman sum *by*

$$K_n(a) = \sum_{x \in \mathbb{F}_{2^n}^*} (-1)^{\operatorname{tr}(x^{-1}+ax)}.$$

Note that we only sum over the nonzero elements of $\mathbb{F}_{2^n}$ in Definition 8. Other authors sometimes include 0 in the sum and set $0^{-1} = 0$. The following is a classical result; for a proof see [17, Proposition 2.4.2].

Lemma 2. *Let $S_m = \{x \in \mathbb{F}_{2^{2m}} : x^{2^m+1} = 1\}$, and let $a \in \mathbb{F}_{2^{2m}}$. Let $u \in \mathbb{F}_{2^m}$ be the unique element in $\mathbb{F}_{2^m}$ such that $a = u \cdot s$ for some $s \in S$. Then*

$$\widehat{1_{S_m}}(a) = -K_m(u).$$

3   A New Construction of Locally-APN Functions

In this section, we present a new construction of locally-APN functions, that uses a spread and a Sidon set.

Definition 9. *Let $m \geq 2$, and let $S = \{s_1, \ldots, s_{2^m+1}\}$ be a Sidon set in $\mathbb{F}_2^{2m}$. Let $P = \{U_1, \ldots, U_{2^m+1}\}$ be a spread of $\mathbb{F}_2^{2m}$, and write $U_i^* := U_i \setminus \{0\}$ for all i. Let $z \in \mathbb{F}_2^{2m}$. We define a function $f_{P,S,z} \colon \mathbb{F}_2^{2m} \to \mathbb{F}_2^{2m}$ by*

$$f_{P,S,z}(x) = \begin{cases} s_i & \text{if } x \in U_i^*, \\ z & \text{if } x = 0. \end{cases}$$

Theorem 1. *Write $f := f_{P,S,z}$. If $m \geq 2$, then f is locally-APN and the differential multiplicities of f with respect to a and b are given by*

$$\delta_f(a,b) = \begin{cases} 2^m - 2 + 2\gamma_f(a,b) & \text{if } b = 0, & (1.1) \\ 2 + 2\gamma_f(a,b) & \text{if } b = s_i + s_j \in S^{(2)}, a \notin U_i^* \cup U_j^*, & (1.2) \\ 2\gamma_f(a,b) & \text{if } b = s_i + s_j \in S^{(2)}, a \in U_i^* \cup U_j^*, & (1.3) \\ 2\gamma_f(a,b) & \text{if } b \notin S^{(2)} \cup \{0\}, & (1.4) \end{cases}$$

where

$$\gamma_f(a,b) = \begin{cases} 1 & \text{if } f(a) = b + z, \\ 0 & \text{if } f(a) \neq b + z. \end{cases}$$

Proof. For any nonzero $x \in \mathbb{F}_2^{2m}$, denote by U_x the set $U_i \in P$ with $x \in U_i$. Consider the equation

$$f(x + a) + f(x) = b \tag{2}$$

for some $a, b \in \mathbb{F}_2^{2m}$ with $a \neq 0$. Note that $x = 0$ and $x = a$ solve (2) if and only if $f(a) = b + z$, i.e. $\gamma_f(a,b) = 1$. Thus, we exclude $x \in \{0, a\}$ in the remainder of the proof.

We determine the number of solutions $x \notin \{0, a\}$ of (2) depending on a and b. First, let $b = 0$. Then (2) implies $f(x + a) = f(x)$ or, equivalently, $U^*_{x+a} = U^*_x$. This is satisfied if and only if $x \in U^*_a \setminus \{a\}$, i.e. for $2^m - 2$ distinct x. Second, let $b \neq 0$ and $b \notin S^{(2)}$. Then (2) has no solutions, since $f(x + a) + f(x) \in S^{(2)}$.

Eventually, let $b \in S^{(2)}$. Then $b = s_i + s_j$ for unique distinct $i, j \in \{1, \ldots, 2^m + 1\}$. Consequently, a solution x of (2) has to satisfy $x \in U^*_i$ and $x + a \in U^*_j$ or the other way round. If $a \in U^*_i$, then $x \in U^*_i$ if and only if $x + a \in U^*_i$, and if $a \in U^*_j$, then $x + a \in U^*_j$ if and only if $x \in U^*_j$. Both are contradictions. Thus, (2) has no solutions if $a \in U^*_i \cup U^*_j$. If $a \notin U^*_i \cup U^*_j$, however, take into account that $U_i \oplus U_j = \mathbb{F}_2^{2m}$. Thus, there exist unique $x \in U^*_i$ and $y \in U^*_j$ with $x + y = a$ or equivalently $y = x + a$ for every nonzero a. Consequently, (2) has two solutions, x and $x + a$, in this case.

Summarizing the above results yields (1.1)–(1.4). Now observe that for a fixed nonzero $a \in \mathbb{F}_2^{2m}$, we have $\delta_f(a, b) > 2$ only for $b = 0$ and, possibly, for $b \in S^{(2)}$ if $b = f(a) + z$. It follows that f is locally-APN. $\qquad\square$

Remark 1. Theorem 1 includes the locally-APN power function $x^{k(2^m - 1)}$, where $\gcd(k, 2^m + 1) = 1$, that Canteaut, and Charpin [5] and Hu et al. [14] studied. More precisely: for k coprime to $2^m + 1$, define the ordered set $S_{m,k} = \{s^k : s \in S_m\}$ (since S_m is a subgroup, $S_{m,k}$ is a permutation of S_m). Then $f_{P_c, S_{m,k}, 0}(x) = x^{k(2^m - 1)}$, in particular $f_{P_c, S_m, 0}(x) = x^{2^m - 1}$.

The construction presented above works for arbitrary m: in $\mathbb{F}_{2^{2m}}$, we always have the classical spread P_c and the Sidon set S_m of size $2^m + 1$ from Proposition 4. Note that our construction is very flexible. We can choose P, S, and z arbitrarily, and also the assignment of the elements of S to the U^*_i is arbitrary. So for fixed P and z, any permutation of S gives a new function.

Regarding the spread P, note that each spread corresponds to a translation plane, and there are numerous constructions of translation planes, see e. g. [3]. For $m = 2$ and $m = 3$, there is, up to equivalence, only the classical spread. For $m \geq 4$, however, there are many more.

Regarding the Sidon set S, note that S_m is currently the only known infinite family of Sidon sets of size $2^m + 1$. There are two other constructions to obtain this family: one was previously known and uses Goppa codes; the other one was introduced by Nagy [18] who constructed the Sidon set as an ellipse in $\mathrm{AG}(2, 2^m)$. Nagy [18] showed that both these Sidon sets are equivalent to S_m. According to Czerwinski and Pott [11], S_m is the unique Sidon set of size $2^m + 1$ if $m = 2$ or $m = 3$. Computational results by Ingo Czerwinski indicate that this also holds for $m = 4$ but that there are more Sidon sets of size $2^m + 1$ if $m = 5$.

Remark 2. Note that the total number of locally-APN functions originating from Theorem 1 is quite large because of the flexibility in this construction we pointed out above. Our computations in small dimensions hint that many of the functions are CCZ-inequivalent. We leave it as an open problem to determine the number of inequivalent functions in the family.

4   The Differential Spectrum and the Algebraic Degree of the Locally-APN Function $f_{P,S,z}$

In this section, we determine the exact differential spectrum of the locally-APN functions $f_{P,S,z}$ introduced in Definition 9, and we determine their algebraic degree.

Theorem 2. *The function $f_{P,S,z}$ on $\mathbb{F}_2^{2m}$ has the following differential spectrum $\mathbb{S} := \mathbb{S}_{f_{P,S,z}}$, where $\mathbb{S}$ is a multiset and the multiplicity of $\delta \in \mathbb{S}$ is denotet by ω_δ.*

(C1) If $z \in S$, then $\mathbb{S} = \{2^m, 2^m - 2, 2, 0\}$ with

$$\omega_{2^m} = 2^m - 1,$$
$$\omega_{2^m - 2} = 2^m(2^m - 1),$$
$$\omega_2 = 2^{m-1}(2^{2m} + 1)(2^m - 1),$$
$$\omega_0 = (2^m - 1)\left(2^{2m} + (2^m + 1)^2(2^{m-1} - 1)\right).$$

(C2) If $z \notin S$ and $S \cup \{z\}$ is Sidon, then $\mathbb{S} = \{2^m - 2, 2, 0\}$ with

$$\omega_{2^m - 2} = 2^{2m} - 1,$$
$$\omega_2 = (2^{2m} - 1)(2^{2m-1} - 2^{m-1} + 1),$$
$$\omega_0 = (2^{2m} - 1)(2^{2m-1} + 2^{m-1} - 2).$$

(C3) If $z \notin S$ and $S \cup \{z\}$ is not Sidon, then $\mathbb{S} = \{2^m - 2, 4, 2, 0\}$ with

$$\omega_{2^m - 2} = 2^{2m} - 1,$$
$$\omega_4 = (2^m - 1)\,3\text{mult}_{S,3}(z),$$
$$\omega_2 = (2^m - 1)\left(2^{m-1}(2^{2m} - 1) + 1 - 6\text{mult}_{S,3}(z)\right),$$
$$\omega_0 = (2^m - 1)\left((2^m + 1)(2^{2m-1} + 2^{m-1} - 2) + 3\text{mult}_{S,3}(z)\right),$$

where $\text{mult}_{S,3}(z)$ is as defined in Definition 6.

Proof. Write $f := f_{P,S,z}$, in brief. First, we introduce a helpful notation. We denote the number of (a, b) in each of the cases (1.1)–(1.4) such that $f(x + a) + f(x) = b$ has δ solutions by $\omega_{\delta,(1.1)}, \ldots, \omega_{\delta,(1.4)}$, respectively. From (1.1)–(1.4), it is clear that

$$\omega_{2^m} = \omega_{2^m,(1.1)}, \qquad \omega_{2^m - 2} = \omega_{2^m - 2,(1.1)},$$
$$\omega_4 = \omega_{4,(1.2)}, \qquad \omega_2 = \omega_{2,(1.2)} + \omega_{2,(1.3)} + \omega_{2,(1.4)}, \qquad (3)$$
$$\omega_0 = \omega_{0,(1.3)} + \omega_{0,(1.4)}.$$

Next, we determine the total number of $(a, b) \in \mathbb{F}_2^{2m} \setminus \{0\} \times \mathbb{F}_2^{2m}$ in each of the cases of (1.1)–(1.4) and denote these numbers by $n_{(1.1)}, \ldots, n_{(1.4)}$, respectively.

First, if $b = 0$, then b is uniquely determined and we have $2^{2m} - 1$ choices for a, so $n_{(1.1)} = 2^{2m} - 1$. Second, if $b = s_i + s_j \in S^{(2)}$ and $a \notin U_i^* \cup U_j^*$, we have

$\binom{2^m+1}{2} = (2^m+1)2^{m-1}$ choices for b and $(2^{2m}-1)-2(2^m-1) = (2^m-1)^2$ choices for a. Consequently, $n_{(1.2)} = (2^{2m} - 1)(2^m - 1)2^{m-1}$. Third, if $b = s_i + s_j \in S^{(2)}$ and $a \in U_i^* \cup U_j^*$, we have $(2^m + 1)2^{2m-1}$ choices for b as before, but now $2(2^m - 1)$ choices for a, so $n_{(1.3)} = (2^{2m} - 1)2^m$. Fourth, if $b \notin S^{(2)} \cup \{0\}$, we have $2^{2m} - \binom{2^m+1}{2} - 1 = (2^m - 1)2^{m-1} - 1$ choices for b and, as a is arbitrary, $2^{2m} - 1$ choices for a. Combining these, we obtain $n_{(1.4)} = (2^{2m} - 1)(2^m + 1)(2^{m-1} - 1)$. We summarize these results for better readability:

$$n_{(1.1)} = 2^{2m} - 1, \qquad n_{(1.2)} = (2^{2m} - 1)(2^m - 1)2^{m-1},$$
$$n_{(1.3)} = (2^{2m} - 1)2^m, \qquad n_{(1.4)} = (2^{2m} - 1)(2^m + 1)(2^{m-1} - 1). \tag{4}$$

We now study the Cases (C1)–(C3) from Theorem 2 separately. From (1.1)–(1.4), we see that $\delta_f(a,b)$ depends on a, b and $\gamma_f(a,b)$. Thus, in each of (C1)–(C3), we determine for each case (1.1) to (1.4) the number of (a,b) such that $\gamma_f(a,b) = 1$, i.e. $f(a) = b+z$. For readability, we label the combination of Case (C1) and (1.1) simply (C1.1) etc.

□

Case (C1). Let $z \in S$. We determine $\omega_{\delta,(1.1)}, \ldots, \omega_{\delta,(1.4)}$ in this case. Plugging these into (3) gives the values of ω_δ in (C1) in the theorem.

(C1.1). Let $b = 0$. Then $\gamma_f(a,b) = 1$ if and only if $f(a) = z$. Since $z \in S$, there are $2^m - 1$ choices for a, such that $f(a) = z$; they form one of the sets U_i^* in our spread P. Consequently, $\omega_{2^m,(1.1)} = 2^m - 1$ and $\omega_{2^m-2,(1.1)} = n_{(1.1)} - \omega_{2^m,(1.1)} = 2^m(2^m - 1)$.

(C1.2). Let $b = s_i + s_j \in S^{(2)}$ and let $a \notin U_i^* \cup U_j^*$. We show that $\gamma_f(a,b) = 0$ for all admissible (a,b) in this case. Suppose there are $b \in S^{(2)}$ and $a \notin U_i^* \cup U_j^*$ such that $f(a) = b+z$ or, equivalently,

$$f(a) = s_i + s_j + z. \tag{5}$$

Note that $a \notin U_i^* \cup U_j^*$ implies $f(a) \in S \setminus \{s_i, s_j\}$. If $z \in \{s_i, s_j\}$, say $s_i = z$, then (5) reduces to $f(a) = s_j$, which is a contradiction. If $z \notin \{s_i, s_j\}$, (5) leads to a contradiction, as well, since the sum of three distinct Sidon elements is no Sidon element, see Proposition 1. It follows that $\omega_{4,(1.2)} = 0$ and $\omega_{2,(1.2)} = n_{(1.2)}$.

(C1.3). Let $b = s_i + s_j \in S^{(2)}$ and let $a \in U_i^* \cup U_j^*$. As in (C1.2), this leads to (5), but now $f(a) \in \{s_i, s_j\}$. Without loss of generality, assume $f(a) = s_i$, which then implies $z = s_j$. So to satisfy (5), we have 2^m choices for b as we can choose $s_i \in S \setminus \{z\}$ and $s_j = z$ is uniquely determined. For each b, we have $2^m - 1$ choices for $a \in U_i$. In summary, we have $\omega_{2,(1.3)} = 2^m(2^m - 1)$ and $\omega_{0,(1.3)} = n_{(1.3)} - 2^m(2^m - 1) = 2^{m-1}(2^m - 1)(2^{2m} + 1)$.

(C1.4). Let $b \notin S^{(2)} \cup \{0\}$. We show that $\gamma_f(a,b) = 0$ for all admissible (a,b) from this case. Suppose $\gamma_f(a,b) = 1$. Then $f(a) + z = b$, but as $f(a), z \in S$, this means $b = 0$ or $b \in S^{(2)}$ which contradicts our assumption. So $\omega_{2,(1.4)} = 0$ and $\omega_{0,(1.4)} = n_{(1.4)}$.

Case (C2). Now let $z \notin S$ such that $S \cup \{z\}$ is Sidon. We proceed as in (C1).

(C2.1). Let $b = 0$. As in (C1.1), $\gamma_f(a, b) = 1$ implies $f(a) = z$, which now contradicts $z \notin S$. Consequently $\omega_{2^m, (1.1)} = 0$ and $\omega_{2^m - 2, (1.1)} = n_{(1.1)}$.

(C2.2). Let $b = s_i + s_j \in S^{(2)}$ and $a \notin U_i^* \cup U_j^*$. As in (C1.2), $\gamma_f(a, b) = 1$ now implies (5). As $S \cup \{z\}$ is Sidon and $f(a) \in S$, this is a contradiction, see Proposition 1. It follows that $\omega_{4, (1.2)} = 0$ and $\omega_{2, (1.2)} = n_{(1.2)}$.

(C2.3). Let $b = s_i + s_j \in S^{(2)}$ and $a \in U_i^* \cup U_j^*$. As in (C1.3), $\gamma_f(a, b) = 1$ now implies $s_j = z$ which contradicts our assumption of (C2). So $\omega_{2, (1.3)} = 0$ and $\omega_{0, (1.3)} = n_{(1.3)}$.

(C2.4). Let $b \notin S^{(2)} \cup \{0\}$. Then $\gamma_f(a, b) = 1$ implies $b = f(a) + z$. We have $2^{2m} - 1$ choices for a, and fixing a uniquely determines b. The condition of (C2)) then guarantees that $b \notin S^{(2)} \cup \{0\}$. So $\omega_{2, (1.4)} = 2^{2m} - 1$ and $\omega_{0, (1.4)} = n_{(1.4)} - \omega_{2, (1.4)} = (2^{2m} - 1)(2^{2m-1} - 2^{m-1} - 2)$.

Case (C3). Let $z \notin S$ such that $S \cup \{z\}$ is not Sidon. We first remark that $\text{mult}_{S,3}(z) > 0$ for all z in this case, since $S \cup \{z\}$ not being Sidon implies there are distinct $s_1, s_2, s_3 \in S$ such that $s_1 + s_2 = s_3 + z$ or, equivalently, $z = s_1 + s_2 + s_3$. We now proceed as before.

(C3.1). Let $b = 0$. By the same reasoning as in (C2.1), $\omega_{2^m, (1.1)} = 0$ and $\omega_{2^m - 2, (1.1)} = n_{(1.1)}$.

(C3.2). Let $b = s_i + s_j \in S^{(2)}$ and $a \notin U_i^* \cup U_j^*$. If $\gamma_f(a, b) = 1$, we obtain (5) or equivalently $z = s_i + s_j + f(a)$, where $s_i, s_j, f(a) \in S$ are distinct. So the number of admissible (a, b) depends on the number of representations of z as the sum of three distinct elements of S, in short on $\text{mult}_{S,3}(z)$. Suppose $z = s_1 + s_2 + s_3$ is such a representation of z in $S^{(3)}$. This representation gives three possible values of b, namely $s_1 + s_2$, $s_1 + s_3$ and $s_2 + s_3$. For each of these choices $f(a)$ is uniquely determined, and so we have $2^m - 1$ choices of a. Consequently, $\omega_{4, (1.2)} = (2^m - 1)3\text{mult}_{S,3}(z)$ and $\omega_{2, (1.2)} = n_{(1.2)} - \omega_{4, (1.2)} = (2^m - 1)(2^{2m} - 1)2^{m-1} - 3\text{mult}_{S,3}(z)$.

(C3.3). Let $b = s_i + s_j \in S^{(2)}$ and $a \in U_i^* \cup U_j^*$. By the same reasoning as in (C2.3), we have $\omega_{2, (1.3)} = 0$ and $\omega_{0, (1.3)} = n_{(1.3)}$.

(C3.4). Let $b \notin S^{(2)} \cup \{0\}$. If $\gamma_f(a, b) = 1$, then $b + z = f(a)$, so we need $b + z \in S$. For a given z, there are $|S| = 2^m + 1$ choices of b such that $b + z \in S$. In (C3.2), we determined that there are $3\text{mult}_{S,3}(z)$ choices of b with $b + z \in S$ and $b \in S^{(2)}$. In the present case, $b \notin S^{(2)}$, so we have $2^m + 1 - 3\text{mult}_{S,3}(z)$ valid choices of b. Each of these choices uniquely determines $f(a)$ and therefore leads to $2^m - 1$ possible values of a. In summary, we have $\omega_{2, (1.4)} = (2^m - 1)(2^m + 1 - 3\text{mult}_{S,3}(z))$ and $\omega_{0, (1.4)} = n_{(1.4)} - \omega_{2, (1.4)} = (2^m - 1)(2^{3m-1} - 5 \cdot 2^{m-1} - 2 + 3\text{mult}_{S,3}(z))$.

Remark 3. As mentioned in Remark 1, for k satisfying $\gcd(k, 2^m + 1) = 1$, we have $f_{P_c, S_{m,k}0}(x) = x^{k(2^m - 1)}$. We can obtain the differential spectrum of this function from Theorem 2, (C2) if m is even, and (C3) if m is odd. In (C3), we then have $\text{mult}_{S,3}(0) = \frac{2^m + 1}{3}$, see Proposition 5. Blondeau, Canteaut, and

Charpin [5, Theorem 7] and Hu et al. [14, Theorem 1] determined the differential spectra of these functions, and our proof of Theorem 2 can be seen as a new proof of their results. $\qquad\square$

Corollary 1 is an immediate consequence of Theorem 2.

Corollary 1. *The function $f_{P,S,z}$ has differential uniformity*

$$\delta(f_{P,S,z}) = \begin{cases} 2^m & \text{if } z \in S, \\ 2^m - 2 & \text{if } z \notin S. \end{cases}$$

If $S = S_m$, we can also precisely determine $\operatorname{mult}_{S_m,3}(z)$, which plays a prominent role in Theorem 2, in terms of Kloosterman sums.

Proposition 6. *For $S_m = \left\{ x \in \mathbb{F}_{2^{2m}} : x^{2^m+1} = 1 \right\}$ and $z \in \mathbb{F}_{2^{2m}}^* \backslash S_m$ we have*

$$\operatorname{mult}_{S_m,3}(z) = \frac{1}{3 \cdot 2^{2m+1}} \left((2^m + 1)^3 + \sum_{u \in \mathbb{F}_{2^m}^*} K_m^3(u) K_m(uu_z) \right),$$

where $u_z \in \mathbb{F}_{2^m}$ is the unique element in $\mathbb{F}_{2^m}$ such that $z = u_z \cdot s$ for some $s \in S_m$.

Proof. Write $S := S_m$. By Proposition 3, we have

$$\operatorname{mult}_{S,3}(z) = \frac{1}{3 \cdot 2^{m+1}} \widehat{(\widehat{1_S})^3}(z). \tag{6}$$

With Definition 7 the right-hand side of (6) equals

$$\frac{1}{3 \cdot 2^{m+1}} \sum_{x \in \mathbb{F}_{2^{2m}}} (-1)^{\operatorname{tr}(xz)} (\widehat{1_S})^3(x). \tag{7}$$

Since $\mathbb{F}_{2^{2m}}^*$ is the direct product of $\mathbb{F}_{2^m}^*$ and S, we can represent any $x \in \mathbb{F}_{2^{2m}}^*$ in the form $x = us$ for unique $u \in \mathbb{F}_{2^m}^*$ and $s \in S$. For $x = 0$, we have $(-1)^{\operatorname{tr}(0)} = 1$ and $\widehat{1_S}(0) = 2^m + 1$. Thus, we rewrite (7) as

$$\frac{1}{3 \cdot 2^{m+1}} \left((2^m + 1)^3 + \sum_{u \in \mathbb{F}_{2^m}^*} \sum_{s \in S} (-1)^{\operatorname{tr}(usz)} (\widehat{1_S})^3(u) \right). \tag{8}$$

It follows from Definition 7 that for $s \in S$ we have $\widehat{1_S}(ys) = \widehat{1_S}(y)$ for all $y \in \mathbb{F}_{2^{2m}}$. Moreover, $\sum_{s \in S} (-1)^{\operatorname{tr}(usz)} = \widehat{1_S}(uz)$. Thus, (8) equals

$$\frac{1}{3 \cdot 2^{m+1}} \left((2^m + 1)^3 + \sum_{u \in \mathbb{F}_{2^m}^*} (\widehat{1_S})^3(u) \widehat{1_S}(uz) \right). \tag{9}$$

We eventually use Lemma 2 to express the Fourier transforms in (9) in terms of Kloosterman sums. To do so for $\widehat{1_S}(uz)$, we need to determine the unique element $u' \in \mathbb{F}_{2^m}^*$ such that $uz = u's'$ for some $s' \in S$. Write $z = u_z s_z$ for unique $u_z \in \mathbb{F}_{2^m}^*$ and $s_z \in S$ and multiply this equation by u. Then $u' = uu_z$, and the proposition follows. $\qquad\square$

Next, we study the algebraic degree of $f_{P,S,z}$. We need the following well-known lemma (e. g. [7, Theorem 8]) about the algebraic degree of the indicator function of an affine subspace. We add a proof for completeness.

Lemma 3. *Let A be a k-dimensional affine subspace of $\mathbb{F}_2^n$. Its indicator function $1_A \colon \mathbb{F}_2^n \to \mathbb{F}_2$ has algebraic degree $n - k$.*

Proof. Let $e_1, \ldots, e_n$ be the standard basis vectors of $\mathbb{F}_2^n$. It is easy to confirm that the support of the Boolean function $\prod_{i=1}^k x_i$ of algebraic degree k is the $(n - k)$-dimensional affine subspace $A := \sum_{i=1}^k e_i + \mathrm{pan}\{e_{k+1}, \ldots, e_n\}$. Since any $(n - k)$-dimensional affine subspace A' of $\mathbb{F}_2^n$ can be obtained from A by an affine transformation, and since the algebraic degree of a function is an affine invariant, the lemma holds. $\square$

Theorem 3. *Let $Z := \sum_{s \in S} s$. The function $f_{P,S,z}$ on $\mathbb{F}_2^{2m}$ has algebraic degree*

$$\deg_{alg}(f_{P,S,z}) = \begin{cases} m & \text{if } z = Z, \\ 2m & \text{if } z \neq Z. \end{cases}$$

Proof. Write $f := f_{P,S,z}$. We use the notation from Theorem 1: we have a spread $P = \{U_1, \ldots, U_{2^m+1}\}$, a Sidon set S with $|S| = 2^m + 1$ and an element $z \in \mathbb{F}_2^{2m}$, and we write $U_i^* := U_i \setminus \{0\}$ for all i. Note that the indicator function $1_{U_i} \colon \mathbb{F}_2^{2m} \to \mathbb{F}_2$ of any U_i has algebraic degree m according to Lemma 3, and we have $1_{U_i}(0) = 1$.

Denote the coordinate functions of f by $f_1, \ldots, f_{2m}$. Since f is constant on every U_i^* so are $f_1, \ldots, f_{2m}$. Consequently, for $x \neq 0$, every coordinate function is the sum of some of the indicator functions of the subspaces from our spread or, in other words, for every $j = 1, \ldots, 2m$, there exists $I_j \subseteq \{1, \ldots, 2^m + 1\}$ such that $f_j(x) = \sum_{i \in I_j} 1_{U_i}(x)$ for $x \neq 0$.

It remains to consider the case $x = 0$. Recall that $f(0) = z$. First, suppose $z = Z$ and write $Z = (Z_1, \ldots, Z_{2m})$. We show that, in this case, also $f_j(0) = \sum_{i \in I_j} 1_{U_i}(0)$. Note that

$$Z = \sum_{s \in S} s = (2^m - 1) \sum_{s \in S} s = \sum_{x \in \mathbb{F}_2^{2m} \setminus \{0\}} f(x).$$

Thus,

$$f_j(0) = Z_j = \begin{cases} 0 & \text{if } |I_j| \text{ is even}, \\ 1 & \text{if } |I_j| \text{ is odd}, \end{cases}$$

and $f_j = \sum_{i \in I} 1_{U_i}$. So f_j is the sum of functions of algebraic degree m and, therefore, $\deg_{alg}(f_j) = m$ for all $j = 1, \ldots, 2m$. Consequently, f has algebraic degree m.

Now let $z \neq Z$. Define $g \colon \mathbb{F}_2^{2m} \to \mathbb{F}_2^{2m}$ such that $f = f_{P,S,Z} + g$ by

$$g(x) = \begin{cases} z + Z & \text{if } x = 0, \\ 0 & \text{if } x \neq 0, \end{cases}$$

and denote its coordinate functions by $g_1, \ldots, g_{2m}$. We have seen before that $\deg_{alg}(f_{P,S,Z}) = m$. By the definition of g, there is least one $j \in \{1, \ldots, 2m\}$ such that $g_j(0) = 1$ and $g_j(x) = 0$ if $x \neq 0$. Then g_j is the indicator function of the trivial subspace, and, according to Lemma 3, $\deg_{alg}(g_j) = 2m$. Consequently, g and hence f have algebraic degree $2m$. $\qquad\qquad\square$

5   Conclusion and Open Questions

In this paper, we introduced an extended definition of locally-APN functions, which also covers non-power functions. We presented a powerful construction of such functions using spreads and Sidon sets, and we determined the differential spectrum and the algebraic degree of these functions. The following questions still remain open:

1. We have seen in Remark 1 that our construction covers several known locally-APN power functions. Can we also find a more general construction covering the remaining known locally-APN power functions mentioned in Sect. 1?
2. As already mentioned in Remark 2: what is the number of inequivalent functions originating from Theorem 1?
3. In Theorem 2, we showed that the differential spectrum of a function from Theorem 1 can depend on $\mathrm{mult}_{S,3}(z)$. To our knowledge, this property of a Sidon set has not been widely studied yet. Can we determine $\mathrm{mult}_{S,3}(z)$ for other Sidon sets except for S_m?
4. Can we construct more functions with cryptographically relevant properties by partitioning $\mathbb{F}_2^n$ into subspaces or by taking a Sidon set as the image set?

Acknowledgments. The authors thank Darrion Thornburgh for valuable comments about k-sums of Sidon elements and pointing out the connection to the Fourier transform and Kloosterman sums. They also thank Ingo Czerwinski for helpful discussions on Sidon sets and providing them with non-classical examples and computational results.

The first author is funded by the Deutsche Forschungsgemeinschaft (DFG, German Research Foundation) – 541511634.

References

1. Anbar, N., Meidl, W.: Bent partitions. Des. Codes Cryptogr. **90**(4), 1081–1101 (2022). https://doi.org/10.1007/s10623-022-01029-z
2. Beierle, C., Langevin, P., Leander, G., Polujan, A., Rasoolzadeh, S.: Millions of inequivalent quadratic APN functions in eight variables (2025). arXiv: 2508.04644 [math.CO]
3. Biliotti, M., Jha, V., Johnson, N.L.: Foundations of Translation Planes. Monographs and Textbooks in Pure and Applied Mathematics, vol. 243, pp. xvi+542. Marcel Dekker, Inc., New York (2001). https://doi.org/10.1201/9781482271003
4. Blondeau, C., Canteaut, A., Charpin, P.: Differential properties of power functions. Int. J. Inf. Coding Theory **1**(2), 149–170 (2010). https://doi.org/10.1504/IJICOT. 2010.032132
5. Blondeau, C., Canteaut, A., Charpin, P.: Differential properties of $x \mapsto x^{2^t-1}$. IEEE Trans. Inform. Theory **57**(12), 8127–8137 (2011). https://doi.org/10.1109/ TIT.2011.2169129
6. Blondeau, C., Nyberg, K.: Perfect nonlinear functions and cryptography. Finite Fields Appl. **32**, 120–147 (2015). https://doi.org/10.1016/j.ffa.2014.10.007
7. Carlet, C.: Boolean Functions for Cryptography and Coding Theory, pp. xiv+562 Cambridge University Press, Cambridge (2020). https://doi.org/10.1017/ 9781108606806
8. Carlet, C.: On APN functions whose graphs are maximal Sidon sets. In: LATIN 2022: Theoretical Informatics. LNCS, vol. 13568, pp. 243–254. Springer, Cham (2022). https://doi.org/10.1007/978-3-031-20624-5_15
9. Carlet, C., Mesnager, S.: On those multiplicative subgroups of $\mathbb{F}_{2^n}^*$ which are Sidon sets and/or sum-free sets. J. Algebraic Combin. **55**(1), 43–59 (2022). https://doi. org/10.1007/s10801-020-00988-7
10. Charpin, P., Peng, J.: Differential uniformity and the associated codes of cryptographic functions. Adv. Math. Commun. **13**(4), 579–600 (2019). https://doi.org/ 10.3934/amc.2019036
11. Czerwinski, I., Pott, A.: Sidon sets, sum-free sets and linear codes. Adv. Math. Commun. **18**(2), 549–566 (2024). https://doi.org/10.3934/amc.2023054
12. Czerwinski, I., Pott, A.: On large Sidon sets. J. Combin. Theory Ser. A **220**, 15 (2026). https://doi.org/10.1016/j.jcta.2025.106129. Paper No. 106129
13. Dillon, J.F.: Elementary Hadamard difference-sets. Thesis (Ph.D.)–University of Maryland, College Park. ProQuest LLC, Ann Arbor, MI, p. 126 (1974)
14. Hu, Z., Li, N., Xu, L., Zeng, X., Tang, X.: The differential spectrum and boomerang spectrum of a class of locally-APN functions. Des. Codes Cryptogr. **91**(5), 1695–1711 (2023). https://doi.org/10.1007/s10623-022-01161-w
15. Lavrauw, M., Polverino, O.: Finite semifields. In: Storme, L., De Beule, J., (eds.) Current Research Topics in Galois Geometry. Mathematics Research Developments. Nova Science, pp. 127–155 (2011)
16. Li, K., Kaleyski, N.: Two new infinite families of APN functions in trivariate form. IEEE Trans. Inform. Theory **70**(2), 1436–1452 (2024). https://doi.org/10.1109/ tit.2023.3313148
17. Mesnager, S.: Bent Functions. Fundamentals and Results. Springer, Cham (2016). https://doi.org/10.1007/978-3-319-32595-8
18. Nagy, G.P.: Sidon sets, thin sets, and the nonlinearity of vectorial Boolean functions. J. Combin. Theory Ser. A **212**, 21 (2025). https://doi.org/10.1016/j.jcta. 2024.106001. Paper No. 106001

19. Nyberg, K.: Differentially uniform mappings for cryptography. In: Helleseth, T. (ed.) EUROCRYPT 1993. LNCS, vol. 765, pp. 55–64. Springer, Heidelberg (1994). https://doi.org/10.1007/3-540-48285-7_6
20. Pott, A.: Almost perfect and planar functions. Des. Codes Cryptogr. **78**(1), 141–195 (2016). https://doi.org/10.1007/s10623-015-0151-x
21. Thornburgh, D.: Uniform exclude distributions of Sidon sets (2024) arXiv: 2407.11783 [math.CO]
22. Thornburgh, D.: On generalizing cryptographic results to Sidon sets in $\mathbb{F}_2^n$ (2025). arXiv: 2501.11184 [math.CO]
23. Xie, X., Mesnager, S., Li, N., He, D., Zeng, X.: On the Niho type locally-APN power functions and their boomerang spectrum. IEEE Trans. Inform. Theory **69**(6), 4056–4064 (2023). https://doi.org/10.1109/tit.2022.3232362

On the Resilience Order of Weightwise Almost Perfectly Balanced Functions

Martin Grenouilloux[1] [iD], Chunlei Li[1(✉)] [iD], and Pierrick Méaux[2] [iD]

[1] University of Bergen, Bergen, Norway
`{martin.grenouilloux,chunlei.li}@uib.no`
[2] University of Luxembourg, Esch-sur-Alzette, Luxembourg
`pierrick.meaux@uni.lu`

Abstract. The recent development of Fully Homomorphic Encryption (FHE) witnessed the emergence of a new generation of tailored cryptographic primitives designed to meet its specific criteria. Among promising candidates for FHE constructions stands out the FLIP cipher, which employs Boolean functions that are evaluated only on specific subsets of $\mathbb{F}_2^n$. In this article, we study Weightwise Almost Perfectly Balanced (WAPB) functions, which are almost balanced on each of these subsets. While WAPB functions have been of great interest for new constructions recently, some aspects, such as resilience remain poorly understood. As such, we take a first step at characterizing the resilience of WAPB functions, through their properties as correctors. We highlight its close connection with the restricted Walsh transform and uncover an algebraic relation between Krawtchouk matrices and Vandermonde matrices, which reduces the problem of determining the corrector order of a WAPB function to a particular instance of the Prouhet-Tarry-Escott problem. This reduction helps us show that for infinitely many integers n, WAPB functions in n variables have corrector order tightly upper bounded by the Hamming weight of n minus one. We conjecture that this observation holds for any positive integer n, which is verified for n up to 62.

Keywords: WAPB function · resilience order · Krawtchouk polynomial · Vandermonde matrix · Prouhet-Tarry-Escott problem

1 Introduction

With the design of FLIP [33], established cryptographic criteria for Boolean functions, which are traditionally used to assess the security of filtered Linear Feedback Shift Registers (LFSRs), combined LFSRs, or more general stream ciphers, no longer directly apply. In this cipher, the relevant properties of the filtering Boolean function are not defined on the entire $\mathbb{F}_2^n$ but instead on a specific, publicly known subset of inputs determined by the execution of the cipher. In particular, the relevant properties of the function are only required on

L. Batina and F. Özbudak (Eds.): WAIFI 2026, LNCS 16611, pp. 199–220, 2026.
https://doi.org/10.1007/978-3-032-27574-5_13

these subsets. This shift has triggered a generalization of standard attacks and the development of new criteria for Boolean functions restricted to subsets or partitions of $\mathbb{F}_2^n$, beginning with the study of restricted cryptographic criteria [7]. So far, these criteria have been regarded for general subsets [4,7], subsets of fixed Hamming weight e.g. [7,14,29,37], and affine spaces [6].

In FLIP, the inputs to the filtering function always have a fixed and known Hamming weight, and tailored attacks to this constraint strengthens them considerably. This phenomenon reflects a broader trend: additional information about a cipher's internal values may dramatically simplify attacks, as observed in algebraic side-channel cryptanalysis [39] or lattice reduction with side information [8]. From the FLIP analysis emerged a systematic study of the partition of $\mathbb{F}_2^n$ into $n+1$ slices, extending the analysis from a single relevant slice to all slices simultaneously, where a slice is the set of all vectors of a given Hamming weight, i.e., $\mathsf{E}_{k,n} = \{x \in \mathbb{F}_2^n \mid \mathsf{w}_\mathsf{H}(x) = k\}$ for $k = 0, 1, \ldots, n$. Notably, Hamming weight is also a widely used leakage model in side-channel attacks [13,23,41], even though in practice approximate weight leakages are often considered more realistic [2,20,38].

A fundamental security requirement for Boolean functions used in stream ciphers is balancedness: the output should take each value equally often to avoid exploitable statistical biases. In the context of slices, [7] introduced Weightwise Perfectly Balanced (WPB) functions and Weightwise Almost Perfectly Balanced (WAPB) functions, which are balanced and almost balanced on every slice $\mathsf{E}_{k,n}$, respectively. The qualifier "almost" accounts for the case when n is not a power of two where the slices always have odd cardinality (which prevents perfect balancedness on any slice). In the FLIP cipher, the filtering function is evaluated on inputs of a fixed and known Hamming weight, corresponding to a single slice of $\mathbb{F}_2^n$ (typically of weight close to $n/2$). Accordingly, the cryptographic behaviour of the function on this slice is of primary importance. It is worth noting that the WAPB property is a stronger requirement, as it imposes (almost) balancedness on all slices simultaneously. Such a strengthening can be seen as an extremal or canonical condition (in the similar spirit as bent functions are studied as maximally nonlinear objects for potential cryptographic applications). Moreover, considering several slices can be relevant beyond the nominal setting of cryptanalysis. Certain attacks may involve restrictions of the function or inputs whose Hamming weights deviate from the central slice. For instance, in the case of FLIP, if some variables are not involved in the filtering function, guess-and-determine strategies lead to the analysis of subfunctions whose inputs lie in neighboring slices. More generally, leakage models such as Hamming weight measurements naturally provide information across multiple slices. The aforementioned aspects motivate the study of WAPB functions functions whose properties remain controlled on a range of slices, rather than on a single one.

Since 2017, many works have introduced new constructions of WPB or WAPB functions, studying their global and restricted cryptographic parameters. Recursive families and secondary constructions were first presented in [7]. Constructions based on field representations yield 2-rotation symmetric WPB functions with high weightwise nonlinearity [26,36], and recently 2-π symmetric WAPB functions for all n [12]. WPB functions obtained by comparing the weights of two halves of the input achieve optimal algebraic immunity [35,42]. Comparing the two halves using total orders leads to WAPB functions for all n [32]. Other approaches modify near-WPB linear or quadratic functions to impose balancedness on all slices [34]. Such a general idea was later generalized to obtain WAPB functions from low-degree functions [19,25,46–48]. Secondary constructions have also been explored, such as direct sums [49], Siegenthaler constructions combined with symmetric functions [15] and t-concatenation [22]. Moreover, several works have modified the support of non-WPB functions to provide WPB or WAPB functions with bounded parameters for the weightwise nonlinearity [14], the global nonlinearity [17], or the algebraic immunity [16]. Heuristic search approaches, notably evolutionary algorithms, have been applied as well [30,31,45]. Additional constructions appear in [10,11,22].

Despite this growing body of work, one central cryptographic parameter has remained largely unexplored for WAPB functions: resilience. Generalizing the notion of balancedness, a Boolean function is said t-resilient if it is uncorrelated to all affine functions combining at most t variables. The resilience has been introduced to estimate the complexity of correlation attacks on stream ciphers [40], and remains relevant when analyzing Boolean functions under restricted input distributions. Previous studies on WAPB functions have characterized their nonlinearity [17], restricted nonlinearity [14], algebraic immunity (both global [16] and weightwise [4]), algebraic degree [18], yet no general result on the resilience of WAPB functions had been established.

The resilience of a WAPB function is particularly intriguing since it is related to both global balancedness and balancedness on all the parts of a particular partition. Since the slices and the hyperplanes corresponding to affine relations are not related, one might expect that enforcing balancedness on all slices would not significantly constrain other properties such as resilience. However, our results show that this intuition does not hold. Even the simplest case of WPB functions provides a stark example: although they are balanced on every slice, they cannot be 1-resilient. More generally, WAPB functions have a low resilience order (the maximum value of t such that the function is t-resilient). We can exhibit n-variable WAPB functions with a resilience order equal to $\mathsf{w_H}(n) - 1$, and demonstrate for infinite values of n that no WAPB function can be $\mathsf{w_H}(n)$-resilient.

To analyze the resilience of WAPB functions, we adopt a perspective that is uncommon in Boolean functions used for cryptography, we study their properties as correctors e.g. [21,24,27] since the corrector order provides an upper bound on the resilience order. A t-corrector is a function for which the sum of Walsh coefficients over each slice $\mathsf{E}_{k,n}$ for $k = 0, 1, \ldots, t$ is null. We show (Theorem 1) that the corrector order of a Boolean function is tightly connected to a system of linear equations involving the restricted Walsh transforms on the slices of the function. A key component of our study is an algebraic connection between Krawtchouk matrices and Vandermonde matrices (Theorem 2), which enables us to transform the system of linear equations into a simpler one (Theorem 3) that is associated with a particular Vandermonde matrix. The link between correctors and restricted Walsh transform provides a general framework to study resilience through corrector properties, and is of independent interest for investigating the number and structure of t-corrector Boolean functions. Furthermore, for WAPB functions, those equations are highly simplified since the restricted Walsh transform of these functions takes only the value -1, 0 or 1. This feature allows us to reduce the determination of the maximum corrector order (and hence an upper bound on the resilience) of WAPB functions to a particular instance of the Prouhet-Tarry-Escott problem (see e.g. [5]). Our last main result (Theorem 4) shows that for infinitely many values of n, n-variable WAPB functions have corrector order upper bounded by $\mathsf{w_H}(n) - 1$. The existence of WAPB functions with resilience $\mathsf{w_H}(n) - 1$ confirms the tightness of this upper bound. Following experimental verification for n up to 62 (as well as for other integers n with $\mathsf{w_H}(n)$ up to 5), we propose a conjecture (Conjecture 1) that all n-variable WAPB functions have corrector order tightly upper bounded by $\mathsf{w_H}(n) - 1$.

2   Preliminaries

Let $\mathbb{N}$ be the set of all non-negative integers and $\mathbb{N}^* = \mathbb{N} \setminus \{0\}$. For a positive integer n, we denote by $[n]$ the set of all integers from 1 to n, i.e. $\{1, \ldots, n\}$ and use the notation $[0, n]$ for the set $\{0, 1, \ldots, n\}$. For readability, we use the notation $+$ instead of $\oplus$ for addition in $\mathbb{F}_2$. For a vector $v \in \mathbb{F}_2^n$, we denote its Hamming weight by $\mathsf{w_H}(v)$, defined as $\mathsf{w_H}(v) = |\{i \in [n] \mid v_i = 1\}|$. The inner product between two vectors $u, v \in \mathbb{F}_2^n$ is given by $u \cdot v = \sum_{i=1}^{n} u_i v_i$. For $k \in [0, n]$, the k-th slice of $\mathbb{F}_2^n$ is the set $\mathsf{E}_{k,n} = \{x \in \mathbb{F}_2^n \mid \mathsf{w_H}(x) = k\}$. The support of a positive integer a in binary expression is defined as $\mathsf{supp}(a) = \{i \mid \sum_{i \geq 0} 2^i = a\}$. For two integers $a, b \in \mathbb{N}$, we denote by $a \preceq b$ when $\mathsf{supp}(a) \subseteq \mathsf{supp}(b)$.

2.1   Boolean Functions and Cryptographic Properties

The set of all Boolean functions from $\mathbb{F}_2^n$ to $\mathbb{F}_2$ will be denoted by $\mathcal{B}_n$.

Definition 1 (Balancedness). *A Boolean function $f \in \mathcal{B}_n$ is said to be balanced if $|\mathsf{supp}(f)| = 2^{n-1} = |\mathsf{supp}(f + 1)|$, where $\mathsf{supp}(f)$ denotes the support of f, namely, the set $\{x \in \mathbb{F}_2^n \mid f(x) = 1\}$.*

Definition 2 (Walsh transform). *Let $f \in \mathcal{B}_n$ be a Boolean function. Its Walsh transform W_f at $a \in \mathbb{F}_2^n$ is defined as:*

$$W_f(a) = \sum_{x \in \mathbb{F}_2^n} (-1)^{f(x) + a \cdot x}.$$

Definition 3 (Resilience). *A Boolean function $f \in \mathcal{B}_n$ is said to be t-resilient if for all $a \in \mathbb{F}_2^n$ such that $0 \leq \mathsf{w_H}(a) \leq t$ we have $W_f(a) = 0$. We denote by $\mathsf{res}(f)$ the maximum value of t such that f is t-resilient.*

Definition 4 (Corrector). *A Boolean function $f \in \mathcal{B}_n$ is said to be a t-corrector if for all $0 \le k \le t$ it satisfies:*

$$\sum_{a \in \mathsf{E}_{k,n}} W_f(a) = 0.$$

We denote by $\mathsf{corr}(f)$ the maximum value of t such that f is a t-corrector.

According to the definitions, it is clear that $\mathsf{corr}(f) \ge \mathsf{res}(f)$ for any $f \in \mathcal{B}_n$. We shall use this fact to prove the non-existence of t-resilient WAPB functions by investigating the maximum corrector order of WAPB functions, which will be introduced in the next subsection.

2.2 Restricted Walsh Transform and WAPB Functions

The Boolean hypercube is partitioned into $n+1$ slices, where each slice consists of vectors sharing the same Hamming weight. We refer to properties that hold slice-wise as weightwise properties. The n-variable symmetric Boolean functions are precisely those that are constant on each slice.

Definition 5 (Restricted Walsh transform). *Let $f \in \mathcal{B}_n$, $S \subset \mathbb{F}_2^n$, its Walsh transform restricted to S at $a \in \mathbb{F}_2^n$ is defined as:*

$$W_{f,S}(a) = \sum_{x \in S} (-1)^{f(x)+a \cdot x}.$$

For $S = \mathsf{E}_{\ell,n}$ we denote $W_{f,\mathsf{E}_{\ell,n}}(a)$ by $\mathcal{W}_{f,\ell}(a)$, and for $a = 0_n$ we denote $\mathcal{W}_{f,\ell}(a)$ by $\mathcal{W}_{f,\ell}$.

Definition 6. *For $f \in \mathcal{B}_n$, we define its restricted Walsh vector in $\mathbb{Z}^{n+1}$ as $\mathsf{rw}_f = (\mathcal{W}_{f,0}, \mathcal{W}_{f,1}, \cdots, \mathcal{W}_{f,n-1}, \mathcal{W}_{f,n})^{\mathsf{T}}.$*

Definition 7 (Weightwise (Almost) Perfectly Balanced Function[7]). *Let $m \in \mathbb{N}^*$ and f be a Boolean function in $n = 2^m$ variables. It will be called weightwise perfectly balanced (WPB) if f is balanced on $\mathsf{E}_{\ell,n}$ for every $\ell \in [1, n-1]$ and*

$$f(0, \cdots, 0) = 0, \quad and \quad f(1, \cdots, 1) = 1.$$

The set of WPB functions in 2^m variables is denoted by $\mathcal{WPB}_m$.
 When n is not a power of 2, some other weights than $\ell = 0$ and n give slices of odd cardinality. We consider the generalization for any $n \in \mathbb{N}^*$, in this case we call $f \in \mathcal{B}_n$ weightwise almost perfectly balanced (WAPB) if:

$$|\mathsf{supp}_\ell(f)| = \begin{cases} |\mathsf{E}_{\ell,n}|/2 & \text{if } |\mathsf{E}_{\ell,n}| \text{ is even,} \\ (|\mathsf{E}_{\ell,n}| \pm 1)/2 & \text{if } |\mathsf{E}_{\ell,n}| \text{ is odd.} \end{cases}$$

The set of WAPB functions in n variables is denoted by $\mathcal{WAPB}_n$.

Below we recall Lucas's Theorem.

Lemma 1. *For a prime p and non-negative integers n, k with $n \geq k$, the following congruence relation holds:*

$$\binom{n}{k} \equiv \prod_{i=0}^{r} \binom{n_i}{k_i} \pmod{p},$$

where $n = n_r p^r + \cdots + n_1 p + n_0$ and $k = k_r p^r + \cdots + k_1 p + k_0$ are the p-ary expansions of n and k, respectively.

This implies that $|\mathsf{E}_{\ell,n}| = \binom{n}{\ell}$ is odd if and only if $\ell \preceq n$, i.e., $\mathsf{supp}(\ell) \subseteq \mathsf{supp}(n)$. By the definition of WAPB functions, we easily have the following property of WAPB functions:

Property 1 (WAPB functions and restricted Walsh transform). Let $n \in \mathbb{N}^*$, $f \in \mathcal{B}_n$ is WAPB if and only if:

$$\forall \ell \in [0, n], \ \mathcal{W}_{f,\ell} = \begin{cases} 0 & \text{if } \ell \npreceq n \\ \pm 1 & \text{if } \ell \preceq n. \end{cases}$$

In particular, when $n = 2^m$ with $m \in \mathbb{N}^*$, $f \in \mathcal{B}_n$ is WPB if and only if:

$$\mathcal{W}_{f,0} = 1, \quad \mathcal{W}_{f,n} = -1, \quad \text{and} \quad \forall \ell \in [1, n-1], \mathcal{W}_{f,\ell} = 0.$$

We make use of Krawtchouk polynomials to characterize the corrector order of WAPB functions. We give the necessary preliminaries here and refer to e.g. [28, Ch.5, §7] for more details.

Definition 8 (Krawtchouk Polynomials). *The Krawtchouk polynomial of degree k, with $0 \leq k \leq n$ is given by:*

$$\mathsf{K}_k(\ell, n) = \sum_{j=0}^{k} (-1)^j \binom{\ell}{j} \binom{n-\ell}{k-j},$$

where the binomial coefficient $\binom{m}{r} = 0$ when $m < r$ or $r < 0$ by convention.

Krawtchouk polynomials can be characterized by the generating series:

$$(1+z)^{n-\ell}(1-z)^{\ell} = \sum_{k=0}^{\infty} \mathsf{K}_k(\ell, n) z^k.$$

In addition, there is an interesting connection between Krawtchouk polynomials and linear functions restricted on the slice $\mathsf{E}_{k,n}$:

$$\mathsf{K}_k(\ell, n) = \sum_{u \in \mathsf{E}_{k,n}} (-1)^{u \cdot v}, \quad \forall v \in \mathsf{E}_{\ell,n}. \tag{1}$$

Property 2 (Krawtchouk polynomials relation). Let $n \in \mathbb{N}^*$ and $k \in [0, n]$. The following relations hold:

- $\mathsf{K}_k(n - \ell, n) = (-1)^k \mathsf{K}_k(\ell, n)$,
- $\mathsf{K}_{n-k}(x, n) = (-1)^x \mathsf{K}_k(x, n)$,
- if n is even and k is odd, $\mathsf{K}_k(n/2, n) = 0$,
- if n is even, $\mathsf{K}_{n/2}(1, n) = 0$.
- $\binom{n}{\ell} \mathsf{K}_k(\ell, n) = \binom{n}{k} \mathsf{K}_\ell(k, n)$.

Property 3 ([9, Prop.4]). Let $n \in \mathbb{N}^*$ and $k \in [0, n]$, the following hold:

$$(k + 1)\mathsf{K}_{k+1}(\ell, n) = (n - 2\ell)\mathsf{K}_k(\ell, n) - (n - k + 1)\mathsf{K}_{k-1}(\ell, n).$$

3   Characterizations of *t*-Correctors

The following theorem characterizes the t-corrector property in terms of the restricted Walsh transform. This allows us to reformulate WAPB functions of corrector order t in terms of Krawtchouk polynomials. Furthermore, by Theorem 2, the following characterization will be converted to a linear system defined by a Vandermonde matrix, which facilitates our investigation of corrector orders of WAPB functions.

Theorem 1. *Let $n \in \mathbb{N}^*$ and $t \in \mathbb{N}$ such that $t \leq n$. A Boolean function $f \in \mathcal{B}_n$ has $\mathsf{corr}(f) = t$ if and only if t is the maximum integer such that the restricted Walsh transform satisfies*

$$\sum_{\ell=0}^{n} \mathcal{W}_{f,\ell}\, \mathsf{K}_k(\ell, n) = 0 \quad \text{for } k = 0, 1, \ldots, t. \tag{2}$$

Proof. Note that

$$\sum_{a \in \mathsf{E}_{k,n}} W_f(a) = \sum_{a \in \mathsf{E}_{k,n}} \sum_{x \in \mathbb{F}_2^n} (-1)^{f(x)+a\cdot x} = \sum_{x \in \mathbb{F}_2^n} (-1)^{f(x)} \left(\sum_{a \in \mathsf{E}_{k,n}} (-1)^{a\cdot x} \right)$$

$$= \sum_{\ell=0}^{n} \sum_{x \in \mathsf{E}_{\ell,n}} (-1)^{f(x)} \left(\sum_{a \in \mathsf{E}_{k,n}} (-1)^{a\cdot x} \right) = \sum_{\ell=0}^{n} \sum_{x \in \mathsf{E}_{\ell,n}} (-1)^{f(x)}\, \mathsf{K}_k(\ell, n)$$

$$= \sum_{\ell=0}^{n} \mathcal{W}_{f,\ell}\, \mathsf{K}_k(\ell, n),$$

where the last second equality follows from (1). The desired statement is then proved by Definition 4. $\qquad \square$

Let $n \in \mathbb{N}^*$ and $t \in \mathbb{N}$ such that $t \leq n$, we define the Krawtchouk matrix $\mathbf{K}$ as an $(n+1) \times (n+1)$ matrix with entries given by

$$\forall k \in [0, n], \ell \in [0, n], \quad \mathbf{K}_{k,\ell} = \mathsf{K}_k(\ell, n).$$

Denote by $\mathbf{K}^{(t+1)}$ the submatrix formed by the leading $(t+1)$ rows of $\mathbf{K}$. By Theorem 1 we see that f is a t-corrector if and only if:

$$\mathbf{K}^{(t+1)} \cdot \mathsf{rw}_f = \mathbf{0}_{t+1}, \tag{3}$$

where $\mathbf{0}_{t+1}$ denotes the column vector of $(t+1)$ zeros. The following theorem provides an interesting algebraic relation between the above Krawtchouk matrix $\mathbf{K}$ and the Vandermonde matrix defined by $0, 1, \ldots, n$ via a non-singular lower triangular matrix[1]. This connection facilitates the study of the above equation in (3).

Theorem 2. *Let*

$$\alpha_k = -\frac{n-k}{2}, \beta_k = \frac{n}{2}, \gamma_k = -\frac{k}{2}, \quad 0 \leq k \leq n,$$

and for $1 \leq r \leq n$ define $r \times (r+1)$ matrices $\mathbf{H}_r$ as follows:

$$\mathbf{H}_r = \begin{bmatrix} \beta_0 & \gamma_1 & & & & \\ \alpha_0 & \beta_1 & \gamma_2 & & & \\ & \alpha_1 & \beta_2 & \gamma_3 & & \\ & & \ddots & \ddots & \ddots & \\ & & & \alpha_{r-2} & \beta_{r-1} & \gamma_r \end{bmatrix}. \tag{4}$$

Let $\mathbf{K}$ be the Krawtchouk matrix with $\mathbf{K}_{k,\ell} = \mathsf{K}_k(\ell, n)$ and $\mathbf{V}$ be the Vandermonde matrix with $V_{k,\ell} = \ell^k$, where k, ℓ with $0 \leq k, \ell \leq n$ index the rows and columns, respectively. Then the matrix $\mathbf{M}$ satisfying

$$\mathbf{M} \cdot \mathbf{K} = \mathbf{V},$$

is an invertible lower triangular matrix whose rows are given by the recursion:

$$\mathbf{M}_{0,0} = 1, \quad \mathbf{M}_{r,[0,r]} = \mathbf{M}_{r-1,[0,r-1]}\mathbf{H}_r \quad for\ r = 1, 2, \ldots, n, \tag{5}$$

where $\mathbf{M}_{r,[0,r]} = [\mathbf{M}_{r,0}, \ldots, \mathbf{M}_{r,r}]$ denotes the leading $(r+1)$ entries in the r-th row of M; in other words, $\mathbf{M}_{r,[0,r]} = [1] \cdot \mathbf{H}_1 \cdot \mathbf{H}_2 \cdots \mathbf{H}_r, \quad r = 2, \ldots, n.$

Proof. Note that both the Krawtchouk matrix and the Vandermonde matrix invertible by definition. Property 3 of the Krawtchouk polynomial is crucial for us to show that the matrix $\mathbf{M}$ satisfying $\mathbf{M} \cdot \mathbf{K} = \mathbf{V}$ is an invertible lower triangular matrix.

---

[1] Surprisingly, it appears that the relation between these two well-studied matrices has not been documented in the literature.

We will proceed by induction on r to gradually determine the non-zero entries in each row of $\mathbf{M}$ denoted as the vector $\mathbf{M}_{r,[0,r+1]}$. Property 3 can be equivalently stated as

$$\ell \cdot \mathsf{K}_k(\ell, n) = -\frac{n-(k-1)}{2}\mathsf{K}_{k-1}(\ell, n) + \frac{n}{2}\mathsf{K}_k(\ell, n) - \frac{k+1}{2}\mathsf{K}_{k+1}(\ell, n),$$

which can be rewritten as

$$
\begin{aligned}
\ell \cdot \mathsf{K}_k(\ell, n) &= \alpha_{k-1}\mathsf{K}_{k-1}(\ell, n) + \beta_k\mathsf{K}_k(\ell, n) + \gamma_{k+1}\mathsf{K}_{k+1}(\ell, n) \\
&= \begin{bmatrix} \alpha_{k-1} & \beta_k & \gamma_{k+1} \end{bmatrix} \cdot \begin{bmatrix} \mathsf{K}_{k-1}(\ell, n) \\ \mathsf{K}_k(\ell, n) \\ \mathsf{K}_{k+1}(\ell, n) \end{bmatrix},
\end{aligned}
\tag{6}
$$

where $\alpha_{k-1} = -\frac{n-(k-1)}{2}$, $\beta_k = \frac{n}{2}$ and $\gamma_{k+1} = -\frac{k+1}{2}$ as defined in Theorem 2.

We start with the case of $r = 0$. Since $\mathsf{K}_0(\ell, n) = 1$ for all ℓ, n, it follows that $\mathbf{M}_{0,[0,n]} = \begin{bmatrix} 1 & 0 & \cdots & 0 \end{bmatrix}$ is lower triangular and verifies $\mathbf{M}_{0,[0,n]} \cdot \mathbf{K} = \mathbf{V}_{0,[0,n]}$.

Now we move to $r = 1$. In this case we need to determine $\mathbf{M}_{1,[0,2]}$ that yields ℓ^1, which corresponds to Row 1 in $\mathbf{V}$. We can expand ℓ^1 using Eq. (6) as follows:

$$
\begin{aligned}
\ell^1 &= \ell \cdot \mathsf{K}_0(\ell, n) \\
&= \beta_0\mathsf{K}_0(\ell, n) + \gamma_1\mathsf{K}_1(\ell, n) \\
&= \underbrace{[1]}_{\mathbf{M}_{0,[0,1]}} \cdot \underbrace{\begin{bmatrix} \beta_0 & \gamma_1 \end{bmatrix}}_{\mathbf{H}_1} \cdot \begin{bmatrix} \mathsf{K}_0(\ell, n) \\ \mathsf{K}_1(\ell, n) \end{bmatrix}.
\end{aligned}
$$

As such, $\mathbf{M}_{1,[0,2]} = \begin{bmatrix} \beta_0 & \gamma_1 \end{bmatrix}$ and $\mathbf{M}_{1,[0,n]} \cdot \mathbf{K} = \mathbf{V}_{1,[0,n]}$. The second row also confirms that $\mathbf{M}$ is lower triangular.

Consider $r = 2$. We want to determine $\mathbf{M}_{2,[0,3]}$ that yields ℓ^2 in $\mathbf{V}$. Similarly, we expand ℓ^2 using Eq. (6) as follows:

$$
\begin{aligned}
\ell^2 = \ell \cdot \ell &= \ell\Big(\gamma_1\mathsf{K}_1(\ell, n) + \beta_0\mathsf{K}_0(\ell, n)\Big) \\
&= \beta_0\ell\mathsf{K}_0(\ell, n) + \gamma_1\ell\mathsf{K}_1(\ell, n) \\
&= \beta_0 \begin{bmatrix} \beta_0 & \gamma_1 \end{bmatrix} \cdot \begin{bmatrix} \mathsf{K}_0(\ell, n) \\ \mathsf{K}_1(\ell, n) \end{bmatrix} + \gamma_1 \begin{bmatrix} \alpha_0 & \beta_1 & \gamma_2 \end{bmatrix} \cdot \begin{bmatrix} \mathsf{K}_0(\ell, n) \\ \mathsf{K}_1(\ell, n) \\ \mathsf{K}_2(\ell, n) \end{bmatrix} \\
&= \underbrace{\begin{bmatrix} \beta_0 & \gamma_1 \end{bmatrix}}_{\mathbf{M}_{1,[0,2]}} \cdot \underbrace{\begin{bmatrix} \beta_0 & \gamma_1 & 0 \\ \alpha_0 & \beta_1 & \gamma_2 \end{bmatrix}}_{\mathbf{H}_2} \cdot \begin{bmatrix} \mathsf{K}_0(\ell, n) \\ \mathsf{K}_1(\ell, n) \\ \mathsf{K}_2(\ell, n) \end{bmatrix}.
\end{aligned}
$$

As such, $\mathbf{M}_{2,[0,3]} = \mathbf{M}_{1,[0,2]} \cdot \mathbf{H}_2 = \begin{bmatrix} (\beta_0^2 + \gamma_1\alpha_0) & (\gamma_1\beta_0 + \gamma_1\beta_1) & (\gamma_1\gamma_2) \end{bmatrix}$ is lower triangular. Thus, we have

$$\mathbf{M}_{2,[0,n]} \cdot \mathbf{K} = \mathbf{M}_{2,[0,3]} \begin{bmatrix} \mathsf{K}_0(\ell, n) \\ \mathsf{K}_1(\ell, n) \\ \mathsf{K}_2(\ell, n) \end{bmatrix} = \mathbf{V}_{2,[0,n]}.$$

<u>Induction:</u> we shall prove that if the statement is true for $r - 1$, then it is true for r.

Again, we use Eq. (6) to expand ℓ^r as follows:

$$\ell^r = \ell \cdot \ell^{r-1} = \ell\Big(\sum_{k=0}^{r-1} \mathbf{M}_{r-1,i} \cdot \mathsf{K}_k(\ell,n) \Big)$$

$$= \begin{bmatrix} \mathbf{M}_{r-1,0} \ \mathbf{M}_{r-1,1} \ \cdots \ \mathbf{M}_{r-1,r-1} \end{bmatrix} \cdot \begin{bmatrix} \ell\mathsf{K}_0(\ell,n) \\ \ell\mathsf{K}_1(\ell,n) \\ \vdots \\ \ell\mathsf{K}_{r-1}(\ell,n) \end{bmatrix}$$

$$= \begin{bmatrix} \mathbf{M}_{r-1,0} \ \mathbf{M}_{r-1,1} \ \cdots \ \mathbf{M}_{r-1,r-1} \end{bmatrix} \cdot \begin{bmatrix} \beta_0 & \gamma_1 & & \\ \alpha_0 & \beta_1 & \gamma_2 & \\ & \ddots & \ddots & \ddots \\ & & \alpha_{r-2} & \beta_{r-1} & \gamma_r \end{bmatrix} \cdot \begin{bmatrix} \mathsf{K}_0(\ell,n) \\ \mathsf{K}_1(\ell,n) \\ \vdots \\ \mathsf{K}_r(\ell,n) \end{bmatrix}$$

$$= \mathbf{M}_{r-1,[0,r]} \cdot \mathbf{H}_r \cdot \begin{bmatrix} \mathsf{K}_0(\ell,n) \\ \mathsf{K}_1(\ell,n) \\ \vdots \\ \mathsf{K}_r(\ell,n) \end{bmatrix},$$

which implies $\mathbf{M}_{r,[0,r]} = \mathbf{M}_{r-1,[0,r]} \cdot \mathbf{H}_r$ and $\mathbf{M}_{r,[0,n]} \cdot \mathbf{K} = \mathbf{V}_{r,[0,n]}$, and that $\mathbf{M}$ is lower triangular. The desired statement follows. $\square$

Denote by $\mathbf{M}_{[0,t],[0,t]}$ the $(t+1)$-order submatrix formed by the leading $(t+1)$ rows and columns of $\mathbf{M}$. The matrix $\mathbf{M}_{[0,t],[0,t]}$ is invertible since the matrix $\mathbf{M}$ is an invertible lower triangular matrix by Theorem 2. The fact that $\mathbf{M} \cdot \mathbf{K} = \mathbf{V}$ with $\mathbf{M}$ being an invertible lower triangular matrix implies that

$$\mathbf{M}_{[0,t],[0,t]} \cdot \mathbf{K}^{(t+1)} = \mathbf{V}^{(t+1)},$$

where $\mathbf{V}^{(t+1)}$ is the submatrix formed by the leading $(t + 1)$ rows of $\mathbf{V}$. The above equality together with (3) enables us to reformulate Theorem 1 as follows.

Theorem 3. *Let $n \in \mathbb{N}^*$ and $t \in \mathbb{N}$ such that $t \leq n$. A Boolean function $f \in \mathcal{B}_n$ has $\mathsf{corr}(f) = t$ if and only if t is the maximum integer such that the restricted Walsh transform vector rw_f satisfies*

$$\begin{bmatrix} 1 & 1 & 1 & \cdots & 1 \\ 0 & 1 & 2 & \cdots & n \\ & & \cdots & \cdots & \\ 0 & 1 & 2^t & \cdots & n^t \end{bmatrix} \cdot \begin{bmatrix} \mathcal{W}_{f,0} \\ \mathcal{W}_{f,1} \\ \vdots \\ \mathcal{W}_{f,n} \end{bmatrix} = \mathbf{0}_{t+1}. \tag{7}$$

Furthermore, due to Property 1 that any WAPB function f has its restricted Walsh transform $\mathcal{W}_{f,\ell} = 0$ when $\ell \not\preceq n$, it suffices to investigate the values of $\mathcal{W}_{f,\ell}$ when $\ell \preceq n$. Assume $\mathsf{w}_{\mathsf{H}}(n) = t$ and denote by rw_f^* the vector of length 2^t obtained by keeping all nonzero coefficients in rw_f. Then we have the following result.

Corollary 1. *Let $n \in \mathbb{N}^*$ with $\mathsf{w_H}(n) = t$. Let $\mathbf{W}$ denote the submatrix formed by the ℓ-th column in $\mathbf{V}^{(t+1)}$ for all integers ℓ satisfying $\ell \preceq n$. Then for any n-variable WAPB function f with $\mathsf{corr}(f) = t$, the 2^t-length vector rw_f^* satisfies the following system of equations in $\mathbf{y} = (y_0, y_1, \ldots, y_{2^t-1})^\mathsf{T} \in \{\pm 1\}^{2^t}$:*

$$\mathbf{W} \cdot \mathbf{y} = \mathbf{0}_{t+1}.$$

The above characterization of t-correctors in terms of a Vandermonde matrix will be used to show the non-existence of WPB functions and WAPB functions of certain corrector order t in the next section.

4   Corrector Order of WAPB Functions

Section 3 characterizes t-correctors in terms of restricted Walsh transforms and Vandermonde matrices. This section will further discuss correction order of WAPB functions by studying the nonzero coefficients in their restricted Walsh transform vectors.

4.1   Non-existence of 1-Resilient WPB Functions

Note that the slice $\mathsf{E}_{k,n}$ has even cardinality if and only $\binom{n}{k}$ is even, equivalently, $k \not\preceq n$ according to Lemma 1. By definition, WPB functions are balanced on each $\mathsf{E}_{k,n}$ for any $k \in [1, n-1]$, indicating that they exist only for the case where $n = 2^m$ for certain positive integer m. The following proposition shows the non-existence of 1-resilient WPB functions.

Proposition 1. *Let f be an n-variable WPB function with $n = 2^m$ for $m \in \mathbb{N}^*$. Then $\mathsf{res}(f) = \mathsf{corr}(f) = 0$.*

Proof. By Definition 7, a WPB function is balanced, hence 0-corrector and 0-resilient. Below we show a WPB function cannot be 1-corrector, thus not 1-resilient.

By Theorem 3 we see that $\mathsf{corr}(f) = 1$ if and only if

$$\mathbf{V}^{(2)} \cdot \mathsf{rw}_f = \mathbf{0}_2,$$

where the restricted Walsh vector rw_f satisfies that $\mathcal{W}_{f,\ell} = 0$ for any $\ell \in [1, n-1]$ since f is balanced on $\mathsf{E}_{\ell,n}$. Excluding these columns in $\mathbf{V}^{(2)}$ reduces the above system to

$$\begin{bmatrix} 1 & 1 \\ 0 & n \end{bmatrix} \cdot \begin{bmatrix} \mathcal{W}_{f,0} \\ \mathcal{W}_{f,n} \end{bmatrix} = \begin{bmatrix} 0 \\ 0 \end{bmatrix},$$

which clearly contradicts the definition of WPB functions in Definition 7. $\square$

4.2 Corrector Order of n-Variable WAPB Functions when $\mathsf{w_H}(n) = 2$ and $\mathsf{w_H}(n) = 3$

We state two propositions for WAPB functions below, where the proof for Proposition 3 is placed in Appendix A.

Proposition 2. *Let $n \in \mathbb{N}^*$ with $\mathsf{w_H}(n) = 2$. Then any WAPB function f has $\mathsf{corr}(f) \leq 1$. Furthermore, the equality is reached if and only if the restricted Walsh vector rw_f satisfies*

$$\mathcal{W}_{f,\ell} = \mathcal{W}_{f,0} \cdot (-1)^{\mathsf{w_H}(\ell)}, \forall \ell \preceq n.$$

Proof. By Corollary 1 it suffices to consider the solution of the linear system $\mathbf{W} \cdot \mathbf{y} = \mathbf{0}_{t+1}$ for $t = 1$ and $t = 2$. More precisely, letting $n = 2^{i_1} + 2^{i_2}$ with positive integers $i_1 < i_2$, we need to study the following two linear systems:

$$\begin{bmatrix} 1 & 1 & 1 & 1 \\ 0 & 2^{i_1} & 2^{i_2} & (2^{i_1} + 2^{i_2}) \end{bmatrix} \cdot \mathbf{y} = \begin{bmatrix} 0 \\ 0 \end{bmatrix}, \tag{8}$$

and

$$\begin{bmatrix} 1 & 1 & 1 & 1 \\ 0 & 2^{i_1} & 2^{i_2} & (2^{i_1} + 2^{i_2}) \\ 0 & 2^{2i_1} & 2^{2i_2} & (2^{i_1} + 2^{i_2})^2 \end{bmatrix} \cdot \mathbf{y} = \begin{bmatrix} 0 \\ 0 \\ 0 \end{bmatrix}, \tag{9}$$

where $\mathbf{y} = (y_{00}, y_{01}, y_{10}, y_{11}) \in \{\pm 1\}^4$. It can be verified that the solutions of (8) are given by $\mathbf{y} = \pm(1, -1, -1, 1)$. This is equivalent to saying that $\mathcal{W}_{f,\ell} = \mathcal{W}_{f,0} \cdot (-1)^{\mathsf{w_H}(\ell)}$ for any $\ell \preceq n$. Furthermore, these solutions do not satisfy (9), which implies that there exists no WAPB function f with $\mathsf{corr}(f) = 2$ when $\mathsf{w_H}(n) = 2$. $\qquad\qquad\square$

Proposition 3. *Let $n \in \mathbb{N}^*$ with $\mathsf{w_H}(n) = 3$. Then any WAPB function f has $\mathsf{corr}(f) \leq 2$. Furthermore, the equality is reached if and only if the restricted Walsh vector rw_f satisfies*

$$\mathcal{W}_{f,\ell} = \mathcal{W}_{f,0} \cdot (-1)^{\mathsf{w_H}(\ell)}, \forall \ell \preceq n.$$

4.3 Corrector Order of n-Variable WAPB Functions for $\mathsf{w_H}(n) > 3$

We shall consider general cases by following the discussion for the cases where $\mathsf{w_H}(n) = 2$ and $\mathsf{w_H}(n) = 3$. In order to discuss the linear systems in a unified way, we need the following lemma whose proof is given in Appendix B.

Lemma 2. *Assume $n \in \mathbb{N}^*$ has $\mathsf{supp}(n) = \{i_1, i_2, \ldots, i_t\}$ and $x_s = 2^{i_s}$ for $s = 1, 2, \ldots, t$. For $1 \leq u \leq t$, the equation $\sum_{\ell \preceq n} \ell^u y_\ell = 0$ can be re-written as*

$$\sum_{s=1}^{t} \sum_{\substack{\{j_1,\ldots,j_s\} \subseteq [t] \\ u_{j_1} + \cdots + u_{j_s} = u \\ u_{j_1},\ldots,u_{j_s} \geq 1}} \frac{u!}{u_{j_1}! \cdots u_{j_s}!} \left(\sum_{\substack{\{i_{j_1},\ldots,i_{j_s}\} \subseteq \mathsf{supp}(\ell) \\ \ell \preceq n}} y_\ell \right) x_{j_1}^{u_{j_1}} \ldots x_{j_s}^{u_{j_s}} = 0. \tag{10}$$

Example 1. Let $\mathsf{supp}(n) = \{i_1, i_2, i_3\}$, $u = 2$ and $x_t = 2^{i_t}$ for $t = 1, 2, 3$. Then

$$\sum_{\ell \preceq n} \ell^2 y_\ell = \sum_{s=1}^{t} \sum_{\substack{\{j_1,\ldots,j_s\} \subseteq [t] \\ (u_{j_1},\ldots,u_{j_s}) \in P_s^*(u)}} \binom{u}{u_{j_1},\ldots,u_{j_s}} \left(\sum_{\substack{\{i_{j_1},\ldots,i_{j_s}\} \subseteq \mathsf{supp}(\ell) \\ \ell \preceq n}} y_k \right) x_{j_1}^{u_{j_1}} \ldots x_{j_s}^{u_{j_s}}$$

$$= \sum_{j=1}^{3} \left(\sum_{2^{i_j} \preceq \ell \preceq n} y_\ell \right) x_j^2 + 2 \sum_{j_1, j_2 \in [3]} \left(\sum_{2^{i_{j_1}} + 2^{i_{j_2}} \subseteq \ell \preceq n} y_\ell \right) x_{j_1} x_{j_2},$$

where the second equality has only two summations. Denoting y_ℓ for $\ell = b_1 2^{i_1} + b_2 2^{i_2} + b_3 2^{i_3}$ as $y_{b_1 b_2 b_3}$ with $b_1, b_2, b_3 \in \{0,1\}$, we obtain the following expansion

$$\begin{aligned}
\sum_{k \preceq n} y_k k^2 = {} & (y_{100} + y_{110} + y_{101} + y_{111})x_1^2 + (y_{010} + y_{110} + y_{011} + y_{111})x_2^2 \\
& + (y_{001} + y_{101} + y_{011} + y_{111})x_3^2 \\
& + 2 \left((y_{110} + y_{111})x_1 x_2 + (y_{101} + y_{111})x_1 x_3 + (y_{011} + y_{111})x_2 x_3 \right).
\end{aligned}$$

Below we shall use the representation of Lemma 2 to investigate the corrector order of n-variable WAPB functions for infinitely many integers n.

Recall that a superincreasing sequence of integers $\mathbf{s} = [s_1, s_2, s_3, \ldots]$ satisfies $s_{r+1} > \sum_{i=1}^{r} s_i$ for all integers $r \geq 1$. It is clear that every subset sum of a superincreasing sequence is unique. Furthermore, we shall call S an m-gap superincreasing sequence if S is a superincreasing sequence and

$$\min \left\{ \left| \sum_{i=1}^{h} a_i - \sum_{i=1}^{h} a_i' \right| : (a_1, \ldots, a_h) \neq (a_1', \ldots, a_h') \in A^h, h \geq 1 \right\} = m, \quad (11)$$

where A is the set of all entries in $\mathbf{s}$.

Theorem 4. *Let $n \in \mathbb{N}^*$ with $\mathsf{supp}(n) = \{i_1, \ldots, i_t\}$ with $i_1 < \cdots < i_t$ and $t > 3$. Suppose $[i_1, \ldots, i_t]$ forms a t-gap superincreasing sequence. Then any n-variable WAPB function f has $\mathsf{corr}(f) \leq t - 1$. Furthermore, the equality is reached if and only if the restricted Walsh vector rw_f satisfies*

$$W_{f,\ell} = W_{f,0} \cdot (-1)^{\mathsf{w_H}(\ell)}, \forall \ell \preceq n.$$

Proof. From Theorem 3 we study the correction order of such functions by considering solutions to the linear systems $\mathbf{W} \cdot \mathbf{y} = \mathbf{0}_t$ and $\mathbf{W} \cdot \mathbf{y} = \mathbf{0}_{t+1}$ where $\mathbf{y} = (y_{0\ldots00}, y_{0\ldots01}, y_{0\ldots10}, \cdots, y_{1\ldots10}, y_{1\ldots11}) \in \{\pm 1\}^{2^t}$.

$$\begin{bmatrix} 1 & 1 & 1 & 1 & \cdots & 1 \\ 0 & 2^{i_1} & 2^{i_2} & (2^{i_1} + 2^{i_2}) & \cdots & (2^{i_1} + \cdots + 2^{i_t}) \\ \vdots & \vdots & \vdots & \vdots & & \vdots \\ 0 & 2^{(t-1)i_1} & 2^{(t-1)i_2} & (2^{i_1} + 2^{i_2})^{t-1} & \cdots & (2^{i_1} + \cdots + 2^{i_t})^{t-1} \end{bmatrix} \cdot \mathbf{y} = \begin{bmatrix} 0 \\ 0 \\ \vdots \\ 0 \end{bmatrix}. \quad (12)$$

For each y_i we can write $i = \sum_{\delta=0}^{t-1} i_\delta 2^\delta$. Let us introduce the notation $\mathbf{y}_\Delta = \{y_i \mid i_\delta = 1, \forall \delta \in \Delta\}$ for some $\Delta \subseteq \{0, 1, \ldots, t-1\}$. Setting the starting row of $\mathbf{W}$ from zero, we will proceed by induction on Rows 1 to Row $t-1$ of $\mathbf{W}$.

We want to prove that all sum coefficients must be null for each linear relation induced by the first r rows. Row 0 forces $\mathbf{y}$ to be a balanced vector over $\{\pm 1\}$.

Initialization. Let $r = 1$. Row 1 induces the equation below

$$\left(\sum_{y_i \in \mathbf{y}_{\{1\}}} y_i\right) \cdot 2^{i_1} + \left(\sum_{y_i \in \mathbf{y}_{\{2\}}} y_i\right) \cdot 2^{i_2} + \cdots + \left(\sum_{y_i \in \mathbf{y}_{\{t\}}} y_i\right) \cdot 2^{i_t} = 0.$$

For ease of presentation we denote the sum coefficients of 2^{i_j} by s_j. Then, the equation from Row 1 becomes

$$s_1 \cdot 2^{i_1} + s_2 \cdot 2^{i_2} + \cdots + s_t \cdot 2^{i_t} = 0. \tag{13}$$

Since $\mathbf{y}$ is balanced, it is clear that any s_j can take for values any even number or its opposite in the range $[0, 2^{t-1}]$ because $|\mathbf{y}_\Delta| = 2^{t-|\Delta|}$. From the initial condition, $[i_1, \ldots, i_t]$ is a t-gap superincreasing sequence and since $s_i < 2^t$ and because the gap between any two i_j is at least t, there doesn't exist any $s_j \neq 0$ satisfying Eq. (13). Consider $r = 2$. From Row 2 one gets the following equation after re-arranging the terms

$$\left(\sum_{y_i \in \mathbf{y}_{\{1\}}} y_i\right) \cdot 2^{2i_1} + \left(\sum_{y_i \in \mathbf{y}_{\{2\}}} y_i\right) \cdot 2^{2i_2} + \cdots + \left(\sum_{y_i \in \mathbf{y}_{\{t\}}} y_i\right) \cdot 2^{2i_t}$$

$$+ 2\left(\sum_{y_i \in \mathbf{y}_{\{1,2\}}} y_i\right) \cdot 2^{i_1+i_2} + \cdots + 2\left(\sum_{y_i \in \mathbf{y}_{\{t-1,t\}}} y_i\right) \cdot 2^{i_{t-1}+i_t} = 0.$$

We use the previously defined notations and knowledge that $s_1 = s_2 = \cdots = s_t = 0$ to simplify it into

$$2s_{1,2} \cdot 2^{i_1+i_2} + 2s_{1,3} \cdot 2^{i_1+i_3} + \cdots + 2s_{t-1,t} \cdot 2^{i_{t-1}+i_t} = 0. \tag{14}$$

Similarly as above, we see that any s_j can take for values any even number or its opposite in the range $[0, 2^{t-2}]$ and we make use of the initial condition that $[i_1, \ldots, i_t]$ is a t-gap superincreasing sequence and bring the same argument to prove there does not exist any $s_{i,j} \neq 0$ satisfying Eq. (14).

Induction: we shall prove that if the statement is true for $r - 1$, then it is true for r, $r \leq t - 1$. From Row r and using Lemma 2 one has

$$\sum_{\ell \preceq n} \ell^r y_\ell = \sum_{m=1}^{t} \sum_{\substack{\{j_1,\ldots,j_m\}\subseteq[t] \\ r_{j_1}+\cdots+r_{j_m}=r \\ r_{j_1},\ldots,r_{j_m}\geq 1}} \frac{r!}{r_{j_1}!\cdots r_{j_m}!} \left(\sum_{\substack{\{i_{j_1},\ldots,i_{j_m}\}\subseteq supp(\ell) \\ \ell \preceq n}} y_\ell\right) x_{j_1}^{r_{j_1}} \ldots x_{j_m}^{r_{j_m}} = 0.$$

Let us denote the multinomial coefficient by $\mu_{1,\ldots,m} = \frac{r!}{r_{j_1}!\cdots r_{j_m}!}$ and the previously defined notations $s_{1,\ldots,m} = \sum(y_\ell)$. We rewrite the above equation into its rearranged form

$$\mu_1\left(s_1 2^{i_1} + \cdots + s_t 2^t\right) + \mu_{1,2}\left(s_{1,2} 2^{i_1+i_2} + \cdots + s_{t-1,t} 2^{i_{t-1}+i_t}\right) + \cdots +$$

$$\mu_{1,2,\ldots,r}\left(s_{1,2,\ldots,r} 2^{i_1+i_2+\cdots+i_r} + \cdots + s_{t-r,t-r+1,\cdots,t} 2^{i_{t-r}+i_{t-r+1}+\cdots+i_t}\right) = 0.$$

We know from Row $r-1$ that $s_1, \ldots, s_t, s_{1,2}, \ldots, s_{t-1,t}, \ldots, s_{t-(r-1),\ldots,t-1,t} = 0$. We then reformulate the above equation into

$$s_{1,2,\ldots,r} 2^{i_1+i_2+\cdots+i_r} + \cdots + s_{t-r,t-r+1,\cdots,t} \cdot 2^{i_{t-r}+i_{t-r+1}+\cdots+i_t} = 0. \tag{15}$$

In a similar fashion, we see that any s_j can take for values any even number or its opposite in the range $[0, 2^{t-r}]$ and we make use of the initial condition that $[i_1, \ldots, i_t]$ is a t-gap superincreasing sequence and bring the same argument to prove there does not exist any $s_j \neq 0$ satisfying Eq. (15).

Finally, expanding $\mathbf{W} \cdot \mathbf{y} = \mathbf{0}_{t+1}$ yields $s_{1,\ldots,t} = y_{1\ldots1}$ that can't satisfy Eq. (12) with knowledge of previous s_j, bringing the argument of impossibility for such a t-corrector. As such, the restricted Walsh vector is uniquely determined by $\mathcal{W}_{f,0}(-1)^{\mathsf{w_H}(\ell)}, \forall \ell \preceq n$. This concludes the proof. $\qquad\square$

Based on the different theoretic results of this section such as Proposition 4 and experimental results (Sect. 5), we introduce the following conjecture on the corrector and resilience order of WAPB functions.

Conjecture 1. Let $n \in \mathbb{N}^*$ with $\mathsf{w_H}(n) = t$. Then there exist WAPB functions f with $\mathrm{corr}(f) = t - 1$ and the restricted Walsh transforms rw_f satisfy

$$\mathcal{W}_{f,\ell} = (-1)^{\mathsf{w_H}(\ell)}, \forall \ell \preceq n,$$

Furthermore, there exists no WAPB function f with $\mathrm{corr}(f) = t$.

Remark 1. Regarding the resilience upper bound of $\mathsf{w_H}(n) - 1$, we know that for all n there exist WAPB functions with resilience order $\mathsf{w_H}(n) - 1$: when we write n as $n = 2^{i_1} + 2^{i_2} + \cdots + 2^{i_t}$ where $i_1 < \cdots < i_t$, the direct sum of WPB functions in 2^{i_s} variables [49] have resilience order exactly $\mathsf{w_H}(n) - 1$. Such WAPB functions are exactly characterized in the conjecture, namely, their Walsh transforms rw_f are given by $\mathcal{W}_{f,\ell} = (-1)^{\mathsf{w_H}(\ell)}, \forall \ell \preceq n$.

Remark 2. Recall that the Prouhet-Tarry-Escott problem [44] for a positive integer r asks for two disjoint multi-sets S_0 and S_1 of the same number of integers satisfying $\sum_{a \in S_0} a^s = \sum_{b \in S_1} b^s$, $s = 0, 1, \ldots, r$. The Prouhet-Tarry-Escott problem can be phrased as a r-regular partition of an integer set [3]. More concretely, for a set S of even number of integers, a partition $\{S_0, S_1\}$ of S, i.e., $S_0 \cap S_1 = \emptyset$ and $S_0 \cup S_1 = S$, is r-regular if $\sum_{a \in S_0} a^s = \sum_{b \in S_1} b^s$, $s = 0, 1, \ldots, r$. When the set $S = \{0, 1, \ldots, 2^t - 1\}$, the Prouhet-Thue-Morse sequence [1] gives a $(t-1)$-regular partition of S as $S_i = \{0 \le a < 2^t - 1 : \mathsf{w_H}(a) \equiv i \mod 2\}, i = 1, 2$. Such a partition $\{S_0, S_1\}$ was conjectured in the second part of [3, Conjecture 5.3] to be the unique $(t-1)$-regular partition of $S = \{0, 1, \ldots, 2^t - 1\}$. One can easily verify that this partition is not t-regular.

Observe that Conjecture 1 is a generalized version of the second part of Conjecture 5.3 in [3]. For a positive integer n of the form $n = 2^{i_1} + 2^{i_2} + \cdots + 2^{i_t}$ with $i_1 < \cdots < i_t$, if in Corollary 1 the system of equations $\mathbf{W} \cdot \mathbf{y} = \mathbf{0}_{t+1}$ has a solution $\mathbf{y} = (y_0, y_1, \ldots, y_{2^t-1})^\mathsf{T} \in \{\pm 1\}^{2^t}$, then it can be used to derive a partition $\{S_0, S_1\}$ of the set $S = \{a_1 2^{i_1} + a_2 2^{i_2} + \cdots + a_t 2^{i_t} : a_1, a_2, \ldots, a_t \in \mathbb{F}_2\}$, where S_0, S_1 is derived based on the sign of y_a, where $a = a_1 + 2a_2 + \cdots + 2^{t-1}a_t$. That is to say, the existence of a WAPB function f with $\mathrm{corr}(f) = t$ implies the existence of a t-regular partition of the set S.

5   Experimental Results

As illustrations, we provide an open-source implementation validating our main results in SageMath [43] available at https://www.github.com/wapbres/wapbres. We give a basis of code to test Theorems 1, 2 and 3 for small parameters. We also verified that Conjecture 1 holds for all dimensions n up to $n = 62$. Our approach consists in exhausting all balanced vectors $\mathbf{y} \in \{\pm 1\}^{w_\mathsf{H}(n)}$ and computing t as in Theorem 3. Precisely there are $\binom{2^{w_\mathsf{H}(n)}}{2^{w_\mathsf{H}(n)-1}}$ such vectors for which we keep track of the largest t-corrector. We give a brief summary of the performance of our method in Table 1. All execution times are in seconds and performed on a single core 13th Gen Intel(R) Core(TM) i5-1345U 4.7 GHz. We see that for values of n with $w_\mathsf{H}(n) = 5$, searching through roughly 2^{32} vectors takes around 2h30, indicating the infeasibility of verifying the case of Hamming weight 6 which induces a search space of roughly 2^{64} vectors.

Table 1. Correction order of WAPB functions in $n \leq 62$ variables.

$w_\mathsf{H}(n)$	1	2	3	4	5
$\mathrm{corr}(f)$	0	1	2	3	4
time (s)	0.02	0.03	0.06	0.21	8916.26

6   Conclusion

This paper investigated the resilience order of WAPB functions from the perspective of their corrector orders. Our main contributions are twofold. First, we give a characterization of the corrector orders of a Boolean function by expressing its restricted Walsh transform vector in terms of linear equations over integers. We established an interesting algebraic relation between the Krawchouk matrix and the Vandermonde matrix, which transforms the linear equations into new ones associated with certain Vandermonde matrix. Second, our analysis of those new equations showed that for infinitely many integers n, the maximum corrector order of n-variable WAPB functions is tightly upper bounded by the Hamming weight of n minus one.

Acknowledgment. The work of Pierrick Méaux was funded by the European Research Council (ERC) under the Advanced Grant program (reference number: 787390).

Disclosure of Interests. The authors have no competing interests to declare that are relevant to the content of this article.

A   Proof of Proposition 3

Proof. Similarly as in the proof for Proposition 2, we consider the solutions to $\mathbf{W} \cdot \mathbf{y} = \mathbf{0}_{t+1}^{\mathsf{T}}$ for $t = 2$ and $t = 3$.

Let $n = 2^{i_1} + 2^{i_2} + 2^{i_3}$ with positive integers $i_1 < i_2 < i_3$, this reduces to studying the solutions for

$$\begin{bmatrix} 1 & 1 & 1 & 1 & 1 & 1 & 1 & 1 \\ 0 & 2^{i_1} & 2^{i_2} & (2^{i_1}+2^{i_2}) & 2^{i_3} & (2^{i_1}+2^{i_3}) & (2^{i_2}+2^{i_3}) & (2^{i_1}+2^{i_2}+2^{i_3}) \\ 0 & 2^{2i_1} & 2^{2i_2} & (2^{i_1}+2^{i_2})^2 & 2^{2i_3} & (2^{i_1}+2^{i_3})^2 & (2^{i_2}+2^{i_3})^2 & (2^{i_1}+2^{i_2}+2^{i_3})^2 \end{bmatrix} \cdot \mathbf{y} = \begin{bmatrix} 0 \\ 0 \\ 0 \end{bmatrix} \tag{16}$$

and

$$\begin{bmatrix} 1 & 1 & 1 & 1 & 1 & 1 & 1 & 1 \\ 0 & 2^{i_1} & 2^{i_2} & (2^{i_1}+2^{i_2}) & 2^{i_3} & (2^{i_1}+2^{i_3}) & (2^{i_2}+2^{i_3}) & (2^{i_1}+2^{i_2}+2^{i_3}) \\ 0 & 2^{2i_1} & 2^{2i_2} & (2^{i_1}+2^{i_2})^2 & 2^{2i_3} & (2^{i_1}+2^{i_3})^2 & (2^{i_2}+2^{i_3})^2 & (2^{i_1}+2^{i_2}+2^{i_3})^2 \\ 0 & 2^{3i_1} & 2^{3i_2} & (2^{i_1}+2^{i_2})^3 & 2^{3i_3} & (2^{i_1}+2^{i_3})^3 & (2^{i_2}+2^{i_3})^3 & (2^{i_1}+2^{i_2}+2^{i_3})^3 \end{bmatrix} \cdot \mathbf{y} = \begin{bmatrix} 0 \\ 0 \\ 0 \\ 0 \end{bmatrix} \tag{17}$$

where $\mathbf{y} = (y_{000}, y_{001}, y_{010}, y_{011}, y_{100}, y_{101}, y_{110}, y_{111}) \in \{\pm 1\}^8$.

Let us first study the solutions of (16). We will iterate through the rows of the matrix to gradually build necessary conditions on its solutions. We refer to Row k as the row indexed by the exponent k for $0 \le k < 3$ in $\mathbf{W}$.

Row 0 indicates that $\mathbf{y}$ must be a balanced vector, namely, half of its entries equal 1 and the other half equal -1. Row 1 yields the following equation:

$$(y_{001} + y_{011} + y_{101} + y_{111}) \cdot 2^{i_1} +$$
$$(y_{010} + y_{011} + y_{110} + y_{111}) \cdot 2^{i_2} +$$
$$(y_{100} + y_{101} + y_{110} + y_{111}) \cdot 2^{i_3} = 0$$

For ease of presentation we denote the sum coefficients of $2^{i_1}, 2^{i_2}, 2^{i_3}$ by s_1, s_2, s_3, respectively. Then the equation from Row 1 becomes

$$s_1 \cdot 2^{i_1} + s_2 \cdot 2^{i_2} + s_3 \cdot 2^{i_3} = 0. \tag{18}$$

By construction, the sums s_1, s_2, s_3 can only take for value any $\pm 4, \pm 2$ or 0. By further applying the constraint of balancedness from Row 0, all possible $\binom{8}{4}$ balanced $\mathbf{y}$ exhibit only the following possible options for the multiset (that allows for same elements) $\{s_1, s_2, s_3\}$:

$$\{0, 0, 0\}, \{0, 0, \pm 2\}, \{0, \pm 2, \pm 2\}, \{\pm 2, \pm 2, \pm 2\}, \{0, 0, \pm 4\}.$$

This means that each term on the left hand of Eq. (18) is either zero or a power of 2. Since $i_1 < i_2 < i_3$, it implies that Eq. (18) can be true only when $s_1 = s_2 = s_3 = 0$.

From Row 2 one gets the following equation (after re-arranging the terms):

$$
\begin{aligned}
(y_{001} + y_{011} + y_{101} + y_{111}) \cdot 2^{2i_1} + \\
(y_{010} + y_{011} + y_{110} + y_{111}) \cdot 2^{2i_2} + \\
(y_{100} + y_{101} + y_{110} + y_{111}) \cdot 2^{2i_3} + \\
(y_{011} + y_{111}) \cdot 2 \cdot 2^{i_1+i_2} + \\
(y_{101} + y_{111}) \cdot 2 \cdot 2^{i_1+i_3} + \\
(y_{110} + y_{111}) \cdot 2 \cdot 2^{i_2+i_3} = 0.
\end{aligned}
\tag{19}
$$

After applying the preceding constraints, namely, balancedness of $\mathbf{y}$ from Row 0 and $s_1 = s_2 = s_3 = 0$ from Row 1, we obtain

$$
2s_{1,2} \cdot 2^{i_1+i_2} + 2s_{1,3} \cdot 2^{i_1+i_3} + 2s_{2,3} \cdot 2^{i_2+i_3} = 0,
$$

where $s_{1,2}, s_{1,3}, s_{2,3}$ correspond to the sum coefficient of $2^{i_1+i_2}, 2^{i_2+i_3}$ and $2^{i_1+i_3}$, respectively. We now discuss the values of $s_{1,2}, s_{1,3}$ and $s_{2,3}$. Assume $s_{1,2} = \pm 2$. Then $y_{011} = y_{111} = \pm 1$.

By considering both the condition of balancedness and that $s_1 = s_2 = s_3 = 0$, we necessarily have $y_{101} = y_{110} = -y_{011} = -y_{111}$. This doesn't satisfy Eq. (19). The same logic applies if $y_{101} = y_{111}$ or if $y_{110} = y_{111}$. The only combination that works is $y_{011} = y_{101} = y_{110} = -y_{111}$, thereby giving the following solutions to Eq. (19)

$$
\mathbf{y} = \pm(1, -1, -1, -1, 1, 1, 1, -1).
$$

Furthermore, these solutions do not satisfy the last equation in (17). That is to say, there is not WAPB function f with $\mathsf{corr}(f) = 3$ when $\mathsf{w_H}(n) = 3$. $\qquad\square$

B   Proof of Lemma 2

Proof. Note that the expansion of $(x_1 + \cdots + x_s)^u$ is given by

$$
(x_1 + \cdots + x_s)^u = \sum_{\substack{u_1 + \cdots + u_s = u \\ u_1, \ldots, u_s \geq 0}} \frac{u!}{u_1! \cdots u_s!} x_1^{u_1} \ldots x_s^{u_s},
$$

where $\frac{u!}{u_1! \cdots u_s!}$ is the multinomial coefficient. In the above sum $(u_1, \ldots, u_s)$ can be seen as a non-negative partition of the integer u. For simplicity, we denote by $P_s(u)$ the set of all such partitions of u. Observe that for any $(u_1, \ldots, u_s) \in P_s(u)$, the integers $u_1, \ldots, u_s, 0$ form an $(s+1)$-tuple in $P_{s+1}(u)$. This implies the inclusive relation

$$
P_s(u) \subset P_{s+1}(u), \quad s = 1, 2, \ldots.
$$

On the other hand, for an $(s+1)$-tuple $(y_1, \ldots, y_{s+1})$ in $P_{s+1}(u)$ with (at least) one zero coordinate, it will degenerate to an s-tuple in $P_s(u)$ after one zero coordinate is removed. Hence, for $s \leq u$, we can partition the set $P_s(u)$ as

$$P_s(u) = P_s^*(u) \sqcup P_{s-1}^*(u) \sqcup \cdots \sqcup P_1^*(u),$$

where $P_r^*(u)$ contains those r-tuples $(u_1, \ldots, u_r)$ with $u_1, \ldots, u_r \geq 1$.

Now we consider the expansion of the sum $\sum_{\ell \preceq n} y_\ell \ell^u = 0$ with $n = 2^{i_1} + \cdots + 2^{i_t}$ and $u > 0$. By denoting $x_j = 2^{i_j}$ and re-arranging the terms in its expansions, we obtain

$$
\begin{aligned}
\sum_{\ell \preceq n} \ell^u y_\ell &= \sum_{s=1}^{t} \sum_{\substack{supp(\ell)=\{i_{j_1},\ldots,i_{j_s}\} \\ \{j_1,\ldots,j_s\} \subseteq [t]}} (x_{j_1} + \cdots + x_{j_s})^u y_\ell \\
&= \sum_{s=1}^{t} \sum_{\substack{supp(\ell)=\{i_{j_1},\ldots,i_{j_s}\} \\ \{j_1,\ldots,j_s\} \subseteq [t]}} y_\ell \sum_{(u_1,\ldots,u_s) \in P_s(u)} \frac{u!}{u_{j_1}! \cdots u_{j_s}!} x_{j_1}^{u_1} \cdots x_{j_s}^{u_{j_s}} \\
&= \sum_{s=1}^{t} \sum_{\substack{supp(\ell)=\{i_{j_1},\ldots,i_{j_s}\} \\ \{j_1,\ldots,j_s\} \subseteq [t] \\ (u_{j_1},\ldots,u_{j_s}) \in P_s(u)}} \frac{u!}{u_{j_1}! \cdots u_{j_s}!} y_\ell x_{j_1}^{u_1} \cdots x_{j_s}^{u_{j_s}}
\end{aligned}
\tag{20}
$$

where the last equality is obtained by re-arranging the summands for $s = 1, \ldots, t$ according to the partition $P_s(u) = P_s^*(u) \sqcup P_{s-1}^*(u) \sqcup \cdots \sqcup P_1^*(u)$. $\square$

References

1. Allouche, J.P., Shallit, J.: The ubiquitous Prouhet-Thue-morse sequence. In: Ding, C., Helleseth, T., Niederreiter, H. (eds.) Sequences and Their Applications, pp. 1–16. Springer, London (1999). https://doi.org/10.1007/978-1-4471-0551-0_1
2. Belaïd, S., Fouque, P.A., Gérard, B.: Side-channel analysis of multiplications in GF(2128). In: Advances in Cryptology - ASIACRYPT 2014, pp. 306–325. Springer, Heidelberg (2014). https://doi.org/10.1007/978-3-662-45608-8_17
3. Bolker, E.D., Offner, C., Richman, R., Zara, C.: The Prouhet-Tarry-Escott problem and generalized Thue-Morse sequences. J. Comb. **7**(1), 117–133 (2016). https://doi.org/10.4310/JOC.2016.v7.n1.a5
4. Bonamino, L., Méaux, P.: Computing the restricted algebraic immunity, and application to weightwise perfectly balanced functions. In: Kim, Y., Miyaji, A., Tibouchi, M. (eds.) Cryptology and Network Security - 24th International Conference, CANS 2025, Osaka, Japan, 17–20 November 2025, Proceedings. Lecture Notes in Computer Science, vol. 16351, pp. 142–169. Springer (2025). https://doi.org/10.1007/978-981-95-4434-9_7
5. Borwein, P.: The Prouhet-Tarry-Escott Problem, pp. 85–95. Springer, New York (2002). https://doi.org/10.1007/978-0-387-21652-2_11

6. Carlet, C., Feukoua, S., Salagean, A.: The stability of the algebraic degree of Boolean functions when restricted to affine spaces. Des. Codes Cryptogr. **93**(11), 4799–4832 (2025). https://doi.org/10.1007/S10623-025-01702-Z

7. Carlet, C., Méaux, P., Rotella, Y.: Boolean functions with restricted input and their robustness; application to the FLIP cipher. IACR Trans. Symmetric Cryptol. **2017**(3) (2017). https://doi.org/10.46586/tosc.v2017.i3.192-227

8. Dachman-Soled, D., Ducas, L., Gong, H., Rossi, M.: LWE with side information: attacks and concrete security estimation. In: Micciancio, D., Ristenpart, T. (eds.) CRYPTO 2020. LNCS, vol. 12171, pp. 329–358. Springer, Cham (2020). https://doi.org/10.1007/978-3-030-56880-1_12

9. Dalai, D.K., Maitra, S., Sarkar, S.: Basic theory in construction of boolean functions with maximum possible annihilator immunity. Des. Codes Cryptogr. **40**(1), 41–58 (2006). https://doi.org/10.1007/s10623-005-6300-x

10. Dalai, D.K., Mallick, K.: A Class of Weightwise Almost Perfectly Balanced Boolean Functions with High Weightwise Nonlinearity. BFA 2023 (2023)

11. Dalai, D.K., Mallick, K.: A class of weightwise almost perfectly balanced Boolean functions. Adv. Math. Commun. **18**(2), 480–504 (2024). https://doi.org/10.3934/amc.2023048

12. Dalai, D.K., Mallick, K., Méaux, P.: Weightwise almost perfectly balanced functions, construction from a permutation group action view. IACR Cryptol. ePrint Arch. 2068 (2024). https://eprint.iacr.org/2024/2068, to appear in Designs, Codes and Cryptography 2026

13. Duval, S., Méaux, P., Momin, C., Standaert, F.X.: Exploring crypto-physical dark matter and learning with physical rounding: towards secure and efficient fresh re-keying. IACR Trans. Cryptogr. Hardw. Embed. Syst. **2021**(1), 373–401 (2020). https://doi.org/10.46586/tches.v2021.i1.373-401

14. Gini, A., Méaux, P.: On the weightwise nonlinearity of weightwise perfectly balanced functions. Discret. Appl. Math. **322**, 320–341 (2022). https://doi.org/10.1016/j.dam.2022.08.017

15. Gini, A., Méaux, P.: Weightwise almost perfectly balanced functions: secondary constructions for all n and better weightwise nonlinearities. In: Isobe, T., Sarkar, S. (eds.) Progress in Cryptology - INDOCRYPT 2022. LNCS, vol. 13774, pp. 492–514. Springer, Cham (2022). https://doi.org/10.1007/978-3-031-22912-1_22

16. Gini, A., Méaux, P.: On the algebraic immunity of weightwise perfectly balanced functions. In: Aly, A., Tibouchi, M. (eds.) Progress in Cryptology - LATINCRYPT 2023. LNCS, vol. 14168, pp. 3–23. Springer, Cham (2023). https://doi.org/10.1007/978-3-031-44469-2_1

17. Gini, A., Méaux, P.: Weightwise perfectly balanced functions and nonlinearity. In: Hajji, S.E., Mesnager, S., Souidi, E.M. (eds.) Codes, Cryptology and Information Security - C2SI 2023. LNCS, vol. 13874, pp. 338–359. Springer, Cham (2023). https://doi.org/10.1007/978-3-031-33017-9_21

18. Gini, A., Méaux, P.: S_0-equivalence classes, a new direction to find better weightwise perfectly balanced functions, and more. Cryptogr. Commun. **16**(6), 1211–1234 (2024). https://doi.org/10.1007/S12095-024-00719-W

19. Guo, X., Su, S.: Construction of weightwise almost perfectly balanced Boolean functions on an arbitrary number of variables. Discret. Appl. Math. **307**, 102–114 (2022). https://doi.org/10.1016/j.dam.2021.10.011

20. Hoffmann, C., Méaux, P., Momin, C., Rotella, Y., Standaert, F.X., Udvarhelyi, B.: Learning with physical rounding for linear and quadratic leakage functions. In: Advances in Cryptology – CRYPTO 2023, pp. 410–439 (2023). https://doi.org/10.1007/978-3-031-38548-3_14

21. Hu, Y., Zhang, Y., Wu, W.: Construction of correctors using resilient fragmentary Boolean functions. Des. Codes Cryptogr. **94**(1), 6 (2026). https://doi.org/10.1007/S10623-025-01771-0

22. Işık, L., Rodríguez-Aldama, R., Šehović, A.: Weightwise (almost) perfectly balanced functions: t-concatenation and the general maiorana-mcfarland class. IACR Cryptol. ePrint Arch. 2130 (2025). https://eprint.iacr.org/2025/2130

23. Joye, M., Olivier, F.: Side-channel analysis. In: van Tilborg, H.C.A. (ed.) Encyclopedia of Cryptography and Security. Springer (2005)

24. Lacharme, P.: Analysis and construction of correctors. IEEE Trans. Inf. Theory **55**(10), 4742–4748 (2009). https://doi.org/10.1109/TIT.2009.2027483

25. Li, J., Su, S.: Construction of weightwise perfectly balanced Boolean functions with high weightwise nonlinearity. Discret. Appl. Math. **279**, 218–227 (2020). https://doi.org/10.1016/j.dam.2020.01.020

26. Liu, J., Mesnager, S.: Weightwise perfectly balanced functions with high weightwise nonlinearity profile. Des. Codes Cryptogr. **87**(8), 1797–1813 (2019). https://doi.org/10.48550/arXiv.1709.02959

27. Luo, S., Wang, W., Zhang, Q., Song, Z.: Constructions of plateaued correctors with high correction order and good nonlinearity via walsh spectral neutralization technique. Des. Codes Cryptogr. **92**(12), 4531–4548 (2024). https://doi.org/10.1007/S10623-024-01497-5

28. MacWilliams, F., Sloane, N.: The Theory of Error-Correcting Codes, 2nd edn. North-holland Publishing Company (1978)

29. Maitra, S., Mandal, B., Martinsen, T., Roy, D., Stănică, P.: Tools in analyzing linear approximation for boolean functions related to FLIP. In: Chakraborty, D., Iwata, T. (eds.) INDOCRYPT 2018. LNCS, vol. 11356, pp. 282–303. Springer, Cham (2018). https://doi.org/10.1007/978-3-030-05378-9_16

30. Mandujano, S., Ku Cauich, J.C., Lara, A.: Studying special operators for the application of evolutionary algorithms in the seek of optimal boolean functions for cryptography. In: Pichardo Lagunas, O., Martínez-Miranda, J., Martínez Seis, B. (eds.) Advances in Computational Intelligence, pp. 383–396. Springer, Cham (2022). https://doi.org/10.1007/978-3-031-19493-1_30

31. Mariot, L., Picek, S., Jakobovic, D., Djurasevic, M., Leporati, A.: Evolutionary construction of perfectly balanced boolean functions. In: 2022 IEEE Congress on Evolutionary Computation (CEC), pp. 1–8. IEEE Press (2022). https://doi.org/10.48550/arXiv.2202.08221

32. Méaux, P.: Weightwise (almost) perfectly balanced functions based on total orders. In: Eichlseder, M., Gambs, S. (eds.) Selected Areas in Cryptography - SAC 2024, pp. 112–138. Springer, Cham (2025). https://doi.org/10.1007/978-3-031-82841-6_5

33. Méaux, P., Journault, A., Standaert, F.-X., Carlet, C.: Towards stream ciphers for efficient FHE with low-noise ciphertexts. In: Fischlin, M., Coron, J.-S. (eds.) EUROCRYPT 2016. LNCS, vol. 9665, pp. 311–343. Springer, Heidelberg (2016). https://doi.org/10.1007/978-3-662-49890-3_13

34. Mesnager, S., Su, S.: On constructions of weightwise perfectly balanced Boolean functions. Cryptogr. Commun. (2021). https://doi.org/10.1007/s12095-021-00481-3

35. Mesnager, S., Su, S., Li, J.: On concrete constructions of weightwise perfectly balanced functions with optimal algebraic immunity and high weightwise nonlinearity. Boolean Funct. Appl. (2021)

36. Mesnager, S., Su, S., Li, J., Zhu, L.: Concrete constructions of weightwise perfectly balanced (2-rotation symmetric) functions with optimal algebraic immunity and high weightwise nonlinearity. Cryptogr. Commun. **14**(6), 1371–1389 (2022). https://doi.org/10.1007/s12095-022-00603-5
37. Mesnager, S., Zhou, Z., Ding, C.: On the nonlinearity of Boolean functions with restricted input. Cryptogr. Commun. (2018). https://doi.org/10.1007/s12095-018-0293-6
38. Oren, Y., Renauld, M., Standaert, F.-X., Wool, A.: Algebraic side-channel attacks beyond the hamming weight leakage model. In: Prouff, E., Schaumont, P. (eds.) CHES 2012. LNCS, vol. 7428, pp. 140–154. Springer, Heidelberg (2012). https://doi.org/10.1007/978-3-642-33027-8_9
39. Renauld, M., Standaert, F.-X.: Algebraic side-channel attacks. In: Bao, F., Yung, M., Lin, D., Jing, J. (eds.) Inscrypt 2009. LNCS, vol. 6151, pp. 393–410. Springer, Heidelberg (2010). https://doi.org/10.1007/978-3-642-16342-5_29
40. Siegenthaler, T.: Correlation-immunity of nonlinear combining functions for cryptographic applications. IEEE Trans. Inf. Theory **30**(5), 776–780 (1984). https://doi.org/10.1109/TIT.1984.1056949
41. Standaert, F.: Introduction to side-channel attacks. In: Verbauwhede, I.M.R. (ed.) Secure Integrated Circuits and Systems, pp. 27–42. Integrated Circuits and Systems, Springer (2010). https://doi.org/10.1007/978-0-387-71829-3_2
42. Tang, D., Liu, J.: A family of weightwise (almost) perfectly balanced Boolean functions with optimal algebraic immunity. Cryptogr. Commun. **11**(6), 1185–1197 (2019). https://doi.org/10.1007/s12095-019-00374-6
43. The Sage Developers: SageMath, the Sage Mathematics Software System (Version 10.7) (2025). https://www.sagemath.org
44. Wright, E.M.: Prouhet's 1851 solution of the Tarry-Escott problem of 1910. Am. Math. Monthly **66**(3), 199–201 (1959). https://doi.org/10.1080/00029890.1959.11989269
45. Yan, L., et al.: IGA: an improved genetic algorithm to construct weightwise (almost) perfectly balanced boolean functions with high weightwise nonlinearity. In: Proceedings of the 2023 ACM Asia Conference on Computer and Communications Security, ASIA CCS 2023, pp. 638–648. Association for Computing Machinery, New York, NY, USA (2023). https://doi.org/10.1145/3579856.3590337
46. Zhang, R., Su, S.: A new construction of weightwise perfectly balanced Boolean functions. Adv. Math. Commun. **17**(4), 757–770 (2023). https://doi.org/10.3934/AMC.2021020
47. Zhao, Q., Jia, Y., Zheng, D., Qin, B.: A new construction of weightwise perfectly balanced functions with high weightwise nonlinearity. Mathematics **11**(5) (2023). https://doi.org/10.3390/math11051193
48. Zhao, Q., Li, M., Chen, Z., Qin, B., Zheng, D.: A unified construction of weightwise perfectly balanced Boolean functions. Discret. Appl. Math. **337**, 190–201 (2023). https://doi.org/10.1016/j.dam.2023.04.021
49. Zhu, L., Su, S.: A systematic method of constructing weightwise almost perfectly balanced Boolean functions on an arbitrary number of variables. Discret. Appl. Math. **314**, 181–190 (2022). https://doi.org/10.1016/j.dam.2022.02.017

A Notion on S-Boxes for a Partial Resistance to Some Integral Attacks

Claude Carlet[1,2]([envelope]) [iD]

[1] Department of Informatics, University of Bergen, 5005 Bergen, Norway
claude.carlet@gmail.com
[2] Department of Mathematics, University of Paris 8, 93526 Saint-Denis, France

Abstract. Recently, the notion of kth-order sum-freedom of a vectorial function $F : \mathbb{F}_2^n \to \mathbb{F}_2^m$ has been introduced, generalizing that of almost perfect nonlinearity (which corresponds to $k = 2$) and having some relation to resistance against integral attacks on block ciphers, by preventing the propagation of the division property of k-dimensional affine spaces. In the present paper, we show that this notion, which is rarely satisfied by vectorial functions, can be weakened while retaining the same behavior with respect to the division property. This leads us to the notion of kth-order t-degree-sum-freedom, whose strength decreases as t increases, and which coincides with kth-order sum-freedom when $t = 1$: for every k-dimensional affine space A, there exists a non-negative integer j of 2-weight at most t such that $\sum_{x \in A}(F(x))^j \neq 0$. We show that t can always be taken smaller than or equal to $\min(k, m)$ under some "reasonable" condition on F (satisfied in particular by all injective functions). This makes the new notion more interesting theoretically and practically than sum-freedom (which is an all-or-nothing notion and which in practice disqualifies almost all functions). The parameter t in the new notion quantifies more precisely the behavior of any "reasonable" function. A quality of this parameter is its simplicity. We also show that t is greater than or equal to $\frac{k}{\deg(F)}$, where $\deg(F)$ is the algebraic degree of F, and we derive two other lower bounds. We study power functions, for which we prove upper bounds. Among them, we study the multiplicative inverse function (used as an S-box in the AES), for which we characterize the kth-order t-degree-sum-freedom by the coefficients of the subspace polynomials of k-dimensional vector subspaces (deducing the exact minimal value of t when k divides n) and we prove that its kth-order t-degree-sum-freedom is equivalent to its $(n - k)$th-order t-degree-sum-freedom.

Keywords: vectorial function · S-box · almost perfect nonlinear · kth-order sum-free · integral attack · division property

1 Introduction

A (vectorial) (n, m)-function $F : \mathbb{F}_2^n \to \mathbb{F}_2^m$ is called kth-order sum-free [7] if, for every k-dimensional affine space A in $\mathbb{F}_2^n$, we have $\sum_{x \in A} F(x) \neq 0$. This is

L. Batina and F. Özbudak (Eds.): WAIFI 2026, LNCS 16611, pp. 221–238, 2026.
https://doi.org/10.1007/978-3-032-27574-5_14

equivalent to saying that the kth-order derivatives $D_{a_1} \ldots D_{a_k} F(x)$ (where the first order derivative is defined as $D_a F(x) = F(x) + F(x + a)$ and the kth-order derivative is its iteration) never vanish when $a_1, \ldots, a_k$ are linearly independent over $\mathbb{F}_2$. Almost perfect nonlinearity corresponds to $k = 2$.

There is a relation between this notion and integral attacks [12]. Todo [16] has introduced, in the framework of these cryptanalyses, the notion of division property of a set, and Boura-Canteaut [3] have translated it into the language of Reed-Muller codes (there also exists a survey [11]). A set $X \subseteq \mathbb{F}_2^n$ has the *division property* at an order l if its indicator has an algebraic degree of at most $n - l$. Integral attacks practically lead to studying the propagation of the division property through rounds, which needs to study it through substitution boxes (S-boxes, which are the nonlinear components in rounds). It is shown in [7] that kth-order sum-freedom makes it impossible the propagation of the division property of k-dimensional affine spaces through the S-box. Since the division property is often (but not always) investigated by cryptanalysts by focussing on affine spaces, the study of kth-order sum-freedom is useful for designers, helping them to protect ciphers against such kinds of integral attacks, and for cryptanalysts, letting them know which affine spaces can be considered in integral attacks. However, the functions satisfying kth-order sum-freedom for a given k are rare, and even if a function satisfies it for some value of k, it may not satisfy it for other values. Fortunately, we show in the present paper that this criterion can be generalized into a weaker version depending on some parameter $t \geq 1$ (kth-order sum-freedom corresponding to $t = 1$), which is practically more useful for designers and attackers (but is still more difficult to study), since it is satisfied by any reasonable vectorial function for any k and some t.

In the present paper, we begin the study of this new notion, called kth-order t-degree-sum-freedom. The condition for such a property to be satisfied by F is that, for every k-dimensional affine space A, there exists a non-negative integer j of 2-weight at most t such that $\sum_{x \in A} (F(x))^j \neq 0$, where the 2-weight of a non-negative integer equals the Hamming weight of its binary expansion. For general k, we show that we can take $t \leq \min(k, m)$ under a reasonable assumption on F, which is satisfied by all injective functions and probably by many other functions (we give an example of an infinite class of non-injective functions satisfying it). It is nice that the all-or-nothing sum-free property can thus be relaxed into a notion which has the same relevance to cryptology, is satisfied at some level by all "reasonable" vectorial functions, with a rather simple definition if we consider functions valued in finite fields. We also show that we necessarily have $t \geq \frac{k}{d}$, where $d = \deg(F)$ is the algebraic degree of F. This generalizes the fact that a function of algebraic degree d cannot be kth-order sum-free for $k > d$. We improve this bound with two other lower bounds: one that is valid when F is a permutation, by means of the algebraic degree of the compositional inverse of F, and another that is valid for every vectorial function and can be tighter in some cases, by means of the algebraic degree of the indicator of the graph of the function. We focus then on power functions, for which we prove upper bounds on the minimum value of t, for a given k. We study specifically the multiplicative inverse function (used as an S-box in the AES), and characterize the minimum value of t by means of the coefficients of the subspace polynomials of k-dimensional vector spaces in $\mathbb{F}_{2^n}$ (which allows us to completely clarify the situation when k divides n). We show that this function is kth-order t-degree-sum-free if and only if it is $(n-k)$th-order t-degree-sum-free, which allows to strengthen the upper bounds found.

2   Preliminaries

Given two positive integers n, m, we call (n, m)-*function* (*vectorial function* if we do not wish to specify n, m) any function $F : \mathbb{F}_2^n \to \mathbb{F}_2^m$. If $m = 1$, we speak of an n-variable Boolean function and we denote it by a lowercase symbol f. We can endow the domain or the co-domain of such a function (or both) with the structure of a finite field, since any finite field $\mathbb{F}_{2^n}$ of characteristic 2 is an n-dimensional vector space over $\mathbb{F}_2$, and given a basis $(\alpha_1, \ldots, \alpha_n)$, we have the correspondence $(x_1, \ldots, x_n) \mapsto \sum_{i=1}^{n} x_i \alpha_i$.

Two (n, m)-functions F, G are called affine (resp. EA) equivalent if (see e.g. [15]) $G = A \circ F \circ A'$ (resp. $G = A \circ F \circ A' + A''$), where A (resp. A') is an affine automorphism of $\mathbb{F}_2^m$ (resp. $\mathbb{F}_2^n$) and A'' is affine. They are more generally called CCZ equivalent if the graph of F, that is, $\{(x, F(x)); x \in \mathbb{F}_2^n\}$, can be mapped to the graph of G by an affine automorphism $\mathcal{A}$ of $\mathbb{F}_2^n \times \mathbb{F}_2^m$. Denoting $\mathcal{A} = (\mathcal{A}_1, \mathcal{A}_2)$, the mapping $F_1 : x \in \mathbb{F}_2^n \mapsto \mathcal{A}_1(x, F(x))$ is then a permutation of $\mathbb{F}_2^n$, and denoting $F_2(x) = \mathcal{A}_2(x, F(x))$, we have $G = F_2 \circ F_1^{-1}$ (see e.g. [6]).

We call F a kth-order sum-free function if, for every k-dimensional affine subspace A of $\mathbb{F}_2^n$ (or of $\mathbb{F}_{2^n}$), we have $\sum_{x \in A} F(x) \neq 0$ [7]. The (n, n)-functions that are second-order sum-free are also called *almost perfect nonlinear* (APN).

When viewing a vectorial function as defined over $\mathbb{F}_2^n$, we can represent it by its (unique) *algebraic normal form* $F(x) = \sum_{I \subseteq \{1,\ldots,n\}} a_I \prod_{i \in I} x_i$ with $a_I \in \mathbb{F}_2^m$ (or $a_I \in \mathbb{F}_{2^m}$). This allows to define its *algebraic degree* $\max\{|I|; a_I \neq 0\}$ where $|\ldots|$ denotes the size.

When viewing a vectorial function as defined over $\mathbb{F}_{2^n}$ and valued in this same field (which includes the possibility that it is valued in a subfield of $\mathbb{F}_{2^n}$ and thus allows us to consider not only (n, n)-functions but also (n, m)-functions, where m divides n), we can represent it by its (unique) *univariate representation* $F(x) = \sum_{i=0}^{2^n - 1} \delta_i x^i$, $\delta_i \in \mathbb{F}_{2^n}$. The algebraic degree of F equals then $\max\{w_2(i); \delta_i \neq 0\}$, where $w_2(i)$ is the 2-weight of i (i.e., the Hamming weight of its binary expansion). Function F is called a *power function* if $F(x) = x^d$, for some exponent $d \in \mathbb{Z}/(2^n - 1)\mathbb{Z}$.

As we already recalled in introduction, a subset X of $\mathbb{F}_2^n$ or $\mathbb{F}_{2^n}$ satisfies the *division property at the order l* if the n-variable Boolean function equal to its indicator (taking the value 1 on X and 0 elsewhere) has algebraic degree at most $n - l$ (see [3]).

3   A Weakening of the Sum-Freedom Notion

To show why sum-freedom can be weakened, let us briefly recall why the kth-order sum-freedom of an S-box avoids the propagation of the division property of k-dimensional affine spaces through it. We first need to say what is the image of a set that must be considered as the result of the processing of a set X (supposed to have the division property) through the S-box: if the S-box F is a permutation, or is more generally injective, then the image of X by F to be considered is the classic one $F(X) = \{F(x); x \in X\} = \{y \in \mathbb{F}_2^m; X \cap F^{-1}(y) \neq \emptyset\}$. If not, then the image to be considered is (see [11]):

$$F((X)) := \{y \in \mathbb{F}_2^m; X \cap F^{-1}(y) \text{ has an odd size}\}$$

(we use a specific notation to avoid any confusion between $F(X)$ and $F((X))$ when F is not injective). What is shown in [7] is that if $\sum_{x \in X} F(x) \neq 0$, then $F((X))$ does not have the division property at the order 2.

Let us recall why this is true and show that the nicely simple notion of sum-freedom is in fact too demanding in most cases, as a property implying the non-propagation of the division property. Let X have an even size (so that it has at least the division property at the order 1). The mapping $X \mapsto F((X))$ preserves the even size, and we have $\sum_{x \in X} F(x) = \sum_{y \in F((X))} y$. The fact that the sum $\sum_{y \in F((X))} y$ is nonzero is equivalent to the property that the indicator of $F((X))$ has at least algebraic degree $n - 1$ (this is a particular case of [6, Corollary 2]; see more details in [7, Subsection 3.2]). Then no propagation of the division property is possible for k-dimensional affine spaces when F is kth-order sum-free, since $F((X))$ only satisfies the division property at the order 1. But we do not need the division property to drop to order 1 for the integral attack to be made impossible, we only need that the division property falls to a small enough level.

The propagation of the division property has been studied in [3,11] through a representation of the S-box by its algebraic normal form. We will not develop here this approach. We shall identify the vector space $\mathbb{F}_2^m$ with the field $\mathbb{F}_{2^m}$ (see Sect. 2). We do so because:

- it is often simpler to address the propagation of the division property in fields than in vector spaces, because it translates into the nullity of some power sums,
- in many block ciphers such as the AES, S-boxes are naturally defined over fields and valued in them,
- most of the important (infinite classes of) vectorial functions for cryptography are defined over fields and valued in them,
- in particular, many important functions for cryptography are power functions over finite fields, and no infinite class of (for instance) APN functions is known by its algebraic normal form.

3.1 Preliminaries on Reed-Muller Codes

We know (see [6,14]) that, for every $1 \leq d \leq m$, the dual of the Reed-Muller code $RM(d-1,m)$, of order $d-1$ and length 2^m, equals the Reed-Muller code $RM(m-d,m)$, of order $m-d$ and of the same length, that is, any m-variable Boolean function f has algebraic degree strictly less than d if and only if, for every Boolean function g of algebraic degree at most $m-d$, we have $\sum_{y\in\mathbb{F}_{2^m}} f(y)g(y) = 0$, or more generally, for every (m,r)-function G (with $r \geq 1$), of algebraic degree at most $m-d$, we have $\sum_{y\in\mathbb{F}_{2^m}} f(y)G(y) = 0$. Hence, by contraposition:

Lemma 1. *Let m be any positive integer, and let $0 \leq d \leq m$. Any nonzero m-variable Boolean function f has algebraic degree at least d if and only if there exists a Boolean function g of algebraic degree at most $m-d$, such that $\sum_{y\in\mathbb{F}_{2^m}} f(y)g(y) \neq 0$, or equivalently there exists, for some $r \geq 1$, an (m,r)-function G of algebraic degree at most $m-d$, such that $\sum_{y\in\mathbb{F}_{2^m}} f(y)G(y) \neq 0$.*

In particular, for every nonzero m-variable Boolean function f, there exists an m-variable Boolean function g (which has of course an algebraic degree of at most m), such that $\sum_{y\in\mathbb{F}_{2^m}} f(y)g(y) \neq 0$ (taking $d = 0$ in Lemma 1, or directly choosing for Boolean function g the indicator of a singleton $\{a\}$, where $f(a) = 1$).

Moreover, when the domain of the Boolean function is endowed with the structure of the field $\mathbb{F}_{2^m}$, we can specify Lemma 1 in a way that will be convenient in our framework. We recall that the 2-weight of j is the Hamming weight of the binary expansion of j.

Lemma 2. *[6, Corollary 2]Let m be any positive integer, and let $0 \leq d \leq m$. Any nonzero m-variable Boolean function f has algebraic degree at least d if and only if there exists a non-negative integer j whose 2-weight satisfies $w_2(j) \leq m-d$, and such that $\sum_{y\in\mathbb{F}_{2^m}} y^j f(y) \neq 0$.*

3.2 Weakening the Notion of Sum-Freedom

From the results recalled above, we deduce the following proposition, by taking $d = m-t$, and observing that if f is the indicator of $F((X))$ in $\mathbb{F}_{2^m}$, where F is an (n,m)-function, then for every (m,r)-function G, we have $\sum_{y\in\mathbb{F}_{2^m}} G(y)f(y) = \sum_{x\in X}(G\circ F)(x)$, and in particular, for $j \geq 1$, $\sum_{y\in\mathbb{F}_{2^m}} y^j f(y) = \sum_{x\in X}\left(F(x)\right)^j$.

Proposition 1. *For any positive integers n, m, t, let $F : \mathbb{F}_2^n \to \mathbb{F}_{2^m}$ be any (n,m)-function (where $\mathbb{F}_2^n$ can be identified with $\mathbb{F}_{2^n}$ or not) and X any set in $\mathbb{F}_2^n$. The set $F((X))$ fails to have the division property at the order $t+1$ (i.e. the m-variable Boolean function equal to the indicator of $F((X))$ has an algebraic degree at least $m-t$) if and only if some non-negative integer j exists such that $w_2(j) \leq t$ and $\sum_{x\in X}\left(F(x)\right)^j \neq 0$, which is equivalent to: some (m,r)-function G (with $r \geq 1$) of algebraic degree at most t exists such that $\sum_{x\in X}(G\circ F)(x) \neq 0$.*

Since any k-dimensional affine space A, with $k \geq 1$, has an even size, the set $F((A))$ has also an even size, and $\sum_{x\in A}\left(F(x)\right)^0$ then equals 0. This leads to the definition:

Definition 1. *Let $F : \mathbb{F}_2^n \to \mathbb{F}_{2^m}$ be an (n,m)-function. Let $1 \leq k \leq n$ and $1 \leq t \leq m$. Then F is called kth-order t-degree-sum-free if, for every k-dimensional affine space A, there exists a positive integer j whose 2-weight is at most t and such that $\sum_{x \in A} \left(F(x)\right)^j \neq 0$.*

This is equivalent to the fact that, for some $r \geq 1$, there exists a vectorial (m,r)-function G of algebraic degree at most t such that $\sum_{x \in X}(G \circ F)(x) \neq 0$.

If an (n,m)-function F is kth-order t-degree-sum-free, the propagation through the S-box of the division property at the order $t+1$ of any k-dimensional affine space fails, since according to Proposition 1, for every k-dimensional affine space A, the algebraic degree of the indicator of $F((A))$ is at least $m - t$.

For any new property of (n,m)-functions F, we need to determine the equivalence relations preserving it. For every $k \geq 2$ and $t \geq 1$, it is easily seen that the property of being kth-order t-degree-sum-free for an (n,m)-function is EA invariant. But it is in general not CCZ invariant, since if F is a (non-affine) permutation, it can be kth-order t-degree-sum-free while its compositional inverse (which is always CCZ equivalent) is not.

The larger t, the weaker the notion of kth-order t-degree-sum-freedom (if a function is kth-order t-degree-sum-free, then it is kth-order t'-degree-sum-free for every $t' \geq t$). The classic kth-order sum-freedom corresponds to $t = 1$. Indeed, the only possibility for F to be kth-order 1-degree sum-free is that $\sum_{x \in A}(F(x))^{2^i} = (\sum_{x \in A} F(x))^{2^i} \neq 0$ for some i, that is, $\sum_{x \in A} F(x) \neq 0$.

Let us see now that any (n,m)-function satisfying some reasonable condition is kth-order t-degree-sum-free for some t smaller than or equal to m.

Lemma 3. *Every (n,m)-function such that, for any k-dimensional affine space A, the set $F((A))$ is non-empty (in particular, every injective (n,m)-function) is kth-order m-degree-sum-free.*

Indeed, we recalled after Lemma 1 the existence, for every nonzero m-variable function f, of an m-variable Boolean function g of algebraic degree at most m such that $\sum_{y \in \mathbb{F}_{2^m}} f(y)g(y) \neq 0$. Taking for f the indicator of $F((A))$, this proves the existence of a non-negative integer j of 2-weight at most m such that $\sum_{y \in \mathbb{F}_{2^m}} y^j f(y) = \sum_{x \in A} \left(F(x)\right)^j \neq 0$. The value $t = m$ satisfies then the condition of Definition 1. This allows to give the following:

Definition 2. *Let n, m be two positive integers and let $2 \leq k \leq n$. Let F be any (n,m)-function such that, for any k-dimensional affine space A, the set $F((A))$ is non-empty (e.g. F is injective). We call kth-order sum-free min-degree of F the smallest value of $t \leq m$ such that F is kth-order t-degree-sum-free.*

Remark. The condition on F in Definition 2 is necessary, since if it is not satisfied for some A, the kth-order sum-free min-degree cannot exist. Take for instance $F(x) = G(x + x^2)$, then for every j, the sum of the values of $(F(x))^j$ over an affine space stable under translation by 1 equals 0.

It would be nice to characterize precisely what is the set of all the (n, m)-functions satisfying this condition. For $k = 1$, it is clearly the set of injective functions. For $k \geq 2$, it is a strict superset of this latter set. For $k = 2$, it is the set of those functions F such that, for every nonzero $a \in \mathbb{F}_2^n$, $D_a F(x)$ takes zero value on at most one pair $\{x, x + a\}$ (among which we have all APN functions). For $k = n$, it is the set of those functions whose image multiset has some points having an odd multiplicity. For $k \in \{3, \ldots, n-1\}$, we leave this question open, but we give an example of an infinite class of non-bijective (m, m)-functions satisfying it for every even k. We take for F any power (m, m)-function $F(x) = x^d$ such that $\gcd(d, 2^m - 1) = 3$ (as are all APN power functions over $\mathbb{F}_{2^m}$ for m even, see [6, Proposition 165]). Let A be a k-dimensional vector subspace of $\mathbb{F}_{2^m}$ and suppose that $F((A)) = \emptyset$. Let us denote by w a primitive element of $\mathbb{F}_4$. The pre-images by F are the singleton $\{0\}$ and the 3-sets $u\mathbb{F}_4^*$, where $u \neq 0$. Since $F((A)) = \emptyset$, then A must not contain 0, and for every $a \in A$, we must have either $aw \in A$ (and $aw^2 \notin A$ which is in fact automatically implied by $0 \notin A$ since $a + aw + aw^2 = 0$) or $aw^2 \in A$ (and $aw \notin A$). Since if $b = aw^2$ then $a = bw$, we have then that A is the disjoint union of a set S of size 2^{k-1} and of the set $wS = \{wx; x \in S\}$. The elements of S are the elements x of A such that $wx \in A$. Hence, $S = A \cap w^2 A$ is an affine space. Since $A = S \cup wS$ is an affine space and S is an affine hyperplane of A, the $\mathbb{F}_2$-vector space E underlying S is then stable under the multiplication by w. Then E is a vector space over $\mathbb{F}_4$ and its dimension $k - 1$, as an $\mathbb{F}_2$-vector space, is then even. This proves that if we take k even, then F satisfies the condition of Definition 2.

Note that, since for m odd, all APN power (m, m)-functions are bijective, all APN power functions satisfy the condition in Definition 2 for any k if m is odd and for any even k if m is even. $\diamond$

3.3 An Upper Bound on the kth-Order Sum-Free Min-Degree

Proposition 2. *Let n, m be two positive integers and let $2 \leq k \leq n$. Let F be any (n, m)-function such that $F((A)) \neq \emptyset$ for every k-dimensional affine space A. Then F has kth-order sum-free min-degree at most $\min(k, m)$, with equality if and only if there exists a k-dimensional affine space A on which F is injective and whose image is an affine space, or $F((A))$ equals $\mathbb{F}_{2^m}$ (with $k \geq m$).*

Proof. Since we know that we can take $t \leq m$, we just have to show, for proving the upper bound, that we can take $t \leq k$. For every k-dimensional affine space A, the set $F((A))$ has size at most 2^k and its indicator f has then an algebraic degree at least $m - k$ (indeed, we know, see [6,14], that any nonzero m-variable Boolean function of algebraic degree at most d has Hamming weight at least 2^{m-d}). There exists then, according to Lemma 2, an integer j of 2-weight at most k such that $\sum_{y \in \mathbb{F}_{2^m}} y^j f(y) = \sum_{x \in A} (F(x))^j \neq 0$. This proves the bound. Let us now determine when this bound is an equality.

If $m \geq k$, then the kth-order sum-free min-degree of F equals k if and only if there exists a k-dimensional affine space A such that, for every non-negative integer j such that $w_2(j) < k$, we have $\sum_{y \in \mathbb{F}_{2^m}} y^j f(y) = \sum_{x \in A} (F(x))^j = 0$, that is, the algebraic degree of the indicator function of $F((A))$ is at most $m - k$, that is, given the size of $F((A))$, equals $m - k$, which is equivalent to the fact that $F((A))$ is a k-dimensional affine space (see [6,14]).

If $m \leq k$, then the kth-order sum-free min-degree of F equals m if and only if $F((A))$ equals $\mathbb{F}_{2^m}$, since otherwise, the algebraic degree of the indicator of $F((A))$ is at least 1 and the kth-order sum-free min-degree of F is then at most $m - 1$. $\qquad\square$

Remark. The fact that t can be taken smaller than or equal to m is, as we saw, not surprising, but the fact that it can be taken smaller than or equal to k makes all the interest of the notion. The argument for proving this property is nicely simple, but this simplicity hides the difficulty to determine concretely the values of j such that $\sum_{x \in A} (F(x))^j \neq 0$, even for quadratic functions. $\qquad\diamond$

We shall see that the bound of Proposition 2 is tight, at least for $m = n$.

3.4   Lower Bounds on the kth-Order Sum-Free Min-Degree

We still assume that F satisfies the condition in Definition 2. After the upper bound on the value of t in the previous subsection, we show in the present subsection a lower bound. Denoting by $\deg\left(F(x)\right)$ the algebraic degree of function $F(x)$, we have, for every non-negative integer j:

$$\deg\left((F(x))^j\right) \leq w_2(j)\,\deg(F),$$

since the algebraic degree of the function $G : x \in \mathbb{F}_{2^m} \mapsto x^j$ equals $w_2(j)$ and we have, for every (n, m)-function F, and any (m, m)-function G: $\deg(G \circ F) \leq \deg(G)\deg(F)$. Let A be any k-dimensional affine space. If $\deg\left((F(x))^j\right) < k$, then we have $\sum_{x \in A}(F(x))^j = 0$. We deduce:

Proposition 3. *Let F be any non-constant (n, m)-function that is kth-order t-degree sum-free. We have:*

$$t \geq \left\lceil \frac{k}{\deg(F)} \right\rceil. \tag{1}$$

Indeed, if $t < \frac{k}{\deg(F)}$ then, for every j such that $w_2(j) \leq t$, we have $\deg\left((F(x))^j\right) \leq \deg(F)\,w_2(j) < k$.

Alternate Lower Bounds. Other upper bounds exist on the algebraic degree of the composition of functions, which are often better than the so-called naive bound $\deg(G \circ F) \leq \deg(G)\deg(F)$, and we shall directly deduce lower bounds on t which will be often better.

The first bound, proved in [4], on the algebraic degree of composed functions, is valid under a very strong hypothesis[1]: the Walsh transform $W_F(u, v) =$

---

[1] This bound is then essentially useful for identifying a feature that should be avoided when choosing an S-box; we give it for being complete on the state of the art.

$\sum_{x\in\mathbb{F}_2^n}(-1)^{v\cdot F(x)+u\cdot x}$ (where $\cdot$ denotes by abuse of notation an inner product in $\mathbb{F}_2^n$ and an inner product in $\mathbb{F}_{2^m}$) has all its values divisible by 2^l, for a large enough value of l. Then we have $\deg(G\circ F)\leq n-l+\deg(G)$, and therefore:

$$t\geq k-n+l.$$

The second upper bound, proved in [2], on the algebraic degree of composed functions, does not apply to all functions either, but its assumption, which is that F is a permutation (with $m=n$, then), is always satisfied when dealing with the model of block ciphers called Substitution-Permutation networks. This bound is:

$$\deg(G\circ F)\leq n-1-\left\lfloor\frac{n-1-\deg(G)}{\deg(F^{-1})}\right\rfloor. \tag{2}$$

It is always less that n (when G itself has an algebraic degree less than n), which is an important advantage over the naive bound, and it outperforms the latter in many cases where this one is less than n. Bound (2) implies:

Proposition 4. *Let F be any (n,n)-permutation and let $\deg(F^{-1})$ be the algebraic degree of the compositional inverse of F. Then, if F is kth-order t-degree sum-free, we have:*

$$t\geq n-1-(n-1-k)\deg(F^{-1}), \tag{3}$$

i.e., the kth-order sum-free min-degree of F is at least this number.

Indeed, according to (2), if $t<n-1-(n-1-k)\deg(F^{-1})$ then, for every j such that $w_2(j)\leq t$, we have $\deg\left((F(x))^j\right)<n-1-\left\lfloor\frac{(n-1-k)\deg(F^{-1})}{\deg(F^{-1})}\right\rfloor=k$.

Corollary 1. *Let n and k be positive integers such that $2\leq k\leq n-2$ and F an (n,n)-permutation. If $\deg(F^{-1})<\frac{n-2}{n-1-k}$, then F is not kth-order sum-free.*

Indeed, (3) implies $t>1$, a contradiction.

A third bound, proved in [5], is (with the naive bound) the only known upper bound valid for all (n,m)-functions:

$$\deg(G\circ F)\leq\deg(1_{\mathcal{G}_F})+\deg(G)-m, \tag{4}$$

where $\deg(1_{\mathcal{G}_F})$ is the algebraic degree of the $(n+m)$-variable Boolean function equal to the indicator of the graph $\mathcal{G}_F=\{(x,F(x));x\in\mathbb{F}_2^n\}$. This bound is in general tighter than the naive bound, and it can be tighter than (2), as seen in [5]. Bound (4) implies:

Proposition 5. *Let F be any (n,m)-function and let $\deg(1_{\mathcal{G}_F})$ be the algebraic degree of the indicator of the graph $\mathcal{G}_F=\{(x,F(x));x\in\mathbb{F}_2^n\}$ of F. Then, if F is kth-order t-degree sum-free, we have:*

$$t\geq m+k-\deg(1_{\mathcal{G}_F}), \tag{5}$$

i.e., the kth-order sum-free min-degree of F is at least this number.

Indeed, according to Relation (4), if $t < m + k - \deg(1_{\mathcal{G}_F})$ then, for every j such that $w_2(j) \leq t$, we have $\deg\left((F(x))^j\right) < k$.

Remark. We have $m + k - \deg(1_{\mathcal{G}_F}) \leq k$ since $\deg(1_{\mathcal{G}_F}) \geq m$, and this is coherent with Proposition 2. $\diamond$

Corollary 2. *Let n, m and k be positive integers such that $2 \leq k \leq n-1$ and F an (n, m)-function. If $\deg(1_{\mathcal{G}_F}) \leq m + k - 2$, then F is not kth-order sum-free.*

Indeed, Relation (5) implies $t \geq 2$, a contradiction with $t = 1$.

This corollary does not allow to prove that the multiplicative inverse function is not kth-order sum-free for some $k \in \{3, \ldots, n-3\}$, since we have for this function that $\deg(1_{\mathcal{G}_{F_{inv}}}) = 2n$ is strictly larger than $n + k - 2$ for every $2 \leq k \leq n$. The fact that $\deg(1_{\mathcal{G}_{F_{inv}}}) = 2n$ comes from the relation $1_{\mathcal{G}_{F_{inv}}}(x, y) = (y + x^{2^n - 2} + 1)^{2^n - 1} = \sum_{i=0}^{2^n - 1}(x(y + 1))^i = (x(y + 1) + 1)^{2^n - 1}$.

We summarize the lower bounds above in a table:

Bound	Condition(s)
$t \geq \left\lceil \dfrac{k}{\deg(F)} \right\rceil$	$F : (n, m)$-function, $\deg(F) > 0$
$t \geq n - 1 - (n - 1 - k)\deg(F^{-1})$	$F : (n, n)$-permutation
$t \geq m + k - \deg(1_{\mathcal{G}_F}),$	$F : (n, m)$-function

4   The Particular Case of $k = 2$ and $m = n$

Since APN functions play an important role in cryptography, the case $k = 2$ and $m = n$ seems particularly interesting to study. When F is an APN (n, n)-function, we have $t = 1$ for $k = 2$. When F is not APN and satisfies the condition in Definition 2 (i.e., for every nonzero $a \in \mathbb{F}_2^n$, $D_a F(x)$ takes value 0 on at most one pair $\{x, x+a\}$, which is a weaker condition than being APN), Proposition 2 tells us that we have $t = 2$ for $k = 2$ (i.e. for every affine plane A, there is a value of j that has 2-weight at most 2 such that $\sum_{x \in A}(F(x))^j \neq 0$). The number of those A such that j must have 2-weight 2 quantifies the level of the non-APNness of F. It does it in a different way from the differential uniformity (equal to the maximum number of solutions of the equations $F(x) + F(x+a) = b$, with $a \neq 0$).

A non-APN (n, n)-function has what [13] calls *vanishing flats* (which are in fact vanishing planes, since they are the affine planes $\{x, y, z, x + y + z\}$, with x, y, z distinct, over which F sums to 0). These vanishing flats are those affine planes A for which the minimum 2-weight of j such that $\sum_{x \in A}(F(x))^j \neq 0$ equals 2. Note that the vanishing flats of a vectorial (n, m)-function F are by definition the projections over $\mathbb{F}_2^n$, parallel to $\mathbb{F}_2^m$, of the affine planes included in the graph of F. The next lemma is straightforward.

Lemma 4. *Let F be an (n, m)-function satisfying the condition in Definition 2 and $P = \{x, y, z, x + y + z\}$ a vanishing flat of F. Then we have that $F(P) =$*

$\{F(x), F(y), F(z), F(x+y+z)\}$ *is an affine plane and, given an integer j, we have $\sum_{x \in P}(F(x))^j \neq 0$ if and only if $F(P)$ is not a vanishing flat of the power function $G(x) = x^j$ over $\mathbb{F}_{2^m}$.*

See more in [9].

5 Some General Results on Power Functions

An (n,n)-function $F(x) = x^d$, where $x \in \mathbb{F}_{2^n}$ and $d \in \mathbb{Z}/(2^n-1)\mathbb{Z}$, is kth-order t-degree-sum-free if and only if, for every k-dimensional affine space A, there exists a non-negative integer j of 2-weight at most t, such that $\sum_{x \in A} x^{dj} \neq 0$. We shall see that the study of the kth-order sum-free min-degree of power functions x^l, where l is a multiple of d, gives information on that of x^d. This is quite positive, but a determination of the kth-order sum-free min-degree of all power functions seems elusive, even for $k = 2$ (for which this determination includes as a very partial sub-problem the determination of all APN power functions, which is currently only in the state of a conjecture). Note also that Lemma 4 does not generalize to vanishing k-flats for $k > 2$.

5.1 The Influence of the kth-Order Sum-Free Min-Degree of "Multiples"

Proposition 6. *Let $d \in \mathbb{Z}/(2^n - 1)\mathbb{Z}$ and let l be a nonzero multiple of d in $\mathbb{Z}/(2^n - 1)\mathbb{Z}$:*

$$0 \not\equiv l \equiv dr \pmod{2^n - 1}.$$

If the power function x^l is kth-order t-degree-sum-free for some k and t, then $F(x) = x^d$ is kth-order $(w_2(r)\,t)$-degree-sum-free.

Moreover, if r is invertible in $\mathbb{Z}/(2^n - 1)\mathbb{Z}$ and r' is its inverse, then $F(x)$ is kth-order $\left(n - 1 - \left\lfloor \frac{n-1-t}{w_2(r')} \right\rfloor\right)$-degree-sum-free.

Proof. Let A be any k-dimensional vector subspace of $\mathbb{F}_{2^n}$ and j a non-negative integer of 2-weight at most t such that $\sum_{x \in A} x^{lj} \neq 0$. Then we have $\sum_{x \in A} x^{drj} \neq 0$ and since we have $w_2(rj) \leq w_2(r)\,w_2(j)$, this proves the first assertion. When r is invertible, Bound (2) says that $w_2(rj) \leq n - 1 - \left\lfloor \frac{n-1-w_2(j)}{w_2(r')} \right\rfloor$. This proves the second assertion. $\square$

5.2 An Upper Bound on the kth-Order Sum-Free Min-Degree of Any Power Permutation

We deduce now from Proposition 6 an improvement for power permutations (in some cases) of the upper bound $\min(k, m)$ on the kth-order sum-free min-degree.

Proposition 7. *Let n, k be positive integers such that $k \leq n-1$. Let $F(x) = x^d$ be any power permutation over $\mathbb{F}_{2^n}$. Let d' be the inverse of d modulo $2^n - 1$, and let $r = d'(2^k - 1) \pmod{2^n - 1}$. Then F is kth-order $(w_2(r))$-degree-sum-free.*

Let r' equal the inverse of r in $\mathbb{Z}/(2^n - 1)\mathbb{Z}$ if $\gcd(k, n) = 1$, and equal to $2^n - 1$ otherwise. Then F is kth-order $\left(n - 1 - \left\lfloor \frac{n-2}{w_2(r')} \right\rfloor\right)$-degree-sum-free.

Proof. We know from [7] that the power function $x^{2^k - 1}$ is kth-order sum-free, that is, kth-order 1-degree sum-free and we have: $d(d'(2^k - 1)) \equiv 2^k - 1 \pmod{2^n - 1}$. Proposition 6 with $l = 2^k - 1$ and $t = 1$ completes the proof. $\square$

There are many cases where $\min\left(w_2(r), n - 1 - \left\lfloor \frac{n-2}{w_2(r')} \right\rfloor\right) < k$. For instance, when $\gcd(k, n) = 1$ and $d' = \frac{l}{2^k - 1}$ (i.e. $d = \frac{2^k - 1}{l}$), where l is co-prime with n and has 2-weight strictly smaller than k, we have $w_2(d'(2^k - 1)) < k$.

5.3   A Generalization

The existence of any class of kth-order sum-free power functions, other than $x^{2^k - 1}$, would give another upper bound similar to that of Proposition 7. The only change (up to a multiplication by a power of 2) on the exponent $2^k - 1 = \sum_{i=0}^{k-1} 2^i$ which seems to preserve the kth-order sum-freedom of the power function $x^{2^k - 1}$ is replacing its exponent by $\sum_{i=0}^{k-1} 2^{ij} = \frac{2^{kj} - 1}{2^j - 1}$ for some j co-prime with n. This corresponds to changing the Frobenius $x \mapsto x^2$ into another generator of the Galois group, its power $x \mapsto x^{2^j}$. The kth-order sum-freedom of the resulting power function comes from the fact recalled in [7] that, if $L_0, \ldots, L_{k-1}$ are linear, then for every $a_1, \ldots, a_k$ in $\mathbb{F}_{2^n}$, we have $D_{a_1} \ldots D_{a_k}\left(\prod_{i=0}^{k-1} L_i\right)(x) = \sum_{\sigma \in G_k} \prod_{i=1}^{k} L_{\sigma(i)}(a_i)$, where G_k is the set of bijections from $\{1, \ldots, k\}$ to $\{0, \ldots, k-1\}$, and we obtain another so-called Moore exponent set (see [1, Bottom of page 2]). Applying Proposition 6 with $l = \frac{2^{kj} - 1}{2^j - 1}$, we have $r = d'(\sum_{i=0}^{k-1} 2^{ij})$, which may have another 2-weight than $d'(2^k - 1)$. The next proposition generalizes Proposition 7.

Proposition 8. *Let n, k be positive integers such that $k \leq n-1$. Let $F(x) = x^d$ be any power permutation over $\mathbb{F}_{2^n}$. Let j be any positive integer co-prime with n. Let d' be the inverse of d modulo $2^n - 1$, $r = d'(\sum_{i=0}^{k-1} 2^{ij}) \pmod{2^n - 1}$ and if $\gcd(\sum_{i=0}^{k-1} 2^{ij}, 2^n - 1) = 1$, let r' the inverse of r in $\mathbb{Z}/(2^n - 1)\mathbb{Z}$, and otherwise $r' = 2^n - 1$. Then F is kth-order w-degree-sum-free where:*

$$w = \min\left(w_2(r), n - 1 - \left\lfloor \frac{n - 2}{w_2(r')} \right\rfloor\right).$$

6   The Case of Multiplicative Inverse Function

The multiplicative inverse function

$$F_{inv}(x) = x^{2^n - 2} = x^{-1}, \quad x \in \mathbb{F}_{2^n},$$

is, as we recalled above, second-order 1-degree-sum-free (i.e. APN) if and only if $n \geq 3$ is odd. For $n \geq 4$ even, it is then 2nd-order 2-sum-free, according to Proposition 2 (the fact that $F_{inv}((A)) \neq \emptyset$ for every affine plane A is obvious since F_{inv} is a permutation). This settles the case $k = 2$, but by curiosity, let us see for which values of j we have $\sum_{x \in A} F_{inv}(x) \neq 0$ for a given affine plane A. We know from [7] that for every affine space A that is not a vector space (i.e. which does not include the zero vector), we have $\sum_{x \in A} F(x) \neq 0$. In the case of a 2-dimensional vector space, that is, $A = \{0, a, b, a + b\}$ with a and b linearly independent over $\mathbb{F}_2$, we have $\frac{1}{a} + \frac{1}{b} + \frac{1}{a+b} = \frac{a^2+b^2+ab}{ab(a+b)} = \frac{a\left(\left(\frac{b}{a}\right)^2 + \frac{b}{a} + 1\right)}{b(a+b)}$. As already observed in [13], the only affine planes over which the inverse function sums to 0 are then the vector spaces of the form $a\,\mathbb{F}_4$, since the equation $x^2 + x + 1 = 0$ has for solutions the two primitive elements of $\mathbb{F}_4$. The plane $a\,\mathbb{F}_4$ is mapped by F_{inv} into $\frac{1}{a}\,\mathbb{F}_4$, and for $w_2(j) = 2$ we have $\sum_{x \in \frac{1}{a}\mathbb{F}_4} x^j = \frac{1}{a^j} \sum_{x \in \mathbb{F}_4} x^j \neq 0$ if and only if j is a nonzero multiple of 3.

On the basis of computer investigations, we conjecture in [8] that, for every $k \in \{3, \ldots, n-3\}$ and every $n \geq 6$, the inverse function is not kth-order sum-free (see also [10]). It is then useful to study the values of t for which it is kth-order t-degree-sum-free.

6.1   A General Upper Bound on the kth-Order Sum-Free Min-Degree for $k \geq 2$

Since F_{inv} has an algebraic degree of $n - 1$, Relation (1) does not give any information, but we have:

Proposition 9. *For every $2 \leq k \leq n$, the multiplicative inverse (n, n)-function is kth-order w-degree-sum-free with $w = \min\left(n - k, n - 1 - \left\lfloor \frac{n-2}{n-k'} \right\rfloor\right)$, where k' equals the inverse of k modulo n if $\gcd(k, n) = 1$ and equals 0 otherwise.*

Proof. We apply Proposition 7 with $d' = 2^n - 2$ and $r = (2^n - 2)(2^k - 1)$ $\pmod{2^n - 1} = 2^n - 2^k \pmod{2^n - 1}$. We have:

- $w_2(r) = n - k$,
- if $\gcd(k, n) = 1$, then the inverse of $2^k - 1$ modulo $2^n - 1$ equals $2^{k(k'-1)} + 2^{k(k'-2)} + \cdots + 2^k + 1$, since $(2^k - 1)(2^{k(k'-1)} + 2^{k(k'-2)} + \cdots + 2^k + 1) = 2^{kk'} - 1 \equiv 1 \pmod{2^n - 1}$. The inverse r' of the opposite $2^n - 2^k$ of $2^k - 1$ modulo $2^n - 1$ equals then $(2^n - 1) - \left(2^{k(k'-1)} + 2^{k(k'-2)} + \cdots + 2^k + 1\right)$ and has 2-weight $n - k'$. Then, $n - 1 - \left\lfloor \frac{n-2}{w_2(r')} \right\rfloor$ equals $n - 1 - \left\lfloor \frac{n-2}{n-k'} \right\rfloor$.

This completes the proof. $\qquad\square$

The bound $w \leq n - k$ in Proposition 9 gives an information when $n - k < k$, that is, $k > \frac{n}{2}$. For $k < \frac{n}{2}$, we will have a result thanks to Corollary 4 below.

We have $n - 1 - \left\lfloor \frac{n-2}{n-k'} \right\rfloor < n - k$ when $\left\lfloor \frac{n-2}{n-k'} \right\rfloor \geq k$, that is, $n - 2 \geq k(n - k')$ (which can happen, when k is not too large and k' is not too small).

Note that for $k = n - 2$, we obtain $w = 2$ (since $n - 2 \geq k(n - k')$ is impossible), which we shall get again thanks to Corollary 4 below.

6.2 Relation with Subspace Polynomials and Consequences

Subspace polynomials (see below) play an important role with respect to the kth-order sum-freedom of the multiplicative inverse function (see [7,8]). In the present subsection, we use them, but we need to develop a different approach for characterizing its kth-order sum-free min-degree.

Note that when we studied the sum-freedom of this function, we observed a difference between the sums $\sum_{x \in A} F_{inv}(x)$ when A is a vector subspace of $\mathbb{F}_2^n$, and when it is another affine subspace. We will see this is also the case for the sums $\sum_{x \in A} (F_{inv}(x))^j$ when considering the kth-order sum-free min-degree.

Subspace Polynomials. Let E_k be a k-dimensional vector space and let $L_{E_k}(x) = \prod_{u \in E_k} (x + u)$. Such a polynomial is called a subspace polynomial. It is a linearized polynomial and we write then $L_{E_k}(x) = \sum_{i=0}^{k} b_{k,i} x^{2^i}$ ($b_{k,0} \neq 0$, $b_{k,k} = 1$); see more in [8].

Consequence on the Sums $\sum_{x \in E_k \setminus \{0\}} x^{-j}$: Since for every $x \in E_k$, we have $x = \frac{1}{b_{k,0}} \sum_{i=1}^{k} b_{k,i} x^{2^i}$, we have for every nonzero $x \in E_k$ and every non-negative integer j, by dividing this equality by x^{j+1}, that $x^{-j} = \frac{1}{b_{k,0}} \sum_{i=1}^{k} b_{k,i} x^{2^i-1-j}$. We deduce that:

$$\sum_{x \in E_k \setminus \{0\}} x^{-j} = \frac{1}{b_{k,0}} \sum_{i=1}^{k} b_{k,i} \left(\sum_{x \in E_k \setminus \{0\}} x^{2^i-1-j} \right).$$

Hence, since for every (n,n)-function F of algebraic degree less than k, we have $\sum_{x \in E_k} F(x) = 0$ and we also have $\sum_{x \in E_k \setminus \{0\}} x^0 = 1$, we deduce:

Lemma 5. *Let n and k be any positive integers such that $k \leq n$ and let E_k be any k-dimensional vector space and $L_{E_k}(x) = \prod_{u \in E_k} (x+u) = \sum_{i=0}^{k} b_{k,i} x^{2^i}$. Let*

$$r = \min\{i; 1 \leq i \leq k \text{ and } b_{k,i} \neq 0\}.$$

Then $\sum_{x \in E_k \setminus \{0\}} x^{-j}$ equals 0 for every $1 \leq j < 2^r - 1$ and is nonzero for $j = 2^r - 1$.

This result extends results from [7,8] (which say, taking $j = 1$, that $\sum_{x \in E_k \setminus \{0\}} x^{-1}$ equals 0 if $r > 1$ and is nonzero if $r = 1$).

Consequence on the Sums $\sum_{x \in E_k \setminus \{0\}} (x + a)^{-j}$; $a \notin E_k$: Let $A_k = a + E_k$ be a k-dimensional affine space not containing 0. We have from [7] that $j = 1$ is such that $\sum_{x \in A_k} x^{-j} \neq 0$ and we do not need to consider larger values of j, but by curiosity, let us see how the method used above for vector spaces can be adapted. We have $A_k = L_{E_k}^{-1}(b)$ for some nonzero $b = L_{E_k}(a)$. Then, for every $x \in A_k$, we have $b = \sum_{i=0}^{k} b_{k,i} x^{2^i}$ and then $\sum_{x \in A_k} x^{-j} = \frac{1}{b} \sum_{i=0}^{k} \left(b_{k,i} \sum_{x \in A_k} x^{2^i-j} \right)$. We find for $j = 1$ that since $\sum_{x \in A_k} x^{2^i-1}$ equals 0 for $i < k$ (since x^{2^i-1} has an algebraic degree of i) and is nonzero for $i = k$ (since $\sum_{x \in A_k} x^{2^k-1}$ equals the value at x of a kth-order derivative of the function x^{2^k-1} and we know from [7] that this constant derivative is nonzero). This provides a different way of proving the known result from [7].

We deduce from the observations above (and from the fact that $j = 2^r - 1$ has a 2-weight of r):

Proposition 10. *Let n and k be positive integers such that $k \leq n$ and let $\mathcal{E}_{k,n}$ be the set of all k-dimensional vector subspaces of $\mathbb{F}_{2^n}$ (Grassmannian). The kth-order sum-free min-degree of the multiplicative inverse function F_{inv} over $\mathbb{F}_{2^n}$ equals:*

$$\max_{E_k \in \mathcal{E}_{k,n}} \min\{i; 1 \leq i \leq k \text{ and } b_{k,i} \neq 0\},$$

where $L_{E_k}(x) = \prod_{u \in E_k}(x + u) = \sum_{i=0}^{k} b_{k,i} x^{2^i}$.

A case where we know that F_{inv} is not kth-order sum-free is when k divides n, since $\sum_{x \in \mathbb{F}_{2^k}} x^{-1}$ equals 0, and the next step is determining the kth-order sum-free min-degree of F_{inv} in such a case.

Remark. Proposition 10 has been proven by a method that might seem computational in nature, but which relies in fact on the deep properties of subspace polynomials and power functions over finite fields. The two next corollaries will illustrate this further. $\diamond$

Corollary 3. *If k divides n, then the kth-order sum-free min-degree of the multiplicative inverse function F_{inv} over $\mathbb{F}_{2^n}$ equals k.*

Indeed, for $E_k = \mathbb{F}_{2^k}$, we have $L_{E_k}(x) = x^{2^k} + x$. Hence, according to Proposition 10, the kth-order sum-free min-degree is at least k, and according to Proposition 2, it equals then k.

We can see that the bound of Proposition 2 is tight, at least for $m = n$.

A more general case (but not the most general case) where we know that F_{inv} is not kth-order sum-free is when $\gcd(k,n) > 1$, see [8]. In such a case, we have $L_{\mathbb{F}_{2^{\gcd(k,n)}}}(x) = x^{2^{\gcd(k,n)}} + x$, and according to Lemma 5, we have then that $\sum_{x \in \mathbb{F}_{2^{\gcd(k,n)}}} x^{-j}$ equals 0 when $j < 2^{\gcd(k,n)} - 1$ and is nonzero when $j = 2^{\gcd(k,n)} - 1$. However, if k does not divide n, this does not address dimension k. Let E_k be a $\frac{k}{\gcd(k,n)}$-dimensional vector subspace of $\mathbb{F}_{2^n}$ over $\mathbb{F}_{2^{\gcd(k,n)}}$, then E_k equals the disjoint union of multiplicative cosets $w\mathbb{F}_{2^{\gcd(k,n)}}^*$ where w ranges over a basis S of E_k over $\mathbb{F}_{2^{\gcd(k,n)}}$, and we have then that $\sum_{x \in E_k} x^{-j} = (\sum_{w \in S} w^{-j})(\sum_{x \in \mathbb{F}_{2^{\gcd(k,n)}}^*} x^{-j})$. This product equals 0 if and only if $\sum_{w \in S} w^{-j} = 0$ or $\sum_{x \in \mathbb{F}_{2^{\gcd(k,n)}}^*} x^{-j} = 0$. Hence the kth-order sum-free min-degree of F_{inv} is larger than or equal to $\gcd(k,n)$. It seems difficult to determine in which cases it equals $\gcd(k,n)$ and in which cases it is larger.

The next corollary generalizes to the sum-free min-degree the equality, proved in [8], between the kth-order sum-freedom of the inverse function and its $(n - k)$th-order sum-freedom.

Corollary 4. *For any $n \geq 2$ and any $k \in \{2, \ldots, n-2\}$, the kth-order sum-free min-degree of the multiplicative inverse (n, n)-function equals its $(n-k)$th-order sum-free min-degree.*

Proof. It is recalled in [8] that if E_k is any k-dimensional vector subspace of $\mathbb{F}_{2^n}$ and $E_{n-k} = L_{E_k}(\mathbb{F}_{2^n})$ then E_{n-k} has dimension $n-k$ and we have the following relation in $\mathbb{F}_{2^n}[x]$:

$$L_{E_{n-k}} \circ L_{E_k}(x) = L_{E_k} \circ L_{E_{n-k}}(x) = x^{2^n} + x.$$

Writing $L_{E_k}(x) = \sum_{i=0}^{k} b_{k,i} x^{2^i}$ and $L_{E_{n-k}}(x) = \sum_{i=0}^{n-k} b_{n-k,i} x^{2^i}$, we have $L_{E_{n-k}} \circ L_{E_k}(x) = \sum_{i=0}^{n-k} \sum_{j=0}^{k} b_{n-k,i}(b_{k,j})^{2^i} x^{2^{i+j}}$. We have then, by considering the coefficient of x^{2^r}:

$$\forall r \in \{1, \ldots, n-1\}, \quad \sum_{i=0}^{r} b_{n-k,i}(b_{k,r-i})^{2^i} = 0. \tag{6}$$

We know from Proposition 10 that, if t is the kth-order sum-free min-degree of the inverse function, then $b_{k,0}$ and $b_{k,t}$ are nonzero and $b_{k,1} = \cdots = b_{k,t-1} = 0$. The relations corresponding to Relation (6) for $r = 1, \ldots, t-1$ imply $b_{n-k,1} = \cdots = b_{n-k,t-1} = 0$ and the tth relation implies $b_{n-k,t} \neq 0$. Proposition 10 completes the proof. $\qquad \square$

Corollary 4 and Proposition 9 give:

Corollary 5. *For every $2 \leq k \leq n$, the multiplicative inverse (n, n)-function is kth-order t-degree-sum-free with $t = \min\left(k, n-k, n-1-\left\lfloor \frac{n-2}{n-k'} \right\rfloor, n-1-\left\lfloor \frac{n-2}{k'} \right\rfloor\right)$, where k' equals the inverse of k modulo n (and $n-k'$ is the inverse of $n-k$ modulo n) if $\gcd(k, n) = 1$ and equals 0 (resp. n) otherwise.*

Remark. The kth-order sum-free min-degree of F_{inv} is then at most this minimum (that we denote by t). For $k \leq \frac{n}{2}$, we have often $t = k$ and we do not get then more information than with Proposition 2. But for $k > \frac{n}{2}$, the bound of Corollary 5 is strictly better than that of Proposition 2. We do not know whether the kth-order sum-free min-degree of F_{inv} is always strictly larger than 1 for $k \in [3, n-3]$, but we conjecture it. Determining the exact kth-order sum-free min-degree of F_{inv} seems challenging. $\qquad \diamond$

Conclusion

We have introduced a notion on vectorial functions extending that of sum-freedom (which itself extended the notion of almost perfect nonlinearity).

This has led to the parameter t that we called kth-order sum-free min-degree, which quantifies to which extent an S-box F prevents from the propagation of the division property of k-dimensional affine spaces, in an integral attack (the smaller the value of t, the better the prevention). We proved that all the vectorial (n, m)-functions having the property that, for every k-dimensional affine space A, the set $F((A)) = \{y \in \mathbb{F}_2^m; A \cap F^{-1}(y) \text{ has an odd size}\}$ is non-empty (in particular, all injective vectorial functions), have a kth-order sum-free min-degree smaller than or equal to $\min(k, m)$. We could also prove several lower bounds on the kth-order sum-free min-degree of vectorial functions.

We leave open the determination, for every $k \in \{3, \ldots, n-1\}$, of the vectorial functions having the property above. We identified an infinite class of (non-necessarily-bijective) (m, m)-functions satisfying this property for even k, including all APN power functions.

A vectorial function is kth-order sum-free if and only if its kth-order sum-free min-degree equals 1. Functions being kth-order sum-free seem rare.

We leave open the problem of determining the kth-order sum-free min-degree t of the multiplicative inverse function for $k \in \{3, \ldots, n-3\}$, and more generally of all the functions x^{2^r-1} for $k \in \{2, \ldots, n-2\} \setminus \{r\}$, and of other cryptographic functions.

Finally, we could show, thanks to a simple and general characterization by subspace polynomials of the kth-order sum-free min-degree of the multiplicative inverse function, that it equals k when k divides n and that, in general, it equals its $(n-k)$th-order sum-free min-degree, thus extending a result on its kth-order sum-freedom.

References

1. Bartoli, D., Zhou, Y.: Asymptotics of Moore exponent sets. J. Comb. Theory Ser. A **175**, 105281 (2020)
2. Boura, C., Canteaut, A.: On the influence of the algebraic degree of F^{-1} on the algebraic degree of $G \circ F$. IEEE Trans. Inf. Theory **59**(1), 691–702 (2013)
3. Boura, C., Canteaut, A.: Another view of the division property. In: Robshaw, M., Katz, J. (eds.) CRYPTO 2016. LNCS, vol. 9814, pp. 654–682. Springer, Heidelberg (2016). https://doi.org/10.1007/978-3-662-53018-4_24
4. Canteaut, A., Videau, M.: Degree of composition of highly nonlinear functions and applications to higher order differential cryptanalysis. In: Proceedings of EURO-CRYPT 2002, Lecture Notes in Computer Science, vol. 2332, pp. 518–533 (2002). https://hal.inria.fr/inria-00072221/document
5. Carlet, C.: Graph indicators of vectorial functions and bounds on the algebraic degree of composite functions. IEEE Trans. Inf. Theory **66**(12), 7702–7716 (2020)
6. Carlet, C.: Boolean Functions for Cryptography and Coding Theory, 562 p. Monograph in Cambridge University Press (2021)
7. Carlet, C.: Two generalizations of almost perfect nonlinearity. J. Cryptol. **38**(2) (2025). (Topical Collection on Advances in Boolean Functions with Applications in Cryptography). Available at Cryptology ePrint Archive 2024/841
8. Carlet, C.: On the vector subspaces of $\mathbb{F}_{2^n}$ over which the multiplicative inverse function sums to zero. Special issue in memory of Kai-Uwe Schmidt of Des. Codes Cryptogr. **93**(4), 1237–1254 (2025). Available at Cryptology ePrint Archive 2024/1007
9. Carlet, C.: A notion on S-boxes for a partial resistance to some integral attacks. Cryptology ePrint Archive 2024/1693

10. Carlet, C., Hou, X.-D.: More on the sum-freedom of the multiplicative inverse function. Des. Codes Crypt. **93**(11), 4981–4997 (2025)
11. Hebborn, P., Leander, G., Udovenko, A.: Mathematical aspects of division property. Cryptogr. Commun. **15**(4), 731–774 (2023)
12. Knudsen, L., Wagner, D.: Integral cryptanalysis. In: Daemen, J., Rijmen, V. (eds.) FSE 2002. LNCS, vol. 2365, pp. 112–127. Springer, Heidelberg (2002). https://doi.org/10.1007/3-540-45661-9_9
13. Li, S., Meidl, W., Polujan, A., Pott, A., Riera, C., Stănică, P.: Vanishing flats: a combinatorial viewpoint on the planarity of functions and their application. IEEE Trans. Inf. Theory **66**(11), 7101–7112 (2020)
14. MacWilliams, F.J., Sloane, N.J.: The Theory of Error-Correcting Codes. North Holland (1977)
15. Özbudak, F., Sınak, A., Yayla, O.: On verification of restricted extended affine equivalence of vectorial boolean functions. In: Koç, Ç.K., Mesnager, S., Savaş, E. (eds.) WAIFI 2014. LNCS, vol. 9061, pp. 137–154. Springer, Cham (2015). https://doi.org/10.1007/978-3-319-16277-5_8
16. Todo, Y.: Structural evaluation by generalized integral property. In: Oswald, E., Fischlin, M. (eds.) EUROCRYPT 2015. LNCS, vol. 9056, pp. 287–314. Springer, Heidelberg (2015). https://doi.org/10.1007/978-3-662-46800-5_12

Boolean Arithmetic over $\mathbb{F}_2$ from Group Commutators

Marc Joye(✉) (iD)

Zama, Paris, France
`marc@zama.ai`

Abstract. This paper studies efficient realizations of arithmetic over the binary field $\mathbb{F}_2$ in nonabelian groups using only intrinsic group operations, namely multiplication and inversion. The constructions rely on commutators to implement Boolean computation within the group structure. Two complementary approaches are presented: a realization of a universal Boolean gate (NAND) and direct realizations of the field operations XOR and AND. These approaches apply to finite nonabelian simple groups and can be implemented using a small number of group operations. Explicit realizations are provided in the alternating groups A_5 and A_6. For the smallest nonabelian simple group A_5, these constructions achieve state-of-the-art efficiency in the number of group operations.

Keywords: Arithmetic over $\mathbb{F}_2$ · Boolean computation · Commutators · Group-based computation · Fully homomorphic encodings

1 Introduction

The expressive power of nonabelian groups as computational media has been recognized for several decades. In algebraic complexity, Barrington' s theorem [1] shows that polynomial-size branching programs of width 5 over a fixed nonabelian simple group capture exactly the class NC^1. Earlier work of Maurer and Rhodes [9] (see also [6]) already demonstrated that Boolean functions can be realized using finite nonabelian simple groups. These results illustrate that noncommutativity provides a rich computational structure even when only a small set of algebraic operations is available.

This observation raises a natural question: to what extent can arithmetic and Boolean computation be realized directly inside group operations? At first sight this appears unlikely, since a group provides essentially a single binary operation (together with inversion), whereas arithmetic requires both addition and multiplication. Nevertheless, suitable encodings of bits as group elements may allow both operations to be recovered from the intrinsic group structure. In particular, arbitrary Boolean circuits can then be evaluated using only the operations of group multiplication and group inversion.

L. Batina and F. Özbudak (Eds.): WAIFI 2026, LNCS 16611, pp. 239–249, 2026.
https://doi.org/10.1007/978-3-032-27574-5_15

To describe this phenomenon, it is convenient to adopt the viewpoint of a *fully homomorphic encoding* inside a group. Informally, elements of a ring are represented by group elements in such a way that appropriate combinations of intrinsic group operations implement the desired arithmetic operations. When the message space is $\mathbb{F}_2 = \{0,1\}$, this corresponds to realizing the Boolean operations XOR and AND on encoded bits using only group operations.

Related Work. Early examples of arithmetic over $\mathbb{F}_2$ in groups include the embedding of $(\mathbb{F}_2, +, \cdot)$ into $\mathrm{SL}(3,2)$ by Ben-Or and Cleve [2], the realization inside S_7 studied by Rappe [13], and the construction in the alternating group A_5 given by Khamsemanan, Ostrovsky, and Skeith [5]. Subsequent work of Nuida gave realizations in the symmetric group S_5 [11], with further refinements in [10]. More recently, Guillot et al. [4] described an explicit realization in the group S_6, building on earlier constructions. Related perspectives appear in works investigating group-theoretic approaches to fully homomorphic encryption [3–5,7,11].

Our Techniques and Contributions. The present work studies how Boolean computation and arithmetic over $\mathbb{F}_2$ can be realized efficiently using only intrinsic operations of certain nonabelian groups, namely multiplication and inversion. Our constructions rely on commutators as a basic algebraic primitive. A key structural property supporting this framework is Ore's theorem, proved in [8], which asserts that every element of a finite nonabelian simple group is a commutator. In particular, the commutator map is sufficiently expressive to reach any element of the group, suggesting that commutator-based constructions can serve as building blocks for Boolean computation inside such groups. Related results appear in [5, Lemma 6.3], where it is shown that, for any distinct elements α, β with $\beta \neq e$ in a finite nonabelian simple group G, the element α can be expressed as a product of functions of the form $(x, y) \mapsto \left[g\,x\,g^{-1}, h\,y\,h^{-1} \right]$ evaluated at (β, β). While this result allows arbitrary products of such commutators, the constructions considered here focus on very small products, with the main case being a single commutator (i.e., $k = 1$) and, in one instance, a product of two commutators.

We present two complementary approaches: (i) a commutator-based realization of a universal Boolean gate (NAND), and (ii) direct realizations of the arithmetic operations XOR and AND over $\mathbb{F}_2$. Both approaches rely solely on multiplication and inversion in the underlying group and apply to finite nonabelian simple groups.

We illustrate these approaches through explicit realizations in small alternating groups. For the smallest nonabelian simple group, namely the alternating group A_5, we obtain a commutator-based realization of the NAND gate requiring 8 group operations. This improves the operation count of the recent construction of Nuida [10], which uses compression functions over groups and requires 10 group multiplications to realize a NAND gate. In addition, we obtain a direct commutator-based realization of the AND gate in A_5 requiring 8 group operations, a functionality not explicitly provided in [10]. Its companion XOR gate over $\mathbb{F}_2$ achieves the same operation count as the construction in [10], namely

10 group operations. We also give a different realization of the AND gate that uses only group multiplications, requiring 9 group multiplications in total; its companion XOR gate is then realized by a single group multiplication.

Our framework also yields particularly efficient realizations in larger simple groups. In the group A_6 (and similarly in S_6), we obtain an AND gate requiring only 5 group multiplications and an XOR gate requiring a single group multiplication. This viewpoint provides a conceptual explanation for the encoding over S_6 described in [4], which is obtained there by modifying an S_7 construction by hand.

Outline of the Paper. The remainder of this paper is organized as follows. Section 2 introduces the required group-theoretic notation and the notion of a homomorphic encoding over a group. Section 3 presents a commutator-based realization of a NAND gate, including an explicit construction in A_5. Section 4 gives realizations of the XOR and AND gates over $\mathbb{F}_2$ with parameters in A_5 and A_6. Section 5 concludes the paper.

2 Background and Notation

Unless specified otherwise, all groups considered in this paper are finite, written multiplicatively, and their identity element is denoted by e.

2.1 Elements of Group Theory

We recall several standard notions from group theory; a general reference is [14].

Basic Notions. Let G be a group. A subgroup $N \leq G$ is said to be *normal*, written $N \trianglelefteq G$, if it is invariant under conjugation, that is, if $g\,N\,g^{-1} = N$ for all $g \in G$. Equivalently, $g\,n\,g^{-1} \in N$ for all $g \in G$ and $n \in N$. The *center* of G is the subgroup $Z(G) := \{z \in G \mid z\,g = g\,z \text{ for all } g \in G\}$, consisting of all elements that commute with every element of G. For $g, h \in G$, the (group) *commutator* of g and h is

$$[g, h] := g\,h\,g^{-1}\,h^{-1} \ .$$

Thus $[g, h] = e$ if and only if g and h commute.

Finally, a nontrivial group G is called *simple* if its only normal subgroups are $\{e\}$ and G itself.

Permutation Groups. For a positive integer n, the *symmetric group* S_n is the group of all permutations of the set $\{1, \ldots, n\}$, with group multiplication given by composition. Its elements are bijections $\sigma\colon \{1, \ldots, n\} \to \{1, \ldots, n\}$, and $\sharp S_n = n!$. Permutations in S_n are written in *cycle notation*. A cycle $(i_1\, i_2\, \ldots\, i_k)$ denotes the permutation sending $i_1 \mapsto i_2, i_2 \mapsto i_3, \ldots, i_k \mapsto i_1$, and fixing all other elements of $\{1, \ldots, n\}$. The identity permutation is denoted by $()$. Disjoint cycles commute, and every permutation can be written uniquely (up to the order of disjoint cycles) as a product of disjoint cycles. A 2-cycle is called a *transposition*.

The *alternating group* A_n is the subgroup of S_n consisting of all even permutations, that is, permutations that can be written as a product of an even number of transpositions. It is a normal subgroup of S_n of index 2, and hence $\sharp A_n = n!/2$. For $n \geq 5$, the group A_n is nonabelian and simple.

2.2 Fully Homomorphic Encodings over Groups

We formalize the notion of realizing arithmetic inside a group.

Definition 1. *Let $(\mathcal{M}, +, \cdot)$ be a commutative ring with 1 and let G be a group. A* fully homomorphic encoding over G *is an injective map*

$$\mathrm{Ecd} \colon \mathcal{M} \longrightarrow G$$

together with efficiently computable operations

$$\mathrm{Add}, \mathrm{Mult} \colon G \times G \to G$$

such that for all $m_1, m_2 \in \mathcal{M}$,

$$\mathrm{Add}\big(\mathrm{Ecd}(m_1), \mathrm{Ecd}(m_2)\big) = \mathrm{Ecd}(m_1 + m_2)$$

and

$$\mathrm{Mult}\big(\mathrm{Ecd}(m_1), \mathrm{Ecd}(m_2)\big) = \mathrm{Ecd}(m_1 \cdot m_2) \ .$$

We illustrate this notion with the realization in $G = A_6$ described in Sect. 4.4. There the message space is $\mathcal{M} = \mathbb{F}_2$, identified with the set $\{0, 1\}$ equipped with addition $\oplus$ (XOR) and multiplication $\wedge$ (AND). The encoding is given by

$$\mathrm{Ecd}(0) = () \ , \qquad \mathrm{Ecd}(1) = (1\,2)(3\,4) \ .$$

If $x = \mathrm{Ecd}(\beta)$ and $x' = \mathrm{Ecd}(\beta')$ for $\beta, \beta' \in \mathbb{F}_2$, then

$$\mathrm{Add}(x, x') = \mathrm{XOR}(x, x') \ , \qquad \mathrm{Mult}(x, x') = \mathrm{AND}(x, x') \ ,$$

where XOR is realized by group multiplication and AND by a commutator-based construction.

3 Boolean Computation via a NAND Gate

We denote the Boolean NAND operator by $\bar{\wedge}$, defined by

$$u \,\bar{\wedge}\, v = \neg(u \wedge v) = 1 - uv \ , \qquad u, v \in \{0, 1\} \ .$$

Since the NAND gate is functionally complete, it suffices to realize NAND in order to obtain universal Boolean computation. In this section, we present a construction in which NAND is implemented by a *single commutator* in a finite nonabelian simple group, and we give an explicit instantiation in A_5.

3.1 Commutator-Based NAND Gate

Let G be a finite nonabelian simple group. Select two distinct nontrivial elements $z_0, z_1 \in G$ that encode the bits

$$0 \longleftrightarrow z_0 \ , \qquad\qquad 1 \longleftrightarrow z_1 \ .$$

The requirement that both encoding elements be nontrivial is essential: if $z_0 = e$, then any commutator involving z_0 is trivial, so NAND cannot be realized as a single commutator.

Fix elements $g, h \in G$ and define

$$\mathrm{NAND}(x, y) := \left[g\,x\,g^{-1}, h\,y\,h^{-1}\right] \ , \qquad x, y \in \{z_0, z_1\} \ . \tag{1}$$

This map depends only on intrinsic group operations.

Correctness Conditions. Let

$$a_i = g\,z_i\,g^{-1} \ , \qquad b_j = h\,z_j\,h^{-1} \ .$$

The map Eq. (1) realizes the Boolean NAND gate under the encoding $0 \leftrightarrow z_0$ and $1 \leftrightarrow z_1$, provided that

$$[a_i, b_j] = z_{i \,\bar{\wedge}\, j} \ , \qquad i, j \in \{0, 1\} \ .$$

Thus the construction reduces to finding four elements a_0, a_1, b_0, b_1 in prescribed conjugacy classes whose commutator table matches the Boolean NAND table. Ore's theorem guarantees that every element of G is a commutator; hence z_0 and z_1 both occur in the image of the commutator map. The present construction strengthens this requirement by imposing simultaneous constraints on four specific commutators arising from fixed conjugacy classes.

Universality. Once such elements z_0, z_1, g, h are found, the commutator map

$$\mathrm{NAND}(x, y) = \left[g\,x\,g^{-1}, h\,y\,h^{-1}\right]$$

realizes the Boolean NAND operation under the given encoding. Since NAND is functionally complete, arbitrary Boolean functions can then be implemented by composing this operation.

3.2 An Explicit NAND Gate in A_5

We now give a concrete realization that implements a Boolean NAND gate entirely within the group law of $G = A_5$.

Define

$$z_0 = (1\,2\,3\,4\,5) \ , \quad z_1 = (1\,3\,5\,4\,2) \ , \quad g = (2\,4)(3\,5) \ , \quad h = (1\,5\,3) \ .$$

The elements z_0 and z_1 are both 5-cycles and therefore lie in the same conjugacy class of A_5. Moreover, g is an involution ($g^{-1} = g$), while h has order 3 (so $h^{-1} = (1\,3\,5)$).

Let $a_i = g\,z_i\,g^{-1}$ and $b_j = h\,z_j\,h^{-1}$. A direct computation in A_5 yields the commutator table

$$
\begin{array}{c|cc}
[a_i, b_j] & j = 0 & j = 1 \\
\hline
i = 0 & z_1 & z_1 \\
i = 1 & z_1 & z_0
\end{array}
$$

which coincides with the Boolean NAND table under the encoding $0 \leftrightarrow z_0$ and $1 \leftrightarrow z_1$.

Expanding the commutator operator, the NAND gate can be evaluated using 8 group operations in A_5 as

$$
\mathrm{NAND}(x, y) = U \cdot V \cdot (V \cdot U)^{-1} \quad \text{where} \quad
\begin{cases}
U = (2\,4)(3\,5) \cdot x \cdot (2\,4)(3\,5) \\
V = (1\,5\,3) \cdot y \cdot (1\,3\,5)
\end{cases} .
$$

4   XOR and AND Gates over $\mathbb{F}_2$

Let G be a finite nonabelian simple group. We now turn from universal Boolean computation to the direct realization of the arithmetic gates over $\mathbb{F}_2$, namely

$$
x \oplus y = x + y \pmod{}2 \quad \text{and} \quad x \wedge y = x \cdot y \pmod 2 .
$$

In contrast to the NAND construction of Sect. 3, it is here natural to adopt the canonical encoding

$$
0 \longleftrightarrow e \,, \qquad\qquad 1 \longleftrightarrow z_1 \,.
$$

so that each bit is represented by an element of the set $\{e, z_1\} \subseteq G$.

4.1   Commutator-Based AND Gate

Fix $g, h \in G$ and define

$$
\mathrm{AND}(x, y) := \left[g\,x\,g^{-1}, h\,y\,h^{-1} \right] \,, \qquad x, y \in \{z_0, z_1\} \,.
$$

Let $a_i = g\,z_i\,g^{-1}$ and $b_j = h\,z_j\,h^{-1}$. Under the encoding $0 \leftrightarrow e$ and $1 \leftrightarrow z_1$, this realizes the Boolean AND gate provided

$$
[a_i, b_j] = z_{i \wedge j} \,, \qquad i, j \in \{0, 1\} \,.
$$

Because $z_0 = e$, we automatically have $a_0 = b_0 = e$, and therefore

$$
[a_0, b_0] = e \,, \qquad\qquad [a_0, b_1] = e \,, \qquad\qquad [a_1, b_0] = e \,,
$$

which match the three zero-outputs of the AND table. The only nontrivial constraint is

$$
[a_1, b_1] = z_1 \,,
$$

that is, two conjugates of z_1 must have commutator equal to z_1.

Remark 1. Suppose z_1 has order 2. If a_1, b_1 are conjugates of z_1, they are involutions as well, and hence

$$[a_1, b_1] = a_1 \, b_1 \, a_1^{-1} \, b_1^{-1} = a_1 \, b_1 \, a_1 \, b_1 = (a_1 \, b_1)^2 \ .$$

Thus the condition $[a_1, b_1] = z_1$ becomes $(a_1 \, b_1)^2 = z_1$, so $T := a_1 \, b_1$ must have order 4. In particular, G must contain two involutions whose product has order 4, and hence a subgroup isomorphic to the dihedral group D_8.

When the condition of Remark 1 is not satisfied, it may still be possible to realize the AND gate by using a product of two commutators. For instance, one may consider constructions of the form

$$\mathrm{AND}(x, y) := \left[g_1 \, x \, g_1^{-1}, h_1 \, y \, h_1^{-1} \right] \cdot \left[g_2 \, x \, g_2^{-1}, h_2 \, y \, h_2^{-1} \right] \ ,$$

which can satisfy the required Boolean table even when a single commutator does not suffice. Products of such commutator maps also appear in structural expressiveness results such as [5, Lemma 6.3].

4.2   XOR Gate over $\mathbb{F}_2$

We present two XOR realizations, depending on the order of z_1.

XOR via Group Multiplication. With the same encoding and z_1 of order 2, group multiplication directly implements addition in $\mathbb{F}_2$:

$$z_0 \cdot z_0 = z_0 \ , \qquad z_0 \cdot z_1 = z_1 \cdot z_0 = z_1 \ , \qquad z_1 \cdot z_1 = z_0 \ .$$

Hence

$$\mathrm{XOR}(x, y) := x \cdot y$$

realizes the XOR gate.

Commutator-Based XOR Gate. If G contains a subgroup isomorphic to D_8, then it contains an element $z_1 = [a_1, b_1]$ of order 2, and the above one-multiplication implementation should be preferred.

If G has no involutions, then it contains an element of order 3. Indeed, by the classification of finite simple groups, the only nonabelian finite simple groups whose order is not divisible by 3 are the Suzuki groups $\mathrm{Sz}(2^{2n+1})$, and these contain dihedral subgroups of order 8, hence involutions (so the contrapositive implies that a simple group without involutions must have order divisible by 3).

When $z_0 = e$, a direct commutator-based realization of the form $\mathrm{XOR}(x, y) = \left[g \, x \, g^{-1}, h \, y \, h^{-1} \right]$ is impossible, since XOR requires $[a_0, b_1] = [a_1, b_0] = z_1$, but $a_0 = e$ implies $[a_0, b_1] = e \neq z_1$. If z_1 has order 3, however, XOR can be obtained from the commutator-based AND gate via

$$\mathrm{XOR}(x, y) := x \cdot y \cdot \mathrm{AND}(x, y) \ ,$$

since $z_1^3 = e$ and a direct verification shows that the Boolean XOR table is satisfied.

246   M. Joye

4.3 Fully Homomorphic Encoding of $\mathbb{F}_2$

Assume that $z_0 = e$ and $z_1 \in G$ are chosen so that the above construction realizes the AND gate, and that XOR is implemented either by group multiplication (when z_1 has order 2) or via $\mathrm{XOR}(x, y) = x \cdot y \cdot \mathrm{AND}(x, y)$ when z_1 has order 3. Then the map

$$\mathrm{Ecd}(0) = e \;, \qquad \mathrm{Ecd}(1) = z_1$$

together with Add = XOR and Mult = AND forms a fully homomorphic encoding of $\mathbb{F}_2$ over G.

4.4 Explicit Realizations

We now give explicit realizations in small simple groups.

A First Explicit realization in A_5. In A_5, involutions are double transpositions. The product of two double transpositions has order 1, 2, 3, or 5, never 4. Hence $(a_1 b_1)^2$ cannot be an involution, and therefore no element $z_1 = [a_1, b_1]$ of order 2 arises in this manner (see Remark 1).

The XOR realization via group multiplication therefore does not apply. We give below concrete parameters for a commutator-based implementation in group $G = A_5$.

Define

$$z_0 = () \;, \qquad z_1 = (1\,2\,3) \;, \qquad g = (1\,5\,4) \;, \qquad h = (1\,2)(3\,4) \;.$$

Here z_1 is a 3-cycle and thus has order 3. A direct computation yields $a_1 = g\, z_1\, g^{-1} = (2\,3\,5)$ and $b_1 = h\, z_1\, h^{-1} = (1\,4\,2)$, and in turn

$$[a_1, b_1] = a_1\, b_1\, a_1^{-1}\, b_1^{-1} = (1\,2\,3) = z_1 \;.$$

Because $z_0 = ()$ (that is, $z_0 = e$), the remaining three entries of the AND table are automatically satisfied. Hence

$$\mathrm{AND}(x, y) = \left[g\, x\, g^{-1},\, h\, y\, h^{-1} \right]$$

realizes the Boolean AND gate in A_5 under the encoding $0 \leftrightarrow ()$, $1 \leftrightarrow (1\,2\,3)$. Since z_1 has order 3, XOR is obtained via

$$\mathrm{XOR}(x, y) = x \cdot y \cdot \mathrm{AND}(x, y) \;.$$

Regarding the operation count, as in the NAND realization, the AND gate requires 8 group operations (7 group multiplications and 1 group inversion), and the XOR gate requires 2 additional group multiplications.

Another realization of the AND gate requiring 9 group multiplications in the multiplication-only model (i.e., disallowing explicit inversion) is described below, together with a matching XOR gate implemented by a single group multiplication.

Remark 2. The same obstruction holds in the symmetric group S_5. Its involutions are either transpositions or double transpositions, and in both cases the product of two conjugate involutions never has order 4. Consequently, S_5 likewise admits no realization with $z_1 = [a_1, b_1]$ of order 2, and one must again rely on the commutator-based constructions for AND and XOR.

A Second Explicit Realization in A_5. We now present an explicit realization of the above type in the group $G = A_5$, using a product of two commutators. Define

$$z_0 = ()\ , \qquad z_1 = (1\,2)(3\,4)\ , \qquad g_1 = (1\,5)(2\,3)\ , \qquad h_1 = (1\,2)(4\,5)\ ,$$
$$g_2 = (2\,5)(3\,4)\ , \qquad h_2 = g_2\, g_1^{-1}\, h_1 = (1\,4)(3\,5)\ .$$

The choice of h_2 is made so that $g_1^{-1} h_1 = g_2^{-1} h_2 = (1\,3\,2\,5\,4)$, which allows a common factor to be extracted in the evaluation of the two commutators. Writing

$$T = x \cdot (1\,3\,2\,5\,4) \cdot y\ ,$$

the AND gate can then be evaluated as

$$\mathrm{AND}(x,y) = \big((1\,5)(2\,3) \cdot T \cdot (1\,2)(4\,5)\big)^2 \cdot \big((2\,5)(3\,4) \cdot T \cdot (1\,4)(3\,5)\big)^2\ .$$

A direct verification shows that this map realizes the Boolean AND gate under the encoding $0 \leftrightarrow ()$ and $1 \leftrightarrow (1\,2)(3\,4)$. Since z_1 has order 2, the companion XOR gate is simply given by

$$\mathrm{XOR}(x,y) = x \cdot y\ .$$

This realization uses only group multiplications. More precisely, the AND gate requires 9 group multiplications, while XOR is implemented by a single group multiplication.

An Explicit Realization in A_6. The situation changes in the group $G = A_6$, which contains a subgroup isomorphic to D_8 (and therefore so does A_n for all $n \geq 6$, since A_6 embeds in A_n as the subgroup fixing $\{7, \ldots, n\}$ pointwise).
Define

$$z_0 = ()\ , \qquad z_1 = (1\,2)(3\,4)\ , \qquad g = (2\,3)(5\,6)\ , \qquad h = (1\,5)(2\,6)\ .$$

All elements lie in A_6, and z_1 is an involution. A direct computation yields $g\, z_1\, g^{-1} = (1\,3)(2\,4)$ and $h\, z_1\, h^{-1} = (3\,4)(5\,6)$. These are involutions whose product, $(1\,3)(2\,4) \cdot (3\,4)(5\,6) = (1\,3\,2\,4)(5\,6)$, is a cycle of type $(4,2)$, and hence has order 4. Consequently,

$$\langle (1\,3)(2\,4), (3\,4)(5\,6) \rangle \cong D_8\ .$$

Moreover, one has $\big[g\, z_1\, g^{-1}, h\, z_1\, h^{-1}\big] = [(1\,3)(2\,4), (3\,4)(5\,6)] = (1\,2)(3\,4) = z_1$. Thus the requirement for realizing the Boolean AND gate in A_6 under the encoding $0 \leftrightarrow ()$, $1 \leftrightarrow (1\,2)(3\,4)$ is satisfied.

Since z_1 has order 2, the XOR gate is obtained directly via group multiplication: $\mathrm{XOR}(x,y) = x \cdot y$.

Since z_0 and z_1 are both involutions, so are their conjugates. In turn, the commutator-based AND becomes $\mathrm{AND}(x,y) = g\,x\,g^{-1}\,h\,y\,h^{-1}\,g\,x\,g^{-1}\,h\,y\,h^{-1}$, for $x, y \in \{z_0, z_1\}$. Noting that $g^{-1}h = (1\,6\,3\,2\,5)$, the AND gate can be evaluated using 5 group multiplications as

$$\mathrm{AND}(x,y) = \big((2\,3)(5\,6) \cdot x \cdot (1\,6\,3\,2\,5) \cdot y \cdot (1\,5)(2\,6)\big)^2 \ .$$

The XOR gate $\mathrm{XOR}(x,y) = x \cdot y$ requires only a single group multiplication.

5   Conclusion

This paper revisits the problem of realizing arithmetic over $\mathbb{F}_2$ within finite nonabelian simple groups using only intrinsic group operations. Two complementary approaches are presented: one based on realizing a universal Boolean gate (NAND), and one based on direct realizations of the arithmetic operations XOR and AND.

Explicit realizations in the alternating groups A_5 and A_6 illustrate different aspects of these approaches. In A_5, compact commutator-based implementations of Boolean gates are obtained. In A_6, the arithmetic operations admit particularly simple realizations, requiring only a very small number of group multiplications. This latter example also sheds light on the encoding over S_6 that appears in recent work.

References

1. Barrington, D.A.M.: Bounded-width polynomial-size branching programs recognize exactly those languages in NC^1. In: 18th ACM Symposium on Theory of Computing (STOC 1986), pp. 1–5 (1986). https://doi.org/10.1145/12130.12131
2. Ben-Or, M., Cleve, R.: Computing algebraic formulas using a constant number of registers. SIAM J. Comput. **21**(1), 54–58 (1992). https://doi.org/10.1137/0221006
3. Grigoriev, D., Ponomarenko, I.: On non-abelian homomorphic public-key cryptosystems. J. Math. Sci. **126**(3), 1158–1166 (2005). https://doi.org/10.1007/s10958-005-0077-3
4. Guillot, P., Hoang Duc, A., Koskas, M., Méhats, F.: Introducing GRAFHEN: GRoup-bAsed fully homomorphic encryption without noise. Cryptology ePrint Archive, Paper 2025/1907 (2025). https://ia.cr/2025/1907
5. Khamsemanan, N., Ostrovsky, R., Skeith, W.E.: On the black-box use of somewhat homomorphic encryption in noninteractive two-party protocol. SIAM J. Discret. Math. **30**(1), 266–295 (2016)
6. Krohn, K., Maurer, W.D., Rhodes, J.: Realizing complex Boolean functions with simple groups. Inf. Control **9**, 190–195 (1966). https://doi.org/10.1016/S0019-9958(66)90229-4
7. Li, J., Wang, L.: Noiseless fully homomorphic encryption. Cryptology ePrint Archive, Paper 2017/839 (2017). https://ia.cr/2017/839

8. Liebeck, M.W., O'Brien, E.A., Shalev, A., Tiep, P.H.: The Ore conjecture. J. Eur. Math. Soc. **12**(4), 939–1008 (2010). https://doi.org/10.4171/JEMS/220

9. Maurer, W.D., Rhodes, J.L.: A property of finite simple non-abelian groups. Proc. Am. Math. Soc. **16**, 552–554 (1965). https://doi.org/10.1090/S0002-9939-1965-0175971-0

10. Nuida, K.: On compression functions over groups with applications to homomorphic encryption. J. Algebra Appl. (to appear). https://doi.org/10.1142/S0219498827500642

11. Nuida, K.: Towards constructing fully homomorphic encryption without ciphertext noise from group theory. In: Takagi, T., Wakayama, M., Tanaka, K., Kunihiro, N., Kimoto, K., Ikematsu, Y. (eds.) International Symposium on Mathematics, Quantum Theory, and Cryptography. MI, vol. 33, pp. 57–78. Springer, Singapore (2021). https://doi.org/10.1007/978-981-15-5191-8_8

12. Ostrovsky, R., Skeith, W.E.: Communication complexity in algebraic two-party protocols. In: Wagner, D. (ed.) CRYPTO 2008. LNCS, vol. 5157, pp. 379–396. Springer, Heidelberg (2008). https://doi.org/10.1007/978-3-540-85174-5_21

13. Rappe, D.K.: Homomorphic cryptosystems and their applications. Ph.D. thesis, Universität Dortmund (2004). https://doi.org/10.17877/DE290R-15728

14. Robinson, D.J.S.: A Course in the Theory of Groups. Graduate Texts in Mathematics, vol. 80, 2nd edn. Springer, Cham (1996). https://doi.org/10.1007/978-1-4419-8594-1

Finite Field Arithmetic, Algorithms, and Implementations

Equality Tests in the Polynomial Modular Number System

Nicolas Méloni[ID], François Palma[(✉)][ID], and Pascal Véron[ID]

Univ Toulon, IMATH, Toulon, France
{nicolas.meloni,francois.palma,pascal.veron}@univ-tln.fr

Abstract. The Polynomial Modular Number System (PMNS) is a non-positional number system that allows for fast modular arithmetic operations. It has shown its effectiveness in both hardware and software contexts, notably outperforming both GMP and OpenSSL from 256 to 8192 bits. Due to its non-positional nature however, equality tests become non-trivial and finding methods to perform them in an optimized fashion remains an open problem. In this work, we present several results, including a new coefficient reduction algorithm that allows for a faster than state of the art equality test without impacting the performance of the overall system.

1 Introduction

Modular arithmetic is core to a wide range of cryptographic protocols. For classical (non post-quantum) cryptography, RSA and ECC are the most notorious examples, with ECC in particular being used anytime an HTTPS connection is established. In terms of post-quantum cryptography, any protocol using euclidean lattices (such as ML-KEM [18]) or elliptic curve isogenies makes use of modular arithmetic.

The Polynomial Modular Number System is a representation system for $\mathbb{Z}/p\mathbb{Z}$ in which elements are polynomials. PMNS have been shown to be very competitive with standard multi-precision arithmetic for ECC-sized integers [2,4,5] but also for larger integer sizes [14,20] up to 8192 bits, particularly when compared with OpenSSL [21] and GMP [27]. PMNS have also been implemented as a faster alternative to regular modular arithmetic in the Zcash library, a bitcoin-based cryptocurrency [8]. In most of these applications, standard operations such as addition, modular multiplication, exponentiation, etc. are the predominant factors in terms of complexity analysis. However, in some cryptosystems [11,17,24], equality tests are performed, for which PMNS are less suited. Namely, due to being a non-positional number system, it is non-trivial to check for equality and instead of simply comparing each coefficient one by one, which is what is done in standard multi-precision arithmetic, the naive method is to perform a polynomial evaluation in $\mathbb{Z}/p\mathbb{Z}[X]$.

L. Batina and F. Özbudak (Eds.): WAIFI 2026, LNCS 16611, pp. 253–270, 2026.
https://doi.org/10.1007/978-3-032-27574-5_16

Recently in [7], a new way to perform equality tests on PMNS has been proposed which necessitates an increase in the number of polynomial coefficients necessary to represent a corresponding integer. For large enough integers (more than 1024 bits) the test becomes less efficient than polynomial evaluation in $\mathbb{Z}/p\mathbb{Z}[X]$.

In Sect. 2 we give an overview on PMNS and how operations are performed in them. Then, in Sect. 3 we focus on the Internal Reduction in PMNS, which is an important step to reduce polynomial coefficients, presenting the state of the art method based on the Montgomery modular reduction and a new algorithm based on the Plantard modular reduction. Afterwards, in Sect. 4 we present the two currently known equality test algorithms for PMNS and two new ones which each offer an interesting performance profile in different use-cases. Finally, in Sect. 5, we give benchmarks of each equality algorithm and analyze the results.

For the sequel of this paper we will consider that a polynomial $P \in \mathbb{Z}[X]$ of degree $n - 1$ can alternatively be considered as a vector in $\mathbb{Z}^n$ and vice versa.

2   Background

In this section, we provide a brief overview of PMNS and the way operations are performed.

2.1   PMNS

Definition 1. *Let $p \geqslant 3$, $n \geqslant 2$, $\gamma \in [1, p - 1]$ and $\rho \in [1, p - 1]$. Let $E \in \mathbb{Z}[X]$ be a monic polynomial of degree n, such that $E(\gamma) \equiv 0 \pmod{p}$. A PMNS is a set $\mathcal{B} \subset \mathbb{Z}[X]$ such that :*

1. $\forall A \in \mathcal{B}$, $\deg(A) < n$,

2. $\forall A(X) = \displaystyle\sum_{i=0}^{n-1} a_i X^i \in \mathcal{B}$, $-\rho < a_i < \rho$ for all i,

3. $\forall a \in \mathbb{Z}/p\mathbb{Z}$, $\exists A \in \mathcal{B}$ such that $A(\gamma) \equiv a \pmod{p}$.

A PMNS $\mathcal{B}$ is defined by any tuple (p, n, γ, ρ, E) that meets these requirements. Requiring all elements of $\mathbb{Z}/p\mathbb{Z}$ to have a corresponding polynomial in the PMNS imposes a number of constraints between the parameters. In particular, let $\mathcal{B}$ defined by (p, n, γ, ρ, E), a PMNS, we need $\mathrm{Card}(\mathcal{B}) \geqslant p$. Since all elements of $\mathcal{B}$ are polynomials of degree at most $n - 1$ whose coefficients are bounded in absolute value by ρ, we get $\mathrm{Card}(\mathcal{B}) = (2\rho - 1)^n$. Therefore a necessary condition on ρ is:

$$\rho > \frac{1}{2} p^{\frac{1}{n}} \tag{1}$$

This does not however guarantee that each integer in $\mathbb{Z}/p\mathbb{Z}$ has at least one polynomial representing it in the PMNS. To establish a sufficient condition, we first need to consider the lattice $\mathcal{L} = \left\{ (x_0, \ldots, x_{n-1}) \in \mathbb{Z}^n : \sum_{i=0}^{n-1} \right.$

$x_i \gamma^i \equiv 0 \pmod{p}\}$. It is the lattice of all vectors of size n for which the associated polynomials vanish in γ modulo p. A canonical basis is the following:

$$
\mathbf{B} = \begin{pmatrix}
p & 0\,0\ldots0\,0 \\
-\gamma \bmod p & 1\,0\ldots0\,0 \\
-\gamma^2 \bmod p & 0\,1\ldots0\,0 \\
\vdots & \ddots\ \ \vdots \\
-\gamma^{n-2} \bmod p\,0\,0\ldots1\,0 \\
-\gamma^{n-1} \bmod p\,0\,0\ldots0\,1
\end{pmatrix}
\tag{2}
$$

Note that the first row of $\mathbf{B}$ corresponds to $(p, 0, \ldots, 0)$ and we have $p \equiv 0 \pmod{p}$. The next $n - 1$ rows are $(-\gamma^i \bmod p, 0, \ldots, 0, 1, 0, \ldots, 0)$ which correspond to the polynomials $X^i - \gamma^i$ and we indeed have that $\gamma^i - \gamma^i \equiv 0 \pmod{p}$. So this is indeed a basis of $\mathcal{L}$.

It has been shown in [1, Theorem 4.2] that as soon as $\rho > \frac{1}{2}\|\mathbf{B}\|_1$ then every element of $\mathbb{Z}/p\mathbb{Z}$ possesses at least one corresponding polynomial in the PMNS. This result holds for any basis of $\mathcal{L}$ and as such, since we want to minimize ρ in order to fit the coefficients in memory, one can use a short basis of $\mathcal{L}$, which can be obtained by applying LLL [13], BKZ [25], or any other such algorithm, although in practice LLL is used.

2.2 Operations

Let $\mathcal{B}$ defined by (p, n, γ, ρ, E), a PMNS. Let $a \in \mathbb{Z}/p\mathbb{Z}$ and $b \in \mathbb{Z}/p\mathbb{Z}$ with $A \in \mathcal{B}$ such that $A(\gamma) \equiv a \pmod{p}$ and $B \in \mathcal{B}$ such that $B(\gamma) \equiv b \pmod{p}$.

Let $\odot$ be either addition or multiplication. To compute $c = a \odot b \pmod{p}$ through our PMNS $\mathcal{B}$, we compute $C(X) = A(X) \odot B(X) \pmod{E(X)}$. In the case of polynomial multiplication, the degree would grow which explains why reducing modulo $E(X)$, which is a monic polynomial of degree n, is necessary. Since E vanishes in γ modulo p, $C(\gamma) \equiv A(\gamma) \odot B(\gamma) \pmod{p}$. The modular reduction by E is called the *external reduction*. To bound $\|A \times B \pmod{E}\|_\infty$ relative to $\|A\|_\infty$ and $\|B\|_\infty$, [10] introduced a constant w whose value depends on E. Let $E(X) = X^n + e_{n-1}X^{n-1} + \cdots + e_1 X + e_0$, let

$$
\mathcal{E} = \begin{pmatrix}
-e_0 & -e_1 & \cdots & -e_{n-1} \\
\cdots & \cdots & \cdots & \cdots \\
\vdots & \vdots & \vdots & \vdots \\
\cdots & \cdots & \cdots & \cdots
\end{pmatrix}
\begin{matrix}
\leftarrow X^n \bmod E \\
\leftarrow X^{n+1} \bmod E \\
\ \\
\leftarrow X^{2n-2} \bmod E
\end{matrix}
,
\tag{3}
$$

then $w = \|(1, 2, \ldots, n) + (n-1, n-2, \ldots, 1)\cdot\mathcal{E}'\|_\infty$ with $\mathcal{E}'$ the matrix verifying $\forall i, j \in \{1, 2, \ldots, n-1\} \times \{1, 2, \ldots, n\}$, $\mathcal{E}'_{i,j} = |\mathcal{E}_{i,j}|$. From this parameter they deduce that $\|A \times B \pmod{E}\|_\infty \leqslant w(\rho - 1)^2$

Another important step is reducing the coefficients. Since we want each coefficient to fit in memory, we must perform an additional step after each multiplication, and sometimes after an addition, called the *internal reduction*. In short, we find a polynomial $\tilde{C}$ such that $\tilde{C}(\gamma) \equiv C(\gamma)$ (mod p) and $\|\tilde{C}\|_\infty < \rho$. This process is similar to finding a close vector to C in $\mathfrak{L}$. We detail this further in Sect. 3.

3   Internal Reduction

The Montgomery-like reduction is the state of the art algorithm for the internal reduction in PMNS. In this section, we first recall the algorithm, and then present 2 other reduction algorithms, including a new one: the Plantard-like reduction.

3.1   Montgomery-Like Internal Reduction

The state of the art internal reduction algorithm is based on the Montgomery modular reduction on integers [16].

The main idea of that algorithm is to replace a costly modular reduction by a non-specific modulus m with a modular reduction and division by ϕ a power of 2, which is much faster. In exchange, the result is congruent to $c\phi^{-1}$ mod m. This extra factor can be handled thanks to the Montgomery domain. Let $a \in \mathbb{Z}/m\mathbb{Z}$ and $b \in \mathbb{Z}/m\mathbb{Z}$. We first compute $a' \equiv a\phi$ mod m and $b' \equiv b\phi$ mod m. It follows that if $c = a' \times b'$ then $c' \equiv (a\phi \times b\phi)\phi^{-1} \equiv ab\phi$ (mod m). This means that the result is still in the Montgomery domain and a sequence of operations can be determined to compute the final result as desired, with potentially a final Montgomery reduction at the end to exit the domain. This translates into the following algorithm for PMNS:

Algorithm 1. Montgomery-like internal reduction [15]

Require: $\mathcal{B}$ defined by (p, n, γ, ρ, E) a PMNS with odd p, $C \in \mathbb{Z}[X]$, $\phi \in \mathbb{N}^*$ and $\mathcal{G}$ a short basis of $\mathfrak{L}$.
Ensure: $C'(\gamma) \equiv C(\gamma)\phi^{-1}$ mod p
  1: $q \leftarrow -C \cdot \mathcal{G}^{-1}$ mod ϕ
  2: $C' \leftarrow (C + (q \cdot \mathcal{G}))/\phi$
  3: **return** C'

Adapting the Montgomery algorithm for PMNS was first proposed in [19]. The version we use here is slightly modified but the principle remains the same and was first proposed in [15]. The idea is to view a polynomial C as a vector in $\mathbb{Z}^n$. Then, we can find a vector $q \in \mathbb{Z}^n$ such that $C + q \cdot \mathcal{G} \equiv 0$ (mod ϕ). Since $\mathcal{G}$ is a basis of $\mathfrak{L}$, which is the lattice described earlier corresponding to the set of all polynomials of degree at most $n - 1$ that vanish in γ mod p, then the result of $q \cdot \mathcal{G}$, if viewed as a polynomial, vanishes in γ mod p. It follows that $C + q \cdot \mathcal{G}$ evaluated in γ is congruent to $C(\gamma) \bmod p$. Finding this vector q is then done by simply computing $-C \cdot \mathcal{G}^{-1} \pmod{\phi}$. It follows that $C + q \cdot \mathcal{G} \equiv C + (-C \cdot \mathcal{G}^{-1}) \cdot \mathcal{G} \equiv C - C \equiv 0 \pmod{\phi}$. As can be seen in Eq. (2), $\det(\mathbf{B}) = p$. Thus, for odd p, $\mathcal{G}$ is always invertible mod ϕ, which is a power of 2.

Similarly to the Montgomery reduction on integers, the result has an extra ϕ^{-1} to deal with. This can be handled similarly to integers by using the Montgomery domain. Since integers have to be converted into polynomials to perform operations in the PMNS anyway, the conversion to the Montgomery domain can be done at the same time for no extra cost.

3.2   Plantard-Like Internal Reduction

In this work, we present for the first time an adaptation of [23, Algorithm 8] to PMNS.

In [23], Plantard presents a new modular reduction algorithm for word-sized integers adapted from the Montgomery modular reduction. The improvement comes from the fact that in a modular exponentiation, part of the computations can be done only once, thus leading to a noticeable speedup. We adapt it to PMNS to get Algorithm 2.

Algorithm 2. Plantard-like internal reduction

Require: $\mathcal{B}$ defined by $(p,\, n,\, \gamma,\, \rho,\, E)$ a PMNS, $C \in \mathbb{Z}[X]$, $\phi \in \mathbb{N}^*$ and $\mathcal{G}$ a short basis of $\mathcal{L}$ invertible modulo ϕ^2.
Ensure: $C'(\gamma) \equiv C(\gamma)\phi^{-2} \bmod p$
  1: $Q \leftarrow -C \cdot \mathcal{G}^{-1} \bmod \phi^2$
  2: $S \leftarrow \left\lceil \frac{1}{\phi} Q \right\rfloor$
  3: $C' \leftarrow \left\lceil \frac{1}{\phi} S \cdot \mathcal{G} \right\rfloor$
  4: **return** C'

Remark 1. Unlike the algorithm presented in [23], we perform rounding operations instead of taking the floor. This allows us to remove the addition by 1 which would not translate well into vectorial operations.

Remark 2. Similarly to the Montgomery domain, one can use a Plantard domain here and put an extra ϕ^2 factor (instead of just ϕ for Montgomery) on each value to ensure that the result stays in the domain.

As can be seen, the process is very similar to the Montgomery-like internal reduction, with the main difference being taking $-C \cdot \mathcal{G}^{-1} \bmod \phi^2$ and not mod ϕ. With this step and the rounding steps added, the addition from the Montgomery-like reduction is unnecessary. Let us first establish that this algorithm properly outputs the expected result.

Proposition 1. *If $\frac{\phi}{2}\|\mathcal{G}\|_1 + 2w(\rho-1)^2 < \frac{\phi^2}{2}$ and $\|C\|_\infty \leqslant 2w(\rho-1)^2$, then Algorithm 2 outputs $C'(X)$ such that $C'(\gamma) \equiv C(\gamma)\phi^{-2} \pmod{p}$.*

Proof. At step 3 of the algorithm, we have $C' = \left\lfloor \frac{1}{\phi}\left(\left\lfloor \frac{1}{\phi}Q \right\rfloor\right) \cdot \mathcal{G} \right\rceil$.

Let $Q_1 = \left\lfloor \frac{1}{\phi}Q \right\rceil$ and $Q_0 = Q - \phi Q_1$. Remark that $\|Q_0\|_\infty \leqslant \frac{\phi}{2}$. Then,

$$\left\lfloor \frac{1}{\phi}\left(\left\lfloor \frac{1}{\phi}Q \right\rceil\right) \cdot \mathcal{G} \right\rceil = \left\lfloor \frac{1}{\phi}(Q_1) \cdot \mathcal{G} \right\rceil$$

$$= \left\lfloor \frac{1}{\phi^2}(\phi Q_1) \cdot \mathcal{G} \right\rceil = \left\lfloor \frac{1}{\phi^2}(Q - Q_0) \cdot \mathcal{G} \right\rceil$$

$$= \left\lfloor \frac{1}{\phi^2}(Q \cdot \mathcal{G} - Q_0 \cdot \mathcal{G}) \right\rceil = \left\lfloor \frac{1}{\phi^2}(Q \cdot \mathcal{G} + C) - \frac{1}{\phi^2}(Q_0 \cdot \mathcal{G} + C) \right\rceil$$

Remark that $\|Q_0 \cdot \mathcal{G}\|_\infty \leqslant \frac{\phi}{2}\|\mathcal{G}\|_1$. Furthermore, we assumed that $\|C\|_\infty \leqslant 2w(\rho-1)^2$. Hence, $\|\frac{1}{\phi^2}(Q_0 \cdot \mathcal{G} + C)\|_\infty \leqslant \frac{1}{\phi^2}(\frac{\phi}{2}\|\mathcal{G}\|_1 + 2w(\rho-1)^2)$. Then, having assumed that $\frac{\phi}{2}\|\mathcal{G}\|_1 + 2w(\rho-1)^2 < \frac{\phi^2}{2}$, it follows that $\|\frac{1}{\phi^2}(Q_0 \cdot \mathcal{G} + C)\|_\infty < \frac{1}{2}$ and therefore $\left\lfloor \frac{1}{\phi^2}(Q_0 \cdot \mathcal{G} + C) \right\rceil = (0, \ldots, 0)$. Furthermore, since $Q \equiv -C \cdot \mathcal{G}^{-1} \bmod \phi^2$, we have $Q \cdot \mathcal{G} \equiv -C \pmod{\phi^2}$. Or, in other words, $Q \cdot \mathcal{G} + C \equiv 0 \pmod{\phi^2}$. This means that $\left\lfloor \frac{1}{\phi^2}(Q \cdot \mathcal{G} + C) \right\rceil = \frac{1}{\phi^2}(Q \cdot \mathcal{G} + C)$. Thus,

$$\left\lfloor \frac{1}{\phi^2}(Q \cdot \mathcal{G} + C) - \frac{1}{\phi^2}(Q_0 \cdot \mathcal{G} + C) \right\rceil = \frac{1}{\phi^2}(Q \cdot \mathcal{G} + C) - \left\lfloor \frac{1}{\phi^2}(Q_0 \cdot \mathcal{G} + C) \right\rceil$$

$$= \frac{1}{\phi^2}(Q \cdot \mathcal{G} + C).$$

Then, $\left\lfloor \frac{1}{\phi}\left(\left\lfloor \frac{1}{\phi}Q \right\rceil\right) \cdot \mathcal{G} \right\rceil = \frac{1}{\phi^2}(Q \cdot \mathcal{G} + C)$ with each coefficient of $Q \cdot \mathcal{G} + C$ a multiple of ϕ^2.

Finally, $(Q \cdot \mathcal{G} + C)(\gamma) \equiv (Q \cdot \mathcal{G})(\gamma) + C(\gamma) \pmod{p}$, thus $C'(\gamma) \equiv (\frac{1}{\phi^2}(Q \cdot \mathcal{G} + C)(\gamma)) \equiv C(\gamma)\phi^{-2} \pmod{p}$.

The main motivation behind adapting this algorithm to PMNS is the bounds on parameters we can afford to choose with it while maintaining an efficient internal reduction algorithm. From [9, Section 3.2], the Montgomery-like internal reduction requires choosing $\rho \geqslant \|\mathcal{G}\|_1 - 1$. From [1, Theorem 4.2], the theoretical bound on ρ to have a PMNS that can represent all the integers in $\mathbb{Z}/p\mathbb{Z}$ is $\rho > \frac{1}{2}\|\mathcal{G}\|_1$ so we can see that there is a gap between the theory and practical considerations.

Proposition 2. *If $\rho > \lfloor \frac{1}{2}\|\mathcal{G}\|_1 \rfloor$ then Algorithm 2 outputs a polynomial C' such that $\|C'\|_\infty < \rho$.*

Proof. At step 1 of the algorithm, we have $\|Q\|_\infty \leqslant \frac{\phi^2}{2}$ if we perform the modular reduction by taking values between $-\frac{\phi^2}{2}$ and $\frac{\phi^2}{2}$ instead of between 0 and $\phi^2 - 1$

in a similar manner to the Montgomery-like internal reduction. It follows that $\|S\|_\infty \leqslant \frac{\phi}{2}$ at step 2 since $S = \left\lfloor \frac{1}{\phi}Q \right\rceil$. As a result, $\|S \cdot \mathcal{G}\|_\infty \leqslant \frac{\phi}{2}\|\mathcal{G}\|_1$. Thus,

$$\|C'\|_\infty = \left\| \left\lfloor \frac{1}{\phi}S \cdot \mathcal{G} \right\rceil \right\|_\infty \leqslant \left\lfloor \frac{1}{2}\|\mathcal{G}\|_1 \right\rceil.$$

Then, if $\rho > \left\lfloor \frac{1}{2}\|\mathcal{G}\|_1 \right\rceil$, clearly $\|C\|_\infty < \rho$.

This means one can take almost the theoretical bound for PMNS and still have an efficient reduction method. Unfortunately, the Plantard-like internal reduction algorithm is slower for a given n than the Montgomery-like internal reduction due to the modular reduction by ϕ^2 instead of ϕ. This is slightly alleviated by the fact that one can choose a smaller n to construct a PMNS if one uses the Plantard-like internal reduction.

4 Equality Tests

Equality test in a PMNS is non-trivial due to the inherent redundancy in the system (number of elements strictly greater than p). In this section we present various algorithms to perform them efficiently.

4.1 Naive Equality Test

Let A and B be two elements of a PMNS, the first equality test consists of simply checking if $(A - B)(\gamma) \equiv 0 \pmod{p}$ to check if $A(\gamma) \equiv B(\gamma) \pmod{p}$. Since the evaluation in γ is slower than a subtraction, this is faster than evaluating both A and B and performing comparisons afterward.   This first version uses

---

Algorithm 3. Horner equality test

Require: $\mathcal{B}$ defined by (p, n, γ, ρ, E) a PMNS, $A, B \in \mathcal{B} \times \mathcal{B}$ thus $\|A\|_\infty < \rho$ and $\|B\|_\infty < \rho$.
Ensure: True if $A(\gamma) \equiv B(\gamma) \pmod{p}$ else False.
1: $C \leftarrow A(X) - B(X)$
2: $c \leftarrow c_{n-1}$
3: **for** $i = n - 2 \ldots 0$ **do**
4:     $c \leftarrow c \times \gamma + c_i \pmod{p}$
5: **end for**
6: **return** True if $c = 0$ else False

---

the Horner polynomial evaluation. However, in general, $\log_{2^{64}}(\gamma) = \log_{2^{64}}(p)$ so this is not optimal. Indeed, we can instead use the following algorithm, which utilises methods from [6, Algorithm 22], an algorithm of conversion from PMNS to binary.

Algorithm 4. Naive equality test

Require: $\mathcal{B}$ defined by (p, n, γ, ρ, E) a PMNS, $A, B \in \mathcal{B} \times \mathcal{B}$ thus $\|A\|_\infty < \rho$ and $\|B\|_\infty < \rho$, $g_0, g_1, \ldots, g_{n-1} \in (\mathbb{Z}/p\mathbb{Z})^n$ pre-computed $\forall i \in \{1, 2, \ldots, n\}$, $g_i \equiv \gamma^i$ (mod p).

Ensure: True if $A(\gamma) \equiv B(\gamma)$ (mod p) else False.

1: $C \leftarrow A(X) - B(X)$
2: $c \leftarrow c_0$
3: **for** $i = 1 \ldots n - 1$ **do**
4:     $c \leftarrow c + c_i g_i$
5: **end for**
6: $c \leftarrow c \bmod p$
7: **return** True if $c = 0$ else False

The notable improvement here is that in Algorithm 3, c quickly becomes the size of p and every subsequent loop iteration multiplies 2 p-sized integers, which is quite slow. Meanwhile, Algorithm 4 multiplies each c_i, each of which fits on a single machine word, with a p-sized integer instead. Hence the time complexity of Algorithm 3 is in $O(n(\log_{2^{64}}(p))^2)$ while the complexity of Algorithm 4 is in $O(n \log_{2^{64}}(p))$ instead.

Hermite Normal Form. Another potential "naive" test would be a lattice membership test. Indeed, $\mathcal{L}$ is the lattice of all polynomials of degree $n - 1$ that vanish in γ modulo p. So $(A - B) \in \mathcal{L} \iff A(\gamma) \equiv B(\gamma)$ (mod p). The common way [3] to do this is to first find a basis of $\mathcal{L}$ in Hermite Normal Form [12] (HNF). It just so happens that matrix **B** from Eq. 2 is already in HNF. Then, the test consists in augmenting the basis with the vector we want to test and checking if it is a linear combination of the other rows. This can be done, for example, with a Gauss pivot. So let us say we want to test whether $\ell \in \mathcal{L}$ with $\ell = (\ell_0, \ell_1, \ldots, \ell_{n-1})$. We can add ℓ as the last row of B (see Eq. 2). Then clearly the matrix can be transformed into

$$\begin{pmatrix} p & 0\,0 \ldots 0\,0 \\ -\gamma \bmod p & 1\,0 \ldots 0\,0 \\ -\gamma^2 \bmod p & 0\,1 \ldots 0\,0 \\ \vdots & \ddots\ \vdots \\ -\gamma^{n-2} \bmod p & 0\,0 \ldots 1\,0 \\ -\gamma^{n-1} \bmod p & 0\,0 \ldots 0\,1 \\ \sum_{i=0}^{n-1} \ell_i(\gamma^i \bmod p) & 0\,0 \ldots 0\,0 \end{pmatrix}$$

with linear combinations. The only check we need is whether $\sum_{i=0}^{n-1} \ell_i \gamma^i \equiv 0$ (mod p) or not. This is exactly the same as the naive equality test presented earlier which is a polynomial evaluation in γ. As such, we do not need to consider this equality test in particular as it can be performed much faster through Algorithm 4 thanks to the pre-computations. Another possibility is to use the SNF (Smith Normal Form) [26] but it again boils down to a polynomial evaluation in $\gamma \bmod p$.

4.2   Translation Equality Test

This equality test was first proposed in [7]. The main idea is to translate to a specific sub-domain in which, after a modular reduction, we are guaranteed to arrive at the all-0 vector if the result is in $\mathcal{L}$.

Algorithm 5. Translation equality test

Require: $A, B \in \mathbb{Z}_{n-1}[X] \times \mathbb{Z}_{n-1}[X]$ with $\|A\|_\infty < l$ and $\|B\|_\infty < l$ with $l = \frac{1}{2}w(\rho - 1)^2$, $\mathcal{T} = (-u, -u, \ldots, -u)$ with $u = \lceil w(\rho - 1)^2 \|\mathcal{G}^{-1}\|_1 \rceil$.
Ensure: $S = 0 \iff A(\gamma) \equiv B(\gamma) \pmod p$.
1: $C \leftarrow (A - B) + \mathcal{T}$
2: $S \leftarrow$ **Montgomery-like**(C)
3: **return** S

Remark 3. Usually, the Montgomery-like step can be made sub-quadratic through using either the basis matrix of a sub-lattice that has a Toeplitz shape [15, Section 5.2] or by the Karatsuba algorithm. But for the purpose of the equality test, the entire lattice must be used which means the matrix will not have any specific shape, which makes the time complexity at best quadratic.

The main drawback of using this algorithm is that it requires choosing ϕ verifying $\phi \geqslant \lceil 2w(\rho - 1)^2 \|\mathcal{G}^{-1}\|_1 \rceil$, instead of $\phi \geqslant 2w(\rho - 1)$ (from [9, Equation 12]). Having stricter requirements means one needs to take a larger n to find a PMNS that verifies them for a given p. This can be seen in Table 1.

Table 1. Comparison table of the value of n needed for a PMNS using the Montgomery-like reduction compared to one verifying the bounds for the translation equality test.

size of p in bits	256	512	1024	2048	4096	6144	8192
n for $\phi \geqslant 2w(\rho - 1)$	5	9	19	42	84	132	184
n for $\phi \geqslant \lceil 2w(\rho - 1)^2 \|\mathcal{G}^{-1}\|_1 \rceil$	5	10	21	48	99	168	232

As can be seen, while this is not very constraining for the smaller integer sizes, for larger integers one needs to choose an n that is 26% larger, which is both costly in terms of memory size and worse in terms of performance, given the fact that the time complexity of all operations depends on the value of n. There is then an obvious trade-off between the cost of normal modular multiplication and that of equality tests for this method, and since generally equality tests are not as frequently performed as modular multiplications, ideally, the growth of the parameter n should be limited.

4.3   Reduced Equality Test

For a general use-case we present here an equality test that does not require a larger bound on ϕ, unlike Algorithm 5. The main idea is to use the Barrett-like reduction to determine if $(A - B)$ is a linear combination of the lines of $\mathcal{G}$, a basis of $\mathfrak{L}$. Let $T \in \mathfrak{L}$, $\exists v \in \mathbb{Z}^n$ such that $T = v \cdot \mathcal{G}$ so $T \cdot \mathcal{G}^{-1} = v$. If $T \in \mathbb{Z}^n \setminus \mathfrak{L}$, $T \cdot \mathcal{G}^{-1} \in \frac{1}{\det(\mathcal{G})}\mathbb{Z}^n$ so $\exists v \in \mathbb{Z}^n$ such that $T = \frac{1}{p}v \cdot \mathcal{G}$ which means performing a rounding operation on each coefficient of $T \cdot \mathcal{G}^{-1}$ and then multiplying the result by $\mathcal{G}$ can confirm whether $T \in \mathfrak{L}$ or not. Actually computing $T \cdot \mathcal{G}^{-1}$ would be costly due to either requiring floating point operations or needing to perform operations with fractions with large numerators (since p is large) so we approximate it with powers of 2, in a similar manner to the Barrett-like reduction.

Algorithm 6. Reduced equality test

Require: $\mathcal{B}$ defined by (p, n, γ, ρ, E) a PMNS, $A, B \in \mathcal{B} \times \mathcal{B}$ thus $\|A\|_\infty < \rho$ and $\|B\|_\infty < \rho$, $\phi \in \mathbb{N}^*$ and $\mathcal{G}$ a short basis of $\mathfrak{L}$, $\mathcal{G}'$ pre-computed such that $\mathcal{G}' = \lfloor \phi\mathcal{G}^{-1} \rceil$.

Ensure: True if $A(\gamma) \equiv B(\gamma)$ (mod p) else False.

1: $C \leftarrow A(X) - B(X)$

2: $S \leftarrow \left\lfloor \frac{1}{\phi}C \cdot \mathcal{G}' \right\rceil \cdot \mathcal{G}$

3: $i \leftarrow 0$

4: **while** $i < n$ **and** $s_i = c_i$ **do**

5:     $i \leftarrow i + 1$

6: **end while**

7: **if** $i = n$ **then**

8:     **return** True

9: **else**

10:     **return** False

11: **end if**

Proposition 3. *If $\phi > 2n(\rho-1)$, then Algorithm 6 outputs True if $A(\gamma) \equiv B(\gamma)$ (mod p) and False otherwise.*

Proof. Since $\mathcal{G}' = \lfloor \phi\mathcal{G}^{-1} \rceil$, there exists $\varepsilon \in \mathcal{M}_n(\mathbb{R})$ such that $\forall i, j \in \{1, 2, \ldots, n\} \times \{1, 2, \ldots, n\}$, $|\varepsilon_{i,j}| \leqslant \frac{1}{2}$ and $\mathcal{G}' = \phi\mathcal{G}^{-1} + \varepsilon$. As a result we will have $\|\varepsilon\|_1 \leqslant \frac{n}{2}$. Thus,

$$\left\lfloor \frac{1}{\phi}C \cdot \mathcal{G}' \right\rceil = \left\lfloor \frac{1}{\phi}\left(C \cdot (\phi\mathcal{G}^{-1} + \varepsilon)\right) \right\rceil = \left\lfloor C \cdot \mathcal{G}^{-1} + \frac{1}{\phi}C \cdot \varepsilon \right\rceil.$$

Remark that $\|\frac{1}{\phi}C \cdot \varepsilon\|_\infty \leqslant \frac{n(\rho-1)}{\phi}$ because $\|C\|_\infty \leqslant 2(\rho - 1)$. However, we previously assumed that $\phi > 2n(\rho - 1)$. Hence, it follows that $\|\frac{1}{\phi}C \cdot \varepsilon\|_\infty < \frac{1}{2}$. If

$A(\gamma) \equiv B(\gamma) \pmod{p}$, then $C(\gamma) \equiv 0 \pmod{p}$ and $C \in \mathcal{L}$. Thus, $\exists v \in \mathbb{Z}^n$ such that $v \cdot \mathcal{G} = C$. In particular $v = C \cdot \mathcal{G}^{-1}$. Thus

$$\left\lfloor C \cdot \mathcal{G}^{-1} + \frac{1}{\phi} C \cdot \varepsilon \right\rceil = C \cdot \mathcal{G}^{-1} = v.$$

The last step compares C with $\left\lfloor C \cdot \mathcal{G}^{-1} + \frac{1}{\phi} C \cdot \varepsilon \right\rceil \cdot \mathcal{G}$ coefficient by coefficient and as such the algorithm returns True when $A(\gamma) \equiv B(\gamma) \pmod{p}$ and False otherwise.

Remark 4. Since $w \geqslant n$ for all E, the condition for this equality test is met when $\phi \geqslant 2w(\rho - 1)$ such as in the case of the Montgomery-like reduction.

4.4   Extended Equality Test

Another aspect of Algorithm 5 is that it allows for larger operands than normal. One would normally expect to have to test two operands in the PMNS and therefore each bounded by ρ in infinite norm but Algorithm 5 can accept operands up to $\frac{1}{2} w(\rho - 1)^2$ instead. Were this bound slightly larger, this algorithm could be used in the middle of operations before an internal reduction since, as noted in Sect. 2.2, the bound for that is $w(\rho - 1)^2$. However the bound for Algorithm 5 being half that instead means it cannot be used in that case. However, this opens up the possibility of an algorithm allowing an equality test even when the operands are as large as $w(\rho - 1)^2$. This is the motivation behind Algorithm 7.   This is essentially an adaptation of Algorithm 5 with the main

---

Algorithm 7. Plantard-like reduction equality test

Require: $A, B \in \mathbb{Z}_{n-1}[X] \times \mathbb{Z}_{n-1}[X]$ with $\|A\|_\infty \leqslant w(\rho - 1)^2$ and $\|B\|_\infty \leqslant w(\rho - 1)^2$,
$\quad \phi \in \mathbb{N}^*$ and $\mathcal{G}$ a short basis of $\mathcal{L}$ invertible modulo ϕ^2.
Ensure: $C' = 0 \iff A(\gamma) \equiv B(\gamma) \pmod{p}$.
  1: $C \leftarrow A(X) - B(X)$
  2: $C' \leftarrow$ **Plantard-like**(C)
  3: **return** C'

---

differences being the use of the Plantard-like internal reduction instead of the Montgomery-like internal reduction and the removal of the Translation vector in the process. This is important because this means that the larger bound on ϕ is no longer necessary. In exchange, this algorithm is slower than Algorithm 5 given that the Montgomery-like reduction is faster than the Plantard-like reduction. Similarly, as noted in Remark 3, the Plantard-like reduction must be performed using the basis of the full lattice so the time complexity is at best quadratic.

Proposition 4. *If $4w(\rho - 1)^2 + \phi\|\mathcal{G}\|_1 < \phi^2$ and $2w(\rho - 1)^2\|\mathcal{G}^{-1}\|_1 < \phi^2$, then Algorithm 7 returns 0 if and only if $A(\gamma) \equiv B(\gamma) \pmod{p}$.*

Proof. If $A(\gamma) \equiv B(\gamma) \pmod{p}$ then $\exists v \in \mathbb{Z}^n$ such that $v \cdot \mathcal{G} = A - B$. We first subtract B from A before applying Algorithm 2 to the result with $C(X) = A(X) - B(X)$ as the operand. Hence, when we compute $Q = -C \cdot \mathcal{G}^{-1} \bmod \phi^2$ at step 2 of Algorithm 2, we have $Q \equiv -v \pmod{\phi^2}$ and, since $v = (A-B) \cdot \mathcal{G}^{-1}$ and we assumed that $2w(\rho-1)^2 \|\mathcal{G}^{-1}\|_1 < \phi^2$, then $\|v\|_\infty < \phi^2$ and thus $Q = -v$. Then at step 2 of Algorithm 2, we have $\|S\|_\infty \leqslant \|\frac{1}{\phi}v\|_\infty + \frac{1}{2}$. Thus $\|C'\|_\infty \leqslant \|\frac{1}{\phi^2}v \cdot \mathcal{G}\|_\infty + \|\frac{1}{2\phi}\mathcal{G}\|_1 + \frac{1}{2}$ and $v \cdot \mathcal{G} = C$. Hence $\|C'\|_\infty \leqslant \|\frac{1}{\phi^2}C\|_\infty + \|\frac{1}{2\phi}\mathcal{G}\|_1 + \frac{1}{2}$. In particular, $\|C\|_\infty = \|A(X) - B(X)\|_\infty \leqslant 2w(\rho-1)^2$ and thus $\|C'\|_\infty \leqslant \frac{2w(\rho-1)^2}{\phi^2} + \frac{\|\mathcal{G}\|_1}{2\phi} + \frac{1}{2}$. In other words, $\|C'\|_\infty \leqslant \frac{4w(\rho-1)^2 + \phi\|\mathcal{G}\|_1}{2\phi^2} + \frac{1}{2}$ and since we assumed that $4w(\rho-1)^2 + \phi\|\mathcal{G}\|_1 < \phi^2$ then $\|C'\|_\infty < \frac{1}{2} + \frac{1}{2}$. Since $C' \in \mathbb{Z}[X]$, $C' = 0$.

If $A(\gamma) \not\equiv B(\gamma) \pmod{p}$, Algorithm 2 always outputs a non-zero polynomial. (see Proposition 1). $\qquad\blacksquare$

Remark 5. If $\phi \geqslant 2w(\rho-1)$ and $\rho \geqslant \|\mathcal{G}\|_1 - 1$ like in the case of the Montgomery-like internal reduction, then the conditions for this proposition are met. Indeed, $2w(\rho-1)^2\|\mathcal{G}^{-1}\|_1 < \phi^2$ since we always have $\|\mathcal{G}^{-1}\|_1 < 1$ in practice. But also:

$$\phi \geqslant 2w(\rho-1) \iff 2\phi(\rho-1) \geqslant 4w(\rho-1)^2$$
$$\iff 2\phi(\rho-1) + \phi\|\mathcal{G}\|_1 \geqslant 4w(\rho-1)^2 + \phi\|\mathcal{G}\|_1.$$

Then, $2\phi(\rho-1) + \phi\|\mathcal{G}\|_1 \leqslant 2\phi(\rho-1) + \phi(\rho+1) \Leftrightarrow 2\phi(\rho-1) + \phi\|\mathcal{G}\|_1 \leqslant \phi(3\rho-1)$. Since $n > 2$ and $w \geqslant n$, $\phi^2 > \phi(2w(\rho-1)) \geqslant \phi(3\rho-1)$ and thus the condition is indeed met. As such, there is no need to increase the bounds on the parameters.

Moreover, there are some early exit conditions that can be used to significantly speed up the algorithm in the majority of cases. This leads to slightly altered Algorithm 8.

Proposition 5. *If $\|A - B\|_\infty\|\mathcal{G}^{-1}\|_1 < \frac{\phi^2}{2\|\mathcal{G}\|_1}$ then Algorithm 8 never reaches step 9.*

Proof. First let us establish the following: $\|S\|_\infty < \frac{\phi}{2\|\mathcal{G}\|_1} \implies (A - B) \in \mathfrak{L}$, with S the value computed at step 3 of the algorithm. Indeed, if $\|S\|_\infty < \frac{\phi}{2\|\mathcal{G}\|_1}$ then, since $\|S \cdot \mathcal{G}\|_\infty \leqslant \|S\|_\infty\|\mathcal{G}\|_1$, we have $\|S \cdot \mathcal{G}\|_\infty < \frac{\phi}{2}$. Thus, clearly, we get that $\left\|\left\lfloor \frac{1}{\phi}S \cdot \mathcal{G}\right\rceil\right\|_\infty = 0$, thus Algorithm 7 returns 0. Using Proposition 4, we can conclude that $(A - B) \in \mathfrak{L}$.

Secondly, let us establish that $\|Q\|_\infty > \|A - B\|_\infty\|\mathcal{G}^{-1}\|_1 \implies C \notin \mathfrak{L}$ with Q the value computed at step 2 of the algorithm. As noted in the proof of Proposition 4, if $(A - B) \in \mathfrak{L}$, then $\exists v \in \mathbb{Z}^n$ such that $v \cdot \mathcal{G} = A - B$ and thus $Q = -v$. Then, $(A - B) \cdot \mathcal{G}^{-1} = v$. Using norm inequalities, we have

$$\|v\|_\infty = \|(A - B) \cdot \mathcal{G}^{-1}\|_\infty \leqslant \|(A - B)\|_\infty\|\mathcal{G}^{-1}\|_1$$

Thus, if $\|Q\|_\infty > \|A - B\|_\infty\|\mathcal{G}^{-1}\|_1$ then $\|Q\|_\infty > \|v\|_\infty$ and as such $Q \neq -v$ and we conclude that $C \notin \mathfrak{L}$. From there, we can conclude. If $\|A - B\|_\infty\|\mathcal{G}^{-1}\|_1 <$

Algorithm 8. Extended equality test

Require: $A, B \in \mathbb{Z}_{n-1}[X] \times \mathbb{Z}_{n-1}[X]$ with $\|A\|_\infty \leqslant w(\rho-1)^2$ and $\|B\|_\infty \leqslant w(\rho-1)^2$, $\phi \in \mathbb{N}^*$ and $\mathcal{G}$ a short basis of $\mathfrak{L}$ invertible modulo ϕ^2.
Ensure: True if $A(\gamma) \equiv B(\gamma) \pmod{p}$ else False.

1: $C \leftarrow A(X) - B(X)$
2: $Q \leftarrow -C \cdot \mathcal{G}^{-1} \bmod \phi^2$
3: $S \leftarrow \left\lfloor \frac{1}{\phi} Q \right\rceil$
4: **if** $\|S\|_\infty < \frac{\phi}{2\|\mathcal{G}\|_1}$ **then**
5:     **return** True
6: **else if** $\|Q\|_\infty > \|C\|_\infty \|\mathcal{G}^{-1}\|_1$ **then**
7:     **return** False
8: **end if**
9: $C' \leftarrow \left\lfloor \frac{1}{\phi} S \cdot \mathcal{G} \right\rceil$
10: **if** $C'(X) = 0$ **then**
11:     **return** True
12: **else**
13:     **return** False
14: **end if**

$\frac{\phi^2}{2\|\mathcal{G}\|_1}$, either $\|S\|_\infty < \frac{\phi}{2\|\mathcal{G}\|_1}$, which means $(A - B) \in \mathfrak{L}$, or $\|S\|_\infty \geqslant \frac{\phi}{2\|\mathcal{G}\|_1}$, in which case $Q \geqslant \frac{\phi^2}{2\|\mathcal{G}\|_1}$ which implies $\|Q\|_\infty > \|A - B\|_\infty \|\mathcal{G}^{-1}\|_1$ from which we conclude that $(A - B) \notin \mathfrak{L}$.

Remark 6. If A and B are both bounded by ρ, then, experimentally the condition is always met for the Montgomery-like PMNS parameters. Meanwhile, if A and B are instead both bounded by $w(\rho - 1)^2$ there are cases where this condition is not verified in practice, in which case we have to perform step 9 and further.

5   Performance

Clock cycle measurements are performed according to the following recommendation from the relevant Intel white paper [22] with slight adaptation (taking the median instead of the minimum measurement) :

- We deactivate the *Turbo-Boost®*
- We try to minimize potential cache misses by "heating" the cache memory with 501 runs that are not measured
- Then we generate 1001 appropriate data sets for which 501 runs are executed and the clock cycles are measured by interrogating the Time Stamp Counter with calls to the RDTSC instruction.
- The performance is the median value

The source code is available at https://github.com/francoispalma/PMNS/tree/master/equality_test

5.1   Inequality Case

In case of inequality, most of the algorithms can check whether the first coefficient being computed is correct, which means the other coefficients don't need to be computed. In particular, Algorithm 8 becomes incredibly fast for this sub-case (see Tables 2 and 3), making it ideal whenever detecting inequality is important. In all tables, Algorithm 5's performance is denoted as "–" whenever it cannot be performed for this value of n due to its increased bounds on parameters and is why two values of n are needed for each prime sizes above 256 bits, with the second value of n being the lowest value we have found where the translation test's bounds are met.

Table 2. Cycle count for equality tests in the inequality case on operands with each coefficient bounded by $(\rho - 1)$ for primes going from 256 to 8192 bits for each equality test method presented using gcc 12.3.0 on intel processor i9-11900KF, modular multiplication added for comparison

$\log_2(p)$	256	512		1024		2048	
n	5	9	10	20	21	42	48
Multiplication	109	337	391	1739	1759	6678	8308
Naive (alg. 4)	238	403	443	1327	1368	4594	5360
Translation (alg. 5)	47	-	197	-	571	-	2620
Reduced (alg. 6)	48	140	179	746	815	3280	4265
Extended (alg. 8)	**27**	**42**	46	**102**	102	**234**	250

$\log_2(p)$	4096		6144		8192	
n	84	99	132	168	184	232
Multiplication	22901	33533	52131	71721	109754	159108
Naive (alg. 4)	18375	22287	45154	57456	88338	78456
Translation (alg. 5)	-	12245	-	36729	-	116271
Reduced (alg. 6)	13454	19019	34330	55422	72996	144282
Extended (alg. 8)	**421**	480	**661**	933	**995**	1298

As noted earlier, tests on double-sized operands, before an internal reduction, are an interesting option to have available so we also benchmark it. For comparison, the methods that do not allow it perform an internal reduction before the test, the cost of which is added to the performance for fairness. Algorithm 5 could be performed for *some* of those without an internal reduction, in which case the performance is closer to the above tables, but we present the cost with the internal reduction here regardless since there is no guarantee the conditions will be met every time.

Table 3. Cycle count for equality tests in the inequality case on operands with each coefficient bounded by $w(\rho - 1)^2$ for primes going from 256 to 8192 bits for each equality test method presented using gcc 12.3.0 on intel processor i9-11900KF, modular multiplication added for comparison

$\log_2(p)$	256	512		1024		2048	
n	5	9	10	20	21	42	48
Multiplication	109	337	391	1739	1759	6678	8308
Naive (alg. 4)	296	690	727	2589	2580	9767	11567
Translation (alg. 5)	115	-	449	-	1733	-	8681
Reduced (alg. 6)	142	448	542	2037	2052	8242	10728
Extended (alg. 8)	**27**	**49**	56	**120**	121	**250**	280

$\log_2(p)$	4096		6144		8192	
n	84	99	132	168	184	232
Multiplication	22901	33533	52131	71721	109754	159108
Naive (alg. 4)	36279	47036	84037	112742	164476	192809
Translation (alg. 5)	-	36968	-	93962	-	233420
Reduced (alg. 6)	31465	43995	73409	113700	148721	255778
Extended (alg. 8)	**479**	561	**865**	1187	**1262**	1646

5.2 Equality Case

This case can be seen as either the performance for an equality test where the result is that there is equality or a constant-time code for each equality test, in case there is such a need.

As can be seen (Table 4), the Reduced equality test (Algorithm 6) presents itself as the best choice for PMNS on prime sizes going from 256 to 4096 bits. Meanwhile, due to the quadratic nature of these algorithms, on larger sizes the Naive algorithm, which becomes sub-quadratic, is better.

The Extended reduction algorithm meanwhile stays as the dominant choice whenever checking equality before an internal reduction is necessary (Table 5).

268   N. Méloni et al.

Table 4. Cycle count for equality tests in the equality case on operands with each coefficient bounded by $(\rho - 1)$ for primes going from 256 to 8192 bits for each equality test method presented using gcc 12.3.0 on intel processor i9-11900KF, modular multiplication added for comparison

$\log_2(p)$	256	512		1024		2048	
n	5	9	10	20	21	42	48
Multiplication	109	337	391	1739	1759	6678	8308
Naive (alg. 4)	223	419	460	1322	1376	**4655**	5380
Translation (alg. 5)	90	-	363	-	1389	-	7093
Reduced (alg. 6)	**67**	**226**	278	**1164**	1271	5094	6875
Extended (alg. 8)	113	319	395	1591	1742	6912	9039

$\log_2(p)$	4096		6144		8192	
n	84	99	132	168	184	232
Multiplication	22901	33533	52131	71721	109754	159108
Naive (alg. 4)	18583	21979	**45177**	56950	90329	**79073**
Translation (alg. 5)	-	30913	-	96647	-	267993
Reduced (alg. 6)	**13494**	21929	54792	94279	126136	275183
Extended (alg. 8)	27265	37015	68571	132813	158711	260780

Table 5. Cycle count for equality tests in the equality case on operands with each coefficient bounded by $w(\rho - 1)^2$ for primes going from 256 to 8192 bits for each equality test method presented using gcc 12.3.0 on intel processor i9-11900KF, modular multiplication added for comparison

$\log_2(p)$	256	512		1024		2048	
n	5	9	10	20	21	42	48
Multiplication	109	337	391	1739	1759	6678	8308
Naive (alg. 4)	293	676	724	2589	2585	9849	11633
Translation (alg. 5)	153	-	610	-	2560	-	13334
Reduced (alg. 6)	134	434	463	1995	2507	10123	13503
Extended (alg. 8)	**116**	**342**	411	**1522**	1544	**6236**	8556

$\log_2(p)$	4096		6144		8192	
n	84	99	132	168	184	232
Multiplication	22901	33533	52131	71721	109754	159108
Naive (alg. 4)	36560	47060	84241	113194	167162	192897
Translation (alg. 5)	-	55561	-	153392	-	381941
Reduced (alg. 6)	39862	54941	95095	151260	202213	387937
Extended (alg. 8)	**25798**	34740	**71223**	138069	**164919**	278998

6   Conclusion

In this paper we have showcased a new internal reduction algorithm for PMNS inspired by the Plantard modular reduction on integers and used it to construct a very efficient equality test for PMNS in the inequality case and to be used before an internal reduction. We have also proposed an equality test for a more normal use case which nevertheless cannot compete with naive conversion to binary on larger prime sizes (6144 bits and further).

Funding This work has been partially funded by TPM Metropol (AAP2024-CALVERE and AAP2025-APACHE projects).

References

1. Bajard, J.C., Marrez, J., Plantard, T., Véron, P.: On polynomial modular number systems over Z/pZ. Adv. Math. Commun. **18**(3), 674–695 (2024). https://doi.org/10.3934/amc.2022018. URL: https://hal.science/hal-03611829
2. Bouvier, C., Imbert, L.: An alternative approach for SIDH arithmetic. In: Garay, J.A. (ed.) PKC 2021. LNCS, vol. 12710, pp. 27–44. Springer, Cham (2021). https://doi.org/10.1007/978-3-030-75245-3_2
3. Cohen, H.: A Course in Computational Algebraic Number Theory, vol. 138. Graduate Texts in Mathematics. Springer, Heidelberg (1993)
4. Coladon, T., Elbaz-Vincent, P., Hugounenq, C.: MPHELL: a fast and robust library with unified and versatile arithmetics for elliptic curves cryptography.. In: 28th IEEE Symposium on Computer Arithmetic, ARITH 2021, Lyngby, Denmark, 14–16 June 2021, pp. 78– 85. IEEE (2021). https://doi.org/10.1109/ARITH51176.2021.0002
5. Didier, L.-S., Dosso, F.Y., Véron, P.: Efficient modular operations using the Adapted Modular Number System. J. Crypto. Eng. 1–23 (2020)
6. Fangan Yssouf Dosso: Contribution de l'arithmétique des ordinateurs aux implémentations résistantes aux attaques par canaux auxiliaires. Université de Toulon, Theses (2020)
7. Dosso, F.Y., Berzati, A., El Mrabet, N., Proy, J.: Redundancy and equality test in the PMNS, application to Elliptic Curve Diffie–Hellman. Cryptology ePrint Archive, Paper 2023/1231 (2023). https://eprint.iacr.org/2023/1231
8. Dosso, F.Y., Duquesne, S., El Mrabet, N., Gautier, E.: PMNS arithmetic for elliptic curve cryptography. Cryptology ePrint Archive, Paper 2025/467 (2025). https://eprint.iacr.org/2025/467
9. Dosso, F.Y., El Mrabet, N., Méloni, N., Palma, F., Véron, P.: Friendly primes for efficient modular arithmetic using the Polynomial Modular Number System. Cryptology ePrint Archive, Paper 2025/090 (2025). https://eprint.iacr.org/2025/090
10. Dosso, F.Y., Robert, J.M., Véron, P.: PMNS for efficient arithmetic and small memory cost. IEEE Trans. Emerg. Topics Comput. **10**(3), 1263–1277 (2022). https://hal.science/hal-03768546v1/file/TETC3187786.pdf
11. ElGamal, T.: A public key cryptosystem and a signature scheme based on discrete logarithms. In: Blakley, G.R., Chaum, D. (eds.) CRYPTO 1984. LNCS, vol. 196, pp. 10–18. Springer, Heidelberg (1985). https://doi.org/10.1007/3-540-39568-7_2
12. Hermite, C.: Sur l'introduction des variables continues dans la théorie des nombres. fre. J. für die reine und angewandte Mathematik **41**, 191–216 (1851)

13. Lenstra, A.K., Lenstra, H.W., Lovász, L.: Factoring polynomials with rational coefficients. Math. Ann. **261**, 515–534 (1982)
14. Méloni, N., Palma, F., Véron, P.: Multi-precision PMNS with CIOS reduction. In: Selected Areas in Cryptography. Springer, Toronto (2025). https://hal.science/hal-05144945
15. Méloni, N., Palma, F., Véron, P.: PMNS for cryptography: a guided tour. Adv. Math. Commun. (2023). https://doi.org/10.3934/amc.2023033. https://hal.science/hal-04195613
16. Montgomery, P.L.: Modular multiplication without trial division. Math. Comput. **44**(170), 519–521 (1985)
17. National Institute of Standards and Technology. Digital Signature Standard (DSS). FIPS PUB 186-4. NIST (2013). https://doi.org/10.6028/NIST.FIPS.186-4
18. National Institute of Standards and Technology. Module-Lattice-Based Key-Encapsulation Mechanism Standard. Technical report. Federal Information Processing Standards Publications (FIPS) 203. U.S. Department of Commerce, Washington, D.C. (2024). https://doi.org/10.6028/NIST.FIPS.203
19. Negre, C., Plantard, T.: Efficient modular arithmetic in adapted modular number system using lagrange representation. In: Mu, Y., Susilo, W., Seberry, J. (eds.) ACISP 2008. LNCS, vol. 5107, pp. 463–477. Springer, Heidelberg (2008). https://doi.org/10.1007/978-3-540-70500-0_34
20. Noyez, L., El Mrabet, N., Potin, O., Véron, P.: Modular multiplication in the AMNS representation: hardware implementation. In: Selected Areas in Cryptography, Montréal (Québec), France (2024)
21. OpenSSL (1998). https://www.openssl.org/. Accessed 07 July 2022
22. Paoloni, G.: White paper: How to Benchmark Code Execution Times on Intel® IA-32 and IA-64 Instruction Set Architectures. Intel Corporation, Technical report (2010)
23. Plantard, T.: Efficient word size modular arithmetic. IEEE Trans. Emerg. Top. Comput. **9**(3), 1506–1518 (2021). https://doi.org/10.1109/TETC.2021.3073475
24. Schnorr, C.-P.: Efficient signature generation by smart cards. J. Cryptol. **4**(3), 161–174 (1991)
25. Schnorr, C.P., Euchner, M.: Lattice basis reduction: improved practical algorithms and solving subset sum problems. In: International Symposium on Fundamentals of Computation Theory, pp. 68–85. Springer, Heidelberg (1991)
26. Smith, H.J.S.: On systems of linear indeterminate equations and congruences. Phil. Trans. R. Soc. Lond. **151**, 293–326 (1861)
27. The GNU Multiple Precision Arithmetic Library (GMP) (1991). https://gmplib.org/. Accessed 07 July 2022

Period Counting Versus Direct Sampling in Oscillator-Based TRNGs: Architectures and Characteristics

Miguel Alcocer[1], Nathalie Bochard[2], Viktor Fischer[2,3], Ana I. Gómez[1(✉)], and Domingo Gómez-Pérez[4]

[1] Universidad Rey Juan Carlos, Móstoles, Spain
m.alcocer.2022@alumnos.urjc.es, ana.gomez.perez@urjc.es
[2] Laboratoire Hubert Curien UMR 5516 Université Jean Monnet Saint-Etienne CNRS, Saint Etienne, France
{nathalie.bochard,fischer}@univ-st-etienne.fr
[3] Faculty of Information Technology, Czech Technical University in Prague, Thakurova 9, 160 00 Prague 6, Czechia
[4] Universidad de Cantabria, Santander, Spain
domingo.gomez@unican.es

Abstract. Many True Random Number Generators (TRNGs) extract randomness from a jittered clock generated by an oscillator either by directly sampling this clock (in sampling mode) or by counting the number of samples equal to one during a reference interval (in counting mode). This architectural choice has a decisive impact on the statistical quality of the output bit stream. Previous works suggest that counting provides an advantage over sampling mode, arguing that jitter accumulation over many oscillator periods yields higher entropy per output bit. The aim of this work is to study both methods more closely using a unified framework of standard random tests. We analyze experimental data acquired from the two TRNG configurations implemented in a Field Programmable Gate Array (FPGA) in self-timed rings (STR). The results show that the sampling method produces higher correlation values that are consistent with other tests: Markov chains, run length, balance, etc. Additionally, we provide results over free running oscillator-based TRNG architectures and we derive experimental limits of the parameters of the two TRNG configurations.

Keywords: Hardware Random Number Generator · Ring Oscillator · Period Counting · Sampling · Correlation Measure · Markov Chains · Statistical Tests

1 Introduction

Random number generators (RNGs) are essential in cryptography. They are used to generate cryptographic keys, nonces, and padding values in cryptographic protocols. Recently, high quality random numbers are also required in countermeasures again side-channel attacks, for example, using random numbers as masking

© The Author(s), under exclusive license to Springer Nature Switzerland AG 2026
L. Batina and F. Özbudak (Eds.): WAIFI 2026, LNCS 16611, pp. 271–283, 2026.
https://doi.org/10.1007/978-3-032-27574-5_17

to protect against differential power analysis [6]. Security in cryptographic applications is guaranteed by good statistical properties and the unpredictability of exploited random numbers. Unpredictability, which is characterized by an output entropy rate, is usually achieved using some physical random process in the so-called True Random Number Generators (TRNGs).

The entropy at the generator output should be estimated by a stochastic model and verified online using dedicated statistical tests [7]. In addition to security, TRNGs should fulfill several practical requirements: as a part of a cryptographic system-on-chip (SoC), they should be implementable in logic devices such as Application Specific Integrated Circuits (ASICs) and Field Programmable Gate Arrays (FPGAs), should occupy small area and feature sufficiently high bit rate at the output [4,9].

TRNGs using self-timed rings (STR-TRNGs) are popular due to their performance and efficient implementation, which exploits jitter of the clock generated in the freely running oscillator as an entropy source.

Two principal architectures of randomness extraction from a jittered clock signal have been proposed: the method based on *sampling* of the jittered clock signal [4] and the principle based on *counting* of samples equal to one [2].

In the architecture based on sampling, a D flip-flop (DFF) samples the output of one oscillator after a jitter accumulation period at the clock edge of the reference clock generated by a second oscillator. In the second architecture based on counting, a digital counter counts the number of samples of the jittered clock signal during the jitter accumulation period determined by the reference clock, and the Least Significant Bit (LSB) of the counter is used as the output bit.

The theoretical literature has gradually favored counting over sampling [2,9], arguing that the integration of jitter over many oscillator periods yields higher entropy per output bit. However, a rigorous, measurement-driven comparison of both architectures on the same hardware under a common mathematical pseudorandomness framework has not yet been carried out.

In this paper, we evaluate outputs of the two RNG architectures based on sampling and counting using different randomness tests, specifically Markov Chains and Correlation Measure. Additionally, we characterize the boundaries for applicable parameters for a Self Timed Ring (STR) architecture.

The paper is organized as follows. In Sect. 2, we present the mathematical background, which is needed to understand the approach. In Sect. 3, we describe in detail the hardware architecture of the two randomness extraction methods. In Sect. 4, we present the experimental results. In Sect. 5, we discuss the obtained results, and in Sect. 6 we conclude the paper and present future steps.

2   Theoretical Background

We denote by $S = (s_1, \ldots, s_N)$ a binary sequence of length N, i.e. $s_i \in \{0,1\}$ for $i = 1, \ldots, N$. For $d \in \{0,1\}$, we introduce the notation $P(s_i = d)$ – the cardinality of $\#\{i \mid s_i = d\}$ divided by N. This can be seen as an empirical approximation of the probability of ones and zeros. For $c \in \{0,1\}^k, d \in \{0,1\}$, we define also the conditional transition probability

$$P(s_i = d | (s_{i-1}, \ldots, s_{i-k}) = \mathbf{c}) = \frac{\#\{i \mid (s_i, \ldots, s_{i-k}) = (d, \mathbf{c})\}}{\#\{i \mid (s_{i-1}, \ldots, s_{i-k}) = \mathbf{c}\}}$$

as the number of occurrences in S of the values $(d, \mathbf{c})$ divided by the number of occurrences of $\mathbf{c}$.

Example 1. Take $S = (1,0,0,1,0,0,1,1)$, $d = 1$ and $\mathbf{c} = (0,0)$.
The values of $P(s_i = d) = 1/2$,

$$P(s_i = d \mid (s_{i-1}, \ldots, s_{i-k}) = \mathbf{c}) = 2/2 = 1.$$

This already shows an imbalance in the conditional transition probability, as the values should be close to $1/2$.

Let us first recall the definition of some popular statistical tests that we shall exploit in this paper.

The k-th order correlation of S introduced by Mauduit-Sárközy [10] is

$$C_k(S) = \max_{D,M} \frac{1}{M} \left| \sum_{n=1}^{M} (-1)^{s_n + s_{n+d_1} + \cdots + s_{n+d_{k-1}}} \right|, \tag{1}$$

where the maximum is over all $1 \leq M \leq N$ distinct lag vectors $D = (d_1, \ldots, d_{k-1})$ with $0 < d_1 < \cdots < d_{k-1} < N - M$, and indices are taken modulo N. In this paper, we focus on $C_2(S)$ for $k = 2$, as the normalized off peak autocorrelation. Random sequences are expected to satisfy $C_k(S) \ll N$, in other case, we say that S show *peaks* in the correlation measure.

Alon *et al.* in [3, Theorem B] established this result in a formal way. For a random sequence of length N, the following bound holds

$$C_k(S) \leq 5\sqrt{\frac{k \ln N}{N}} \tag{2}$$

with probability close to one as $N \to \infty$. We will denote this bound $\mathcal{B}_k(N)$. For $k = 2$, Schmidt [11] tightened the constant in (2) to $\sqrt{2} \approx 1.41$, but the probability estimate is different and more difficult to calculate. Therefore, we will work with the bound presented in Eq. (2) as our working threshold. Finally, we introduce the following notation for the autocorrelation function

$$R(S, d) = \frac{1}{N-d} \sum_{i=0}^{N} (-1)^{s_i + s_{i+d}}. \tag{3}$$

This represents the correlation measure of the second order (order of 2) for the whole sequence, comparing with the calculation in smaller windows as in Eq. (1). The *normalized balance* of S is

$$B(S) = \frac{1}{N} \sum_{i=1}^{N} s_i - \frac{1}{2} = P(s_i = 1) - \frac{1}{2}. \tag{4}$$

An ideal binary source has $B(S) = 0$.

274   M. Alcocer et al.

Another important measure is the k-bit Markov chain by computing all 2^k conditional transition probabilities $P(s_i \mid (s_{i-1}, \ldots, s_{i-k}) = \mathbf{c})$. An ideal i.i.d. source satisfies $P(s_i = 1 \mid \cdot) = 0.5$ for all contexts. The maximum deviation from this ideal value across all contexts is

$$\Delta_k = \max_{\mathbf{c} \in \{0,1\}^k} \left| P(s_n = 1 \mid \mathbf{c}) - 0.5 \right|. \tag{5}$$

A key structural property of C_k that relates to Δ_k is that it controls the equidistribution of the histogram of k consecutive values in S: for any pattern $\mathbf{c} \in \{0,1\}^k$, the relative frequency deviation satisfies

$$\Gamma_k = \max_{\mathbf{c} \in \{0,1\}^k} \left| \frac{\#\{i \mid (s_{i-1}, \ldots, s_{i-k}) = \mathbf{c}\}}{N} - \frac{1}{2^k} \right| \leq C_k(S), \tag{6}$$

for any S [10, Theorem B]. The correlation measure of order k captures notions studied in different statistical tests. For example $k = 1$ recovers the bit balance, while the values $k = 2, 3, 4$ represent the equidistribution for pairs, triples, and quadruples. They detect patterns that lower order correlations cannot detect. This motivates evaluating C_k for several values of k simultaneously. Recently, the relationship between Δ_k and $C_k(S)$ has been studied, in the sense that the latter can provide an upper bound on the previous [1].

3   Hardware Setup

In this section, we present the hardware architectures used to acquire the data analyzed in this paper.

3.1   Source of Randomness

We use jittered clock signals generated by STRs [5] as a source of randomness.

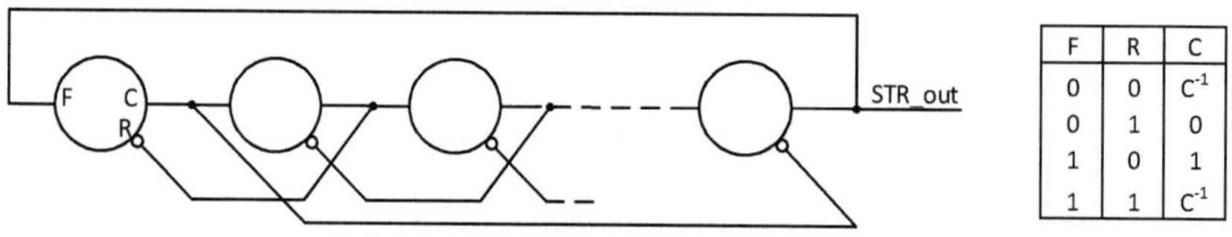

Fig. 1. Self-timed ring architecture and truth table

STRs consist of a set of L Muller cells interconnected in a loop via a handshake protocol determined by the truth table of the Muller cell presented in Fig. 1, where F_i is the forward input, R_i the reverse input, and C_i is the output of the i-th cell. Compared to commonly used ring oscillators, in which only one event (rising or falling edge) moves across the ring, multiple signal edges can circulate in the STRs at the same time. The output frequency of the ring depends on the time that the edge needs to pass across the ring and also on the number of edges present in the ring. The advantage of such a structure over the ring oscillators is that the frequency does not automatically decrease when the size of the ring increases.

Our ultimate goal is to leverage these multiple edges to extract several bits at the same time. However, this feature is not exploited in this paper: to simplify our study, we observe only one STR output as a first step. In this way, the behavior of the STR is close to that of a ring oscillator. Therefore, the conclusions of this paper can be extended to any free running oscillator-based architecture.

3.2   Randomness Extraction from the Jittered Clock

We compare two randomness extraction methods, i.e. time-to-digital conversions. The first method based on sampling of a jittered clock signal is presented in Fig. 2. It uses a D flip-flop to sample the signal coming from one free running oscillator using the clock signal generated by the second oscillator. The output RNG signal is obtained after a time period determined by a period counter. This jitter accumulation time guarantees uncertainty in the output sampled value.

The second method enabling efficient randomness extraction is based on counting the number of samples equal to one during the same jitter accumulation period, with the output value being the least significant bit of the counter, as it is presented in Fig. 3.

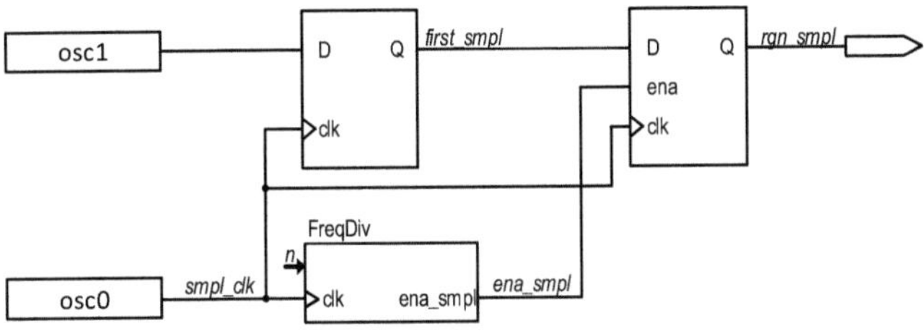

Fig. 2. Elementary oscillator-based TRNG (EO-TRNG) with entropy extraction based on sampling of a jittered clock signal

The two structures were implemented in an Intel Cyclone V FPGA using a hardware platform dedicated to TRNG testing (see [8] for more details). The random bits generated at the *rng_smpl* and *rng_cnt* outputs are stored in an internal memory of the device and transmitted to a PC via a USB connection.

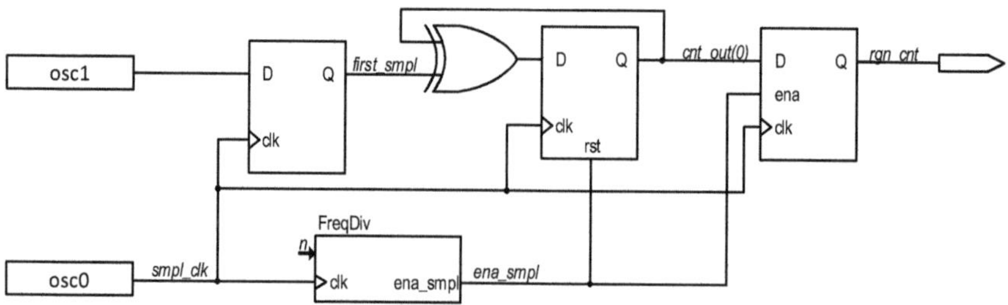

Fig. 3. Oscillator-based TRNG with entropy extraction based on counting samples of a jittered clock signal equal to one

In both cases, the physical source of randomness used to generate truly random numbers is the phase noise of the two freely running oscillators (STRs). This phase noise is the temporal manifestation of electronic noises present at the transistor level. Electronic noise is a mixture of various noise sources with different origins: some of them have global impact to the entire electronic circuit, such as noise coming from the power supply or noises related to processed data, while others are generated locally inside the oscillator.

Global noises should not be used as a source of randomness, as they can be manipulated by a potential attacker. To reduce impact of the global noises on the output signal, we exploit a differential principle: using two equivalent rings, the clock signal used as a time reference in the RNG undergoes the same global variations as the signal from which the phase noise is extracted. In other words, the reference clock signal features the same global variations as the sampled signal, while local noise variations (inside the two rings) remain independent. According to this principle, the impact of global noises on one ring is compensated by the influence of the same global noises on the second ring. The extracted digital noise is thus caused essentially by the local noises existing inside the two oscillators.

Local noise also has several sources that are not statistically equivalent. Among the most common types of noise, the thermal noise comes from the thermal agitation of electrons. This noise is present in all electronic components, it is well understood, and easy to model. It is commonly assumed that its statistical characteristics follow a Gaussian distribution.

Another prevalent noise in electronic circuits is the flicker noise, a low-frequency noise related to impurities present in semiconductor materials and the trapping of charge carriers. Its impact increases with the reduction of integration technology (higher integration density). Unlike thermal noise, flicker noise is more difficult to model. First, it is highly correlated, which contradicts the principle of independence between successive generated bits. Second, its physical characteristics are more difficult to quantify and incorporate into model.

Taking into account contribution of the flicker to the final entropy rate could help to increase output bit rate. However, addressing flicker noise is one of the challenges for this type of generator and it is beyond the scope of this paper.

Here, we focus on analyzing the contribution of thermal noises in free running oscillators, while the main goal of this paper is to compare two methods for extracting phase noise in differential structures based on integrated, free running oscillators.

3.3 Details of the Implemented Design

The experiments presented here are composed of two STR structures of 33 Muller cells.

- *Osc0*: $f_0 \approx 137\,\text{MHz}$
- *Osc1*: $f_1 \approx 192\,\text{MHz}$
- Ratio: $r = f_1/f_0 \approx 1.4015$

Other important intermediate signals are (see Fig. 2 and Fig. 3, respectively):

- *first_smpl*: the first D flip-flop samples *Osc1* signal at rising edges of the *Osc0* reference clock to give an intermediate state without accumulation of jitter. This signal is used as baseline reference for both methods. If the process is stationary and the two oscillators are independent, we obtain a pattern that repeats over time, the shape of which depends on the ratio of the two frequencies.
- *cnt_out(0)*: in the counter mode, the XOR gate and the flip-flop it feeds constitute the last stage of a counter. Placed after the *first_smpl* signal, it counts samples equal to one.
- *ena_smpl*: output of the frequency divider *FreqDiv* that generates an enable signal every n clock period to capture a random bit after a chosen accumulation time. The longer the accumulation time (n), the greater the entropy, but the lower the generator bit rate.
- *rng_smpl* (resp. *rng_cnt*): output of the generator obtained by subsampling the *first_smpl* (resp. *cnt_out(0)*) signal. It gives a new bit after each n clock period (jitter accumulation time).

4 Experimental Results

To validate our hypothesis, binary sequences of length $N = 250\,000$ bytes were acquired on the hardware platform, due to memory restrictions on the acquisition equipment.

We denote S_n^s the binary sequence given by the output signal *rgn_smpl* following the schematics in Fig. 2 with accumulation time n. Similarly, we also consider the binary sequence S_n^c given by the output signal *rgn_cnt* following the schematics in Fig. 3 with accumulation time n. For our experiments, we restrict the accumulation time n to the following set of values $\{150, 300, 600, 1200, 2400\}$.

On the one hand, higher values of n cause a lower output bit rate. On the other hand, smaller values of n cause a smaller entropy rate per bit and increase the dependence between the output bits. Therefore, we analyze the quality of the resulting binary sequences with respect to n, focusing on smaller values. For this purpose, we use the binary measures presented in Sect. 2.

4.1   Accumulation Time

For each binary sequence, we calculate the normalized second order correlation C_2 as defined in Eq. 2 for each value of n, along with $\mathcal{B}_2(N)$. Figure 4 contains the experimental runs for different values of n. Results show very different behaviors with respect to the correlation measure of the second order. While both methods present peaks in the correlation measure of the second order, these peaks are different in magnitude. Both $C_2(S_n^c)$ and $C_2(S_n^s)$ decrease exponentially with respect to n until they reach a certain threshold. Another observation is that $C_2(S_n^s)$ hints non random behavior, as it is greater than $\mathcal{B}_2(N)$ until $n = 2400$, whereas $C_2(S_n^c) \leq \mathcal{B}_2(N)$ for $n \geq 300$.

These findings were further validated and displayed in Fig. 5 on three additional hardware configurations with frequency ratios $r \approx 1.434$, $r \approx 1.496$, and $r \approx 1.616$, using sequences of length $N = 8 \times 10^6$ bits. Across all three configurations, $C_2(S_n^c)$ drops below $\mathcal{B}_2(N)$, confirming the counting mode behavior. Notably, $C_2(S_n^s)$ converges to a persistent plateau at or above $\mathcal{B}_2(N)$, revealing that the output does not pass the correlation bound regardless of the accumulation time.

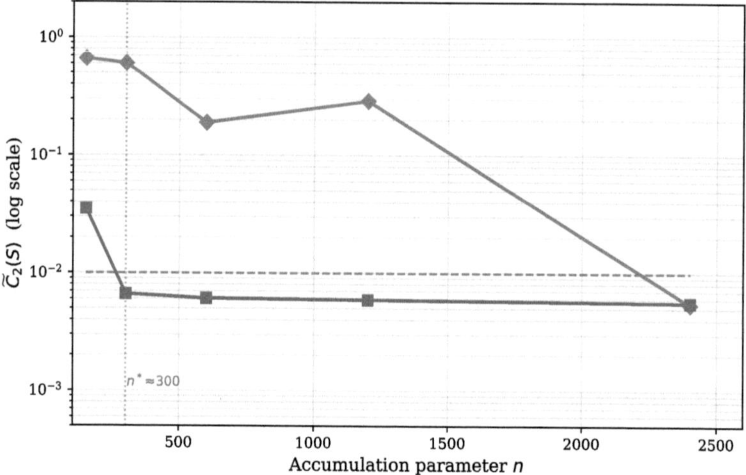

Fig. 4. C_2 (log scale) vs. n. Correlation measure of sequence S_n^c, S_n^s are represented by blue squares and purple diamonds, respectively. The dashed line represents correlation bound $\mathcal{B}_2(N)$ for the corresponding output length $N = 250\,000$. (Color figure online)

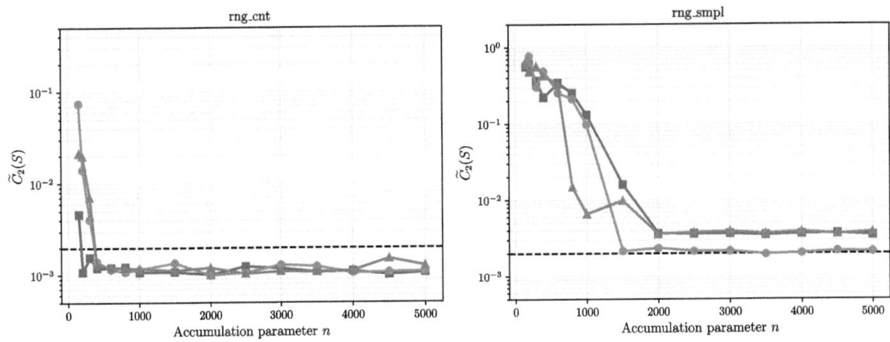

Fig. 5. C_2 (log scale) vs. n for three hardware configurations with frequency ratios $r \approx 1.434$ (blue squares), $r \approx 1.496$ (orange circles), and $r \approx 1.616$ (green triangles). Left panel: sequences; right panel: rng_smpl. The dashed line represents $\mathcal{B}_2(N)$ for $N = 8 \times 10^6$ bits. (Color figure online)

To confirm the previous results, we model S_n^s, S_n^c using Markov chains. For this purpose, it is sufficient to study the joint and conditional probabilities (see Eq. (6) and Eq. (5), respectively). As expected, the deviations from ideal joint and conditional probabilities decrease monotonically with n. At $n = 150$ the conditional deviation $\Delta_4 = 0.066 > 0.05$, exceeding the 5% tolerance threshold. For $n \geq 300$ the deviations for S_n^c lie below this threshold for any $k \leq 4$. As a summary, for $n \geq 300$ and $k = 1, 2, 3, 4$, $\Gamma_k < 0.010$ and $\Delta_k < 0.015$ are below the required tolerance. The deviations for S_n^s present values above the required tolerance until $n = 2400$ for several values of k. Figure 6 shows the deviation from the expected distribution of a random sequence in logarithm scale for Markov joint probability deviation (Γ_k) and Markov conditional probability deviation (Δ_k)

4.2   Edge Case: $n = 300$

As mentioned earlier, it is interesting to investigate separately the first case with a minimum accumulation time n. Previous observations suggest that accumulation time $n \geq 300$ is necessary for this model to pass the statistical tests in the counter design. In this subsection, we will analyze this particular case using different statistics.

First, both binary sequences can be considered without significant bias, due to the balance values $B(S_{300}^s) = 0.014$ and $B(S_{300}^c) = 0.001$.

The *number of k-runs* is the number of consecutive ones of length k. These values should follow a binomial distribution. This means that it is expected that half of the runs have length 1, a quarter of them length 2, and so on. In general, length k is expected for 2^{-k} of them.

For both S_{300}^s and S_{300}^c, the distribution of number of k- approximates the binomial distribution, with shorter runs of average 1.89 REF cycles, that is con-

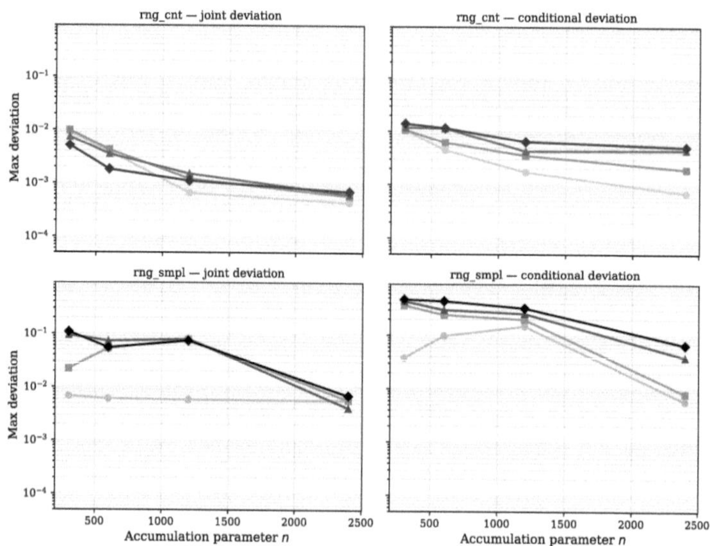

Fig. 6. Markov chain deviation for $S^c_{n \geq 300}$ (top row) and $S^s_{n \geq 300}$ (bottom row) (log scale) versus n. Left panel depicts the maximum joint probability deviation Γ_k, while the right panel depicts the maximum conditional probability deviation Δ_k. The values of the Markov chains order ($k = 1, 2, 3, 4$) are represented by circles, squares, triangles and diamonds with progressively darker shades, respectively.

sistent with the ratio $f_1/f_0 \approx 1.408$. However, S^s_{300} deviates for the expected value 2^{-k}, for $k \geq 1$ with the maximum deviation being at $k = 1$. This comparison is summarized in Fig. 7.

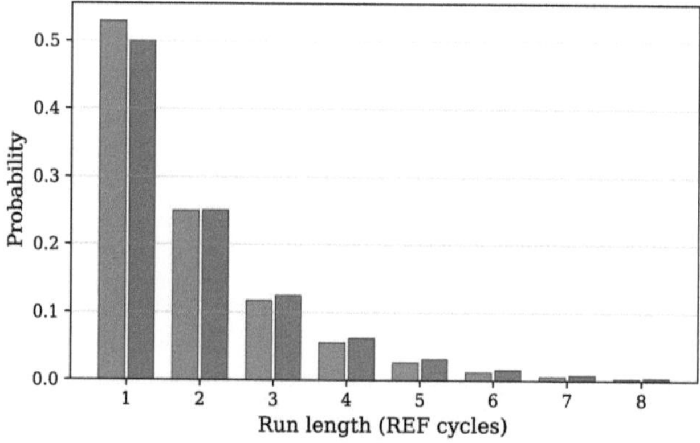

Fig. 7. Proportion of k-runs versus run length k for the binary sequence S^s_{300} (purple) and S^c_{300} (blue). (Color figure online)

We examine the autocorrelation function $R(S, d)$ as defined in Eq. (3). This function is expected to have a constant value without visible peaks. However, the values of $|R(S^s_{300}, d)|$ for $1 < d < 200$ have peaks of the order of 0.60 at $d = 2$, which is close to $\{f_1/f_0\} \approx 2.2$. Additionally, values for $d < 10$ are greater than $\mathcal{B}_2(N) = 0.05$. In contrast, $|R(S^c_{300}, d)|$ does not present significant peaks. Figure 8 summarizes the obtained results.

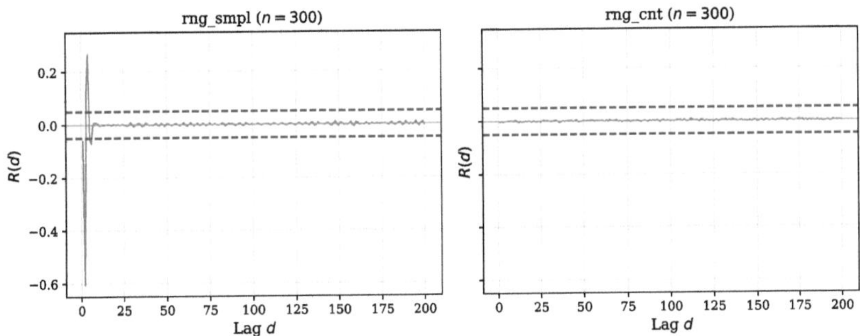

Fig. 8. Autocorrelation function $R(S, d)$ versus lag d. On the left $R(S^s_{300}, d)$ (purple) and on the right $|R(S^c_{300}, d)|$ (blue). Dotted lines represent bound $\mathcal{B}_2(N)$ for $N = 250\,000$ (Color figure online)

Finally, the correlation measure of order k can be used as a randomness test by comparing with the bound in Eq. 2. On the one hand $C_k(S^c_{300}) < \mathcal{B}_k(N)$ for $k \in \{2, 3, 4\}$, thus passing this test. On the other hand $C_k(S^s_{300}) > \mathcal{B}_k(N)$ for $k \in \{2, 4\}$. The specific values for the correlation measure of order k are displayed in Table 1.

Table 1. Values of correlation measure of order k, $C_k(S)$, compared with the $\mathcal{B}_k(N)$ for $k \in \{2, 3, 4\}$ and $N = 250\,000$.

	S^s_{300}			S^c_{300}		
k	C_k	$\mathcal{B}_k$	Accept?	C_k	$\mathcal{B}_k$	Accept?
2	0.604	0.050	✗ REJECT	0.009	0.050	✓ OK
3	0.031	0.061	✓ OK	0.008	0.061	✓ OK
4	0.507	0.071	✗ REJECT	0.010	0.071	✓ OK

5 Discussion

The experiments were carried out on the same device and hardware platform to allow a fair comparison of the quality of the binary output. The results confirm previous expectations, more specifically: First, by increasing the accumulation time n we reduce the correlation on the output, thus reducing unwanted patterns in C_2 as seen in Fig. 4. Second, the counting mode should be preferred over the sampling one when designing oscillator-based TRNGs, as the autocorrelation of the output requires higher accumulation time for sampling in order to obtain comparable results. The counting mode with an accumulation parameter $n \geq 300$ achieves $C_2 \approx 0.009$ below the theoretical bound, whereas the sampling mode requires $n \geq 2100$ for similar values (for $N = 250\,000$ bits). With larger sequences ($N = 8 \times 10^6$ bits), $C_2(S_n^s)$ converges to $\mathcal{B}_2(N)$, i.e., the output fails the correlation test regardless of the accumulation time. Finally, in the counting mode, generated data pass also Markov chains tests, as deviations are not statistically significant for memory depths $k \leq 4$ in Fig. 6.

We have further studied the edge case $n = 300$, that shows that the counting mode performs better that the sampling mode in the performed tests of randomness (Balance, k run length, autocorrelation and higher order correlation of order k). However, some tests are not sufficiently sensitive to detect patterns in the sampling mode as seen in Table 1. For these experiments, sampling mode produces sequences that present structure: $C_2 \approx 0.604$, with peaks present for small d.

6   Conclusions

As a summary these results confirm the theoretical predictions of Allini et al. [2] and Lubicz and Fischer [9], and provide thresholds based on experiments on the accumulation time when counting mode is used ($n \approx 300$). The benefits from this results are two-fold: they show bounds on the bit rate of this generators as well as it helps practitioners to design secure TRNGs.

Although these are initial results, these experimental results open several questions to deepen our understanding of this family of TRNG designs. As the analysis has been performed only on the output, further experiments should be done with internal signals to confirm that this behavior is consistent over different architectures.It remains as future work to reproduce the experiments across different hardware boards, with similar environmental conditions (temperature, voltage,..), due to the time-consuming nature of the setup of the experiments.

Our opinion is that correlation measure has not been considered among practitioners as they relied on several statistical independent tests, each of them measuring a different type of statistical defect. We believe that correlation measure is sensitive enough for many of this defects and can help to discard in earlier stages of hardware designs, only studying this figure of merit. Initial theoretical results support our conclusions [1], however a full validation is still a work in progress

Currently we are evaluating one output of the design, however taking into account several outputs can increase the performance on the TRNG. One unsolved question is what information gives the cross-correlation measure of the outputs about the underlying hardware of the TRNG. In particular, if values of the cross correlation give information about length of the ring oscillators, number of stages, etc.

Finally, some theoretical analysis of the relationships between the stochastic model, the design parameters and the correlation measure should be done in the future to validate the entropy per bit at the output. We believe that our results provide a first step towards formalization of the stochastic model of the generator based on counting.

Acknowledgments. M. Alcocer, A.I Gómez and D. Gómez were financially supported by grant PID2023-151238OA-I00 funded by MICIU/AEI/10.13039/501100011033 and by ERDF, EU.

This work was also partially supported by the French PSPC Regions AAP 1 – AURA 2019 call, funded by "Pôle de compétitivité MINALOGIC", France.

References

1. Alcocer, M., Gómez, A.I., Gomez-Perez, D.: Correlation bounds and Markov analysis for ring-oscillator TRNGs: a joint validation framework (2026). https://arxiv.org/abs/2603.13525
2. Allini, E.N., Skórski, M., Petura, O., Bernard, F., Laban, M., Fischer, V.: Evaluation and monitoring of free running oscillators serving as source of randomness. IACR Trans. Cryptogr. Hardw. Embed. Syst. **2018**(3), 214–242 (2018)
3. Alon, N., Kohayakawa, Y., Mauduit, C., Moreira, C.G., Rödl, V.: Measures of pseudorandomness for finite sequences: typical values. Proc. Lond. Math. Soc. **95**(3), 778–812 (2007)
4. Baudet, M., Lubicz, D., Micolod, J., Tassiaux, A.: On the security of oscillator-based random number generators. J. Cryptol. **24**(2), 398–425 (2011)
5. Cherkaoui, A., Fischer, V., Aubert, A., Fesquet, L.: Comparison of self-timed ring and inverter ring oscillators as entropy sources in FPGAs. In: 2012 Design, Automation & Test in Europe Conference & Exhibition (DATE), pp. 1325–1330 (2012). https://doi.org/10.1109/DATE.2012.6176697
6. Dubey, A., Cammarota, R., Aysu, A.: Maskednet: the first hardware inference engine aiming power side-channel protection. In: 2020 IEEE International Symposium on Hardware Oriented Security and Trust (HOST), pp. 197–208. IEEE (2020)
7. Killmann, W., Schindler, W.: A proposal for: functionality classes for random number generators. BSI Technical Report (2011)
8. Laban, M., Drutarovský, M., Fischer, V., Varchola, M.: Modular evaluation platform for evaluation and testing of physically unclonable functions. In: Radioelektronika 2018 - 28th International Conference, Prague, Czech Republic (2018). https://ujm.hal.science/ujm-01814539
9. Lubicz, D., Fischer, V.: Entropy computation for oscillator-based physical random number generators. J. Cryptol. **37**(2), 15 (2024). https://doi.org/10.1007/s00145-024-09494-6
10. Mauduit, C., Sárközy, A.: On finite pseudorandom binary sequences I: measure of pseudorandomness, the Legendre symbol. Acta Arith. **82**(4), 365–377 (1997)
11. Schmidt, K.U.: The correlation measures of finite sequences: limiting distributions and minimum values. Trans. Am. Math. Soc. **369**, 429–446 (2017). arXiv:1404.0172 (2014)

On the Feasibility of 256-Bit Prime Field Arithmetic via PMNS on 8-Bit Architectures

Daichi Aoki[1,2]([✉]) and Tsuyoshi Takagi[2]

[1] NEC Corporation, Kanagawa, Japan
daichi_aoki@nec.com
[2] The University of Tokyo, Tokyo, Japan
takagi@g.ecc.u-tokyo.ac.jp

Abstract. This paper investigates the design of a Polynomial Modular Number System (PMNS) for 256-bit prime field arithmetic on 8-bit architectures. To fit coefficients into 8-bit words, we target a basis matrix $\mathbf{B}$ with $\|\mathbf{B}\|_1 < 2^8$ and dimension $n > 31$. We first demonstrate that standard LLL reduction is effective, achieving this 8-bit target for a wide range of parameters and approaching the heuristic theoretical limits for small γ. However, for large γ or as the dimension grows, a gap emerges between LLL's output and the theoretical minimum. We prove the existence of a global minimum and propose a local-improvement framework using Integer Linear Programming (ILP). This work establishes LLL as a powerful initializer and provides a principled algorithmic foundation for purpose-specific PMNS optimization.

Keywords: Polynomial modular number system · Lattice basis reduction · LLL algorithm · Integer programming

1 Introduction

Efficient modular arithmetic is a cornerstone of modern public-key cryptography. For widely used systems such as RSA [17] and Elliptic-curve cryptography (ECC) [13], the performance of the underlying modular multiplication is critical. This is particularly challenging on resource-constrained 8-bit architectures, such as AVR microcontrollers, which are prevalent in the Internet of Things (IoT) ecosystem. On such devices, standard multi-precision arithmetic suffers from heavy carry-propagation, making the implementation of ECC based on 256-bit prime fields costly in terms of both time and power [1,9,10,12,15,16].

The Polynomial Modular Number System (PMNS), introduced by Bajard, Imbert and Plantard [3], offers a promising alternative. By representing residues as polynomials with bounded coefficients, PMNS allows modular multiplication to be performed using polynomial multiplication followed by a coefficient reduction step. This structure is highly suitable for constrained environments as it minimizes carry operations and enables efficient execution even on simple 8-bit

L. Batina and F. Özbudak (Eds.): WAIFI 2026, LNCS 16611, pp. 284–299, 2026.
https://doi.org/10.1007/978-3-032-27574-5_18

cores. However, the efficiency of PMNS is governed by the size of the coefficients, which depends on the quality of the lattice basis $\mathbf{B}$ associated with the system.

Earlier work on modular number systems and their variants [8,14] studied non-standard residue representations as alternatives to classical special-prime techniques such as generalized Mersenne numbers. PMNS extended this direction by using polynomial representatives modulo p, thereby exposing a high degree of parallelism and reducing carry propagation in arithmetic [2,3,5,7]. A recent work [4] has clarified the theoretical foundation of PMNS, in particular by establishing general existence results and by relating valid PMNS constructions to basis properties of an associated lattice. The theorem from [4] on PMNS over $\mathbb{Z}/p\mathbb{Z}$ shows that if $\mathbf{B}$ is the lattice basis matrix associated with the system, then any $\rho > \|\mathbf{B}\|_1 /2$ is a sufficient coefficient bound for constructing a PMNS. The relevance of PMNS to cryptography has also become increasingly explicit in recent work, especially in the context of elliptic-curve cryptography when pseudo-Mersenne primes are unavailable or undesirable [6].

For 8-bit architectures, the design constraint is exceptionally stringent. To avoid overflow and maximize performance on an 8-bit ALU, all coefficients must fit within a signed 8-bit word. Specifically, for a 256-bit prime p, the design goal is to find a basis matrix $\mathbf{B}$ such that its l_1-operator norm satisfies $\|\mathbf{B}\|_1 < 2^8$ in a dimension $n > 31$. While several methods exist to construct PMNS bases, they typically rely on generic lattice reduction algorithms like LLL [11]. A fundamental conceptual gap exists here: LLL is designed to minimize the l_2-norm (Euclidean length) of basis vectors, whereas the practical feasibility of PMNS is determined by the l_1-operator norm. It remains unclear how close LLL can bring us to the theoretical limits of 8-bit PMNS, and how to proceed when it fails.

In this paper, we bridge this gap by investigating both the feasibility and the direct optimization of 8-bit PMNS for 256-bit primes. Our contributions are as follows:

1. We provide an experimental study on whether standard LLL reduction can achieve the $\|\mathbf{B}\|_1 < 2^8$ requirement for random 256-bit primes. We show that LLL is effective for a meaningful range of parameters, often approaching heuristic limits.
2. To address the limitation of l_2-based reduction, we reformulate the PMNS basis design as a direct l_1-operator norm minimization problem over the group of unimodular transformations $\mathbf{U} \in \mathrm{GL}_n(\mathbb{Z})$.
3. We prove that this optimization problem always attains its minimum and propose a local-improvement framework based on Integer Linear Programming (ILP) to further refine the bases.

By positioning LLL as a powerful initializer and the ILP framework as a specialized tool for direct l_1 optimization, this work provides a principled foundation for designing PMNS tailored to the most constrained hardware environments.

The remainder of the paper is organized as follows. Section 2 recalls preliminaries on lattices and LLL reduction. Section 3 reviews PMNS and formulates the basis design problem arising in 8-bit PMNS architectures. Section 4 investigates the feasibility of the target bound $\|\mathbf{B}\|_1 < 2^8$ via LLL reduction. Section 5 studies direct l_1-norm optimization over unimodularly equivalent bases and presents exact and local optimization frameworks.

2   Preliminaries

Definition 1 (l_p-norm). *For $1 \leq p \leq \infty$, the l_p-norm of $v = (v_1, \ldots, v_n) \in \mathbb{R}^n$ is defined by*

$$\|v\|_p = \begin{cases} \left(\sum_{i=1}^n |v_i|^p\right)^{1/p}, & 1 \leq p < \infty, \\ \max_{1 \leq i \leq n} |v_i|, & p = \infty. \end{cases}$$

The norm $\|v\|_\infty$ is called the max-norm of v.

Let $\mathbf{A} = (a_{ij}) \in \mathbb{R}^{m \times n}$. The induced l_p-operator norm of $\mathbf{A}$ is defined by

$$\|\mathbf{A}\|_p = \max_{\|x\|_p = 1} \|\mathbf{A}x\|_p.$$

Especially, for $p = 1, \infty$, we have

$$\|\mathbf{A}\|_1 = \max_{1 \leq j \leq n} \sum_{i=1}^m |a_{ij}|, \quad \|\mathbf{A}\|_\infty = \max_{1 \leq i \leq m} \sum_{j=1}^n |a_{ij}|.$$

Definition 2 (Lattice). *A lattice $\mathcal{L}$ is a discrete subgroup of $\mathbb{R}^n$, that is, the set of all the integral combinations of $d \leq n$ linearly independent vectors over $\mathbb{R}$:*

$$\mathcal{L} = \mathbb{Z}b_1 + \cdots + \mathbb{Z}b_d, \quad b_i \in \mathbb{R}^n.$$

A set of d vectors $B = (b_1, \ldots, b_d)$ that generates a lattice $\mathcal{L}$ is called a basis, and each b_i is called a basis vector. The integer d is called the dimension of the lattice and is denoted by $\dim(\mathcal{L})$. Particularly, when $d = n$, the lattice is said to be full-rank.

An $n \times d$ matrix $\mathbf{B}$ with each basis vector b_i as a row is called the basis matrix of the lattice $\mathcal{L}$. Furthermore, the lattice generated by $\mathbf{B}$ is denoted by $\mathcal{L}(\mathbf{B})$. If $d \geq 2$, then any d-dimensional lattice has infinitely many distinct bases. The volume of a lattice is given by $\mathrm{vol}(\mathcal{L}) = \sqrt{\det(\mathbf{B}^T \mathbf{B})}$ and does not depend on the choice of lattice basis.

Definition 3 (Gram-Schmidt Orthogonalization, GSO). *Let $\mathcal{L} \subset \mathbb{R}^n$ be a d-dimensional lattice. For a basis $(b_1, \ldots, b_d)$ of $\mathcal{L} \subset \mathbb{R}^n$, the GSO vectors $b_1^*, \ldots, b_d^* \in \mathbb{R}^n$ are defined as follows:*

$$\begin{cases} b_1^* = b_1, \\ b_i^* = b_i - \sum_{j=1}^{i-1} \mu_{i,j} b_j^* \quad (2 \leq i \leq d), \end{cases}$$

where $\mu_{i,j} = \dfrac{b_i \cdot b_j^*}{\|b_j^*\|^2}$, which we call the GSO coefficient.

Given a basis for a lattice $\mathfrak{L}$, the operation of transforming it into a basis for the same lattice where each basis vector is short and the basis vectors are nearly orthogonal to each other is called lattice basis reduction. LLL [11] is the most famous and representative lattice basis reduction algorithm.

Definition 4 (LLL-reduced basis). *Let $b_1^*, \ldots, b_d^*$ be the GSO vectors of the basis $(b_1, \ldots, b_d)$ for a d-dimensional lattice, and $\mu_{i,j}$ $(1 \le j < i \le d)$ be the GSO coefficients. The basis is said to be LLL-reduced with respect to the parameter $\frac{1}{4} < \delta < 1$ when it satisfies the following two conditions:*

1. *Size-reduction condition: $|\mu_{i,j}| \le \frac{1}{2}$ for all $1 \le j < i \le d$,*
2. *Lovász condition: $\|b_k^*\|^2 \ge (\delta - \mu_{k,k-1}^2) \|b_{k-1}^*\|^2$ for all $2 \le k \le d$.*

Given a basis for a lattice $\mathfrak{L}$, the operation of transforming it into another basis of the same lattice whose vectors are shorter and closer to orthogonal is called lattice basis reduction. Among the many reduction algorithms, LLL is the most classical and widely used one. In this paper, LLL reduction is used as a first-stage method for constructing basis matrices of the PMNS lattice with small l_1-operator norm.

3 Polynomial Modular Number System

Definition 5 (Polynomial Modular Number System). *Let $p \ge 3$, $n \ge 2$, $\gamma \in [1, p-1]$ and $\rho \in [1, p-1]$ be integers. Let $E(X) \in \mathbb{Z}[X]$ be a monic polynomial of degree n that satisfies $E(\gamma) \equiv 0 \pmod{p}$. A Polynomial Modular Number System (PMNS) is a set $\mathfrak{B} \subset \mathbb{Z}[X]$ such that:*

1. *$\forall A(X) \in \mathfrak{B}$, $\deg(A(X)) < n$,*
2. *$\forall A(X) = \sum_{i=0}^{n-1} a_i X^i \in \mathfrak{B}$, $-\rho < a_i < \rho$ for all i,*
3. *$\forall a \in \{0, \ldots, p-1\}$, $\exists A(X) \in \mathfrak{B}$ such that $A(\gamma) \equiv a \pmod{p}$.*

The polynomial $E(X)$ is called the reduction polynomial with respect to p.

A PMNS $\mathfrak{B}$ is characterized by the parameter set (p, n, γ, ρ, E). In this paper, we denote $\mathfrak{B} = (p, n, \gamma, \rho, E)$.

Arithmetic operations using the PMNS are performed in the following steps:

1. **Conversion**: Given $a, b \in \mathbb{Z}/p\mathbb{Z}$, compute $A(X), B(X) \in \mathfrak{B}$ representing a, b,
2. **External reduction**: Compute $C(X) = A(X) \odot B(X) \bmod E(X)$, where $\odot$ denotes the polynomial operation corresponding to multiplication or addition,
3. **Internal reduction**: Find a polynomial $\tilde{C}(X)$ such that $\deg \tilde{C}(X) < n$, $\tilde{C}(\gamma) \equiv C(\gamma) \bmod p$ and $|\tilde{c}_i| < \rho$ for all coefficients $\tilde{c}_i$ of $\tilde{C}(X)$.

Given a PMNS $\mathfrak{B} = (p, n, \gamma, \rho, E)$, we can define the n-dimensional lattice $\mathfrak{L}_{\mathfrak{B}}$ given by:

$$\mathfrak{L}_{\mathfrak{B}} = \left\{ (x_0, \ldots, x_{n-1}) \in \mathbb{Z}^n \,\middle|\, \sum_{i=0}^{n-1} x_i \gamma^i \equiv 0 \bmod p \right\}.$$

Let $\mathbb{Z}_{n-1}[X]$ be the set of polynomials of degree at most $n-1$ in $\mathbb{Z}[X]$. The set $\mathfrak{L}_{\mathfrak{B}}$ can be viewed as a subset of $\mathbb{Z}_{n-1}[X]$ consisting of polynomials vanishing at $\gamma \bmod p$. Then we can perform the internal reduction step in the following steps:

1. Find a polynomial $T(X) \in \mathfrak{L}_{\mathfrak{B}}$ such that $\|C(X) - T(X)\|_\infty < \rho$,
2. Set $\tilde{C}(X) = C(X) - T(X)$.

Theorem 1 provides a condition on ρ for a tuple $\mathfrak{B} = (p, n, \gamma, \rho, E)$ to be a PMNS given p, n, E.

Theorem 1 ([4, **Theorem 4.2**]). *Let $p \geq 3$, $n \geq 2$ and $\gamma \in [1, p-1]$ be integers, $E(X) \in \mathbb{Z}_n[X]$ be a monic polynomial and γ be a root of $E(X)$ in $\mathbb{Z}/p\mathbb{Z}$. Let $\mathbf{B}$ be any basis of the lattice $\mathfrak{L}_{\mathfrak{B}}$ for a tuple $\mathfrak{B} = (p, n, \gamma, \rho, E)$. Then,*

$$\text{for any } \rho > \frac{1}{2}\|\mathbf{B}\|_1 = \frac{1}{2}\max_j\left\{\sum_{i=0}^{n-1}|b_{ij}|\right\},$$

$\mathfrak{B} = (p, n, \gamma, \rho, E)$ *is a PMNS.*

Here $\|\mathbf{B}\|_1$ denotes the induced matrix l_1-norm, i.e., the maximum absolute column sum. Thus, the problem of constructing a PMNS with a small coefficient bound can be reduced to the problem of finding a basis matrix $\mathbf{B}$ with small l_1-operator norm. One basis matrix of $\mathfrak{L}_{\mathfrak{B}}$ is given by

$$\mathbf{B} = \begin{pmatrix} p & 0 & 0 & 0 & \ldots & 0 \\ -\gamma & 1 & 0 & 0 & \ldots & 0 \\ 0 & -\gamma & 1 & 0 & \ldots & 0 \\ \vdots & & \ddots & \ddots & & \vdots \\ 0 & \ldots & \ldots & -\gamma & 1 & 0 \\ 0 & \ldots & \ldots & 0 & -\gamma & 1 \end{pmatrix}. \tag{1}$$

Each row of this matrix corresponds to a polynomial vanishing at $\gamma \bmod p$, and the rows form a basis of $\mathfrak{L}_{\mathfrak{B}}$. Applying Theorem 1 to this basis yields the sufficient condition

$$\rho > \frac{1}{2}\|\mathbf{B}\|_1 = \frac{p+\gamma}{2}.$$

For the 256-bit setting considered in this paper, this bound is far too large to be useful. Therefore, the central task is not merely to construct a basis of $\mathfrak{L}_{\mathfrak{B}}$, but to construct one whose l_1-operator norm is much smaller than that of the canonical basis (1). This is the basis design problem studied in the remainder of the paper.

4 Feasibility of 8-Bit PMNS via LLL Reduction

In this paper, we consider the design of a PMNS with 256-bit modulus for 8-bit architectures. In such a setting, each coefficient of a polynomial in the PMNS is required to fit into an 8-bit word, so our target is to construct parameters (p, n, γ, ρ, E) with

$$\rho \leq 2^7.$$

By Theorem 1, it is sufficient to find a basis matrix $\mathbf{B}$ of the lattice $\mathcal{L}_{\mathfrak{B}}$ such that $\|\mathbf{B}\|_1 < 2^8$, because then one may take $\rho = 2^7$. Moreover, in order that the PMNS represent all elements of $\mathbb{Z}/p\mathbb{Z}$, the number of admissible coefficient vectors must be at least p. Since each coefficient a_i satisfies $-\rho < a_i < \rho$, it has exactly $2\rho - 1$ possible integer values. Therefore, a necessary condition is

$$(2\rho - 1)^n \geq p.$$

Equivalently,

$$n \geq \left\lceil \frac{\log p}{\log(2\rho - 1)} \right\rceil.$$

In particular, when $\rho = 2^7$, this becomes $n \geq \left\lceil \frac{\log p}{\log 255} \right\rceil$. However, this bound only reflects the coefficient range and does not capture the effect of the parameter γ. To obtain a more informative γ-dependent estimate, let $A(\gamma) = \sum_{i=0}^{n-1} a_i \gamma^i$ with $|a_i| < \rho$. Then the absolute value of $A(\gamma)$ is bounded by

$$|A(\gamma)| \leq (\rho - 1) \sum_{i=0}^{n-1} \gamma^i = (\rho - 1)\frac{\gamma^n - 1}{\gamma - 1}.$$

Hence all values represented by such polynomials lie in the interval

$$\left[-(\rho - 1)\frac{\gamma^n - 1}{\gamma - 1}, \ (\rho - 1)\frac{\gamma^n - 1}{\gamma - 1} \right].$$

Therefore, a heuristic necessary condition for potentially covering all residue classes modulo p is

$$2(\rho - 1)\frac{\gamma^n - 1}{\gamma - 1} + 1 \geq p.$$

For large γ, this suggests

$$2(\rho - 1)\gamma^{n-1} \gtrsim p,$$

which gives the heuristic estimate

$$n \gtrsim \frac{\log_2 p - \log_2(2(\rho - 1))}{\log_2 \gamma} + 1.$$

The practical design problem is to determine, for a given 256-bit prime p, parameter pairs (n, γ) for which the lattice $\mathcal{L}_{\mathfrak{B}}$ admits a basis matrix whose l_1-operator norm is smaller than 2^8. According to the above discussion, larger n is required when γ is small (e.g., $n > 84$ for $\gamma = 8$) compared to when γ is large (e.g., $n > 63.25$ for $\gamma = 16$). The primary goal of our experiments in this section is to investigate how closely the standard LLL algorithm can approach this theoretical limit and successfully find a basis satisfying the 8-bit constraint.

Throughout this section, we fix the reduction polynomial to be

$$E(X) = X^n - \lambda, \quad \lambda = \gamma^n \bmod p.$$

Although more efficient choices of E for the external reduction step are possible, our purpose here is to isolate the basis design problem arising from the internal reduction step. Under these settings, the 8-bit PMNS design problem reduces to constructing a basis matrix of $\mathcal{L}_\mathfrak{B}$ satisfying $\|\mathbf{B}\|_1 < 2^8$.

A natural starting point is the canonical basis matrix

$$\mathbf{B} = \begin{pmatrix} p & 0 & 0 & 0 & \ldots & 0 \\ -\gamma & 1 & 0 & 0 & \ldots & 0 \\ 0 & -\gamma & 1 & 0 & \ldots & 0 \\ \vdots & & \ddots & \ddots & & \vdots \\ 0 & \ldots & \ldots & -\gamma & 1 & 0 \\ 0 & \ldots & \ldots & 0 & -\gamma & 1 \end{pmatrix},$$

which is a basis matrix of $\mathcal{L}_\mathfrak{B}$. However, this basis is far from sufficient for 8-bit PMNS design. Indeed, the maximum column sum of $\mathbf{B}$ is $p + \gamma$, so $\|\mathbf{B}\|_1 = p + \gamma$, and Theorem 1 only yields the condition

$$\rho > \frac{p + \gamma}{2},$$

which is useless for a 256-bit prime modulus. Therefore, some form of lattice basis reduction is indispensable. In this section, we first study whether LLL reduction alone is sufficient to achieve feasibility, before introducing any dedicated optimization.

4.1   Experimental Setting

For each pair of parameters (n, γ), we sampled 100 random 256-bit prime moduli p. For each sampled prime, we generated the canonical basis matrix $\mathbf{B}$ given by (1), applied LLL reduction, and checked whether the reduced basis matrix satisfies $\|\mathbf{B}\|_1 < 2^8$. The value plotted in Fig. 1 is the proportion of sampled primes for which this inequality holds. In other words, the "success probability" in Fig. 1 means the empirical success rate, over 100 independently sampled 256-bit primes, that LLL reduction produces a basis matrix suitable for 8-bit PMNS design. The figure considers the parameter ranges

$$n \in [30, 90], \quad \gamma \in \{8, 16, 32, 64, 128, 256, 512\}.$$

All experiments were carried out in SageMath 10.6, and used its default LLL implementation with the parameter $\delta = 0.99$. The primes were sampled uniformly from the set of 256-bit primes using SageMath's random prime generation functionality. This range of n covers the practically relevant regime near the minimum admissible dimension $n > 31$. Figure 1 is therefore intended to show, for each γ, where the target inequality $\|\mathbf{B}\|_1 < 2^8$ is likely to be achieved by LLL reduction alone.

4.2   Gap Analysis: Theoretical Limits vs. Practical Reduction

Figure 1 shows the empirical success probability of LLL reduction achieving $\|\mathbf{B}\|_1 < 2^8$ for several choices of (n, γ). A notable feature of Fig. 1 is that the success probability is typically concentrated in a specific dimension window for each fixed γ, rather than improving monotonically as n increases. To understand this behavior, we compare these results with the heuristic lower bound derived in Sect. 4. For a 256-bit prime p and the target $\rho = 2^7$, the condition $n \gtrsim \frac{\log_2 p - \log_2(2(\rho-1))}{\log_2 \gamma} + 1$ approximately becomes

$$n \gtrsim \frac{256 - \log_2 254}{\log_2 \gamma} + 1 \approx \frac{248.011}{\log_2 \gamma} + 1.$$

For instance, when $\gamma = 16$, this heuristic suggests $n \gtrsim \frac{248.011}{4} + 1 \approx 63.0$. Our experimental results for $\gamma = 16$ show that the success probability indeed starts to rise around $n = 63$, suggesting that the standard LLL algorithm is highly effective at finding bases close to this theoretical limit.

However, this efficiency has limits. In cases where $\gamma = 256$ or $\gamma = 512$, where the heuristic suggests that dimensions as low as $n = 32$ or $n = 28.5$ might be sufficient, our experiments with LLL reduction failed to yield any valid bases in the range $n \in [30, 90]$. This discrepancy indicates that the simple heuristic estimate does not capture the full complexity of basis construction under the 8-bit constraint as γ grows larger.

The observed decrease in success probability as n grows beyond the optimal window can be explained by two factors. First, according to the definition of the l_1-operator norm $\|\mathbf{B}\|_1 = \max_j \sum_i |b_{ij}|$, an increase in the dimension n directly increases the number of summands in each column. Even if individual coefficients remain small, the cumulative sum is more likely to exceed the 2^8 threshold in higher dimensions. Second, in higher-dimensional lattices, the length of the shortest vectors found by reduction algorithms tends to increase due to the exponential approximation factors inherent in lattice reduction.

Furthermore, to investigate whether more powerful reduction techniques could close the remaining gap to the theoretical bound, we conducted the same experiments using the BKZ algorithm with a block size equal to the dimension n. In our additional experiments (not shown here), BKZ with block size n does not improve the success probability over LLL. This suggests that the gap between the theoretical lower bound and the experimental results is not due to the suboptimality of the reduction algorithm itself, but rather to the inherent difficulty of satisfying the 8-bit constraint at the minimum dimension. Therefore, these results suggest the necessity of directly optimizing the l_1-operator norm to reach the theoretical limits of PMNS design.

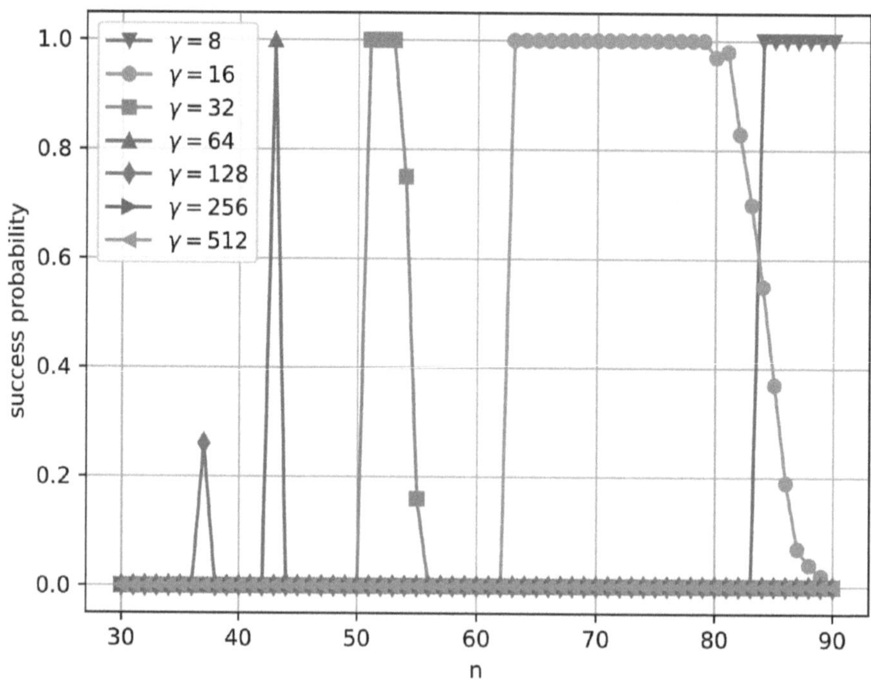

Fig. 1. Empirical success probability, over 100 randomly sampled 256-bit primes, that the LLL-reduced basis satisfies $\|\mathbf{B}\|_1 < 2^8$ for each pair (n, γ).

4.3   Interpretation: Feasibility vs. Optimization

The above observations answer the first design question:

- Can an 8-bit PMNS basis with $\|\mathbf{B}\|_1 < 2^8$ be obtained from the canonical lattice basis by standard lattice reduction?

Our experiments indicate that the answer is yes. Therefore, LLL reduction already establishes the feasibility of the design goal. In particular, for $\gamma = 32, 64, 128$ in Fig. 1, successful instances already appear at dimension close to the theoretical minimum, although the empirical success probability becomes sensitive to the parameter choice in that regime.

However, this does not eliminate the need for further optimization. The point is that LLL is not designed to minimize the quantity $\|\mathbf{B}\|_1$ itself. Rather, it aims at producing a basis consisting of relatively short and nearly orthogonal vectors. As a consequence, although LLL can make the basis norm small enough to cross the target threshold, there is no reason to expect that the obtained basis is optimal with respect to the actual PMNS objective. The quantity $\|\mathbf{B}\|_1$ is fundamental here, since Theorem 1 directly translates a bound on $\|\mathbf{B}\|_1$ into a valid choice of ρ. Once feasibility is confirmed, the more refined question is therefore the following:

- How small can we make $\|\mathbf{B}\|_1$, or equivalently, how small can we make the required dimension n while still keeping $\|\mathbf{B}\|_1 < 2^8$?

This second question is important for at least two reasons. First, smaller n is desirable for implementation efficiency. Second, even when $\|\mathbf{B}\|_1 < 2^8$ already holds, a larger margin below the threshold leads to a more robust parameter choice.

For these reasons, we regard LLL reduction not as the final design tool, but as a feasibility method and an initializer. This motivates the next section, where we study the optimization problem

$$\min_{\mathbf{U}\in\mathrm{GL}_n(\mathbb{Z})} \|\mathbf{U}\mathbf{B}\|_1 ,$$

that is, the minimization of the l_1-operator norm over all unimodularly equivalent bases of the same lattice. This formulation targets the quantity that governs the PMNS bound, and allows us to search for bases that are better adapted to 8-bit implementation than the basis returned by LLL alone.

5 l_1-Operator Norm Optimization over Unimodular Transformations

In the previous section, we showed that LLL reduction alone can already produce basis matrices satisfying the target inequality $\|\mathbf{B}\|_1 < 2^8$ for a range of parameters. This establishes the feasibility of 8-bit PMNS design by standard lattice reduction. However, LLL is not designed to minimize $\|\mathbf{B}\|_1$ itself. Its objective is to output a basis consisting of relatively short and nearly orthogonal vectors, whereas the PMNS condition is governed directly by the l_1-operator norm through Theorem 1. Therefore, once feasibility has been confirmed, it is natural to study the following optimization problem: among all bases of the same lattice, find one whose l_1-operator norm is as small as possible. Since any other basis of the same lattice is obtained from a given basis by a unimodular transformation, this leads to optimization over $\mathrm{GL}_n(\mathbb{Z})$. The relevance of $\|\mathbf{B}\|_1$ comes from the fact that $\rho > \|\mathbf{B}\|_1 / 2$ is sufficient for PMNS construction, and for 8-bit design the target is $\|\mathbf{B}\|_1 < 2^8$.

5.1 Problem Formulation

Let $\mathbf{B}$ be a basis matrix of a full-rank lattice $\mathcal{L}$. For any unimodular matrix $\mathbf{U} \in \mathrm{GL}_n(\mathbb{Z})$, the matrix $\mathbf{U}\mathbf{B}$ is again a basis matrix of the same lattice. We therefore consider the function

$$f_{\mathbf{B}} : \mathrm{GL}_n(\mathbb{Z}) \to \mathbb{R}, \quad f_{\mathbf{B}}(\mathbf{U}) = \|\mathbf{U}\mathbf{B}\|_1$$

and aim to solve

$$\min_{\mathbf{U}\in\mathrm{GL}_n(\mathbb{Z})} \|\mathbf{U}\mathbf{B}\|_1 .$$

This formulation directly captures the quantity relevant to the PMNS design. As a first step, we show that this optimization problem is well posed.

Lemma 1. *Let $\mathbf{B} \in \mathrm{GL}_n(\mathbb{R})$ and $\mathbf{U} \in \mathbb{R}^{n \times n}$. The following holds:*

$$\left\|\mathbf{B}^{-1}\right\|_1^{-1} \|\mathbf{U}\|_1 \leq \|\mathbf{U}\mathbf{B}\|_1 \leq \|\mathbf{B}\|_1 \|\mathbf{U}\|_1 .$$

Proof. By the submultiplicativity of the operator norm $\|\cdot\|_1$, we obtain

$$\|\mathbf{U}\mathbf{B}\|_1 \leq \|\mathbf{U}\|_1 \|\mathbf{B}\|_1 ,$$
$$\|\mathbf{U}\|_1 = \left\|(\mathbf{U}\mathbf{B})\mathbf{B}^{-1}\right\|_1 \leq \|\mathbf{U}\mathbf{B}\|_1 \left\|\mathbf{B}^{-1}\right\|_1 .$$

$\square$

It follows that any unimodular matrix achieving a sufficiently small value of $\|\mathbf{U}\mathbf{B}\|_1$ must lie in a bounded subset of $\mathbb{Z}^{n \times n}$. Since integer matrices in a bounded region form a finite set, the minimum of $f_{\mathbf{B}}$ is always attained. More precisely, we show that there exists a minimizer $\mathbf{U}^\star \in \mathrm{GL}_n(\mathbb{Z})$, and that every entry of $\mathbf{U}^\star = (u_{ij}^\star)$ satisfies $|u_{ij}^\star| \leq \|\mathbf{B}\|_1 \left\|\mathbf{B}^{-1}\right\|_1$.

Theorem 2. *Fix $\mathbf{B} \in \mathrm{GL}_n(\mathbb{R})$. Consider the function $f_{\mathbf{B}} : \mathrm{GL}_n(\mathbb{Z}) \to \mathbb{R}$ defined by $f_{\mathbf{B}}(\mathbf{U}) = \|\mathbf{U}\mathbf{B}\|_1$. There exists a unimodular matrix $\mathbf{U}^\star = (u_{ij}^\star)$ that attains the minimum of the function $f_{\mathbf{B}}$, and its entries satisfy*

$$|u_{ij}^\star| \leq \|\mathbf{B}\|_1 \left\|\mathbf{B}^{-1}\right\|_1 .$$

Proof. For any $t \in \mathbb{R}_{\geq 0}$, let $\mathcal{S}_t = \{\mathbf{U} \in \mathrm{GL}_n(\mathbb{Z}) \mid f_{\mathbf{B}}(\mathbf{U}) \leq t\}$. From Lemma 1, we have

$$f_{\mathbf{B}}(\mathbf{U}) \leq t \Rightarrow \|\mathbf{U}\|_1 \leq t \left\|\mathbf{B}^{-1}\right\|_1 . \tag{2}$$

Since $\mathbf{U}$ is an integer matrix, there are only finitely many matrices $\mathbf{U}$ satisfying $\|\mathbf{U}\|_1 < t \left\|\mathbf{B}^{-1}\right\|_1$. Therefore, the set $\mathcal{S}_t$ is finite for any t.

Let $M := f_{\mathbf{B}}(\mathbf{I}_n) = \|\mathbf{B}\|_1$. Since $\inf f_{\mathbf{B}}(\mathbf{U}) \leq M < \infty$, the set $\mathcal{S}_M$ is nonempty. A real-valued function defined on a finite set always attains its minimum. Therefore, the following holds:

$$\exists \mathbf{U}^\star \in \mathcal{S}_M \text{ such that } f_{\mathbf{B}}(\mathbf{U}^\star) = \min_{\mathbf{U} \in \mathcal{S}_M} f_{\mathbf{B}}(\mathbf{U}).$$

Moreover, since $\mathcal{S}_M$ contains all candidates for $\mathbf{U}^\star$ in $\mathrm{GL}_n(\mathbb{Z})$, the following also holds:

$$f_{\mathbf{B}}(\mathbf{U}^\star) = \min_{\mathbf{U} \in \mathrm{GL}_n(\mathbb{Z})} f_{\mathbf{B}}(\mathbf{U}).$$

From the above discussion, it follows that a unimodular matrix $\mathbf{U}^\star$ attaining the minimum of the function $f_{\mathbf{B}}$ belongs to the finite set $\mathcal{S}_M$. In particular, setting $t = M$ in (2), we obtain $\|\mathbf{B}\|_1 \left\|\mathbf{B}^{-1}\right\|_1 \geq \|\mathbf{U}^\star\|_1 \geq |u_{ij}^\star|$. $\square$

5.2   Exact Optimization by Exhaustive Search

Theorem 2 immediately yields a conceptually simple exact algorithm. Let

$$R = \|\mathbf{B}\|_1 \, \|\mathbf{B}^{-1}\|_1 \, .$$

Then every entry of an optimal unimodular matrix $\mathbf{U}^\star$ lies in the interval $[-R, R]$. Therefore, one may enumerate all integer matrices $\mathbf{U}$ with entries in this interval, test whether $\det(\mathbf{U}) = \pm 1$, and evaluate $\|\mathbf{UB}\|_1$. The best candidate found in this finite search is an optimal basis of the lattice.

This exhaustive search, described as Algorithm 1, is useful mainly as a theoretical baseline. It proves that the optimization problem is decidable in finite time and provides a reference notion of global optimality. However, its complexity grows extremely quickly: the number of candidates for $\mathbf{U}^\star$ is $(2R + 1)^{n^2}$, which leads to the exponential-time complexity $\exp(O(n^2 \log R))$.

Algorithm 1. Exhaustive Search

Input: A basis matrix $\mathbf{B} \in \mathbb{Z}^{n \times n}$ of a lattice $\mathfrak{L}$
Output: A basis matrix $\mathbf{B}^\star$ of $\mathfrak{L}$ that minimizes $\|\mathbf{B}^\star\|_1$
 1: $\mathbf{B}^\star = \mathbf{B}$, $M = \|\mathbf{B}^\star\|_1$
 2: Enumerate all integer matrices $\mathbf{U}$ with entries in $[-R, R]$
 3: **for** each $\mathbf{U}$ **do**
 4:     **if** $\det \mathbf{U} = \pm 1$ **then**
 5:        $M' = \|\mathbf{UB}\|_1$
 6:        **if** $M' < M$ **then**
 7:           $M \leftarrow M'$
 8:           $\mathbf{B}^\star \leftarrow \mathbf{UB}$
 9: **return** $\mathbf{B}^\star$

5.3   A Local Improvement Principle via Row Updates

To obtain a more practical optimization framework, we next consider local basis transformations. Let $\mathbf{B} = (b_{ij})$ be a basis matrix whose row vectors form an n-dimensional full-rank lattice $\mathfrak{L}$. Let b_i denote the i-th row vector of $\mathbf{B}$. Fix an index i, and update only the i-th row by adding integer combinations of the other rows:

$$b_i' = b_i + \sum_{\substack{1 \le l \le n \\ l \ne i}} k_l b_l.$$

If we define $k = (k_1, \ldots, k_n) \in \mathbb{Z}^n$ with $k_i = 1$, then this update can be written as

$$b_i' = k\mathbf{B}.$$

Because such a row operation corresponds to multiplication by a unimodular matrix, the updated matrix $\mathbf{B}'$ remains a basis matrix of the same lattice $\mathfrak{L}$.

This observation reduces the global optimization problem to a sequence of local basis improvements. The key question is: for a fixed row index i, when does there exist an integer vector k such that replacing b_i with $k\mathbf{B}$ strictly decreases $\|\mathbf{B}\|_1$? The following theorem gives a necessary and sufficient characterization of this event.

Theorem 3. *Let c_j denote the j-th column vector of $\mathbf{B}$. For a fixed index i, let $\mathbf{B}'$ denote the matrix obtained by updating the i-th row of $\mathbf{B}$ as $b'_i = k\mathbf{B}$. Then, $\|\mathbf{B}'\|_1 < \|\mathbf{B}\|_1$ holds if and only if $k = (k_1, \ldots, k_n) \in \mathbb{Z}^n$ with $k_i = 1$ satisfies*

$$(k\mathbf{B})_j \in [-D_j, D_j], \quad \text{for all } j = 1, \ldots, n,$$

where $D_j = \|\mathbf{B}\|_1 - \|c_j\|_1 + |b_{ij}| - 1$.

Proof. For any j, the difference between the norm of the j-th column vector c_j of $\mathbf{B}$ and that of the j-th column vector c'_j of $\mathbf{B}'$ is given as follows:

$$\|c'_j\|_1 - \|c_j\|_1 = \sum_{r=1}^{n} |b'_{rj}| - \sum_{r=1}^{n} |b_{rj}| = |b'_{ij}| - |b_{ij}|.$$

We have the following relation:

$$\|\mathbf{B}'\|_1 < \|\mathbf{B}\|_1 \Leftrightarrow \|c'_j\|_1 < \|\mathbf{B}\|_1 \text{ for all } j$$
$$\Leftrightarrow |b'_{ij}| \leq D_j \text{ for all } j$$
$$\Leftrightarrow (k\mathbf{B})_j \in [-D_j, D_j] \text{ for all } j.$$

$\square$

This characterization is particularly useful because it converts the question of norm reduction into a finite system of linear inequalities over the integers. In other words, the possibility of improving the current basis through a single row update can be checked by solving an integer feasibility problem:

$$\begin{aligned}
\text{Find} \quad & k \in \mathbb{Z}^n \\
\text{subject to} \quad & (k\mathbf{B})_j \in [-D_j, D_j], \quad j = 1, \ldots, n \\
& k_i = 1.
\end{aligned} \tag{3}$$

5.4 ILP-Based Heuristic Framework

The problem (3) can be reformulated as an integer linear programming problem (ILP). Thus, for a fixed row i, one may ask an ILP solver to find an integer vector k satisfying the constraints, and if such a k exists, update the row $b_i \leftarrow k\mathbf{B}$. Repeating this process as long as a norm-reducing row update is found yields a greedy local search procedure.

The outer greedy loop (Algorithm 2) repeatedly chooses a row index i, while the inner procedure (Algorithm 3) checks, via ILP, whether the row can be updated so as to decrease the l_1-operator norm of the basis matrix. Since each successful update strictly decreases $\|\mathbf{B}\|_1$ and the norm is nonnegative, the procedure must terminate. The resulting basis is locally optimal with respect to the chosen update rule, although it is not guaranteed to be globally optimal.

In Algorithm 2, several strategies can be considered for selecting the row i. For example, a column of $\mathbf{B}$ having the largest l_1-norm is selected, and the indices of the top K entries in that column, sorted by decreasing absolute value, are taken as $S \subset \{1, \ldots, n\}$ and used as candidates for i. Moreover, to reduce the computational cost of the ILP solver, only the rows in S are allowed to be used for updating; that is, $k_l = 0$ for $l \notin S$.

At the same time, we emphasize that this ILP-based method should presently be regarded as a proposed optimization framework, rather than as a fully validated practical algorithm. In the current stage of this work, we do not yet have sufficient experimental evidence to support detailed claims about its effectiveness, scalability, or advantage over LLL across a broad parameter range. For this reason, in this paper we position the method primarily as a natural consequence of the problem formulation above: once the design objective is expressed as minimizing $\|\mathbf{UB}\|_1$, the row-update characterization and its ILP reformulation provide a principled route toward direct norm optimization. A thorough implementation study, including solver design, tuning of row-selection strategies, and large-scale empirical comparison with LLL-based baselines, is left for future work.

Algorithm 2. Row-Update Search

Input: A basis matrix $\mathbf{B} \in \mathbb{Z}^{n \times n}$ of a lattice $\mathfrak{L}$
Output: A basis matrix $\mathbf{B}^\star$ of $\mathfrak{L}$
1: $\mathbf{B}^\star = \mathbf{B}$, $M = \|\mathbf{B}^\star\|_1$
2: **repeat**
3:     Choose row i
4:     $\mathbf{B}^\star, M \leftarrow \mathsf{CanReduce\_Update}(\mathbf{B}^\star, M, i)$ $\triangleright$ If M can be reduced, update $b_i^\star, M$
5: **until** there exists no i that reduces M
6: **return** $\mathbf{B}^\star$

Algorithm 3. CanReduce_Update

Input: A basis matrix $\mathbf{B} \in \mathbb{Z}^{n \times n}$, integers $M \geq 0$, $i \in [1, n]$
Output: A basis matrix $\mathbf{B}$ and an integer M
1: Solve the problem (3) with $\mathbf{B}$ and i by ILP solver
2: **if** there exists a solution k **then**
3:     Update $b_i \leftarrow k\mathbf{B}$
4:     Update $M \leftarrow \|\mathbf{B}\|_1$
5: **return** $\mathbf{B}, M$

---

6 Conclusion

We have studied the feasibility and optimization of 8-bit PMNS for 256-bit prime moduli from the viewpoint of lattice basis norms. Using the PMNS condition $\rho > \|\mathbf{B}\|_1 / 2$, we reduced the design problem to the construction of a basis matrix of the lattice $\mathcal{L}_{\mathfrak{B}}$ whose l_1-operator norm is smaller than 2^8. In the 256-bit setting considered here, this corresponds to the practical goal of constructing PMNS parameters with coefficients that fit into 8-bit words and with dimension $n > 31$. This reduction turns the search for suitable PMNS parameters into a concrete lattice basis design problem.

Our first main result is that standard lattice reduction already establishes the feasibility of this goal in a meaningful parameter range. The experiments show that LLL reduction alone can produce basis matrices satisfying $\|\mathbf{B}\|_1 < 2^8$, and that, among the parameter choices for which successful instances occur, the dimension required for feasibility tends to decrease as γ increases. In particular, for $\gamma = 32, 64, 128$ in our experiments, successful instances already appear close to the theoretical minimum dimension, although the empirical success probability becomes more sensitive to the parameter choice in that regime. These results show that LLL reduction is an effective first-stage method for constructing 8-bit PMNS bases.

Our second main result is a direct optimization formulation targeting the quantity that actually governs PMNS construction. Since LLL is not designed to minimize $\|\mathbf{B}\|_1$ itself, we formulated the problem of minimizing $\|\mathbf{UB}\|_1$ over unimodular transformations $\mathbf{U} \in \mathrm{GL}_n(\mathbb{Z})$. We proved that this optimization problem always attains its minimum, derived a finite exact search formulation, and established a local-improvement principle based on row updates. We further showed that the corresponding norm-reduction condition can be expressed as an integer linear programming problem, which yields a natural framework for PMNS-specific basis optimization beyond generic lattice reduction.

From the viewpoint of the overall paper, the roles of these two parts are complementary. Section 4 answers the feasibility question: can the target $\|\mathbf{B}\|_1 < 2^8$ already be reached by standard lattice reduction? Sect. 5 addresses the subsequent design question: once feasibility is known, how should one optimize the basis with respect to the actual PMNS objective? In this sense, LLL should be viewed not as the final design tool, but as an effective feasibility method and initializer, whereas the l_1-optimization approach provides the conceptual foundation for further purpose-specific improvement of PMNS bases.

At the present stage, the proposed optimization procedures should be regarded as a theoretical and algorithmic framework rather than as fully validated practical methods. Their implementation and large-scale empirical evaluation are left for future work. Important next steps include implementing the proposed optimization algorithms, comparing them systematically with LLL-based baselines over a broad parameter range, and studying how the choice of γ, the reduction polynomial, and the row-selection strategy affect the achievable norm and the resulting implementation cost. Such investigations may lead to a more practical methodology for constructing PMNS tailored to constrained architectures.

References

1. Akyildiz, I.F., Su, W., Sankarasubramaniam, Y., Cayirci, E.: Wireless sensor networks: a survey. Comput. Netw. **38**(4), 393–422 (2002)
2. Bajard, J.C., Imbert, L., Plantard, T.: Modular number systems: beyond the Mersenne family. In: Handschuh, H., Hasan, A. (eds.) SAC 2004. LNCS, vol. 3357, pp. 159–169. Springer, Heidelberg (2004)
3. Bajard, J.C., Imbert, L., Plantard, T.: Arithmetic operations in the polynomial modular number system. In: 17th IEEE Symposium on Computer Arithmetic, Cape Cod, MA, USA, pp. 206–213. IEEE Computer Society (2005)
4. Bajard, J.C., Marrez, J., Plantard, T., Véron, P.: On polynomial modular number systems over Z/pZ. Adv. Math. Commun. **18**(3), 674–695 (2024)
5. Coladon, T., Elbaz-Vincent, P., Hugounenq, C.: MPHELL: a fast and robust library with unified and versatile arithmetics for elliptic curves cryptography. In: IEEE 28th Symposium on Computer Arithmetic, Lyngby, Denmark, pp. 78–85. IEEE (2021)
6. Dosso, F.Y., Duquesne, S., El Mrabet, N., Gautier, E.: PMNS arithmetic for elliptic curve cryptography. In: Nitaj, A., Petkova-Nikova, S., Rijmen, V. (eds.) AFRICACRYPT 2025. LNCS, vol. 15651, pp. 164–191. Springer, Cham (2025)
7. Dosso, F.Y., El Mrabet, N., Méloni, N., Palma, F., Véron, P.: Friendly primes for efficient modular arithmetic using the polynomial modular number system. J. Cryptogr. Eng. **15**(3), 18 (2025)
8. Garner, H.L.: The residue number system. IRE Trans. Electron. Comput. **8**(2), 140–147 (1959)
9. Gura, N., Patel, A., Wander, A., Eberle, H., Shantz, S.C.: Comparing elliptic curve cryptography and RSA on 8-bit CPUs. In: Joye, M., Quisquater, J.J. (eds.) CHES 2004. LNCS, vol. 3156, pp. 119–132. Springer, Heidelberg (2004)
10. Hutter, M., Schwabe, P.: Multiprecision multiplication on AVR revisited. J. Cryptogr. Eng. **5**(3), 201–214 (2015)
11. Lenstra, A.K., Lenstra, H.W., Lovász, L.: Factoring polynomials with rational coefficients. Math. Ann. **261**(4), 515–534 (1982)
12. Liu, A., Ning, P.: TinyECC: a configurable library for elliptic curve cryptography in wireless sensor networks. In: Proceedings of the 7th International Conference on Information Processing in Sensor Networks, IPSN 2008, St. Louis, Missouri, USA, pp. 245–256. IEEE Computer Society (2008)
13. Miller, V.S.: Use of elliptic curves in cryptography. In: Williams, H.C. (ed.) CRYPTO 1985. LNCS, vol. 218, pp. 417–426. Springer, Heidelberg (1985)
14. Negre, C., Plantard, T.: Efficient modular arithmetic in adapted modular number system using Lagrange representation. In: Mu, Y., Susilo, W., Seberry, J. (eds.) ACISP 2008. LNCS, vol. 5107, pp. 463–477. Springer, Heidelberg (2008)
15. Oliveira, L.B., Aranha, D.F., Morais, E., Daguano, F., López, J., Dahab, R.: Tiny-Tate: computing the Tate pairing in resource-constrained sensor nodes. In: Sixth IEEE International Symposium on Network Computing and Applications (NCA 2007), Cambridge, MA, USA, pp. 318–323. IEEE Computer Society (2007)
16. Park, D., Chang, N.S., Lee, S., Hong, S.: Fast implementation of NIST P-256 elliptic curve cryptography on 8-bit AVR processor. Appl. Sci. **10**(24), 8816 (2020)
17. Rivest, R.L., Shamir, A., Adleman, L.M.: A method for obtaining digital signatures and public-key cryptosystems. Commun. Assoc. Comput. Mach. **21**(2), 120–126 (1978)

Fast and Energy-Efficient Polynomial Multiplication Using FFT, FFNT, and NTT on GPUs for Fully Homomorphic Encryption

Ali Şah Özcan[1]([✉]), Cihangir Tezcan[2], and Erkay Savaş[1]

[1] Sabancı University, 34956 Istanbul, Turkey
{alisah,erkays}@sabanciuniv.edu
[2] Middle East Technical University, 06800 Ankara, Turkey
cihangir@metu.edu.tr

Abstract. Many cryptographic schemes, such as fully homomorphic encryption (FHE), zero-knowledge proofs, and post-quantum cryptography, rely on fast multiplication of large polynomials. Efficient algorithms for computing the discrete Fourier transform, including the Fast Fourier Transform (FFT), Negacyclic FFT (FFNT), and Number Theoretic Transform (NTT), can be employed to accelerate these multiplications.

In this paper, we introduce an improved version of the FFNT that eliminates pre- and post-processing steps, and we show that it provides competitive performance among FFT- and FFNT-based methods for polynomial multiplication on GPUs, including implementations based on the highly optimized CUDA library cuFFT. We then compare the performance of the proposed FFNT with that of the NTT when both are implemented in CUDA for GPU platforms.

Our results show that FFNT can outperform NTT in both speed and energy efficiency on GPUs with sufficiently strong floating-point hardware for some FHE settings. In particular, for polynomial multiplication, the FFNT kernel achieves up to 60% higher energy efficiency than the NTT kernel on such devices. Furthermore, in a practical setting, our GPU-based implementation of torus FHE (TFHE) using FFNT for NAND gate bootstrapping can be up to 46% faster than its NTT-based counterpart when ample floating-point resources are available.

These results suggest that the choice between NTT- and FFNT-based polynomial multiplication should be guided by the underlying GPU architecture and by the precision and efficiency requirements of the target FHE workload.

Keywords: Fully Homomorphic Encryption · Polynomial Multiplication · Fast Fourier Transform · Number Theoretic Transform · Negacyclic Fast Fourier Transform · Graphics Processing Units

L. Batina and F. Özbudak (Eds.): WAIFI 2026, LNCS 16611, pp. 300–327, 2026.
https://doi.org/10.1007/978-3-032-27574-5_19

1   Introduction

Multiplication of high-degree polynomials is the most time-consuming operation in many cryptographic algorithms such as fully homomorphic encryption (FHE) schemes [9,11,12,18], zero-knowledge proofs [6,7,10], and lattice-based post-quantum cryptography [8,17]. Accelerating resource- and time-intensive polynomial multiplication facilitates the wider adoption of these advanced cryptographic schemes in practical applications. Numerous efficient methods, such as Karatsuba's algorithm [21] or convolution-based Discrete Fourier Transformation (DFT) algorithm [15] exist for fast polynomial multiplication; the latter is widely used in cryptographic algorithms that involve high-degree polynomials, ranging from hundreds to tens of thousands of terms. Generally, the main idea behind DFT-based techniques is to accelerate classical cyclic convolution by using efficient FFT algorithms to perform polynomial multiplication modulo $(X^N - 1)$, which is equivalent to multiplication in the polynomial ring $\mathbb{C}/(X^N - 1)$. In most of the aforementioned cryptographic schemes, however, multiplication in the rings $\mathbb{C}/(X^N + 1)$ or $\mathbb{Z}_q/(X^N + 1)$, based on negacyclic convolution, is required. Although not described in the original torus-based FHE (TFHE) scheme [12], one method for calculating negacyclic convolution using DFT has already been implemented in the TFHE library [13], albeit in an inefficient way.

The Number Theoretic Transform (NTT) [1] is essentially a form of DFT that operates over integers rather than complex numbers. It employs a finite field with characteristic q, typically a prime number, and a primitive N-th root of unity, ω, where $\omega^N \equiv 1 \pmod{q}$ and $\omega^k \not\equiv 1 \pmod{q}$ for $k < N$. Consequently, an integer variant of the FFT algorithm can also be used to compute NTT in $O(N \log N)$ steps. There is also an efficient variant of the NTT algorithm for computing a negacyclic NTT using a primitive $2N$-th root of unity. In this paper, we will use NTT to refer to the integer version of the FFT algorithm, while FFT and FFNT will exclusively denote FFT algorithms that work with complex numbers.

Crandall proposed an efficient method for direct calculation of negacyclic convolution in [16], which was recently improved in [2]. Both methods utilize the Discrete Galois Transform (DGT), which operates over the field $\mathbb{F}_{q^2}$. Thus, DGT is also a number-theoretic variant of FFT, requiring fewer number of iterations than conventional NTT. However, since DGT operates in $\mathbb{F}_{q^2}$ and NTT in $\mathbb{F}_q$, multiplications in NTT are simpler compared to those in DGT. Therefore, comparing the efficiencies of DGT and NTT is not always straightforward.

The algorithm proposed in [23] is a negacyclic version of FFT, appropriately named the Fast Fourier Negacyclic Transform (FFNT). The authors claim to achieve around 24× better performance than the generic NTL library (comparable to Crandall's method) and around 4× better performance than a straightforward approach with DFT. Although these speed-ups are observed for CPUs, to the best of our knowledge, no GPU results have been provided so far. In this work, we provide the best optimizations of FFT, FFNT, and NTT on GPUs and compare their performance for polynomial multiplications together with NVIDIA's cuFFT library.

Since the FFNT algorithm in [23] uses $e^{2\pi i/N}$ as the primitive N-th root of unity, it necessitates pre- and post-processing steps involving a total of $2N$ extra multiplications in $\mathbb{C}$. We show that these additional $2N$ complex multiplications in pre- and post-processing steps can easily be eliminated if we work with $e^{\pi i/N}$, the primitive $2N$-th root of unity.

In many practical homomorphic encryption (HE) schemes such as BGV, BFV and CKKS [9,11,18], ciphertext usually consists of two very large degree polynomials with large coefficients, which requires multi-precision arithmetic. In addition, these HE schemes contain computationally involved operations such as homomorphic multiplication, relinearization, bootstrapping, key switching, and automorphism, which necessitate polynomial arithmetic with large coefficients. During such operations, generally so-called *gadget decomposition* is applied first to reduce the large polynomial multiplication with large coefficients into multiple polynomial multiplications with smaller coefficients.

A common approach for decomposition is to use Residue Number Systems (RNS) [30], which represents large integers with their residues with respect to a set of smaller pairwise co-prime integers. The idea is to reduce large integers using smaller machine-size moduli into single-precision integers and use NTT for the polynomial multiplications. Many GPU and FPGA optimizations are provided in the literature for RNS-variants of NTT.

When using RNS in gadget decomposition, NTT is generally the first choice for polynomial arithmetic since RNS is a number theoretic representation of large integer coefficients. However, FFT can also be used for multiplying polynomials with integer coefficients, which can be more efficient; especially on platforms equipped with efficient floating-point units. Many CPUs and GPUs claim exceptional speed for floating-point computations. Also, FFT is the first choice for polynomial multiplication in the TFHE scheme, where polynomial coefficients are torus elements. These facts further motivate the exploration of using FFT in FHE schemes.

Recently, in [5], the gadget decomposition proposed in [22] was adapted for numbers represented in base 2^k instead of the RNS form, a choice that aligns more naturally with digital hardware operating on binary numbers. This adaptation was reported to simplify and accelerate relinearization operations. Since this representation permits the use of both FFT and NTT for relinearization, identifying which method is more efficient had remained an open question until now. In this work, we show when the FFT outperforms the NTT on GPUs across a wide range of polynomial sizes in terms of execution time and energy efficiency.

Given that computation over encrypted data is reported to be several orders of magnitude slower than processing in the clear [20], there is strong motivation to accelerate fully homomorphic encryption (FHE) schemes using GPU, FPGA, and ASIC platforms. An essential design consideration for such FHE accelerators is energy efficiency, owing to their inherently heavy computational workloads. Although energy efficiency has been extensively studied in the context of machine learning applications [29], it remains largely underexplored for FHE acceleration. Furthermore, as GPU implementations of cryptographic algorithms have been reported to be less efficient than their FPGA and CPU counterparts [31], a systematic investigation into the energy efficiency of GPU-based FHE implementations becomes imperative for advancing practical and sustainable homomorphic computation.

We can summarize our contributions and outcomes of our work as follows:

1. We first present a more efficient version of the negacyclic FFT (New FFNT), which eliminates unnecessary multiplications in the pre- and post-processing steps.
2. Although FFT, FFNT, and NTT can all be used for fast polynomial multiplication in FHE applications, we demonstrate that FFNT and NTT outperform standard FFT-based solutions.
3. Despite the highly optimized FFT implementation in NVIDIA's cuFFT library, it is not an efficient alternative for FHE applications, for the reasons discussed in this paper. However, FHE applications could greatly benefit from an efficient implementation of FFNT integrated into the cuFFT library, enabling faster polynomial multiplications.
4. We argue and demonstrate that whether FFNT- or NTT-based polynomial multiplication is superior depends largely on floating-point arithmetic capacity in the underlying GPU platform.
5. On GPU platforms equipped with a sufficiently large number of 64-bit floating-point units, FFNT can outperform NTT-based solutions. Otherwise, our experiments show that NTT-based polynomial multiplication consistently surpasses FFNT-based implementations by a substantial margin. A similar trend is observed in terms of energy efficiency. Our investigations reveal that energy efficiency is influenced by factors such as execution time, GPU utilization, and the architectural characteristics of the underlying GPU hardware.
6. The digit decomposition method of [5] reduces polynomial multiplication with large coefficients into multiple multiplications with arbitrarily small coefficients at the cost of additional DFT operations. This approach allows leveraging high-speed, low-precision floating-point units in FHE applications. An open problem has been whether FFT, in this context, could be advantageous on GPUs. We show that NTT with 64-bit primes at least matches the performance of the FFNT-based solution and yields significantly better timing results most of the time for $N \in [2^{10}, 2^{16}]$, which is the typical parameter range for FHE schemes. This advantage largely stems from the 53-bit precision limit of the largest floating-point numbers compared to the full 64-bit precision provided by their integer counterparts.
7. We further demonstrate that our NTT-based implementation of the TFHE scheme outperforms other known implementations.

Our optimized CUDA codes for FFT, FFNT, cuFFT, and NTT are publicly available[1].

---

[1] Our CUDA codes for efficient polynomial multiplication via FFT, FFNT, cuFFT, and NTT are publicly available for future improvements and comparisons: https://github.com/Alisah-Ozcan.

2 Preliminaries

Instead of using the schoolbook method for multiplying large degree polynomials, a method based on Discrete Fourier Transform is adopted for efficiency as it reduces the complexity from $O(N^2)$ to $O(N \log N)$. The forward DFT maps a complex-valued vector $\boldsymbol{a}$ into a complex-valued vector $\boldsymbol{A}$ of the same size by

$$A_k = \sum_{n=0}^{N-1} a_n e^{\frac{-2\pi i}{N} kn}, \text{ for } k = 0, \ldots, N-1, \tag{1}$$

where $\omega = e^{\frac{-2\pi i}{N}}$ is a primitive N^{th} root of unity. The inverse transform (iDFT) is the same, except the sign of the exponent e is positive and each sum is divided by N:

$$a_n = \frac{1}{N} \sum_{k=0}^{N-1} A_k e^{\frac{2\pi i}{N} kn}, \text{ for } n = 0, \ldots, N-1. \tag{2}$$

There is a DFT-based algorithm to compute $a(x) \cdot b(x) \bmod (x^N - 1)$, where $a(x), b(x) \in \mathbb{C}/(x^N - 1)$, when DFT is computed with primitive N^{th} of unity:

$$a(x) \cdot b(x) \bmod (x^N - 1) = iDFT\big(DFT(a(x)) \odot DFT(b(x))\big), \tag{3}$$

where $\odot$ denotes pairwise multiplication of vector elements A_k and B_k for $k = 0, \ldots, N-1$, sometimes referred to as the *Hadamard multiplication*. Algorithm 1 that computes DFT is commonly known as Fast Fourier Transform with complexity $O(N \log N)$. We will refer to Algorithm 1 as cyclic FFT as it uses primitive N^{th} of unity ζ and it is used to implement multiplication in $\mathbb{C}/(x^N - 1)$.

In Algorithm 1, the function, $\Gamma(i, n)$ is the bit reverse function, which reverses the binary representation of the n-bit number, i. The operations performed in Steps 10–13 are famously known as the Cooley-Tukey (CT) butterfly operation [15]. Note that there are $N/2$ different powers of ζ used in all executions of Step 11 of the algorithm, which are usually precomputed for efficiency. The precomputed powers of ζ can be stored in a table in bit-reverse order for easy access during the FFT computation. There are also variants of the FFT algorithm that allow online computation of powers of ζ with no need for storage. Algorithm 4.2 in [14], known as the decimation in frequency (DIF) FFT algorithm, uses Gentlemen Sande (GS) butterfly circuit [19] and computes the powers of a primitive N^{th} root of unity ζ on the fly.

Given the cyclic FFT_1 in Algorithm 1, one can compute multiplication $a(x) \cdot b(x) \bmod (x^N + 1)$, where $a(x), b(x) \in \mathbb{C}/(x^N + 1)$, which is often referred to as negacyclic multiplication. To this end, we use pre- and post-processing steps, which involve the multiplication of vector elements with certain powers of primitive $2N^{\text{th}}$ root of unity, ω with $\zeta = \omega^2$; operation is known as twisting and untwisting. The method is illustrated in Algorithm 2, and the transformation employed in the process is commonly referred to as the negacyclic FFT (FFNT).

Algorithm 1. Cyclic FFT (FFT_1)

Require: $a(x) \in \mathbb{C}[x]/(x^N - 1)$ and a primitive N-th root of unity ζ, and N
Ensure: $\mathbf{A} \in \mathbb{C}^N$ where $\mathbf{A} = DFT(a(x))$, in bit-reversed order
1: **for** i from 0 to $N - 1$ **do**
2:     $\mathbf{A}[i] \leftarrow a_i$
3: **end for**
4: $\Delta \leftarrow N$; $m \leftarrow 1$
5: **while** $m < N$ **do**
6:     $\Delta \leftarrow \Delta/2$
7:     **for** i from 0 to $m - 1$ **do**
8:         $j_1 \leftarrow 2 \cdot i\Delta$; $j_2 \leftarrow j_1 + \Delta - 1$
9:         **for** j from j_1 to j_2 **do** $\triangleright$ *CT butterfly*
10:            $t_0 \leftarrow \mathbf{A}[j]$
11:            $t_1 \leftarrow \mathbf{A}[j + \Delta] \cdot \zeta^{\Gamma(i, N/2)}$
12:            $\mathbf{A}[j] \leftarrow (t_0 + t_1)$
13:            $\mathbf{A}[j + \Delta] \leftarrow (t_0 - t_1)$
14:        **end for**
15:    **end for**
16: **end while**
17: **return A**

Algorithm 2. FFNT-based multiplication in $\mathbb{C}/(x^N + 1)$

Require: $a(x), b(x) \in \mathbb{C}[x]/(x^N + 1)$ and a primitive N-th root of unity ζ and primitive $2N$-th root of unity ω, where $\zeta = \omega^2$
Ensure: $c(x) = a(x)b(x) \bmod (x^N + 1)$
1: **for** i from 0 to $N - 1$ **do**
2:     $\tilde{a}_i \leftarrow a_i \cdot \omega^i$ $\triangleright$ *twisting*
3:     $\tilde{b}_i \leftarrow b_i \cdot \omega^i$ $\triangleright$ *twisting*
4: **end for**
5: $\tilde{\boldsymbol{A}} \leftarrow \text{FFT}_1(\tilde{a}(x), \zeta, N)$
6: $\tilde{\boldsymbol{A}} \leftarrow \text{FFT}_1(\tilde{b}(x), \zeta, N)$
7: **for** i from 0 to $N - 1$ **do**
8:     $\tilde{\boldsymbol{C}}_i \leftarrow \tilde{\boldsymbol{A}}_i \cdot \tilde{\boldsymbol{B}}_i$
9: **end for**
10: $\tilde{c}(x) \leftarrow \text{iFFT}_1(\tilde{\boldsymbol{C}}, \zeta^{-1}, N)$
11: **for** i from 0 to $N - 1$ **do**
12:    $c_i \leftarrow \tilde{c}_i \cdot \omega^{-i}$ $\triangleright$ *untwisting*
13: **end for**
14: **return A**

If the coefficients of $a(x)$ are real numbers, twisting them results in a vector of complex numbers of size $N/2$

$$(\tilde{a}_i + j\tilde{a}_{i+N/2})$$

for $j = 0, 1, \ldots, N/2 - 1$ since $\omega^{N/2} = \sqrt{-1} = j$. This operation is referred to as folding, and we obtain Algorithm 3, which is more efficient as it uses half-sized FFT operations [23]. The new algorithm proves to be very useful for FHE schemes which often employ the polynomial rings $\mathbb{R}/(x^N + 1)$ and $\mathbb{Z}/(x^N + 1)$.

Algorithm 3. FFNT-based multiplication in $\mathbb{R}/(x^N+1)$ with half size FFT_1 [23]

Require: $a(x), b(x) \in \mathbb{R}[x]/(x^N+1)$ and a primitive N-th root of unity ζ and primitive $2N$-th root of unity ω, where $\zeta = \omega^2$

Ensure: $c(x) = a(x)b(x) \bmod (x^N + 1)$

1: **for** i from 0 to $N/2 - 1$ **do**
2:     $\tilde{a}_i \leftarrow (a_i + ja_{i+N/2}) \cdot \omega^i$ ▷ *folding and twisting*
3:     $\tilde{b}_i \leftarrow (b_i + jb_{i+N/2}) \cdot \omega^i$ ▷ *folding and twisting*
4: **end for**
5: $\tilde{\boldsymbol{A}} \leftarrow \mathrm{FFT}_1(\tilde{a}(x), \zeta, N/2)$
6: $\tilde{\boldsymbol{B}} \leftarrow \mathrm{FFT}_1(\tilde{b}(x), \zeta, N/2)$
7: **for** i from 0 to $N/2 - 1$ **do**
8:     $\tilde{\boldsymbol{C}}_i \leftarrow \tilde{\boldsymbol{A}}_i \cdot \tilde{\boldsymbol{B}}_i$
9: **end for**
10: $\tilde{c}(x) \leftarrow \mathrm{iFFT}_1(\tilde{\boldsymbol{C}}, \zeta^{-1}, N/2)$
11: **for** i from 0 to $N/2 - 1$ **do**
12:     $c_i \leftarrow \mathrm{Re}(\tilde{c}_i) \cdot \omega^{-i}$
13:     $c_{i+N/2} \leftarrow \mathrm{Im}(\tilde{c}_i) \cdot \omega^{-i}$ ▷ *unfolding and untwisting*
14: **end for**
15: **return A**

Number Theoretic Transform [1] is simply the variant of DFT, which works with vectors (or polynomials) of integers, instead of complex numbers. When the integers are the elements of the field $\mathbb{Z}_q$ for a prime q, then one can find a primitive N^{th} root (or $2N^{th}$ root) of unity such that $\zeta^N \bmod q = 1$ ($\omega^{2N} \bmod q = 1$). For this, we need to have $q \bmod N = 1$ (or $q \bmod 2N = 1$). Then, the efficient FFT algorithms such as FFT_1 in Algorithm 1 or its negacyclic version FFT_2 (see Sect. 3) can be profitably used step by step if the complex arithmetic is substituted with modular arithmetic in $\mathbb{Z}_q$. One can even obtain an integer counterpart of Algorithm 3 for the fast polynomial multiplication in $\mathbb{Z}_q/(x^N+1)$ such as Discrete Galois Transform [16], where the polynomial coefficients are the elements of the extended field $\mathbb{F}_{q^2}$.

For efficiency reasons, it is often advantageous to use special-form primes larger than q to define the NTT for polynomial multiplication. A common choice in FHE applications is the so-called Goldilocks prime $P = 2^{64} - 2^{32} + 1$, since modular reduction by P can be performed efficiently using only addition and subtraction operations. Such primes are sometimes referred to as *carrier primes*.

When NTT arithmetic is carried out modulo P, care must be taken to ensure that the overall computation remains consistent with arithmetic modulo q. In particular, we must prevent intermediate values from exceeding P, as they would otherwise be reduced modulo P. A loose upper bound on the size of a coefficient in the resulting polynomial after multiplication (Eq. 3) is Nq^2. Thus, to preserve correctness, we require that $P > Nq^2$.

Most modern CPUs are equipped with advanced hardware units for fast floating-point arithmetic, often combined with SIMD (Single Instruction, Multiple Data) capabilities to enable the parallel processing of numerous real number operations. As a result, the FFT may be preferred over the NTT for polynomial multiplication in HE schemes. Additionally, FFT is a natural choice for the TFHE scheme, where polynomial coefficients are rational numbers. In some cases, FFT is also recommended for polynomial multiplication when coefficients are integers, leveraging the hardware efficiencies associated with floating-point computations.

Similarly, leading graphics processing unit (GPU) manufacturer NVIDIA® offers optimized FFT implementations for CUDA-enabled devices through their cuFFT library. The cuFFT library supports both real-valued and complex-valued inputs and efficiently computes Discrete Fourier Transforms. FFT is an $(O(N \log N))$ algorithm for any input size, but cuFFT performs faster when the input size is of the form $(2^a \times 3^b \times 5^c \times 7^d)$. Additionally, cuFFT supports 16-bit (half-precision), 32-bit (single-precision), and 64-bit (double-precision) floating-point formats, with FFT computations generally being faster at lower precisions. Since cryptographic schemes often require higher-precision computations, it is important to thoroughly investigate the advantages of using FFT for polynomial multiplication in homomorphic encryption (HE) schemes, including TFHE. Here, one of our primary objectives is to explore and address this key research question.

In this work, we employed two NVIDIA® GPUs with different floating-point arithmetic capacities to demonstrate that NTT and FFNT may exhibit distinct execution characteristics. The specifications of these GPUs are listed in Table 1. The A100 40 GB, high-performance data-center GPU, represents a device with relatively strong double-precision capability, featuring 32 FP64 units per streaming multiprocessor, whereas the RTX 4090 is a high-end GPU optimized for compute-intensive workloads but equipped with only two FP64 units per streaming multiprocessor.

3   A More Efficient Algorithm for FFNT

By replacing only Step 11 of FFT_1, Algorithm 1, with $t_1 \leftarrow \omega^{\Gamma(m+i,N)}$, where ω is the primitive $2N^{\text{th}}$ root of unity, one can obtain the negacyclic version of the FFT, referred to here as FFT_2. Then, the multiplication in $\mathbb{R}/(x^n + 1)$ can be performed without pre- and post-processing steps:

$$a(x) \cdot b(x) \bmod (x^N + 1) = \mathsf{iFFT}_2\big(\mathsf{FFT}_2(a(x), \omega, N) \odot \mathsf{FFT}_2(b(x), \omega, N), \omega^{-1}, N\big). \tag{4}$$

Table 1. The basic specifications of the GPUs used in this work. Clock rates correspond to the maximum boost frequency. Execution units are reported per Streaming Multiprocessor (SM).

Specification	A100 40 GB	GeForce RTX 4090
Cores	6912	16384
Clock Rate	1420 MHz	2550 MHz
Bandwidth	1.56 TB/s	1.01 TB/s
CC	8.0	8.9
Architecture	Ampere	Ada Lovelace
FP32 per SM	64	128
FP64 per SM	32	2
INT32 per SM	64	64

As a result, FFT_2 saves $3N$ multiplication operations while the storage space for powers of ω doubles, as the algorithm needs all powers ω^j for $j = 1, \ldots, N-1$. Consequently, we obtain a more efficient Algorithm 4. Note that folding and unfolding operations do not require any arithmetic; therefore, they are essentially free. We also provide steps of FFT_3 and iFFT_3 in Algorithm 5 and Algorithm 6, respectively, since they are slightly different from FFT_1 and FFT_2. Note that $\omega = e^{-\pi/N}$, which is a primitive $2N^{\mathrm{th}}$ root of unity with $\omega^{2N} = 1$ and $\omega^N = -1$.

Algorithm 4. Multiplication in $\mathbb{R}/(x^N + 1)$ with half size FFTs

---

Require: $a(x), b(x) \in \mathbb{R}[x]/(x^N + 1)$ and a primitive $2N$-th root of unity ω.
Ensure: $c(x) = a(x)b(x) \bmod (x^N + 1)$
1: **for** i from 0 to $N/2 - 1$ **do**
2:     $\tilde{a}_i \leftarrow (a_i + ja_{i+N/2})$ $\triangleright$ *folding*
3:     $\tilde{b}_i \leftarrow (b_i + jb_{i+N/2})$ $\triangleright$ *folding*
4: **end for**
5: $\tilde{A} \leftarrow \mathrm{FFT}_3(\tilde{a}, \omega, N/2)$
6: $\tilde{B} \leftarrow \mathrm{FFT}_3(\tilde{b}(x), \omega, N/2)$
7: **for** i from 0 to $N/2 - 1$ **do**
8:     $\tilde{C}_i \leftarrow \tilde{A}_i \cdot \tilde{B}_i$
9: **end for**
10: $C \leftarrow \mathrm{iFFT}_3(\tilde{C}, \omega^{-1}, N/2)$
11: **for** i from 0 to $N/2 - 1$ **do**
12:     $c_i \leftarrow \mathrm{Re}(C_i); \; c_{i+N/2} \leftarrow \mathrm{Im}(C_i)$ $\triangleright$ *unfolding*
13: **end for**
14: **return A**

Algorithms 3, 5, and 6 can be adapted to the NTT, the integer equivalent of the FFT, by employing the method introduced by Crandall [16], which refers to it as the Discrete Galois Transform (DGT). If a primitive $2N$-th root of unity

---

Algorithm 5. Negacyclic half FFT (FFT$_3$)

Require: $a \in \mathbb{C}^{N/2}$, a primitive $2N$-th root of unity ω, and N
Ensure: $\mathbf{A} \in \mathbb{C}^{N/2}$ (in bit-reversed order)
 1: **for** i from 0 to $N/2 - 1$ **do**
 2:     $\mathbf{A}[i] \leftarrow a[i]$
 3: **end for**
 4: $\Delta \leftarrow N/4$; $m \leftarrow 1$
 5: **while** $m < N/2$ **do**
 6:     **for** i from 0 to $m - 1$ **do**
 7:        $j_1 \leftarrow 2 \cdot i\Delta$; $j_2 \leftarrow j_1 + \Delta - 1$
 8:        **for** j from j_1 to j_2 **do** $\triangleright$ *CT butterfly*
 9:           $t_0 \leftarrow \mathbf{A}[j]$
10:           $t_1 \leftarrow \mathbf{A}[j + \Delta] \cdot \omega^{\Gamma(2m+i,N)}$
11:           $\mathbf{A}[j] \leftarrow (t_0 + t_1)$
12:           $\mathbf{A}[j + \Delta] \leftarrow (t_0 - t_1)$
13:        **end for**
14:     **end for**
15:     $\Delta \leftarrow \Delta/2$; $m \leftarrow 2m$
16: **end while**
17: **return A**

ω exists in $\mathbb{F}_{q^2}$, these algorithms can facilitate fast polynomial multiplication in $\mathbb{Z}_q[X]/(x^N + 1)$. This requires replacing complex number arithmetic with arithmetic in $\mathbb{F}_{q^2}$.

We will now evaluate whether the DGT offers advantages over the classical NTT variant of FFT$_2$ by analyzing the number of $\mathbb{F}_q$ multiplications, as these are the most prevalent and computationally expensive operations. The multiplication in $\mathbb{F}_q$, which is modular multiplication, can be efficiently computed using either Barrett's method [4] or Montgomery algorithm [24] for modular reduction. Each multiplication in $\mathbb{F}_{q^2}$ takes four $\mathbb{F}_q$ multiplications[2]. The $\mathbb{Z}_q[X]/(x^N + 1)$ multiplication based on our variant of the DGT algorithm, which requires no twisting and untwisting operations, performs

$$3N \log \frac{N}{2} \text{ and } 6N \log \frac{N}{2}$$

$\mathbb{F}_q$ multiplication and $\mathbb{F}_q$ addition/subtraction operations, respectively. The NTT-based algorithm, on the other hand, requires

$$1.5N \log N \text{ and } 3N \log N$$

$\mathbb{F}_q$ multiplications and addition/subtraction operations, respectively. Then, the NTT-based algorithm performs fewer number of $\mathbb{F}_q$ multiplications and addition/subtraction operations than the DGT-based algorithm for $N > 2^2$, which

---

[2] Since the Karatsuba algorithm [21] introduces extra single-precision integer addition instructions, whose complexity on general purpose computers are comparable to those of single-precision multiplication instructions we use schoolbook method for $\mathbb{F}_{q^2}$ multiplication.

Algorithm 6. Negacyclic half inverse FFT (iFFT$_3$)

Require: $\mathbf{A} \in \mathbb{C}^{N/2}$ (in bit-reversed order), a primitive $2N$-th root of unity ω, and N
Ensure: $a \in \mathbb{C}^{N/2}$

1: **for** i from 0 to $N/2 - 1$ **do**
2:    $a[i] \leftarrow \mathbf{A}[i]$
3: **end for**
4: $\Delta \leftarrow 1; m \leftarrow N/2$
5: **while** $m > 1$ **do**
6:    $j_1 \leftarrow 0; h \leftarrow m/2$
7:    **for** i from 0 to $h - 1$ **do**
8:       $j_2 \leftarrow j_1 + \Delta - 1$
9:       **for** j from j_1 to j_2 **do** $\triangleright$ *GS butterfly*
10:          $t \leftarrow a[j]$
11:          $a_j \leftarrow 0.5 \cdot (t + a[j + \Delta])$
12:          $a[j + \Delta] \leftarrow 0.5 \cdot ((t - a[j + \Delta]) \cdot \omega^{-\Gamma(2h+i,N)})$
13:       **end for**
14:    **end for**
15:    $\Delta \leftarrow 2\Delta; m \leftarrow m/2$
16: **end while**
17: **return** a

---

is always the case in all HE schemes. Therefore, we prefer the NTT-based algorithm for $\mathbb{Z}_q[X]/(x^N + 1)$ multiplications over the DGT-based algorithm.

Next, we compare the efficiencies of the FFT-based and NTT-based algorithms for both $\mathbb{Z}_q[X]/(x^N + 1)$ and $\mathbb{R}[X]/(x^N + 1)$ multiplications. Modern computers have highly optimized floating-point hardware units responsible for supporting real arithmetic, where IEEE 754 is used as the standard to represent real numbers. The precision of the mantissa (significand) part of the double floating-point type is 52 bits (53 bits with the implicit bit). As a result, the precision of the polynomial coefficients r can be determined by the inequality $2^{53} > Nr^2$, as Nr^2 is the precision of each coefficient of the resulting polynomial after multiplication. This is true for polynomials in both $\mathbb{Z}_q[X]/(x^N + 1)$ and $\mathbb{R}[X]/(x^N + 1)$. For instance, when $N = 2^{13}$, $r < 2^{20}$.

To compute $\mathbb{R}[X]/(x^N + 1)$ multiplications using NTT, we need to employ a so-called *carrier prime* approach. Here, we use a sufficiently large prime q, where $q > Nr^2$. Since q can be as large as a 64-bit value in a 64-bit machine, utilizing NTT actually gives higher precision than the IEEE 754 double floating-point number type. For instance, when $N = 2^{13}$, $r < 2^{25.5}$.

Most modern computers are equipped with dedicated hardware units for integer and floating-point arithmetic: the Arithmetic Logic Unit (ALU) and the Floating-Point Unit (FPU), respectively. Because scientific computations often involve numerous floating-point operations, many platforms include multiple highly optimized FPUs to enhance performance. Additionally, multiplication units within both ALUs and FPUs are among the most complex and resource-intensive operations, often consuming significant time. Let τ_i^m, τ_f^m τ_i^a and τ_f^a denote the latencies of instructions for integer multiplication, floating-point multiplications, integer addition and floating point addition, respectively[3]. Depending on the hardware design and the number of available units, it is possible that $\tau_f < \tau_i$, meaning floating-point arithmetic operations could be performed more quickly than integer operations in some architectures.

Since the FFT involves complex arithmetic, multiplying two complex coefficients in $\mathbb{R}[X]/(x^N + 1)$ requires four floating-point multiplications. In contrast, the corresponding operation in NTT is modular multiplication in $\mathbb{F}_q$, which can be performed with three integer multiplications, assuming Montgomery multiplication is employed and one operand, such as a power of ω, is fixed.

Now, we can estimate the execution time for polynomial multiplication in $\mathbb{R}[X]/(x^N + 1)$ or $\mathbb{Z}_q[X]/(x^N + 1)$ using either FFNT- or NTT-based methods. Focusing on the dominant multiplication operations, the estimated execution time for FFNT-based polynomial multiplication is given by:

$$T_{\text{FFNT}} = 3N \log \frac{N}{2} (\tau_f^m + 2\tau_f^a).\tag{5}$$

For the NTT-based method, the execution time is:

$$T_{\text{NTT}} = 4.5N \log N (\tau_i^m + 2\tau_i^a).\tag{6}$$

Comparing these, the FFNT-based approach can become advantageous if:

$$(\tau_f^m + 2\tau_f^a) < \frac{3 \log N}{2(\log N - 1)}(\tau_i^m + 2\tau_i^a),\tag{7}$$

meaning that floating-point multiplication units must either be sufficiently fast or sufficiently numerous for FFNT to outperform NTT. While many GPU architectures feature floating-point units optimized for the short precision ranges commonly used in AI applications, not all of them provide a sufficient number of 64-bit double-precision units (FP64), which are particularly relevant for FHE implementations. In addition, other factors—such as memory access patterns and microarchitectural characteristics, whose details are often only partially known—can also influence performance. Therefore, experimental evaluation on GPU platforms with different hardware configurations is essential to draw reliable conclusions.

In the subsequent sections, we present comprehensive implementation results for FHE-related operations on two different GPU architectures: one featuring a relatively large number of FP64 units, and another high-end GPU equipped with only a small number of such units.

---

[3] By latency, we typically mean the average execution time of a single multiplication when multiple such operations are conducted in parallel across many available multiplication units.

4   Implementation Results

In this section, we present the implementation results of FFT[4], cuFFT, FFNT[5], and NTT-based[6] solutions on the two GPU devices listed in Table 1. In the first subsection, we compare four FFT- and FFNT-based approaches for polynomial multiplication and show that the newly optimized FFNT implementation generally outperforms the others. We then compare FFNT- and NTT-based solutions for polynomial multiplication and demonstrate that the better-performing algorithm depends on the GPU architecture. Next, we evaluate both approaches in the context of the digit decomposition method proposed in [5], where NTT is almost always faster on both GPU platforms. Finally, we compare FFNT- and NTT-based solutions for TFHE, where FFNT-based methods are typically favored. Our TFHE experiments indicate that the choice of algorithm should be guided by the characteristics of the underlying GPU device.

4.1   Performance of Various FFT Algorithms on GPU

In this section, first we provide timing results for four different methods for polynomial multiplications using four different FFT algorithms. Recall that one polynomial multiplication requires two forward and one inverse FFT operations, and one Hadamard multiplication. These methods are outlined as follows:

- **Straightforward Method:** Here we use the cyclic FFT in Algorithm 1, FFT_1, where the input size of the FFT is $2N$. Inputs are padded with 0's to match the FFT size and the polynomial reduction with $x^N + 1$ is performed at the end separately. This method is included as a reference implementation with no optimization except that only the number of GPU kernels is minimized.

- **cuFFT-based Method:** This is similar to the straightforward method in the sense that cyclic convolution is used since highly optimized functions of NVIDIA's cuFFT library are employed for fast FFT computation. The library includes the dedicated functions `cufftExecR2C()`, `cufftExecC2C()`, `cufftExecD2Z()`, and `cufftExecZ2Z()`. In the function designations, the letters R, C, D, and Z represent single precision real value, single precision complex value, double precision real value, and double precision complex value, respectively. Since we focus on real-valued inputs, in our experiments we used `cufftExecR2C()`, `cufftExecD2Z()`, and related functions. While the FFT computations can be performed extremely fast with the library functions, the fact that the FFT size doubles from N to $2N$ has a significant adverse effect on timing. Additionally, we need separate kernels for FFT, inverse FFT and Hadamard multiplication operations, which also affects the performance considerably since using the cuFFT library allows only limited kernel optimization.

---

[4] https://github.com/Alisah-Ozcan/GPU-FFT.
[5] https://github.com/Alisah-Ozcan/GPU-FFT.
[6] https://github.com/Alisah-Ozcan/GPU-NTT.

- **Algorithm 3:** The algorithm is introduced as FFNT recently in [23] and it is reported that FFNT can be used for polynomial multiplication (i.e., here Algorithm 3) without errors and performs four times faster than a conventional FFT on CPUs. Since it computes the negacyclic FFT and the number of kernels can be optimized, it is expected to be much faster than both the straightforward and the cuFFT-based methods.
- **Algorithm 4:** This method is almost identical to Algorithm 3 except for the fact that twisting and untwisting operations are not performed. As the omitted operations are not the bottleneck of the computations, Algorithm 4 is expected to be only slightly better than Algorithm 3, and its advantage becomes more (or less) important depending on the underlying architecture, as we observe in the timing results presented below.

We implemented all four methods in CUDA using double precision for floating point arithmetic (64-bit) (FP64 units) and ring dimensions in the range of $[2^{10}, 2^{16}]$; meaningful values that cover the majority of FHE applications. We provide the execution timings for a single polynomial multiplication in Table 2 and Table 3 for RTX 4090 and A100 GPUs, respectively. The rightmost column tabulates the speedup values of Algorithm 4 over straightforward, cuFFT and Algorithm 3, in that order.

Table 2. Polynomial multiplication timings using straightforward, cuFFT, Algorithm 3 and Algorithm 4 on an **RTX 4090** in ms.

$\log N$	Straightfwd	$cuFFT$	Algorithm 3	Algorithm 4	Speedup (%)
10	0.018085	0.057696	0.013814	0.013627	32.7/323.4/1.4
11	0.019141	0.123766	0.020032	0.017473	9.5 608.3/14.6
12	0.019935	0.147175	0.021425	0.018457	8.0/697.4/16.1
13	0.021213	0.326779	0.020204	0.019618	8.1/1565.7/.0
14	0.022723	0.118859	0.021674	0.020767	9.4/472.3/4.4
15	0.031040	0.118964	0.022426	0.021839	42.1/444.7/2.7
16	0.060727	0.118882	0.023451	0.023401	159.5/408.0/0.2

On the RTX 4090, the cuFFT-based implementation is generally the slowest among the four methods due to inefficient kernel utilization, whereas Algorithm 4 proves to be the fastest. The speedup ranges from 0.2% to as high as 16% compared to Algorithm 3, the second fastest method, mainly because it requires fewer multiplications.

For the A100, however, the results are more nuanced and mixed due to several contributing factors. First, the highly optimized cuFFT library efficiently utilizes the 32 FP64 units for smaller ring dimensions, $N = 2^{10}$ and $N = 2^{11}$, yielding the best performance in these cases. However, this advantage disappears rapidly for larger ring dimensions because of its high memory demands and the large number of kernel launches required. Consequently, even for smaller ring dimensions such

Table 3. Polynomial multiplication timings using straightforward, cuFFT, Algorithm 3 and Algorithm 4 on an **A100** in ms.

$\log N$	Straightfwd	cuFFT	Algorithm 3	Algorithm 4	Speedup (%)
10	0.056218	0.028544	0.045747	0.046029	$18.1/-61.3/-0.6$
11	0.055194	0.041037	0.054195	0.054221	$1.8/-32.1/-0.05$
12	0.055296	0.058368	0.054246	0.054502	$1.4/6.6/-0.5$
13	0.055194	0.099482	0.054323	0.054195	$1.8/45.5/0.2$
14	0.053197	0.189875	0.054912	0.054810	$-3.0/71.1/0.2$
15	0.055834	0.061184	0.055424	0.052198	$6.5/14.7/5.8$
16	0.064973	0.062387	0.053222	0.053222	$18.1/14.7/0.0$

as $N = 2^{10}$ and $N = 2^{11}$, the cuFFT-based implementation is not preferable, as its kernel usage becomes difficult to optimize when integrated into more involved implementations of FHE schemes.

Regarding the comparison between Algorithm 3 and Algorithm 4 on the A100, the advantage of the latter appears to diminish. This is primarily because the A100 features a large number of FP64 units operating in parallel, making the additional multiplications in the pre- and post-processing steps of Algorithm 3 account for only a negligible portion of the total execution time.

Having established that the latter two algorithms are generally preferable to the former two, we utilized Algorithm 4 and Algorithm 3 to perform multiple polynomial multiplications in batches. The execution times for these algorithms taken on the RTX 4090 and A100 GPUs, are presented in Table 4. In FHE applications, where polynomial multiplications are commonly performed in batches, the timing results clearly demonstrate that Algorithm 4, introduced in this paper, is usually the better FFNT-based algorithm for computing the negacyclic DFT. In the next section, we compare the performance of FFNT and NTT algorithms for polynomial multiplication.

4.2 Comparing NTT and FFNT Based Polynomial Multiplications

We implemented an optimized iterative NTT algorithm, similar to the radix-2 CT algorithm in [27], which is also the integer counterpart of FFT_2. We then compared the execution times of polynomial multiplication based on NTT with those of polynomial multiplication using Algorithm 4, for both single and batch executions. For the NTT, a 64-bit prime modulus was used, whereas Algorithm 4 employed double-precision floating-point arithmetic.

The execution times for single and batch polynomial multiplications are reported in Table 5 and Table 6, respectively. The results indicate that NTT-based polynomial multiplication is significantly faster than the FFNT-based approach on the RTX 4090, whereas the opposite trend is observed on the A100 architecture. It is also important to note that the NTT-based method operates with full 64-bit precision, while the FFNT-based method is constrained by the underlying floating-point representation (IEEE 754 double precision),

Table 4. Batch polynomial multiplication timings in ms with FFNT (**Algorithm** 3) and new FFNT (**Algorithm** 4) on **RTX 4090/ A100** with double precision (64 bits).

$\log N$	Batch Count	Algorithm 3 (RTX 4090/ A100)	Algorithm 4 (RTX 4090/ A100)	Speedup (%) (RTX 4090/ A100)
10	16	0.01619/0.04661	0.01392/0.04715	14.0/−1.1
	32	0.01446/0.04723	0.01383/0.04692	4.3/0.6
	64	0.01499/0.04725	0.01422/0.04697	5.1/0.6
	128	0.01895/0.04910	0.01776/0.04759	6.2/3.1
11	16	0.01840/0.05473	0.01753/0.05450	4.7/0.4
	32	0.01907/0.05609	0.01801/0.05475	5.5/2.3
	64	0.02457/0.05905	0.02309/0.05660	6.0/4.1
	128	0.04003/0.06115	0.03802/0.05946	5.0/2.7
12	16	0.01995/0.05524	0.01908/0.05488	4.3/0.6
	32	0.02602/0.05903	0.02418/0.05703	7.0/3.3
	64	0.04375/0.06149	0.04042/0.06016	7.6/2.1
	128	0.07677/0.08094	0.07255/0.07928	5.4/2.0
13	16	0.02708/0.05893	0.02582/0.05703	4.6/3.2
	32	0.04597/0.06225	0.04348/0.06033	5.4/3.0
	64	0.08218/0.08376	0.07793/0.08153	5.1/2.6
	128	0.15507/0.12585	0.14552/0.12375	6.1/1.6
14	16	0.04899/0.06348	0.04629/0.06225	5.5/1.9
	32	0.08973/0.08668	0.08313/0.08422	7.3/2.8
	64	0.16583/0.13068	0.15591/0.12736	5.9/2.5
	128	0.32673/0.23682	0.30078/0.19898	7.9/15.9
15	16	0.09512/0.08857	0.08898/0.07045	6.4/20.4
	32	0.18024/0.13506	0.16637/0.10849	7.6/19.6
	64	0.34261/0.24476	0.32225/0.20410	5.9/16.6
	128	0.68908/0.40442	0.64001/0.40453	7.1/−0.02
16	16	0.19190/0.11617	0.17859/0.11153	6.9/3.9
	32	0.37072/0.21465	0.34394/0.21140	7.2/1.5
	64	0.72321/0.40732	0.68192/0.40857	5.7/−0.3
	128	1.44600/0.75438	1.36500/0.75430	5.6/0.01

which provides only 52 bits of effective precision—typically the usable range for FHE schemes when floating-point arithmetic is used. In the following section, we demonstrate that the precision offered by the underlying arithmetic units can play a decisive role in the practical implementation of FHE schemes.

4.3   NTT and FFNT for Decomposition in FHE Schemes

Lattice-based fully homomorphic encryption (FHE) schemes are built upon noisy encryption, where plaintexts are masked with random noise to ensure security.

Table 5. Polynomial multiplication timings in ms for new FFNT (Algorithm 4) and NTT on **RTX 4090/A100** with double precision (64 bits).

$\log N$	Algorithm 4 (RTX 4090/ A100)	NTT (RTX 4090/ A100)	Speedup (%) (RTX 4090/ A100)
10	0.01362/0.04602	0.00545/0.05585	150.0%/−21.3%
11	0.01747/0.05422	0.00846/0.06289	106.0%/−16.0%
12	0.01845/0.05450	0.00875/0.06241	111.0%/−14.5%
13	0.01961/0.05419	0.00877/0.06269	124.0%/−15.6%
14	0.02076/0.05481	0.00924/0.06400	125.0%/−16.7%
15	0.02183/0.05219	0.00985/0.06420	122.0%/−23.0%
16	0.02340/0.05322	0.01146/0.06343	104.0%/−19.1%

Many advanced FHE schemes employ the technique of decomposition to control noise growth. This process converts a pair of ciphertext polynomials into a set of shorter polynomials[7], which are then multiplied by an evaluation key; a computationally intensive procedure known as *relinearization*. Similar operations are also applied in key switching, external product, and homomorphic rotations of ciphertext, which are among the most time-consuming operations in FHE cryptography.

The residue number system (RNS) has become the standard approach both for decomposition and for enabling efficient arithmetic in most FHE schemes. In RNS [3], a large integer is represented by its residues with respect to a set of smaller, pairwise coprime moduli. Since the number of Discrete Fourier Transforms (DFTs) required during relinearization grows quadratically with the number of moduli, DFT computations constitute the primary performance bottleneck.

In [22], the costly decomposition operation over a large modulus is replaced with more efficient operations over a ring of integers with a small bound. This approach notably reduces the number of DFTs considerably without incurring additional noise growth. The authors demonstrated the effectiveness of their method by applying it to the relinearization operation. Their CPU-based experiments show that the proposed relinearization technique achieves speedups of $2.3\times$ and $3.3\times$ compared to the previous method when the base ring dimension is 2^{15} or 2^{16}, respectively.

The authors in [5] extended the result of [22] beyond the full-RNS setting and showed that one can completely decouple the big number decomposition from the cyclotomic arithmetic aspects. In other words, a large integer coefficient of a ciphertext polynomial can be decomposed into a set of smaller sized binary digits, referred to as *limbs*. For example, when $N = 2^{15}$ the CKKS scheme requires a coefficient modulus Q, where $\log Q \approx 880$ to achieve more than 100-bit security. One, now, can decompose an 880-bit integer in different ways depending on the limb size, K. For instance, when $K = 5$ and $K = 19$, the number of limbs, the batch size, $\ell = 176$ and $\ell = 47$, respectively.

---

[7] Here, the term short polynomial refers to polynomials with smaller coefficients.

Table 6. Batch polynomial multiplication timings in ms with FFNT and NTT on **RTX 4090/A100** with double precision (64 bits).

$\log N$	Batch Count	Algorithm 4 (RTX 4090/ A100)	NTT (RTX 4090/ A100)	Speedup (%) (RTX 4090/ A100)
10	16	0.0139/0.0472	0.0054/0.0565	159.9/−16.5
	32	0.0138/0.0469	0.0060/0.0562	128.9/−16.6
	64	0.0142/0.0470	0.0063/0.0570	126.3/−17.6
	128	0.0178/0.0476	0.0067/0.0612	164.1/−22.3
11	16	0.0175/0.0545	0.0093/0.0633	89.3/−13.9
	32	0.0180/0.0548	0.0104/0.0652	73.1/−16.0
	64	0.0231/0.0566	0.0118/0.0669	96.4/−15.4
	128	0.0380/0.0595	0.0148/0.0776	157.8/−23.4
12	16	0.0191/0.0549	0.0104/0.0650	83.3/−15.6
	32	0.0242/0.0570	0.0121/0.0676	99.5/−15.6
	64	0.0404/0.0602	0.0166/0.0808	143.7/−25.5
	128	0.0726/0.0793	0.0228/0.1187	218.1/−33.2
13	16	0.0258/0.0570	0.0122/0.0685	112.2/−16.7
	32	0.0435/0.0603	0.0170/0.0837	156.4/−27.9
	64	0.0779/0.0815	0.0236/0.1238	230.8/−34.1
	128	0.1455/0.1237	0.0396/0.2060	267.3/−39.9
14	16	0.0463/0.0623	0.0176/0.0872	163.5/−28.6
	32	0.0831/0.0842	0.0275/0.1309	202.4/−35.7
	64	0.1559/0.1274	0.0423/0.2178	268.2/−41.5
	128	0.3008/0.1990	0.0791/0.4024	280.1/−50.5
15	16	0.0890/0.0705	0.0298/0.1172	198.6/−39.9
	32	0.1664/0.1085	0.0512/0.1833	224.9/−40.8
	64	0.3222/0.2041	0.0858/0.3386	275.6/−39.7
	128	0.6400/0.4045	0.1698/0.6324	277.0/−36.0
16	16	0.1786/0.1115	0.0671/0.1939	166.2/−42.5
	32	0.3439/0.2114	0.1249/0.3581	175.4/−41.0
	64	0.6819/0.4086	0.2229/0.6705	206.0/−39.1
	128	1.3650/0.7543	0.4516/1.2912	202.3/−41.6

The new decomposition technique in [5], not only provides computational efficiency and flexibility, but also significant decrease in the memory footprint for storing relinearization keys across different levels compared to the RNS-based counterparts. They verified both in theory and in practice that the performance of key-switching, external and internal products and automorphisms, the most time- and resource-consuming operations across many FHE schemes, using their representation are similar or faster than the one achieved by [22], and they discussed the high impact of these results for people who work on low-level or hardware optimizations as well as the benefits of the new parametrizations for people currently working on compilers for FHE. They even managed to lower the running time of the gate bootstrapping of TFHE by eliminating 12.5% of its DFTs.

However, the results of [5] were only tested on CPUs and no experiments were performed on GPUs. Moreover, [5] notes that both NTT or FFNT can be used but, leaves it as an open problem to determine the one that provides the best performance. In this work we answer this problem for GPUs.

In the case of FFNT for CKKS $(\log N, \log Q) = (15, 880)$, due to the mantissa size of double-precision floating point variables, the largest limb size K can be chosen is $K = 19$ which means we have $\ell = 47$ limbs and we need to perform 141 DFTs for a polynomial multiplication as it requires two forward and one inverse DFT operations (see Table 7). On the other hand, for the case of NTT, if we chose a 64-bit prime, then K can be as large as $K = 25$ which in turn means $\ell = 36$. This reduces the problem to compute 108 NTTs. An algorithm for calculating the required mantissa size for a given limb size K was provided in [5] and a closer look shows that the required mantissa size obtained from this algorithm is actually $2 \times K + 14$.

Table 7. NTT and **Algorithm** 4 comparison for the method of [5] on various GPUs. Polynomial sizes are $N = 2^{15}$ and $\log Q = 880$-bit.

$\log N$	Method	Precision	K	ℓ	RTX 4090	A100
15	Algorithm 4	52-bit mantissa	19	47	0.243070	**0.182170**
		23-bit mantissa	5	176	**0.157513**	0.339533
15	NTT	253-bit Montgomery	120	8	0.147706	0.342656
		190-bit Montgomery	88	10	0.125258	0.270285
		126-bit Montgomery	56	16	0.095382	0.214989
		64-bit Montgomery	25	36	**0.048721**	**0.201190**

We compare the method of [5] using the new FFNT (Algorithm 4), and NTT in Tables 7 and 8 when $(\log N, \log Q) = (15, 880)$ and $(\log N, \log Q) = (16, 1761)$, respectively. It can be seen that our NTT-based implementations are always faster than FFNT-based Algorithm 4 on RTX 4090 while it is either comparable to or faster than FFNT on A100 (see the values we highlighted in bold in Tables 7 and 8). It is also interesting to note that, for the case of FFNT on RTX 4090, using single precision and performing more DFTs is faster than using double precision and performing a fewer number of DFTs when $(\log N, \log Q) = (15, 880)$ in Table 7. But, the double precision should be preferred for $(\log N, \log Q) = (16, 1761)$ as the number of DFTs becomes prohibitively high as can be observed in Table 8.

For the NTT-based implementation, we experimented with carrier primes of varying sizes, ranging from 64 bits to 253 bits, in order to evaluate the impact of different limb and batch configurations on performance. The results indicate that the 64-bit carrier prime generally provides the best performance, with one exception on the A100, where a 126-bit carrier prime performs better (see Table 8).

Table 8. NTT and **Algorithm** 4 comparison for the method of [5] on various GPUs. Polynomial sizes are $N = 2^{16}$ and $\log Q = 1761$-bit.

$\log N$	Method	Precision	K	ℓ	RTX 4090	A100
16	Algorithm 4	52-bit mantissa	19	93	**1.004000**	**0.565862**
		23-bit mantissa	4	441	1.670000	1.364915
16	NTT	253-bit Montgomery	119	15	0.510726	0.955110
16		190-bit Montgomery	88	21	0.473901	0.854502
16		126-bit Montgomery	56	32	0.347722	**0.696550**
16		64-bit Montgomery	25	71	**0.236247**	0.781466

Consequently, the optimal configuration of the NTT-based implementation achieves speedups of $0.157513/0.048721 = 3.2$ and $1.004000/0.236247 = 4.25$ for $(\log N, \log Q) = (15, 880)$ and $(\log N, \log Q) = (16, 1761)$, respectively, on the RTX 4090.

It is worth noting that the observed increase in speedup values is partly due to the higher precision of the NTT-based method (64-bit), which contrasts with the relatively lower precision of single- and double-precision floating-point arithmetic. This difference in precision leads to a significant increase in batch sizes for the floating-point methods, thereby amplifying the relative advantage of the NTT-based approach.

In conclusion, the main drawback of using FFNT instead of NTT is the limited mantissa size of the floating-point data types. The mantissa sizes of the available CUDA data types are as follows:

- `bfloat16` (CUDA half precision) has 7-bit mantissa, not big enough for our purposes.
- `float16` has 10-bit mantissa, not big enough for our purposes.
- `TF32` (CUDA Tensor float) has 10-bit mantissa, not big enough for our purposes.
- `float32` has 23-bit mantissa $=> K$ can be at most 4 which require excessively many DFTs, which may rarely be advantageous over double precision.
- `float64` has 52-bit mantissa $=> K$ can be at most 19.

Thus, due to the limited mantissa sizes of available floating-point data types, NTT-based solutions may be preferred over FFNT-based ones when employing the method of [5]. In contrast, for the implementation of CKKS and BFV schemes - where the Residue Number System (RNS) is the standard approach for modulus decomposition - NTT-based methods are consistently the more suitable choice.

4.4   NTT and FFNT for TFHE

In this section, we demonstrate that an NTT-based approach with a 64-bit carrier prime can also be applied to fixed-point arithmetic in TFHE, delivering favorable performance on GPUs. To assess which method - NTT or the new FFNT (**Algorithm** 4)- provides a more efficient approach for accelerating polynomial multiplication in the TFHE scheme, we implemented the TFHE scheme in CUDA using a GPU library [28][8] with both techniques. The benchmark used for the evaluation involves full TFHE bootstrapping (including blind rotation and key switching) using both NTT and FFNT arithmetic backends under the identical cryptographic parameters. This setup ensured that the comparison reflects the actual performance of TFHE as a whole, rather than that of isolated polynomial multiplications.

Although we have developed specialized GPU kernels for both FFNT and NTT in other contexts (e.g., for larger polynomial sizes or general-purpose FHE schemes), in our TFHE implementation, the NTT and FFNT routines are invoked as `__device__` functions within a unified bootstrapping kernel. This design choice is motivated by the relatively small ring dimensions used in TFHE - typically $N = 1024$ or $N = 2048$ - which enable a full transform (forward or inverse) to be computed entirely within a single thread block. Under these conditions, there is no need to launch a separate kernel or coordinate multiple blocks for the transform, since all data dependencies can be efficiently resolved within a single block-synchronized execution unit using shared memory.

After extensive testing, we observed that in the context of TFHE bootstrapping the better algorithm is dictated by the underlying GPU architecture, mirroring the results obtained from standalone polynomial multiplication benchmarks. We measured the execution time of the TFHE `NAND` gate bootstrapping operation for various data types using both NTT- and FFNT-based backends. The data type `uintX`, where $X \in 8, 16, 32, 64, 128, 256$, denotes an encrypted integer of X bits, with each bit represented by an LWE ciphertext[9]. Both the NTT and FFNT benchmarks were implemented using optimized CUDA kernels with identical control flow and logic to ensure a fair comparison. The timing results for the RTX 4090 and A100 GPUs are reported in Table 9. On the RTX 4090, the NTT-based implementation achieves speedups ranging from 12.4% to 22.5%, reaffirming its consistent performance advantage over the FFNT-based counterpart. In contrast, on the A100 GPU, the FFNT-based implementation outperforms the NTT-based approach, achieving speedups ranging from 2.8% to 46.1%.

Table 9. TFHE `NAND` bootstrapping timings (in ms) on **RTX 4090/A100**.

Data Type	NTT	Algorithm 4 (RTX 4090/ A100)	NTT Speedup (%) (RTX 4090/ A100)
uint8	8.9013/13.2276	10.8195/12.8568	21.5/−2.8
uint16	8.9735/12.7472	10.8551/12.3928	21.0/−2.8
uint32	9.8358/15.2172	11.0583/12.8195	12.4/−15.8
uint64	12.3934/21.6111	14.0669/14.2551	13.5/−34.0
uint128	19.9196/32.4150	23.8368/19.0849	19.7/−41.1
uint256	35.0424/57.0872	42.9429/30.7688	22.5/−46.1

---

[8] https://github.com/Alisah-Ozcan/HEonGPU.
[9] LWE (Learning With Errors) is a cryptographic primitive in which each ciphertext encrypts a single plaintext bit.

Based on these results, we adopted the NTT arithmetic backend for all subsequent comparative benchmarks against state-of-the-art TFHE implementations reported in the literature for RTX 4090. All experiments were performed using 128-bit LWE security parameters, with an LWE dimension of $n = 630$ and a ring dimension of $N = 1024$. Table 10 presents the measured TFHE NAND gate latency across different data types, comparing our implementation with TFHE-rs (GPU backend) [33] and the recent work of Yu Xiao et al. [32].

Table 10. Comparison of NTT-based TFHE NAND timings (in ms) on RTX 4090 against those of [32,33] across different data types.

Data Type	[33]	[32]	NTT-based TFHE	Speedup (%) [33]/[32]
uint8	31.53	18.63	**12.72**	148.0/46.0
uint16	31.54	18.61	**12.75**	147.0/46.0
uint32	31.55	18.87	**13.60**	132.0/39.0
uint64	32.03	24.23	**15.88**	102.0/53.0
uint128	33.74	29.97	**23.10**	46.0/30.0
uint256	58.32	58.30	**38.24**	52.0/52.0

Table 10 shows that our NTT-based implementation achieves lower NAND gate latency across all data types when compared to both TFHE-rs [33] and the work by Yu Xiao et al. [32]. The performance gains are especially significant for smaller data types such as uint8, uint16, uint32, and uint64 where we observe up to 148% speedup over TFHE-rs [33] and 53% over the work [32]. As the data width increases, our implementation continues to outperform both baselines, achieving consistent reductions in latency.

An important architectural factor affecting these results is the behavior of modern GPUs such as the RTX 4090, which incorporates a large number of Streaming Multiprocessors (SMs). Due to this high degree of parallelism, the device cannot be fully saturated when bootstrapping only a limited number of LWE ciphertexts. Our experiments show that performance begins to plateau once approximately 64 or more LWE bootstraps are processed concurrently. Consequently, executing NAND bootstraps on uint8 (8 LWE), uint32 (32 LWE), or even uint64 data types does not produce significant differences in total latency. In all these cases, the GPU remains underutilized, which explains the near-constant cost observed across small and medium bit-widths and further validates the effectiveness of our bit-parallel evaluation strategy.

5   Power and Energy Characterization of NTT and FFNT

This section presents a detailed characterization of the NTT and FFNT kernels in terms of instantaneous power consumption and total energy cost on modern NVIDIA GPUs.

All measurements were performed under thermally stable conditions with the **graphics clock frequency** and **graphics transfer rate** manually locked to their maximum operating values to prevent dynamic frequency scaling and ensure consistent power readings. Power data were obtained using the NVIDIA Management Library (**NVML**), while execution durations were recorded via CUDA event timing.

5.1   Measurement Methodology

Average instantaneous power (P_{avg}) was recorded through the NVML API at 5 ms intervals during function execution to obtain fine-grained temporal traces of power variation. Each experimental scenario was executed continuously for 10 s to capture steady-state behavior and to average out short-term power fluctuations. The NVML routine `nvmlDeviceGetPowerUsage(device, &power_mW)`, returns instantaneous power values (in mW) from the on-board sensors, encompassing both static leakage and dynamic switching power across GPU core and memory rails [26]. Execution times (t_{exec}) for the function portion were measured using CUDA events, excluding data transfers and initialization overheads. The total energy consumption for each function was computed as

$$E_{\mathrm{total}} = P_{\mathrm{avg}} \times t_{\mathrm{exec}},$$

where t_{exec} is expressed in seconds. All configurations ($\log N \in \{12, 13, 14, 15, 16\}$, $Batch \in \{32, 64, 128\}$) were executed repeatedly, and reported results represent the mean of multiple runs to mitigate both thermal drift and frequency variability.

5.2   Observed Behavior of NTT & FFNT Kernels

Since the RTX 4090 and A100 exhibit opposite performance trends due to their architectural differences, we analyze the results for each architecture in two separate sections.

RTX 4090. The integer-based NTT kernels exhibit extremely high arithmetic intensity and are dominated by integer instructions ($>99\,\%$), as experimentally confirmed by Nsight Compute instruction counters. Each operation consists of modular additions and multiplications that fully utilize the **INT32/INT64 ALUs** and the register files within each Streaming Multiprocessor (SM). As the problem size and batch count increase, the measured average power rises from approximately $260\,\mathrm{W}$ at $\log N = 12$ to about $450\,\mathrm{W}$ at $\log N = 16$, approaching the GPU's thermal design power (TDP) limit (see Table 11).

This growth indicates near-maximum SM occupancy, where all integer pipelines remain active per cycle. At the same time, the high degree of instruction-level parallelism and minimal global memory traffic result in short execution times (10–450 ms). Consequently, despite their high instantaneous power, NTT kernels complete rapidly and consume significantly less total energy compared to their floating-point counterparts.

The FFNT kernels, in contrast, exhibit a markedly different profile as can be observed in Table 11. Each SM on the RTX 4090 integrates only two FP64 units (256 in total across the GPU), resulting in limited throughput. Thus, FP64 kernels sustain lower instantaneous power (150–250 W) but much longer execution times; up to 1.36 s for the largest configuration. Although the GPU operates at a lower average wattage, the extended runtime leads to higher overall energy consumption.

Table 11. Comprehensive power, execution time, and total energy comparison between **NTT** and **FFNT**-based polynomial multiplication kernels on the RTX 4090/A100 (64-bit precision). Power values are obtained from 10 s NVML traces sampled every 5 ms; execution times (in ms) are taken from benchmark results.

Precision	$\log N$	Batch	P_{avg} (W)		E_{total} (mJ)		NTT Eff. (%)
			FFNT	NTT	FFNT	NTT	
			RTX 4090/A100	RTX 4090/A100	RTX 4090/A100	RTX 4090/A100	RTX 4090/A100
64-bit	12	32	147.9/123.2	259.7/171.3	3.57/6.75	3.15/11.12	13.6/−39.3
		64	172.1/155.6	336.9/229.9	6.96/9.41	5.58/18.56	24.5/−49.3
		128	183.4/179.2	419.8/272.3	13.31/14.21	9.57/32.37	39.0/−56.1
	13	32	177.3/150.6	362.7/220.4	7.24/9.09	6.15/15.16	17.7/−66.8
		64	177.3/179.4	418.7/264.3	13.82/14.58	9.86/32.72	40.1/−55.4
		128	184.9/212.9	428.8/309.7	26.90/26.32	16.99/63.80	58.3/−58.7
	14	32	173.4/176.4	418.7/261.7	14.41/14.85	11.51/22.82	25.3/−53.7
		64	180.3/212.8	435.1/306.9	28.12/27.03	18.42/66.80	52.6/−59.5
		128	199.6/241.2	441.4/355.2	60.05/48.05	34.93/142.98	72.0/−66.4
	15	32	177.2/213.7	433.9/303.1	29.49/23.03	22.22/35.53	32.8/−54.3
		64	196.6/244.9	444.8/356.6	63.35/49.98	38.17/120.77	66.0/−58.6
		128	225.6/254.7	446.1/381.1	144.37/103.04	75.74/240.91	90.7/−57.3
	16	32	194.8/247.2	442.2/355.2	67.01/27.57	29.67/68.99	125.8/−60.0
		64	220.7/263.2	446.1/386.8	150.49/107.59	99.42/259.18	51.3/−58.5
		128	247.7/276.2	446.9/396.4	338.16/208.41	201.82/512.31	67.5/−59.3

As summarized in Table 11, NTT kernels achieve higher instantaneous power but shorter runtime, whereas FFNT kernels exhibit the opposite trend. For instance, at $\log N = 16$, Batch = 128, the FFNT-64 kernel ran for 1.36 s at 247 W (338 J total), while the NTT-64 kernel finished in 0.45 s at 447 W (202 J). This means that FFNT consumed about 45% less power on average but required about **1.7× more energy** to perform the same computation.

The results align with the functional-unit composition of the Ada Lovelace architecture. Each SM integrates 128 FP32 cores, 64 INT32 cores, and 2 FP64 cores, along with Tensor, load/store, and scheduling units [25]. Power distribution studies indicate that approximately 70–75% of the total dynamic power arises from SM activity, about 15% from the memory subsystem, and 10–15% from leakage and control circuitry.

In NTT kernels, dense integer arithmetic keeps the INT32 pipelines saturated for most of the execution window, ensuring full utilization of warp schedulers and maximizing computational throughput. Because these kernels have low global-memory bandwidth requirements and exhibit coalesced access patterns, the memory subsystem contributes minimally to total power draw. As a result, even though instantaneous power is high, runtime remains short and overall energy consumption is minimized.

In contrast, FFNT kernels – particularly their FP64 variants – cannot maintain such high utilization. The RTX 4090 provides only two FP64 execution units per SM (256 total), resulting in underutilization of the available computational and memory resources. This imbalance extends execution time substantially while the GPU continues to draw baseline static and clock power. Therefore, FFNT appears thermally "cooler" in instantaneous terms but energetically "less efficient" due to its longer active duration.

As the batch size increases further, both NTT and FFNT kernels are expected to converge toward the GPU's power envelope of 440–450 W, where performance-per-watt improvements begin to saturate. Beyond this point, increasing concurrency mainly adds latency hiding rather than true throughput scaling. Hence, integer-heavy kernels such as NTT better exploit the Ada architecture's compute-to-memory balance than floating-point workloads.

A100. Since NTT kernels are both slower and more power-intensive on A100 GPUs, the advantages of FFNT-based solutions are twofold, resulting in significantly higher energy efficiency compared to the energy efficiency achieved by NTT-based implementations on the RTX 4090 (see Table 11).

As the problem size and batch count increase, the measured average power consumption for FFNT rises from approximately 123 W at $\log N = 12$ to about 276 W at $\log N = 16$, which remains substantially lower than the corresponding values for NTT (cf. 171 W to 396 W). On the A100, each streaming multiprocessor (SM) integrates 64 FP32 cores, 64 INT32 cores, and 32 FP64 cores. The smaller number of FP64 units compared to INT32 units appears to reduce the overall power requirements when FFNT is used for polynomial multiplication.

6 Conclusion

In this work, we investigated efficient GPU implementations of three widely used polynomial multiplication algorithms—FFT, FFNT, and NTT—with a focus on their applicability to cryptographic workloads. We introduced a new optimized variant of FFNT that eliminates the pre- and post-processing steps and demonstrated its performance advantage over both traditional FFT and FFNT implementations, including the highly optimized CUDA cuFFT library.

Our results show that NTT consistently outperforms both FFT and FFNT in terms of execution time and energy efficiency for polynomial sizes commonly used in cryptographic algorithms, including schemes employing approximate arithmetic such as TFHE, on GPU architectures with relatively few double-precision floating-point units, such as the **RTX 4090**. This conclusion is further supported by experiments in a practical homomorphic encryption setting: when integrating NTT and FFNT separately into a TFHE bootstrapping pipeline, the NTT-based implementation achieved lower overall latency, reaffirming its suitability for GPU-accelerated cryptographic computations.

On the other hand, for GPU architectures with a larger number of double-precision floating-point units, such as the **A100 (Ampere)**, we observed the opposite trend, where FFNT-based solutions may be preferable. In the case of TFHE, the FFNT-based implementation clearly outperforms the NTT-based approach. FFNT can also be advantageous for FHE implementations that rely on digit decomposition rather than RNS arithmetic, particularly due to its superior energy efficiency. Nevertheless, for FHE schemes employing RNS arithmetic, NTT-based solutions may still be preferable, as cyclotomic arithmetic significantly reduces the size of the RNS base.

References

1. Agarwal, R., Burrus, C.: Fast convolution using fermat number transforms with applications to digital filtering. IEEE Trans. Acoust. Speech Signal Process. **22**(2), 87–97 (1974). https://doi.org/10.1109/TASSP.1974.1162555
2. Al Badawi, A., Veeravalli, B., Aung, K.M.M.: Efficient polynomial multiplication via modified discrete galois transform and negacyclic convolution. In: Arai, K., Kapoor, S., Bhatia, R. (eds.) FICC 2018. AISC, vol. 886, pp. 666–682. Springer, Cham (2019). https://doi.org/10.1007/978-3-030-03402-3_47
3. Bajard, J.C., Méloni, N., Plantard, T.: Efficient RNS bases for cryptography. In: 17th IMACS World Congress Scientific Computation, Applied Mathematics and Simulation, Paris, France (2005). https://hal-lirmm.ccsd.cnrs.fr/lirmm-00106470
4. Barrett, P.: Implementing the rivest Shamir and Adleman public key encryption algorithm on a standard digital signal processor. In: Odlyzko, A.M. (ed.) CRYPTO 1986. LNCS, vol. 263, pp. 311–323. Springer, Heidelberg (1987). https://doi.org/10.1007/3-540-47721-7_24
5. Belorgey, M.G., Carpov, S., Gama, N., Guasch, S., Jetchev, D.: Revisiting key decomposition techniques for FHE: simpler, faster and more generic. In: Chung, K., Sasaki, Y. (eds.) ASIACRYPT 2024, Part I. LNCS, vol. 15484, pp. 176–207. Springer, Cham (2024). https://doi.org/10.1007/978-981-96-0875-1_6
6. Ben-Sasson, E., Bentov, I., Horesh, Y., Riabzev, M.: Scalable, transparent, and post-quantum secure computational integrity. IACR Cryptol. ePrint Arch. 46 (2018). http://eprint.iacr.org/2018/046
7. Bitansky, N., Canetti, R., Chiesa, A., Tromer, E.: From extractable collision resistance to succinct non-interactive arguments of knowledge, and back again. In: Goldwasser, S. (ed.) Innovations in Theoretical Computer Science 2012, Cambridge, MA, USA, 8–10 January 2012, pp. 326–349. ACM (2012). https://doi.org/10.1145/2090236.2090263
8. Bos, J., et al.: Crystals - kyber: a CCA-secure module-lattice-based KEM. In: 2018 IEEE European Symposium on Security and Privacy (EuroS&P), pp. 353–367 (2018). https://doi.org/10.1109/EuroSP.2018.00032
9. Brakerski, Z., Gentry, C., Vaikuntanathan, V.: (Leveled) fully homomorphic encryption without bootstrapping. ACM Trans. Comput. Theory **6**(3), 13:1–13:36 (2014). https://doi.org/10.1145/2633600

10. Bünz, B., Bootle, J., Boneh, D., Poelstra, A., Wuille, P., Maxwell, G.: Bulletproofs: short proofs for confidential transactions and more. In: 2018 IEEE Symposium on Security and Privacy, SP 2018, Proceedings, 21–23 May 2018, San Francisco, California, USA, pp. 315–334. IEEE Computer Society (2018). https://doi.org/10.1109/SP.2018.00020

11. Cheon, J.H., Kim, A., Kim, M., Song, Y.: Homomorphic encryption for arithmetic of approximate numbers. In: Takagi, T., Peyrin, T. (eds.) ASIACRYPT 2017. LNCS, vol. 10624, pp. 409–437. Springer, Cham (2017). https://doi.org/10.1007/978-3-319-70694-8_15

12. Chillotti, I., Gama, N., Georgieva, M., Izabachène, M.: Faster fully homomorphic encryption: bootstrapping in less than 0.1 seconds. In: Cheon, J.H., Takagi, T. (eds.) ASIACRYPT 2016, Part I. LNCS, vol. 10031, pp. 3–33. Springer, Heidelberg (2016). https://doi.org/10.1007/978-3-662-53887-6_1

13. Chillotti, I., Gama, N., Georgieva, M., Izabachène, M.: TFHE: fast fully homomorphic encryption library (2016). https://tfhe.github.io/tfhe/

14. Chu, E., George, A.: Inside the FFT black box: serial and parallel fast Fourier transform algorithms (2019). https://api.semanticscholar.org/CorpusID:60390207

15. Cooley, J.W., Tukey, J.W.: An algorithm for the machine calculation of complex Fourier series. Math. Comput. **19**(90), 297–301 (1965). http://www.jstor.org/stable/2003354

16. Crandall, R.E.: Integer convolution via split-radix fast galois transform (1999). https://api.semanticscholar.org/CorpusID:37425031

17. Ducas, L., Lepoint, T., Lyubashevsky, V., Schwabe, P., Seiler, G., Stehle, D.: CRYSTALS – Dilithium: digital signatures from module lattices. Cryptology ePrint Archive, Paper 2017/633 (2017). https://eprint.iacr.org/2017/633

18. Fan, J., Vercauteren, F.: Somewhat practical fully homomorphic encryption. IACR Cryptol. ePrint Arch. 144 (2012). http://eprint.iacr.org/2012/144

19. Gentleman, W.M., Sande, G.: Fast Fourier transforms: for fun and profit. In: Proceedings of the November 7–10, 1966, Fall Joint Computer Conference, AFIPS 1966 (Fall), pp. 563–578. Association for Computing Machinery, New York (1966). https://doi.org/10.1145/1464291.1464352

20. Jung, W., et al.: HEAAN demystified: accelerating fully homomorphic encryption through architecture-centric analysis and optimization. CoRR abs/2003.04510 (2020). https://arxiv.org/abs/2003.04510

21. Karatsuba, A., Ofman, Y.: Multiplication of multidigit numbers on automata. Sov. Phys. Dokl. **7**, 595 (1962)

22. Kim, M., Lee, D., Seo, J., Song, Y.: Accelerating HE operations from key decomposition technique. In: Handschuh, H., Lysyanskaya, A. (eds.) CRYPTO 2023. LNCS, vol. 14084, pp. 70–92. Springer, Cham (2023). https://doi.org/10.1007/978-3-031-38551-3_3

23. Klemsa, J.: Fast and error-free negacyclic integer convolution using extended Fourier transform. In: Dolev, S., Margalit, O., Pinkas, B., Schwarzmann, A. (eds.) CSCML 2021. LNCS, vol. 12716, pp. 282–300. Springer, Cham (2021). https://doi.org/10.1007/978-3-030-78086-9_22

24. Montgomery, P.L.: Modular multiplication without trial division. Math. Comput. **44**(170), 519–521 (1985)

25. NVIDIA Corporation: NVIDIA Ada Lovelace Architecture Whitepaper (2023). https://resources.nvidia.com/en-us-tensor-core/nvidia-ada-lovelace-architecture-whitepaper

26. NVIDIA Developer Documentation: NVIDIA Management Library (NVML) API Reference (2024). https://docs.nvidia.com/deploy/nvml-api/

27. Özcan, A.S., Javeed, A., Savas, E.: High-performance number theoretic transform on GPU through radix2-CT and 4-step algorithms. IEEE Access **13**, 87862–87883 (2025). https://doi.org/10.1109/ACCESS.2025.3570024
28. Ozcan, A.S., Savas, E.: HEonGPU: a GPU-based fully homomorphic encryption library 1.0. Cryptology ePrint Archive, Paper 2024/1543 (2024). https://eprint.iacr.org/2024/1543
29. Schoonhoven, R., Veenboer, B., Van Werkhoven, B., Batenburg, K.J.: Going green: optimizing GPUs for energy efficiency through model-steered auto-tuning . In: 2022 IEEE/ACM International Workshop on Performance Modeling, Benchmarking and Simulation of High Performance Computer Systems (PMBS), pp. 48–59. IEEE Computer Society, Los Alamitos (2022). https://doi.org/10.1109/PMBS56514.2022.00010, https://doi.ieeecomputersociety.org/10.1109/PMBS56514.2022.00010
30. Szabó, N., Tanaka, R.: Residue Arithmetic and Its Applications to Computer Technology. McGraw-Hill Series in Information Processing and Computers. McGraw-Hill (1967). https://books.google.com.tr/books?id=Osk-AAAAIAAJ
31. Tezcan, C.: Optimization of advanced encryption standard on graphics processing units. IEEE Access **9**, 67315–67326 (2021). https://doi.org/10.1109/ACCESS.2021.3077551
32. Xiao, Y., et al.: GPU acceleration for FHEW/TFHE bootstrapping. IACR Trans. Cryptogr. Hardw. Embed. Syst. **2025**(1), 314–339 (2024). https://doi.org/10.46586/tches.v2025.i1.314-339, https://tches.iacr.org/index.php/TCHES/article/view/11931
33. Zama: TFHE-RS: a pure rust implementation of the TFHE scheme for Boolean and integer arithmetics over encrypted data (2022). https://github.com/zama-ai/tfhe-rs

From Compact to Fast: Exploring 32 and 64-Bit AES Datapaths for IoT Systems

Doaa Ashmawy$^{(\boxtimes)}$ and Arash Reyhani-Masoleh

Electrical and Computer Engineering Department, Western University,
London, ON, Canada
{dashmawy,areyhani}@uwo.ca

Abstract. This work presents new hardware architectures for AES-128 encryption tailored to resource-constrained IoT devices. We propose 32-bit and 64-bit round-based designs that significantly increase throughput while maintaining an efficient area–performance balance. Each datapath includes an independent key schedule module capable of generating multiple round key bytes per cycle in parallel with encryption, eliminating key-setup latency. Both architectures use the lightweight S-box [11] and are synthesized in STM 65 nm CMOS technology for fair comparison with the 16-bit baseline in [2].

The proposed 32-bit and 64-bit designs reduce encryption latency to 44 and 22 cycles per 128-bit block and increase throughput from 734 Mbps to 1.512 Gbps and 2.950 Gbps, corresponding to improvements of 106% and 301.9%, respectively. These gains are achieved with area overheads of 29.1% and 84.4%. Energy consumption per block decreases from 7.62 nJ to 4.69 nJ and 3.37 nJ at 10 MHz.

Both architectures also improve throughput-per-area, achieving 1.60× and 2.18× higher efficiency than the baseline. Compared to designs in [13], they achieve lower latency with fewer S-boxes, resulting in smaller area and higher throughput. These features make the architectures well suited for low-latency, energy-efficient AES implementations in IoT hardware.

Keywords: AES-128 · 32-bit datapath · 64-bit datapath · key scheduling · ASIC · IoT

1 Introduction

The Advanced Encryption Standard (AES) is a widely used symmetric-key block cipher standardized by the National Institute of Standards and Technology (NIST) in 2001, replacing the earlier Data Encryption Standard (DES) [10]. AES follows a Substitution–Permutation Network (SPN) structure composed of four fundamental transformations: SubBytes (nonlinear substitution using an S-box), ShiftRows (row-wise permutation), MixColumns (linear mixing of columns), and AddRoundKey (key addition). The algorithm processes plaintext in blocks of 128 bits and supports three different key lengths: 128, 192, and

L. Batina and F. Özbudak (Eds.): WAIFI 2026, LNCS 16611, pp. 328–340, 2026.
https://doi.org/10.1007/978-3-032-27574-5_20

256 bits. Prior work has explored AES implementations with 192-bit and 256-bit key sizes [3], while other designs employ 128-bit datapaths to achieve high throughput in non-resource-constrained environments [5,14]. In this work, we focus on AES encryption with a 128-bit key (AES-128) and 32-bit/64-bit datapaths. The cipher key is expanded through a key schedule to generate a set of round keys, with one round key applied during each round of the AES-128 encryption process.

Efficient AES hardware for IoT systems requires careful trade-offs between throughput, area, and power [1,8]. Prior work has focused on low-area 16-bit datapath designs optimized for constrained devices, including a column-wise datapath and key expansion, lightweight serial MixColumns, and compact S-boxes, which achieve higher throughput gains over 8-bit designs [2].

In this work, we propose 32-bit and 64-bit datapath designs for AES-128 encryption to improve throughput, while accepting a corresponding increase in area. The proposed designs employ the lightweight S-box [11] and are implemented using STM 65 nm CMOS technology. For a fair comparison, we also implement the 16-bit baseline design [2] using the same S-box and ASIC library. We design separate key schedule modules for each datapath width, enabling the generation of multiple round-key bytes per clock cycle in parallel with the data path. Wider datapaths process more AES state bytes per cycle, thereby reducing overall encryption latency and increasing throughput for IoT applications with higher data-rate requirements.

1.1 Contributions

The contributions of this work are summarized as follows:

- We propose and implement two AES-128 architectures with 32-bit and 64-bit datapaths that improve throughput, energy consumption, and area efficiency. The proposed designs achieve higher throughput, lower latency, and require fewer S-boxes than recent 32-bit and 64-bit implementations in [13].
- A separate key schedule module is designed for each datapath width to compute round keys concurrently with the datapath.
- We employ the lightweight S-box [11] in the proposed designs, including both the datapath and key schedule modules. We also implement the 16-bit baseline design presented in [2] using the same S-box to ensure a fair comparison with the other two designs. All designs are coded in Verilog and verified using ModelSim© and NIST reference test cases.
- ASIC implementation results using STM 65 nm CMOS technology demonstrate the trade-off between throughput and area for the proposed 32-bit and 64-bit datapaths compared to the 16-bit baseline [2]:
  - Throughput improves by 106% and 301.9%, respectively, at the cost of area increases of 29.1% and 84.4%.
  - Throughput-per-area efficiency improves by 59.4% and 117.7%, respectively.
  - Energy efficiency is also improved, with reductions of 38.5% and 55.8% in energy per 128-bit block, respectively.

The rest of the paper is organized as follows. Section 2 presents background and preliminaries. Sections 3 and 4 describe the proposed AES cores. Section 5 describes the hardware code development, presents ASIC implementation results, and compares them to previous work. Section 6 concludes the paper.

2   Background and Preliminaries

This section provides technical background on AES-128 transformations and summarizes prior work relevant to the proposed 32- and 64-bit AES designs. The 16-bit AES core [2] is used as our baseline for comparison.

2.1   AES Transformations

AES-128 operates on a 128-bit state array through a sequence of transformations: AddRoundKey, SubBytes, ShiftRows, and MixColumns. After the initial round-key addition, rounds 1–9 apply all four transformations, while the final round omits MixColumns.

AddRoundKey. Each round key is XORed with the state array bytes.

SubBytes. Each byte of the state is substituted using the AES S-box, implemented using finite-field arithmetic in $GF(2^8)$. For area efficiency and fair comparison, we employ the lightweight S-box from [11] in the proposed 32- and 64-bit designs as well as in the 16-bit baseline [2].

ShiftRows. Rows 1–3 of the state are cyclically shifted left by 1, 2, and 3 bytes, while row 0 remains unchanged. We employ Järvinen et al. [7] 32-bit and 64-bit permutation blocks to implement this operation, as their compact, column-wise structure matches the proposed designs and avoids additional multiplexers for rearranging plaintext or initial key bytes.

MixColumns. Columns of the state are multiplied by a fixed polynomial over $GF(2^8)$. Our designs utilize Satoh et al. [12] 32-bit MixColumns circuit, which has a low implementation area and short critical path delay. The 32-bit core uses one MixColumns unit, while the 64-bit core instantiates two 32-bit units to process two columns per clock cycle.

2.2   Baseline 16-Bit AES Core [2]

The 16-bit AES hardware architecture [2] features a column-wise datapath with a 16-bit serial MixColumns circuit and a lightweight key derivation block. It employs the Bonus S-box [9] and STM 65 nm CMOS technology, achieving a balanced trade-off between area and throughput. The column-wise organization eliminates the need for additional multiplexers to reorder plaintext or key bytes at the start of encryption, reducing hardware overhead and control complexity.

For experimental comparison, we implemented this baseline in Verilog and used the lightweight S-box [11] for both the datapath and key derivation block. The proposed 32- and 64-bit designs also employ the same lightweight S-box and STM 65 nm technology to maintain fair comparison. By widening the datapath, the new designs reduce encryption latency and substantially increase throughput, making them well-suited for IoT systems with constrained resources and higher data-rate requirements.

3   The Proposed 32-Bit AES-128 Design

This section introduces a novel design for the AES encryption algorithm that makes use of a 32-bit data processing path. The ciphertext becomes available after 44 clock cycles, including 4 cycles to load the input data and apply the input key, plus 4 cycles for each of the ten encryption rounds.

3.1   The Proposed 32-Bit Datapath

The proposed 32-bit AES datapath is presented in Fig. 1. It consists of sixteen byte registers, denoted as B_0–B_{15}, organized into four parallel 32-bit words, four lightweight S-box units, and a 32-bit MixColumns module.

In the proposed datapath, the state is initially combined with the input key using 32 2-input XOR gates and then fed into the 32-bit permutation block [7]. The permutation block outputs four bytes of the state per clock cycle, which are subsequently processed by the four lightweight S-boxes [11] and then by the 32-bit MixColumns unit [12]. Our 32-bit permutation block differs slightly from the original design in [7] by eliminating certain multiplexers, as no decryption is implemented. The control unit coordinates the operation of the datapath by generating the necessary select signals for the multiplexers. It ensures that data flows correctly through the permutation block.

3.2   The Proposed 32-Bit Key Schedule Module

The proposed 32-bit key expansion unit is designed to compute four bytes of the round key per clock cycle. Figure 2 shows the architecture of the proposed 32-bit key schedule module. The key register array consists of sixteen 8-bit registers, R_0 to R_{15}.

The four 8-bit multiplexers to the right of the unit choose between loading the input key or computing the new round key. The first four clock cycles are used to load the input key into the key schedule module. Table 1 clarifies the process of key generation. The key register contents are given for each clock cycle to check for the correct operation of the unit.

Four bytes of the round key are supplied to the main datapath each clock cycle. These four key bytes are read from the four 8-bit key registers $R_{12}, R_{13}, R_{14}, R_{15}$. The key for rounds 0–9 is added using a 32-bit XOR gate placed before the 32-bit permutation block, while the final round key is added

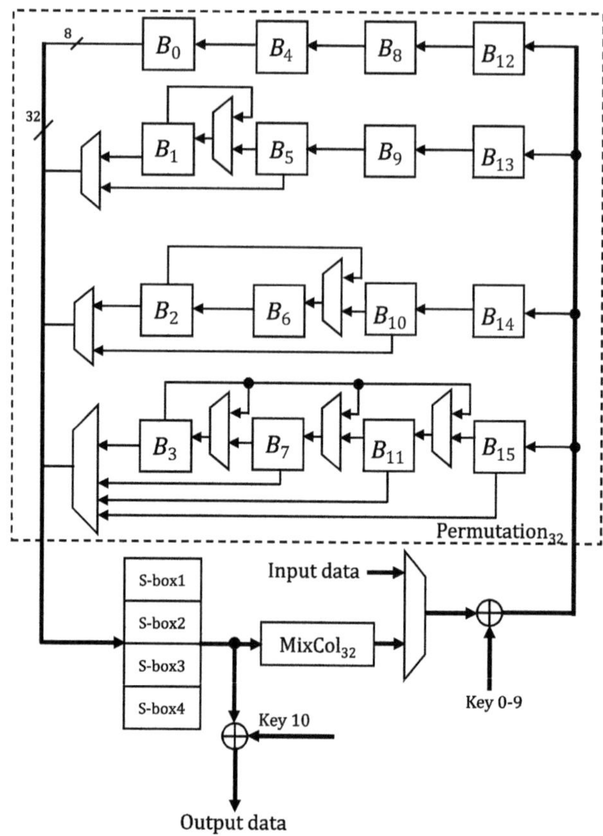

Fig. 1. The proposed 32-bit AES datapath.

Table 1. Operation of the proposed 32-bit key schedule module.

t	R_0	R_4	R_8	R_{12}	R_1	R_5	R_9	R_{13}	R_2	R_6	R_{10}	R_{14}	R_3	R_7	R_{11}	R_{15}	Operations
0	a_0	a_4	a_8	a_{12}	a_1	a_5	a_9	a_{13}	a_2	a_6	a_{10}	a_{14}	a_3	a_7	a_{11}	a_{15}	$b_0 = a_0 \oplus S(a_{13}) \oplus Rcon,\ b_4 = a_4 \oplus b_0$
																	$b_1 = a_1 \oplus S(a_{14}),\ b_5 = a_5 \oplus b_1$
																	$b_2 = a_2 \oplus S(a_{15}),\ b_6 = a_6 \oplus b_2$
																	$b_3 = a_3 \oplus S(a_{12}),\ b_7 = a_7 \oplus b_3$
1	$\mathbf{b_4}$	a_8	a_{12}	$\mathbf{b_0}$	$\mathbf{b_5}$	a_9	a_{13}	$\mathbf{b_1}$	$\mathbf{b_6}$	a_{10}	a_{14}	$\mathbf{b_2}$	$\mathbf{b_7}$	a_{11}	a_{15}	$\mathbf{b_3}$	$b_8 = a_8 \oplus b_4,\ b_9 = a_9 \oplus b_5$
																	$b_{10} = a_{10} \oplus b_6,\ b_{11} = a_{11} \oplus b_7$
2	$\mathbf{b_8}$	a_{12}	b_0	b_4	$\mathbf{b_9}$	a_{13}	b_1	b_5	$\mathbf{b_{10}}$	a_{14}	b_2	b_6	$\mathbf{b_{11}}$	a_{15}	b_3	b_7	$b_{12} = a_{12} \oplus b_8,\ b_{13} = a_{13} \oplus b_9$
																	$b_{14} = a_{14} \oplus b_{10},\ b_{15} = a_{15} \oplus b_{11}$
3	$\mathbf{b_{12}}$	b_0	b_4	b_8	$\mathbf{b_{13}}$	b_1	b_5	b_9	$\mathbf{b_{14}}$	b_2	b_6	b_{10}	$\mathbf{b_{15}}$	b_3	b_7	b_{11}	Shift

$S(a_i)$ denotes the S-box substitution of a_i. Bold bytes indicate values updated in the previous cycle.

to the state using another 32-bit XOR gate located after the four lightweight S-boxes of the main datapath in order to skip MixColumns. In the proposed 32-bit design, the addition of the key for rounds 0–9 is not delayed as the key

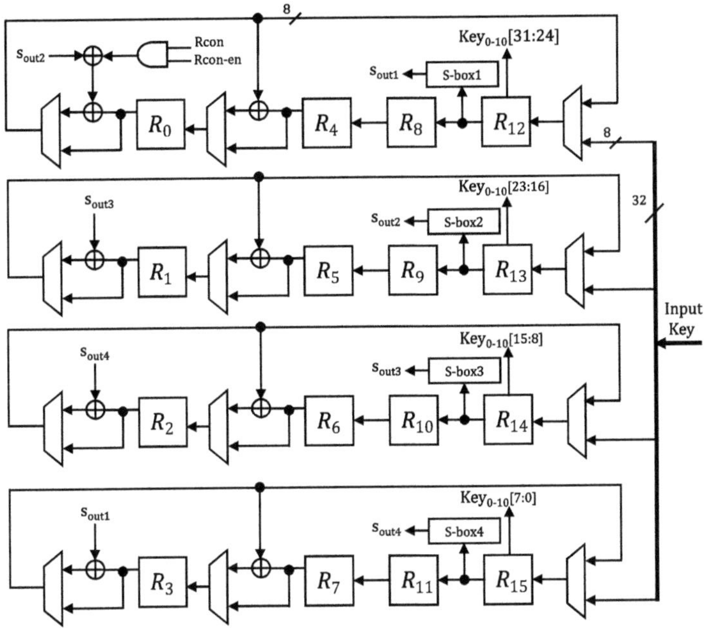

Fig. 2. The proposed 32-bit key schedule module.

bytes are added directly to either the input data for round 0, or to the output of the 32-bit MixColumns for rounds 1–9.

4   The Proposed 64-Bit AES-128 Design

This section presents a new design for the AES encryption algorithm using a 64-bit datapath. The output data becomes available after 22 clock cycles: 2 cycles for loading the input data and applying the input key (Key 0), followed by 2 cycles for each of the ten encryption rounds.

4.1   The Proposed 64-Bit Datapath

Figure 3 outlines the proposed 64-bit datapath.

We incorporate a 64-bit column-wise permutation block similar to the one proposed in [7], with some modifications due to the absence of the decryption function, which eliminates the need for certain multiplexers. The wiring between the output of the 64-bit permutation block and the 8 lightweight S-boxes is such that S-box1, S-box2, ..., S-box8 are connected to $D_0, D_5, D_2, D_7, D_4, D_1, D_6, D_3$, respectively, according to [7]. We use two 32-bit MixColumns in the datapath to output two columns of the state each clock cycle. Two 64-bit XOR gates are used for AddRoundKey. The first XOR gate is

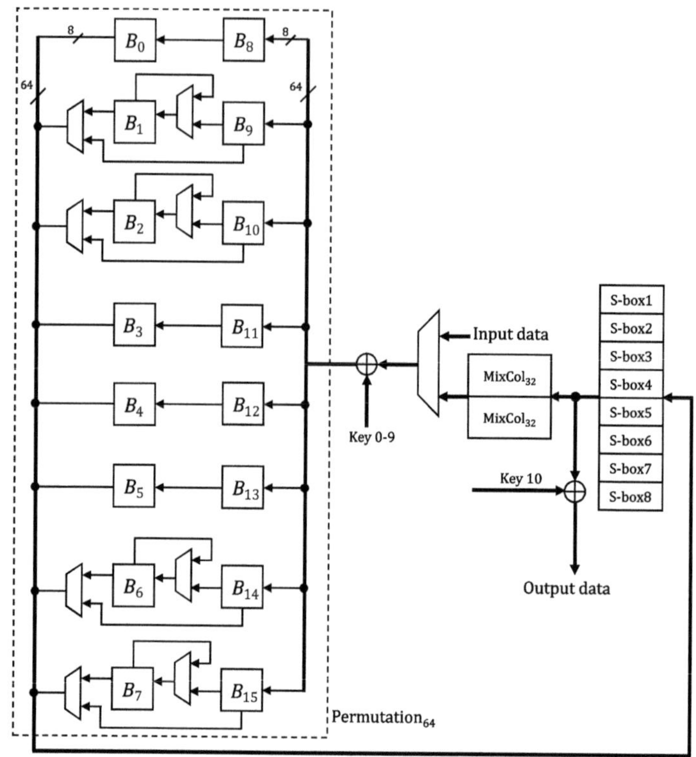

Fig. 3. The proposed 64-bit AES datapath.

placed before the permutation block, to add the key for rounds 1–9. The second 64-bit XOR gate is used for the final key addition with the output of the S-boxes in order to skip MixColumns. Thus, the output data is obtained by adding the outputs of the eight S-boxes in the datapath to the final key.

4.2   The Proposed 64-Bit Key Schedule Module

The architecture of the proposed 64-bit key schedule module, which is capable of computing eight bytes of the key per clock cycle, is illustrated in Fig. 4. To the right of the unit, there are eight 8-bit multiplexers that select either loading the input key or calculating the new key. The first two clock cycles are used to load the input key into the 64-bit key schedule module.

After the loading operation is complete, and according to Table 2, two clock cycles are required to evaluate the new key. Eight bytes of the key for rounds 0–10 are added to the state each clock cycle. The round key bytes are available from the eight 8-bit key registers $R_8, R_9, R_{10}, R_{11}, R_{12}, R_{13}, R_{14}, R_{15}$. No delay is necessary for rounds 0–9 as the key bytes are added directly to either the input data (for round 0) or to the output of the two 32-bit MixColumns units

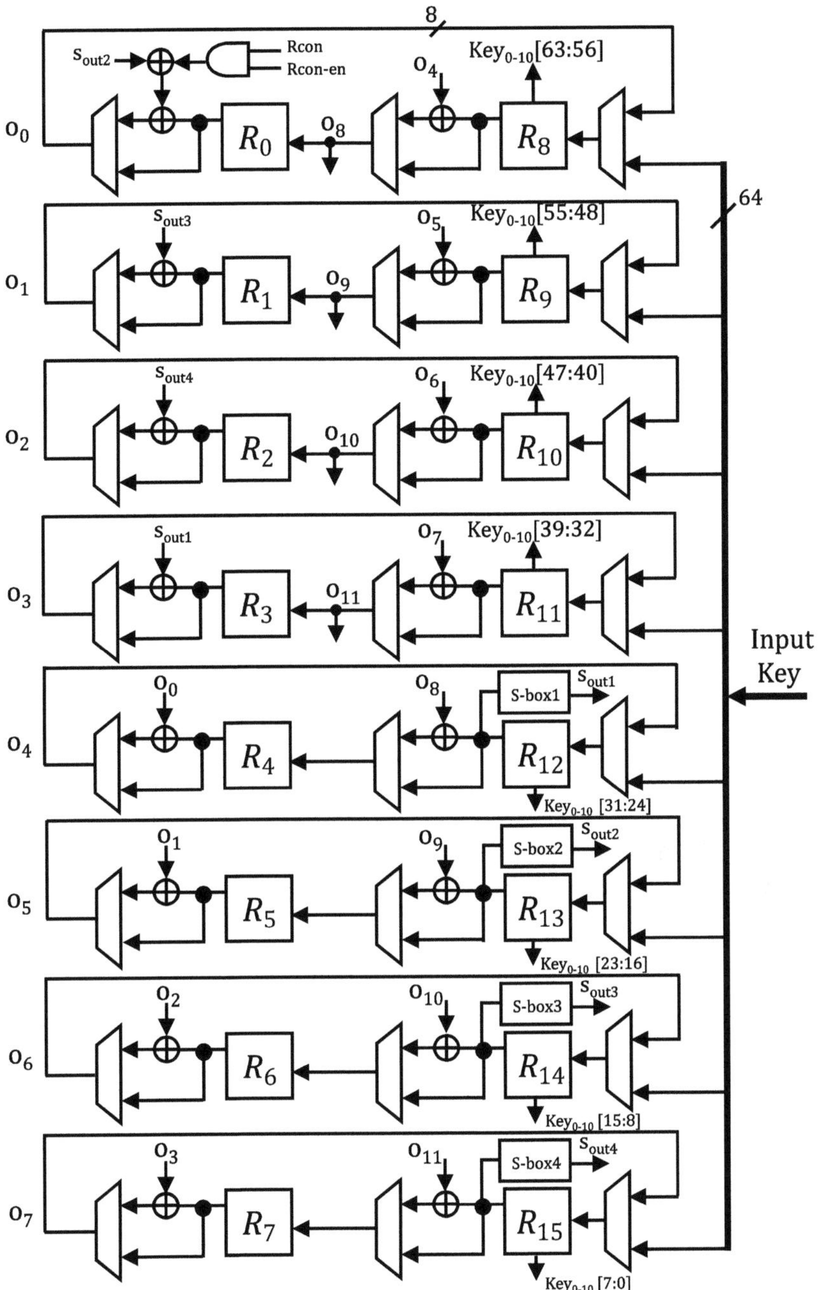

Fig. 4. The proposed 64-bit key schedule module.

(for rounds 1–9) using one 64-bit XOR gate. The final key (round 10) bytes are added to the output of the eight S-boxes using a separate 64-bit XOR gate.

Table 2. Operation of the proposed 64-bit key schedule module.

t	R_0	R_8	R_1	R_9	R_2	R_{10}	R_3	R_{11}	R_4	R_{12}	R_5	R_{13}	R_6	R_{14}	R_7	R_{15}	Operations
0	a_0	a_8	a_1	a_9	a_2	a_{10}	a_3	a_{11}	a_4	a_{12}	a_5	a_{13}	a_6	a_{14}	a_7	a_{15}	$b_0 = a_0 \oplus S(a_{13}) \oplus Rcon, \; b_8 = a_8 \oplus b_4$
																	$b_1 = a_1 \oplus S(a_{14}), \; b_9 = a_9 \oplus b_5$
																	$b_2 = a_2 \oplus S(a_{15}), \; b_{10} = a_{10} \oplus b_6$
																	$b_3 = a_3 \oplus S(a_{12}), \; b_{11} = a_{11} \oplus b_7$
																	$b_4 = a_4 \oplus b_0, \; b_{12} = a_{12} \oplus b_8$
																	$b_5 = a_5 \oplus b_1, \; b_{13} = a_{13} \oplus b_9$
																	$b_6 = a_6 \oplus b_2, \; b_{14} = a_{14} \oplus b_{10}$
																	$b_7 = a_7 \oplus b_3, \; b_{15} = a_{15} \oplus b_{11}$
1	b_8	b_0	b_9	b_1	b_{10}	b_2	b_{11}	b_3	b_{12}	b_4	b_{13}	b_5	b_{14}	b_6	b_{15}	b_7	Shift

$S(a_i)$ denotes the S-box substitution of a_i. Bold bytes indicate values updated in the previous cycle.

5 Performance Evaluation

This section describes the code development process and presents the ASIC implementation results of the new hardware architectures. The proposed 32-bit and 64-bit datapath AES designs and the 16-bit baseline AES [2] were implemented using Verilog. Each core consists of separate datapath and key expansion modules, following the column-wise architecture described in Sects. 3 and 4. Functional verification was performed using ModelSim© and NIST test vectors [10], confirming correct ciphertext generation and the expected encryption cycles of 88, 44, and 22 cycles per block for the 16-, 32-, and 64-bit cores, respectively, and using the same 65 nm technology.

5.1 Hardware Coding Process

We coded separate modules for the datapath and key schedule for each core. A controller unit generates control and status signals to coordinate data movement, key schedule, and the operation of multiplexers in each clock cycle. Testbenches were developed to validate functional correctness. Each testbench instantiated the AES design and applied NIST reference vectors [10], comparing the outputs against expected ciphertext. Simulation was performed at the logic level using ModelSim©.

The key schedule modules for the 32-bit and 64-bit datapath widths employ the same lightweight S-box used in the datapath and generate 4 and 8 key bytes per clock cycle, respectively, in parallel with the datapath operations. The controller manages all multiplexers required to select between loading the input key, performing the RotWord operation, and updating the key register array. For the datapath, the controller coordinates byte-level operations, including loading the input data, executing the ShiftRows transformation, and enabling MixColumns computations.

We also coded the 16-bit baseline design [2] and used the same lightweight S-box in both the datapath and key schedule modules to ensure a fair comparison with the proposed schemes.

5.2   ASIC Implementation Results

ASIC implementation was carried out using the STM 65 nm CMOS standard cell library and Synopsys Design Compiler. The 16-bit baseline [2], the proposed 32-bit design, and the proposed 64-bit core were all synthesized using the lightweight S-box [11] in both the datapath and key schedule module to ensure a fair comparison.

Table 3 compares the ASIC implementation results of the baseline 16-bit AES [2], the proposed 32-bit AES, and the proposed 64-bit AES. The comparison includes area, delay, maximum operating frequency, throughput, power, latency, energy, and throughput-to-area efficiency. The 16-bit baseline occupies 2763 GE and operates at 504.35 MHz, achieving a throughput of 734 Mbps with a latency of 88 cycles. Increasing the datapath width to 32 bits significantly improves performance. The 32-bit architecture reaches 1.512 Gbps throughput, corresponding to a 2.06× improvement over the 16-bit design, while the area increases to 3566 GE (+29.1%). The latency is reduced by half to 44 cycles, and the maximum frequency slightly increases to 519.84 MHz due to a comparable critical path. The 64-bit architecture further improves performance, reaching 2.95 Gbps throughput, which corresponds to a 4.02× increase compared with the 16-bit baseline and 1.95× compared with the proposed 32-bit design. This performance gain is achieved with an area of 5094.75 GE, representing an 84.4% increase relative to the 16-bit architecture and a 42.9% increase compared to the 32-bit design. Despite the wider datapath, the maximum frequency remains comparable (506.99 MHz), while the latency decreases to only 22 cycles.

From an area-efficiency perspective, the throughput-per-area improves significantly with wider datapaths. We define the throughput-per-area metric as the throughput in Kbps divided by the area in GE, where the area in GE is obtained by dividing the ASIC area in μm^2 by the area of a two-input NAND gate in the corresponding library. The area of a two-input NAND gate in STM 65 nm is $2.08\,\mu m^2$, while in NanGate 45 nm it is $0.798\,\mu m^2$. The efficiency increases from 265.6 Kbps/GE for the 16-bit architecture to 424 Kbps/GE for the 32-bit design (+59.4%) and 579.1 Kbps/GE for the 64-bit architecture (+117.7%).

Power consumption scales with datapath width, increasing from $866.5\,\mu W$ for the 16-bit design to $1066.5\,\mu W$ (+23.1%) for the 32-bit architecture and $1532.3\,\mu W$ (+76.8%) for the 64-bit design. However, the throughput improvement outpaces the power increase, leading to better energy efficiency per processed bit. Energy consumption is calculated at a given clock frequency by multiplying the power obtained from the CAD tool by the latency in cycles and dividing by the clock frequency (we use a clock frequency of 10 MHz). As a result, the proposed 32-bit and 64-bit architectures achieve energy reductions

Table 3. ASIC implementation results of 16-bit, 32-bit, and 64-bit AES cores using STM 65 nm and the lightweight S-box [11].

Design	# S-box	Technology	Area (GE)	Delay (ns)	Max Freq (MHz)	Throughput (Mbps)	Power (μW)	Latency (cycles)	Energy @10 MHz (nJ/128-bit)	Efficiency (Kbps/GE)
1-bit [4]	1	45 nm	1974	1.98	505.05	45.91	143	1408	20.10	23.26
8-bit [4]	2	45 nm	2535	1.93	518.13	376.94	192.4	176	3.38	148.72
8-bit [6]	2	45 nm	2805.33	2.95	338.98	246.53	329.12	176	5.79	87.88
16-bit [2]	4	65 nm	2763.25	1.983	504.35	734.00	866.5	88	7.62	265.6
32-bit [13]	12	0.18 μm	-	3.44	290.7	664.5	-	54	-	-
32-bit (new)	8	65 nm	3566.04	1.924	519.84	1512.00	1066.5	44	4.69	424.0
64-bit [13]	24	0.18 μm	-	3.35	298.5	1123.8	-	32	-	-
64-bit (new)	12	65 nm	5094.75	1.972	506.99	2950.00	1532.3	22	3.37	579.1

of 38.5% and 55.8%, respectively, compared to the 16-bit baseline, decreasing the energy per 128-bit block from 7.62 nJ for the 16-bit datapath to 4.69 nJ and 3.37 nJ for the 32-bit and 64-bit datapaths, respectively.

5.3   Comparison to Previous Work Using Different Technologies

Table 3 shows a clear advantage of the proposed 32-bit and 64-bit architectures over prior designs in terms of throughput, latency, energy consumption, and area efficiency. The 1-bit implementation [4] represents a compact design with minimal area (1974 GE) and low power consumption (143 μW), but it achieves limited throughput of only 45.91 Mbps and incurs very high latency of 1408 cycles, resulting in modest area efficiency (23.26 kbps/GE).

The prior 8-bit in [4] achieves 376.94 Mbps throughput with 176 cycles latency, while maintaining reasonable area and power. The alternative 8-bit design in [6] trades higher area and power for improved energy efficiency compared to the 1-bit design [4], though its throughput (246.53 Mbps) remains lower than the other 8-bit architecture [4]. Both 8-bit designs significantly outperform the 1-bit implementation [4] but still suffer from relatively high latency.

The previous 16-bit design [2] further enhances throughput to 734 Mbps while reducing latency to 88 cycles, demonstrating the benefits of increased datapath width. However, this improvement is accompanied by higher area and power consumption, and its area efficiency reaches 265.6 kbps/GE.

Building on this progression, the proposed 32-bit architecture nearly doubles the throughput to 1512 Mbps while halving the latency to 44 cycles compared to the 16-bit design. Although this improvement incurs increases in area and power, the design substantially improves area efficiency from 265.6 to 424 kbps/GE and reduces energy per 128-bit block.

The 64-bit architecture further extends these gains, delivering 2950 Mbps throughput and reducing latency to 22 cycles, representing an additional twofold improvement in throughput over the 32-bit design. Despite higher area and power requirements, the 64-bit design achieves the best overall area efficiency of 579.1 kbps/GE and the lowest energy consumption of 3.37 nJ per 128-bit block among all evaluated architectures.

The proposed architectures are compared with the 32-bit and 64-bit designs in [13]. While [13] reports throughputs of 664.5 Mbps and 1123.8 Mbps for its 32-bit and 64-bit implementations, respectively, the proposed designs significantly improve throughput to 1512 Mbps and 2950 Mbps. A direct comparison of energy metrics and throughput-per-area efficiency is not possible, as power and area in GE are not reported in [13]. It is worth noting that the proposed designs achieve lower latency and require fewer S-boxes than the counterparts in [13] for both 32-bit and 64-bit datapaths. This comparison is independent of the implementation technology.

Overall, the results show a clear trend where increasing datapath width from 1-bit to 64-bit progressively improves throughput and reduces latency, while also enhancing area efficiency. The proposed 32-bit and 64-bit designs provide substantial gains over prior work, achieving a superior balance between performance and area efficiency suitable for high-performance IoT devices.

6   Conclusion

This work presented the design, implementation, and ASIC evaluation of 32-bit and 64-bit AES-128 architectures optimized for IoT applications. Both designs integrate a separate key schedule module that operates in parallel with the datapath, eliminating key-setup latency and enabling maximum throughput without resource sharing. The proposed architectures deliver substantial improvements in performance and efficiency over prior work. Compared to the 32-bit and 64-bit designs in [13], the new implementations achieve significantly higher throughputs of 1.512 Gbps and 2.950 Gbps, respectively, while reducing latency and requiring fewer S-boxes, independent of the underlying technology.

Using the lightweight S-box [11] and STM 65 nm technology for fair comparison with the 16-bit baseline in [2], the proposed 32-bit and 64-bit cores complete AES-128 encryption in 44 and 22 cycles per block—2× and 4× faster than the baseline. Throughput increases from 734 Mbps to 1.512 Gbps and 2.950 Gbps, corresponding to improvements of approximately 2.06× and 4.02×, with area overheads of 29.1% and 84.4%. Despite the larger datapaths, throughput-per-area efficiency improves by 59% and 118%, and energy per block decreases by 38.5% and 55.8% for the 32-bit and 64-bit designs, respectively.

Overall, the proposed architectures offer flexible design points that balance area, throughput, and energy efficiency. While ultra-compact designs such as the 1-bit architecture in [4] minimize hardware footprint, the 32-bit and 64-bit implementations provide significantly higher throughput, improved area efficiency, and lower energy per block. These characteristics make them well suited for high-speed, energy-constrained environments, including real-time secure communication and low-latency wireless encryption in IoT systems.

Acknowledgment. The authors would like to thank the reviewers for their insightful comments, which helped improve this paper. We would like to thank the Canadian Microelectronics Corporation (CMC) Microsystems for providing the CAD tools used in this work. This work was supported in part by the Natural Sciences and Engineering Research Council (NSERC) of Canada under the Discovery Grant.

The authors have no competing interests to declare that are relevant to the content of this article.

References

1. Agwa, S., Yahya, E., Ismail, Y.: Power efficient AES core for IoT constrained devices implemented in 130nm CMOS. In: 2017 IEEE International Symposium on Circuits and Systems (ISCAS), pp. 1–4. IEEE (2017)
2. Ashmawy, D., Reyhani-Masoleh, A.: A new 16-Bit IoT ASIC design for the AES encryption algorithm. In: 2025 IEEE International Symposium on Circuits and Systems (ISCAS), pp. 1–5 (2025)
3. Balli, F., Banik, S.: Six shades of AES. In: International Conference on Cryptology in Africa, pp. 311–329. Springer (2019)
4. Balli, F., Caforio, A., Banik, S.: The area-latency symbiosis: towards improved serial encryption circuits. IACR Trans. Cryptogr. Hardw. Embed. Syst. 239–278 (2021)
5. Gueron, S., Mathew, S.: Hardware implementation of AES using area-optimal polynomials for composite-field representation $GF((2^4)^2)$ of $GF(2^8)$. In: 23nd IEEE Symposium on Computer Arithmetic, ARITH 2016, Silicon Valley, CA, USA, 10–13 July 2016, Proceedings, pp. 112–117 (2016)
6. Hamalainen, P., Alho, T., Hannikainen, M., Hamalainen, T.D.: Design and implementation of low-area and low-power AES encryption hardware core. In: 9th EUROMICRO Conference on Digital System Design (DSD 2006), pp. 577–583 (2006)
7. Järvinen, T., Salmela, P., Hämäläinen, P., Takala, J.: Efficient byte permutation realizations for compact AES implementations. In: 2005 13th European Signal Processing Conference, pp. 1–4. IEEE (2005)
8. Kim, H.K., Sunwoo, M.H.: Low power AES using 8-bit and 32-bit datapath optimization for small internet-of-things (IoT). J. Signal Process. Syst. **91**(11–12), 1283–1289 (2019)
9. Maximov, A., Ekdahl, P.: New circuit minimization techniques for smaller and faster AES SBoxes. IACR Trans. Cryptogr. Hardw. Embed. Syst. 91–125 (2019)
10. NIST Standard: Specification for the Advanced Encryption Standard (AES) (2001)
11. Reyhani-Masoleh, A., Taha, M., Ashmawy, D.: Smashing the implementation records of AES S-box. IACR Trans. Cryptogr. Hardw. Embed. Syst. 298–336 (2018)
12. Satoh, A., Morioka, S., Takano, K., Munetoh, S.: A compact Rijndael hardware architecture with S-box optimization. In: Advances in Cryptology - ASIACRYPT 2001, 7th International Conference on the Theory and Application of Cryptology and Information Security, Gold Coast, Australia, 9–13 December 2001, Proceedings, pp. 239–254 (2001)
13. Sheikhpour, S., Mahani, A., Bagheri, N.: Reliable advanced encryption standard hardware implementation: 32-bit and 64-bit data-paths. Microprocess. Microsyst. **81**, 103740 (2021)
14. Ueno, R., et al.: High throughput/gate AES hardware architectures based on datapath compression. IEEE Trans. Comput. **69**(4), 534–548 (2020)

Combinatorics, Curves, and Dynamical Structures

Constructing a Parameterized Polynomial Map Through Functional Graphs and Applications

Hugo Teixeira[1]($\boxtimes$), Claude Gravel[2], and Daniel Panario[1]

[1] Carleton University, Ottawa, Canada
HugoTeixeira@cmail.carleton.ca, daniel@math.carleton.ca
[2] Toronto Metropolitan University, Toronto, Canada
gravel@torontomu.ca

Abstract. Let q be an odd prime power and let $\mathbb{F}_q$ denote the finite field with q elements. We study the functional graph structure of the polynomial map $F(X) = (aX^q + bX)(X^q - X)^{n-1}$ over the quadratic extension $\mathbb{F}_{q^2}$, building on the dynamical analysis of Brochero and Teixeira. By representing $\mathbb{F}_{q^2}$ as a two-dimensional vector space over $\mathbb{F}_q$, we obtain an explicit coordinate form for F that reveals how the parameters a, b, and n control the preimage structure. We introduce three parameterized families of functions derived from F with controlled preimage properties: the first achieves constant preimage size $\gcd(n, q-1)$ for non-zero elements when $a^2 \neq b^2$; the second incorporates affine shifts to avoid degenerate orbits; and the third projects to a one-dimensional setting. We analyze the cycle structure, connected components, and trees attached to cycles in each case. As an application, we observe that these families satisfy the conditions for ϵ-strongly universal hash functions, suggesting potential use in randomized algorithms and data structures. We aim to connect the rich theory of functional graphs over finite fields with parameterized function families having predictable combinatorial behavior.

Keywords: functional graphs · finite fields · quadratic extensions · polynomial dynamics · preimage structure · iterations of functions

1 Introduction

Given a finite set S and a function $f : S \to S$, the *functional graph* of f is the directed graph $\mathcal{G}_f = (S, E)$ where $(x, y) \in E$ if and only if $f(x) = y$. The study of functional graphs of polynomial maps in the particular case where S is a finite fields has attracted considerable attention due to its applications in cryptography [16,18,19] and biology [2,12,24] and [15, Chapter 17.1]. Classical results concern the iteration of monomials X^d, where the functional graph structure is determined by the multiplicative structure of $\mathbb{F}_q^*$. More generally, Chebyshev polynomials [9,21], Rédei functions [7,20], and various other polynomial families have been analyzed. A survey of over 120 papers on the dynamics of different

L. Batina and F. Özbudak (Eds.): WAIFI 2026, LNCS 16611, pp. 343–357, 2026.
https://doi.org/10.1007/978-3-032-27574-5_21

functions over finite fields is presented in [14]. Many more papers on this topic have been published in recent years; for example [1,4,5,10,11,17,23].

In general, since every vertex in a functional graph has out-degree exactly one, each connected component consists of a unique cycle with trees attached to the cycle vertices. Understanding the functional graph of a map f, that is, characterizing its cycle lengths, the number of cycles of each length, and the structure of the attached trees, provides a precise visual representation and an insight into the behavior of f. Moreover, periodicity, rho-length and permutation properties are easily derived from the functional graph.

In this paper, we focus on the map

$$F(X) = (aX^q + bX)(X^q - X)^{n-1} \tag{1}$$

over the quadratic extension $\mathbb{F}_{q^2}$, where $q = p^s$ is an odd prime power, $a, b \in \mathbb{F}_q$, and $n \geq 2$ is a positive integer. This function was recently analyzed by Brochero and Teixeira [6], who provided a complete characterization of its dynamics. Our goal is to explore how parameterized families of functions derived from F can be constructed with controlled preimage properties, and investigate what they imply for the functional graph structure.

A key tool in our analysis is the representation of $\mathbb{F}_{q^2}$ as a two-dimensional vector space over $\mathbb{F}_q$. Let $\beta \in \mathbb{F}_{q^2}$ be any element such that $\beta \notin \mathbb{F}_q$ and $\beta^2 \in \mathbb{F}_q$. Then, $\{1, \beta\}$ forms a basis of $\mathbb{F}_{q^2}$ over $\mathbb{F}_q$, and every element $X \in \mathbb{F}_{q^2}$ can be uniquely written as

$$X = x + y\beta$$

for some $x, y \in \mathbb{F}_q$. We denote this representation by the ordered pair $\langle x, y \rangle \in \mathbb{F}_q \times \mathbb{F}_q$.

This identification $\mathbb{F}_{q^2} \cong \mathbb{F}_q \times \mathbb{F}_q$ allows us to study the map F both in $\mathbb{F}_{q^2}$ and in $\mathbb{F}_q \times \mathbb{F}_q$. The Frobenius automorphism $X \mapsto X^q$ acts on coordinates as $(x + y\beta)^q = x - y\beta$, since $\beta^q = -\beta$. Hence, we have

$$F(x + y\beta) = (a(x - y\beta) + b(x + y\beta))((x - y\beta) - (x + y\beta))^{n-1}$$
$$= ((a + b)x + (b - a)y\beta)(-2y\beta)^{n-1}.$$

We define the parameters

$$\delta_1 = a + b \quad \text{and} \quad \delta_2 = b - a, \tag{2}$$

which play a central role in the preimage structure of F. When n is odd, using the vector space representation, we can express F in the following way.

$$F(x, y) = (4\beta^2)^{\frac{n-1}{2}} y^{n-1} \langle \delta_1 x, \delta_2 y \rangle. \tag{3}$$

Remark 1. Throughout this paper, we use the notations $F(X)$, $F(x + y\beta)$, and $F(x, y)$ interchangeably to denote the same function, with the understanding that $X = x + y\beta$ corresponds to $\langle x, y \rangle$.

Since $q - 1$ is even (as q is an odd prime power), the value $\gcd(n, q - 1)$ is minimized when n is odd. This impacts directly on the preimage size of cyclic element as we show in Sect. 2. For this reason, we focus on odd values of n throughout this paper and work with the form given in Eq. (3).

We present three parameterized families of functions based on (3):

- **Family 1:** (Sect. 3.1): Functions $h_{\langle \delta_1, \delta_2 \rangle} : \mathbb{F}_q^* \times \mathbb{F}_q^* \to \mathbb{F}_q^* \times \mathbb{F}_q^*$ parameterized by $\langle \delta_1, \delta_2 \rangle \in \mathbb{F}_q^* \times \mathbb{F}_q^*$. When $\gcd(n, q - 1) = 1$, these functions are injective.
- **Family 2:** (Sect. 3.2): Functions $h_{\langle \delta_1, \delta_2, c, d \rangle} : \mathbb{F}_q^* \times \mathbb{F}_q^* \to \mathbb{F}_q^* \times \mathbb{F}_q^*$ with additional shift parameters $c, d \in \mathbb{F}_q^*$ that avoid the degenerate component containing zero. The parameter space has size $(q - 1)^4$.
- **Family 3:** (Sect. 3.3): Functions $h_{\langle a, b \rangle} : \mathbb{F}_q^* \to \mathbb{F}_q^*$ obtained by projection to one coordinate, with parameter space of size $(q - 1)^2$.

For each family, we analyze the preimage structure and relate it to the functional graph properties established in [6]. As an application, we observe that these families satisfy the conditions for ϵ-strongly universal hash functions, which have uses in randomized algorithms, load balancing, and data structures.

We finish this section by providing an overview of the structure of the paper. Section 2 presents the necessary background on functional graphs and recalls key results from [6] needed for this paper. Section 3 introduces our three parameterized families and analyzes their properties. Section 4 discusses the connection to universal hashing and provides computational examples. Section 5 illustrates the structure of some examples of functional graphs. Finally, Sect. 6 concludes with remarks on future directions.

2   Preliminaries

In this section, we recall key results on the functional graph of F from [6].

The following theorems from [6] characterize the functional graph of $F(X) = (aX^q + bX)(X^q - X)^{n-1}$ over $\mathbb{F}_{q^2}$.

Definition 1 (Preimage). *Given a function $f : S \to S$ and an element $\alpha \in S$, the* preimage *of α under f is $f^{-1}(\alpha) = \{x \in S : f(x) = \alpha\}$. The in-degree of α in $\mathcal{G}_f$ equals $|f^{-1}(\alpha)|$.*

Theorem 1 (Preimage Sizes [6, Theorems 3, 6 and 8]). *Let F be as in (1) with $\delta_1 = a + b$ and $\delta_2 = b - a$. For any $\alpha \in \mathbb{F}_{q^2}$:*

1. If $\alpha \neq 0$ and $\delta_1 = 0$, then $|F^{-1}(\alpha)| \in \{0, q - 1, q \gcd(n, q - 1)\}$.
2. If $\alpha \neq 0$ and $\delta_2 = 0$, then $|F^{-1}(\alpha)| \in \{0, q \gcd(n, q - 1)\}$.
3. If $\alpha \neq 0$ and $\delta_1 \delta_2 \neq 0$, then $|F^{-1}(\alpha)| \in \{0, \gcd(n, q - 1)\}$.
4. If $\alpha = 0$, then $|F^{-1}(\alpha)| \in \{q - 1, 2q - 2\}$.

This theorem reveals that the preimage structure depends critically on the parameters δ_1 and δ_2. In particular, when both are non-zero, the preimage size of any non-zero element is either 0 or $\gcd(n, q - 1)$. This uniformity is a key property that we exploit in our parameterized families.

Theorem 2 (Zero Component [6, Lemmas 17, 24 and 33]). *The connected component containing $0 \in \mathbb{F}_{q^2}$ consists only of elements of the form $\langle x, 0 \rangle$ and $\langle 0, y \rangle$. Moreover, any element in this component reaches 0 after at most two iterations of F.*

In the coordinate representation, elements $\langle x, 0 \rangle$ correspond to $x \in \mathbb{F}_q \subset \mathbb{F}_{q^2}$, while elements $\langle 0, y \rangle$ correspond to $y\beta$ with $y \in \mathbb{F}_q^*$. The zero component thus contains exactly $2q - 1$ elements (including zero itself), and these are precisely the elements with $xy = 0$.

Figure 1 illustrates the functional graph of $F(X) = (X^5 + 2X)(X^5 - X)$ over $\mathbb{F}_{25}$. The elements are labeled $0, 1, \ldots, 24$, corresponding to powers of a primitive element. The connected component containing 0 is visible on the bottom right side of the figure.

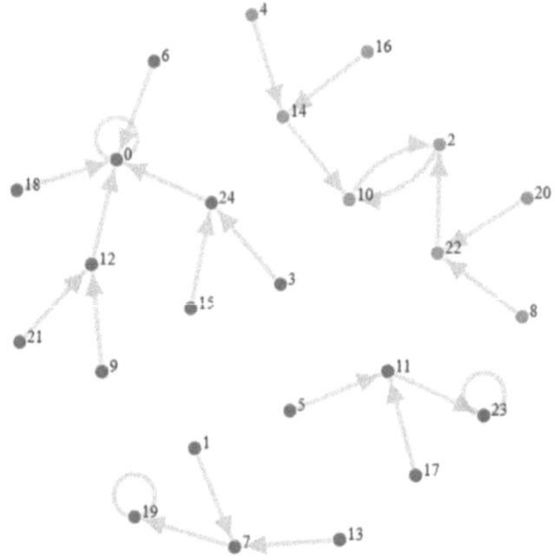

Fig. 1. Functional graph of $F(X) = (X^5 + 2X)(X^5 - X)$ over $\mathbb{F}_{25}$. Elements are labeled 0 to 24. The component on the bottom right contains 0 and the elements mapping to it.

3 Parameterized Families with Controlled Preimage Properties

We now introduce three parameterized families of functions derived from F. Our goal is to understand how the choice of parameters affects the preimage structure and, consequently, the behavior of the functional graph when the parameter is fixed.

3.1 Family 1: Basic Parameterization

Construction 1. *Let $\mathbb{F}_q$ be the finite field with $q = p^s$ elements, where p is an odd prime and s is a positive integer. Let n be a positive odd integer such that $\gcd(n, q - 1) = 1$, and let $\beta \in \mathbb{F}_{q^2} \setminus \mathbb{F}_q$ satisfy $\beta^2 \in \mathbb{F}_q$.*

Define the family $\mathcal{F}_1 = \{h_{\langle \delta_1, \delta_2 \rangle} : \langle \delta_1, \delta_2 \rangle \in \mathbb{F}_q^* \times \mathbb{F}_q^*\}$, *where each function* $h_{\langle \delta_1, \delta_2 \rangle} : \mathbb{F}_q^* \times \mathbb{F}_q^* \to \mathbb{F}_q^* \times \mathbb{F}_q^*$ *is defined by*

$$h_{\langle \delta_1, \delta_2 \rangle}(x, y) = (4\beta^2)^{\frac{n-1}{2}} y^{n-1} \langle \delta_1 x, \delta_2 y \rangle. \tag{4}$$

Theorem 3. *The family* $\mathcal{F}_1$ *has the following properties:*

1. *For any fixed* $\langle \delta_1, \delta_2 \rangle \in \mathbb{F}_q^* \times \mathbb{F}_q^*$, *the function* $h_{\langle \delta_1, \delta_2 \rangle}$ *is a bijection from* $\mathbb{F}_q^* \times \mathbb{F}_q^*$ *to itself.*
2. *For any distinct* $\langle x_1, y_1 \rangle, \langle x_2, y_2 \rangle \in \mathbb{F}_q^* \times \mathbb{F}_q^*$ *and any* $\langle u_1, v_1 \rangle, \langle u_2, v_2 \rangle \in \mathbb{F}_q^* \times \mathbb{F}_q^*$, *there is at most one parameter* $\langle \delta_1, \delta_2 \rangle$ *such that* $h_{\langle \delta_1, \delta_2 \rangle}(x_1, y_1) = \langle u_1, v_1 \rangle$ *and* $h_{\langle \delta_1, \delta_2 \rangle}(x_2, y_2) = \langle u_2, v_2 \rangle$.

Proof. (1) For a fixed $\langle \delta_1, \delta_2 \rangle \in \mathbb{F}_q^* \times \mathbb{F}_q^*$, suppose $h_{\langle \delta_1, \delta_2 \rangle}(x_1, y_1) = h_{\langle \delta_1, \delta_2 \rangle}(x_2, y_2)$ for some $\langle x_1, y_1 \rangle, \langle x_2, y_2 \rangle \in \mathbb{F}_q^* \times \mathbb{F}_q^*$. Then

$$(4\beta^2)^{\frac{n-1}{2}} y_1^{n-1} \langle \delta_1 x_1, \delta_2 y_1 \rangle = (4\beta^2)^{\frac{n-1}{2}} y_2^{n-1} \langle \delta_1 x_2, \delta_2 y_2 \rangle.$$

Comparing components:

$$y_1^{n-1} \delta_1 x_1 = y_2^{n-1} \delta_1 x_2,$$
$$y_1^n \delta_2 = y_2^n \delta_2.$$

Since $\delta_2 \neq 0$, the second equation gives $y_1^n = y_2^n$. Since $\gcd(n, q-1) = 1$ and $y_1, y_2 \in \mathbb{F}_q^*$, we conclude $y_1 = y_2$. Substituting into the first equation and using $\delta_1 \neq 0$, we get $x_1 = x_2$. Thus $h_{\langle \delta_1, \delta_2 \rangle}$ is injective, hence bijective on the finite set $\mathbb{F}_q^* \times \mathbb{F}_q^*$.

(2) We have that $h_{\langle \delta_1, \delta_2 \rangle}(x_1, y_1) = \langle u_1, v_1 \rangle$ and $h_{\langle \delta_1, \delta_2 \rangle}(x_2, y_2) = \langle u_2, v_2 \rangle$ is equivalent to the following system of equations

$$(4\beta^2)^{\frac{n-1}{2}} \delta_1 y_1^{n-1} x_1 = u_1,$$
$$(4\beta^2)^{\frac{n-1}{2}} \delta_2 y_1^n = v_1,$$
$$(4\beta^2)^{\frac{n-1}{2}} \delta_1 y_2^{n-1} x_2 = u_2,$$
$$(4\beta^2)^{\frac{n-1}{2}} \delta_2 y_2^n = v_2.$$

We have that the system is solvable for δ_1 and δ_2 if and only if

$$\frac{u_1}{(4\beta^2)^{\frac{n-1}{2}} y_1^{n-1} x_1} = \frac{u_2}{(4\beta^2)^{\frac{n-1}{2}} y_2^{n-1} x_2} \quad \text{and} \quad \frac{v_1}{(4\beta^2)^{\frac{n-1}{2}} y_1^n} = \frac{v_2}{(4\beta^2)^{\frac{n-1}{2}} y_2^n}.$$

In this case, we have a unique solution. $\qquad\qquad\qquad\qquad\qquad\qquad\qquad\quad\square$

Observation 1. *The constraint* $\gcd(n, q-1) = 1$ *is essential for injectivity. When* $\gcd(n, q-1) = d > 1$, *the equation* $y_1^n = y_2^n$ *has d solutions for y_2 given y_1, leading to non-trivial collisions.*

3.2   Family 2: Shifted Parameterization

As noted in Theorem 2, the functional graph of F has a degenerate component containing elements of the form $\langle x, 0 \rangle$ and $\langle 0, y \rangle$. To avoid this component, we introduce shift parameters.

Definition 2. *For $d \in \mathbb{F}_q^*$ and $x \in \mathbb{F}_q^*$, we define the function $\phi_d : \mathbb{F}_q^* \to \mathbb{F}_q$ by*

$$\phi_d(x) = \begin{cases} x & \text{if } x \neq d, \\ 0 & \text{if } x = d. \end{cases}$$

We observe that ϕ_d is a bijection from $\mathbb{F}_q^*$ to $\mathbb{F}_q \setminus \{d\}$.

Construction 2. *Let $\mathbb{F}_q$, n, and β be as in Construction 1. Define the family $\mathcal{F}_2 = \{h_{\langle \delta_1, \delta_2, c, d \rangle} : \langle \delta_1, \delta_2, c, d \rangle \in (\mathbb{F}_q^*)^4\}$, where each function $h_{\langle \delta_1, \delta_2, c, d \rangle} : \mathbb{F}_q^* \times \mathbb{F}_q^* \to \mathbb{F}_q^* \times \mathbb{F}_q^*$ is defined by*

$$h_{\langle \delta_1, \delta_2, c, d \rangle}(x, y) = (4\beta^2)^{\frac{n-1}{2}} (\phi_d(y) - d)^{n-1} \langle \delta_1(\phi_c(x) - c), \delta_2(\phi_d(y) - d) \rangle. \quad (5)$$

The shifts by c and d ensure that the transformed coordinates $\phi_c(x) - c$ and $\phi_d(y) - d$ are always non-zero for $x, y \in \mathbb{F}_q^*$. This avoids the degenerate behavior associated with the zero component.

Theorem 4. *The family $\mathcal{F}_2$ has the following properties:*

1. *For any fixed parameter $\langle \delta_1, \delta_2, c, d \rangle \in (\mathbb{F}_q^*)^4$, the function $h_{\langle \delta_1, \delta_2, c, d \rangle}$ is a bijection.*
2. *For any distinct $\langle x_1, y_1 \rangle, \langle x_2, y_2 \rangle \in \mathbb{F}_q^* \times \mathbb{F}_q^*$ and any $\langle u_1, v_1 \rangle, \langle u_2, v_2 \rangle \in \mathbb{F}_q^* \times \mathbb{F}_q^*$, the number of parameters $\langle \delta_1, \delta_2, c, d \rangle$ satisfying both $h(x_1, y_1) = \langle u_1, v_1 \rangle$ and $h(x_2, y_2) = \langle u_2, v_2 \rangle$ is at most $q - 1$.*

Proof. (1) The proof is analogous to Theorem 3(1), using the fact that the shifts are bijections.

(2) We write $y_i - d$ for $\phi_d(y_i) - d$, noting that $y_i \neq d$ implies no special case. Writing out the conditions, we get

$$(4\beta^2)^{\frac{n-1}{2}} \delta_1(y_1 - d)^{n-1}(x_1 - c) = u_1, \quad (6)$$

$$(4\beta^2)^{\frac{n-1}{2}} \delta_2(y_1 - d)^n = v_1, \quad (7)$$

$$(4\beta^2)^{\frac{n-1}{2}} \delta_1(y_2 - d)^{n-1}(x_2 - c) = u_2, \quad (8)$$

$$(4\beta^2)^{\frac{n-1}{2}} \delta_2(y_2 - d)^n = v_2. \quad (9)$$

From (7) and (9), if $v_1 \neq v_2$, we can solve for d using

$$\frac{(y_1 - d)^n}{(y_2 - d)^n} = \frac{v_1}{v_2}.$$

Since $\gcd(n, q - 1) = 1$, this gives a unique solution for d. Then δ_2 is determined by (7), and δ_1 and c can be found from the remaining equations.

If $v_1 = v_2$, then $(y_1 - d)^n = (y_2 - d)^n$, which (since $\gcd(n, q-1) = 1$) implies $y_1 - d = y_2 - d$, and hence $y_1 = y_2$. In this case, d is a free parameter (any of $q - 1$ values in $\mathbb{F}_q^*$), and for each choice of d, there is a unique solution for δ_1, δ_2, c. Thus at most $q - 1$ solutions exist.

A similar analysis applies when $u_1(y_2 - d)^{n-1} = u_2(y_1 - d)^{n-1}$. $\qquad\square$

3.3 Family 3: One-Dimensional Projection

By projecting to a single coordinate, we obtain a family over $\mathbb{F}_q^*$ rather than $\mathbb{F}_q^* \times \mathbb{F}_q^*$.

Construction 3. *Let $\mathbb{F}_q$, n, and β be as in Construction 1. Define the family $\mathcal{F}_3 = \{h_{\langle a,b\rangle} : \langle a, b\rangle \in \mathbb{F}_q^* \times \mathbb{F}_q^*\}$, where each function $h_{\langle a,b\rangle} : \mathbb{F}_q^* \to \mathbb{F}_q^*$ is defined by*

$$h_{\langle a,b\rangle}(y) = (4\beta^2)^{\frac{n-1}{2}} a(\phi_b(y) - b)^n. \tag{10}$$

This family can be viewed as the second-coordinate projection of Family 2 with $\delta_2 = a$ and $d = b$.

Theorem 5. *The family $\mathcal{F}_3$ has the following properties:*

1. *For any fixed parameter $\langle a, b\rangle \in \mathbb{F}_q^* \times \mathbb{F}_q^*$, the function $h_{\langle a,b\rangle}$ is a bijection on $\mathbb{F}_q^*$.*
2. *For any distinct $y_1, y_2 \in \mathbb{F}_q^*$ and any $u_1, u_2 \in \mathbb{F}_q^*$ with $u_1 \neq u_2$, there is exactly one parameter $\langle a, b\rangle$ such that $h_{\langle a,b\rangle}(y_1) = u_1$ and $h_{\langle a,b\rangle}(y_2) = u_2$.*

Proof. (1) For fixed $\langle a, b\rangle$, the map $y \mapsto (\phi_b(y) - b)^n$ is a bijection on $\mathbb{F}_q^*$ when $\gcd(n, q - 1) = 1$ (as raising to the n-th power is then a bijection on $\mathbb{F}_q^*$). Multiplication by the non-zero constant $(4\beta^2)^{\frac{n-1}{2}} a$ preserves bijectivity.

(2) The conditions give:

$$(4\beta^2)^{\frac{n-1}{2}} a(y_1 - b)^n = u_1,$$

$$(4\beta^2)^{\frac{n-1}{2}} a(y_2 - b)^n = u_2.$$

Dividing: $(y_1 - b)^n/(y_2 - b)^n = u_1/u_2$. Since $u_1 \neq u_2$ and $\gcd(n, q-1) = 1$, there is a unique n-th root, giving

$$\frac{y_1 - b}{y_2 - b} = \left(\frac{u_1}{u_2}\right)^{\frac{1}{n}},$$

from which b can be uniquely determined (since $y_1 \neq y_2$). Finally, a is determined by the first equation. $\qquad\square$

4   Applications and Computational Examples

4.1   Connection to Universal Hashing

The properties established in Sect. 3 have implications for families hash functions. We briefly recall the relevant definitions.

A *hash function* is a function with inputs in a message space $\mathcal{M}$ that produces fixed-sized outputs in a tag space $\mathcal{T}$. Generally, a hash function may take arrays of different lengths as input; however, we consider hash functions with fixed-size input in this section.

The following properties, though not an exhaustive list, correspond to different levels of security for a given hash function [13].

Definition 3 (Security Properties). *Let $f : \mathcal{M} \to \mathcal{T}$ be a hash function. We define the following security properties:*

- ***Collision resistance:*** *It is computationally infeasible to find $x, x' \in \mathcal{M}$ such that $x \neq x'$ and $f(x) = f(x')$.*
- ***Preimage resistance:*** *For any $y \in \mathcal{T}$, it is computationally infeasible to find $x \in \mathcal{M}$ such that $f(x) = y$.*
- ***Second preimage resistance:*** *Given any $x \in \mathcal{M}$, it is computationally infeasible to find $x' \in \mathcal{M}$ such that $f(x) = f(x')$.*

In many applications, to enhance variability and resistance against brute force cyberattacks, a family of keyed hash functions is employed.

Definition 4 (Hash Family [3]). *A $(|\mathcal{K}|, |\mathcal{M}|)_{|\mathcal{T}|}$ hash family is an array with rows indexed by $\mathcal{K}$ and with columns indexed by $\mathcal{M}$, where each entry is an element taken from a set $\mathcal{T}$. Each row is interpreted as a mapping $f_k : \mathcal{M} \to \mathcal{T}$, for $k \in \mathcal{K}$.*

A hash family is equivalently represented by a four-tuple $(\mathcal{K}, \mathcal{M}, \mathcal{T}, \mathcal{H})$, where $\mathcal{H} = \{f_k : \mathcal{M} \to \mathcal{T} \mid k \in \mathcal{K}\}$.

To further enhance collision resistance, Carter and Wegman [8] define a class of families of hash functions, known as universal hash functions.

Definition 5 (Universal Hash Family [8]). *A hash family $(\mathcal{K}, \mathcal{M}, \mathcal{T}, \mathcal{H})$ is universal if for all distinct $m, m' \in \mathcal{M}$ it holds that*

$$\Pr[h_k(m) = h_k(m')] \leq \frac{1}{|\mathcal{T}|},$$

where the probability is taken over uniform choices of $k \in \mathcal{K}$.

An even stronger property is defined by Stinson [22].

Definition 6 (ϵ-Strongly Universal Hash Family [22]). *A hash family $(\mathcal{K}, \mathcal{M}, \mathcal{T}, \mathcal{H})$ is ϵ-strongly universal if the following conditions are satisfied:*

- *For any $m \in \mathcal{M}$ and $t \in \mathcal{T}$, there are exactly $\frac{|\mathcal{K}|}{|\mathcal{T}|}$ functions $f_k \in \mathcal{H}$ such that $f_k(m) = t$.*
- *For all distinct $m, m' \in \mathcal{M}$ and all $t, t' \in \mathcal{T}$ it holds that*

$$\Pr[h_k(m) = t \text{ and } h_k(m') = t'] \leq \frac{\epsilon}{|\mathcal{T}|},$$

where the probability is taken over uniform choices of $k \in \mathcal{K}$.

When $\epsilon = \frac{1}{|\mathcal{T}|}$, a family of ϵ-strongly universal hash functions is referred to as strongly universal.

From the family of functions constructed in the previous section, we can obtain the following.

Remark 2 We emphasize that universal and strongly universal hash families, as defined by Carter and Wegman [8] and Stinson [22], do *not* require the hash function to be compressing. The universality condition concerns collision probabilities over a random choice of key. In particular, bijective families (where $|\mathcal{M}| = |\mathcal{T}|$) are a well-studied and important special case; see, e.g., [3,22]. Compression is a property relevant to cryptographic hash functions in the sense of Definition 3, but it is distinct from the universality framework employed here.

Theorem 6.   *1. The function family from Construction 1 is universal and 1-strongly universal.*
2. *The function family from Construction 2 is universal and $\frac{1}{q-1}$-strongly universal.*
3. *The function family from Construction 3 is strongly universal (i.e., $\frac{1}{|\mathcal{T}|}$-strongly universal).*

Proof. For Construction 1, from Theorem 3 we have that for any distinct $\langle x_1, y_1 \rangle$ and $\langle x_2, y_2 \rangle$ in $\mathbb{F}_q^* \times \mathbb{F}_q^*$ and any $\langle u_1, v_1 \rangle$ and $\langle u_2, v_2 \rangle$ in $\mathbb{F}_q^* \times \mathbb{F}_q^*$, there is at most one $\langle \delta_1, \delta_2 \rangle$ such that $h_{\langle \delta_1, \delta_2 \rangle}(x_1, y_1) = \langle u_1, v_1 \rangle$ and $h_{\langle \delta_1, \delta_2 \rangle}(x_2, y_2) = \langle u_2, v_2 \rangle$. Hence, the probability is smaller or equal to $\frac{1}{(q-1)^2} = \frac{1}{|\mathcal{T}|}$ and Construction 1 is 1-Strongly Universal. The results for Constructions 2 and 3 follow similarly from Theorems 4, and 5 by computing the probabilities (Table 1). $\square$

Table 1. Summary of the three parameterized families

Family	Domain	Codomain	Parameter Space	ϵ
$\mathcal{F}_1$	$\mathbb{F}_q^* \times \mathbb{F}_q^*$	$\mathbb{F}_q^* \times \mathbb{F}_q^*$	$\mathbb{F}_q^* \times \mathbb{F}_q^*$	1
$\mathcal{F}_2$	$\mathbb{F}_q^* \times \mathbb{F}_q^*$	$\mathbb{F}_q^* \times \mathbb{F}_q^*$	$(\mathbb{F}_q^*)^4$	$1/(q-1)$
$\mathcal{F}_3$	$\mathbb{F}_q^*$	$\mathbb{F}_q^*$	$\mathbb{F}_q^* \times \mathbb{F}_q^*$	$1/(q-1)$

4.2   Computational Aspects

With respect to implementation considerations, we recall that the function families presented in Construction 1, 2 and 3 are given respectively by

$$h_{\langle \delta_1, \delta_2 \rangle}(x, y) = (4\beta^2)^{\frac{n-1}{2}} y^{n-1} \langle \delta_1 x, \delta_2 y \rangle,$$

$$h_{\langle \delta_1, \delta_2, c, d \rangle}(x, y) = (4\beta^2)^{\frac{n-1}{2}} (\phi_d(y) - d)^{n-1} \langle \delta_1(\phi_c(x) - c), \delta_2(\phi_d(y) - d) \rangle,$$

$$h_{(a,b)}(y) = (4\beta^2)^{\frac{n-1}{2}} a(\phi_b(y) - b)^n.$$

We determine the algorithms to compute hash tags for each of the constructions. In these algorithms the computation of y^{n-1} takes $O(\log n)$ multiplications using repeated squaring algorithm.

For Construction 1 we have the following algorithm.

Algorithm 1. Hash computation for Construction 1

Input: Message $(x, y) \in \mathbb{F}_q^* \times \mathbb{F}_q^*$, key $(\delta_1, \delta_2) \in \mathbb{F}_q^* \times \mathbb{F}_q^*$
Output: Hash value $(u, v) \in \mathbb{F}_q^* \times \mathbb{F}_q^*$

1: $k \leftarrow (4\beta^2)^{\frac{n-1}{2}}$ ▷ Precomputed constant
2: $y_{pow} \leftarrow y^{n-1}$ ▷ $O(\log n)$ multiplications
3: $u \leftarrow k \cdot \delta_1 \cdot y_{pow} \cdot x$
4: $v \leftarrow k \cdot \delta_2 \cdot y_{pow} \cdot y$
5: **Return** (u, v)

For Construction 2, we recall that for $x \in \mathbb{F}_q$ and $d \in \mathbb{F}_q^*$ the function ϕ_d is defined as $\phi_d(x) = x$, if $x \neq d$ and $\phi_d(d) = 0$, otherwise. Therefore, the computation of the function $\phi_d(x)$ does not require algebraic operations.

Algorithm 2. Hash computation for Construction 2

Input: Message $(x, y) \in \mathbb{F}_q^* \times \mathbb{F}_q^*$, key $(\delta_1, \delta_2, c, d) \in \mathbb{F}_q^* \times \mathbb{F}_q^* \times \mathbb{F}_q^* \times \mathbb{F}_q^*$
Output: Hash value $(u, v) \in \mathbb{F}_q^* \times \mathbb{F}_q^*$

1: $k \leftarrow (4\beta^2)^{\frac{n-1}{2}}$ ▷ Precomputed constant
2: **if** $x = c$ **then**
3:      $x_{shift} \leftarrow -c$
4: **end if**
5: **if** $x \neq c$ **then**
6:      $x_{shift} \leftarrow x - c$
7: **end if**
8: **if** $y = d$ **then**
9:      $y_{shift} \leftarrow -d$
10: **end if**
11: **if** $y \neq d$ **then**
12:      $y_{shift} \leftarrow y - d$
13: **end if**
14: $y_{pow} \leftarrow y_{shift}^{n-1}$ ▷ $O(\log n)$ multiplications
15: $u \leftarrow k \cdot \delta_1 \cdot y_{pow} \cdot x_{shift}$

16: $v \leftarrow k \cdot \delta_2 \cdot y_{pow} \cdot y_{shift}$
17: **Return** (u, v)

For Construction 3, we take the coordinate v of the return value of Algorithm 2. That is equivalent to the following algorithm.

Algorithm 3. Hash computation for Construction 3

Input: Message $y \in \mathbb{F}_q^*$, key $(a, b) \in \mathbb{F}_q^* \times \mathbb{F}_q^*$
Output: Hash value $u \in \mathbb{F}_q^*$

1: $k \leftarrow (4\beta^2)^{\frac{n-1}{2}}$ ▷ Precomputed constant
2: **if** $y = b$ **then**
3:     $y_{shift} \leftarrow -b$
4: **end if**
5: **if** $y \neq b$ **then**
6:     $y_{shift} \leftarrow y - b$
7: **end if**
8: $y_{pow} \leftarrow y_{shift}^n$ ▷ $O(\log n)$ multiplications
9: $u \leftarrow k \cdot a \cdot y_{pow}$
10: **Return** u

We briefly comment on the cost and memory usage of our algorithms. In terms of the number of field multiplications, all constructions require $O(\log n)$ multiplications that come from the required exponentiation. The storage cost of a key is $O(\log q)$ bits for all constructions. The storage cost of the precomputed elements is $O(\log q)$ bits for all construction. All constructions naturally support parallel processing: multiple messages can be hashed independently, the computation of y^{n-1} can use parallel exponentiation algorithms, and SIMD instructions can process multiple field operations simultaneously.

4.3   Security Considerations

We briefly discuss the security model and potential attack vectors relevant to our constructions.

The security guarantee provided by ϵ-strongly universal hash families is *information-theoretic*: for a uniformly random key, the collision probability is bounded by $\epsilon/|\mathcal{T}|$ regardless of the adversary's computational power. This is fundamentally different from the computational security model of cryptographic hash functions (Definition 3), where one assumes bounded adversaries. In the universal hashing framework, the key is secret and chosen uniformly at random; the adversary sees only the hash outputs.

Regarding algebraic attacks, we note that in the coordinate representation, each function in our families has a simple algebraic form: the map $h_{\langle \delta_1, \delta_2 \rangle}(x, y) = (4\beta^2)^{(n-1)/2} y^{n-1} \langle \delta_1 x, \delta_2 y \rangle$ is a monomial map composed with a diagonal linear transformation. Given a single input-output pair, an attacker could recover the key $\langle \delta_1, \delta_2 \rangle$ by solving a system of two equations over $\mathbb{F}_q$. However, this is not a vulnerability in the universal hashing model: the key is never reused across sessions, and the security bound already accounts for the worst-case over all input pairs.

Concerning the cycle structure, the functional graphs of our families are fully characterized by the results of Brochero and Teixeira [6]. When $\gcd(n, q-1) = 1$, each function is a bijection, so the functional graph consists entirely of cycles with no preperiodic tails. Knowledge of the cycle structure does not provide an advantage beyond what the universality bound already concedes, since the bound holds for all pairs of distinct inputs simultaneously.

Finally, we remark that Gröbner basis techniques, while effective for solving polynomial systems, are not the relevant threat model for universal hashing. The algebraic simplicity of our functions is by design: it enables efficient evaluation and the clean algebraic structure that yields tight universality bounds.

5   Functional Graph Examples

To illustrate the structure of functional graphs arising from our families, we present examples with small parameters. Since the domain and codomain coincide in each family, we can iterate the functions and visualize the resulting functional graphs.

Let $q = 7$ and $n = 5$. We have that 3 is not a square in $\mathbb{F}_q$, hence we take $\beta \in \mathbb{F}_{q^2}$ such that $\beta^2 = 3$. For Algorithm 1, we take $\mathcal{M} = \mathcal{T} = \mathbb{F}_7^* \times \mathbb{F}_7^*$ and, as an example, we fix $\langle \delta_1, \delta_2 \rangle = \langle 2, 4 \rangle$. The functional graph of $h_{\langle 2,4 \rangle}(x, y)$ is given in Fig. 2.

Fig. 2. Functional graph of $h_{\langle 2,4 \rangle}(x, y)$ over $\mathbb{F}_7^* \times \mathbb{F}_7^*$. All connected components are pure cycles (no tails), confirming bijectivity. The coloring distinguishes the six connected components.

For Algorithm 2, we take again $\mathcal{M} = \mathcal{T} = \mathbb{F}_7^* \times \mathbb{F}_7^*$ and fix $\langle \delta_1, \delta_2, c, d \rangle = \langle 2, 4, 2, 3 \rangle$. Therefore, we obtain the functional graph of $h_{\langle 2,4,2,3 \rangle}(x, y)$ given in Fig. 3.

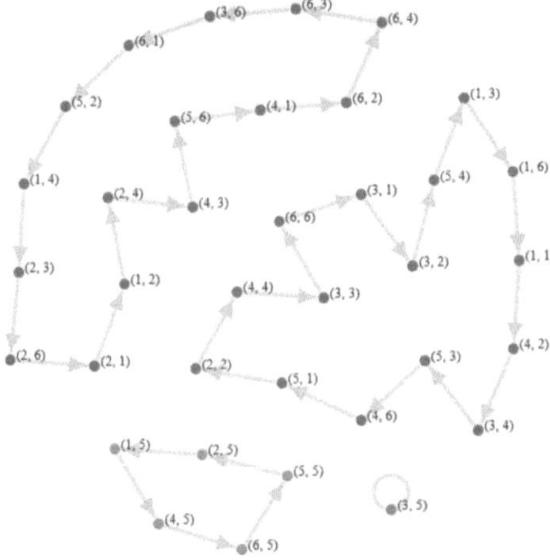

Fig. 3. Functional graph of $h_{\langle 2,4,2,3\rangle}(x, y)$ over $\mathbb{F}_7^* \times \mathbb{F}_7^*$ showing each connected component.

Finally, for Algorithm 3, since the function is defined over $\mathbb{F}_q$ instead of $\mathbb{F}_{q^2}$, we take $q = 49$ to obtain the same number of messages (and vertices in the graph). We take $\mathcal{M} = \mathcal{T} = \mathbb{F}_{49}^*$ and the key $\langle a, b\rangle = \langle 2, 4\rangle$; see Fig. 4.

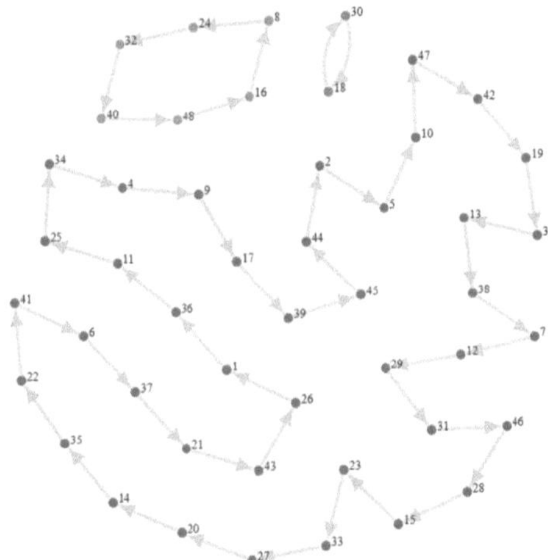

Fig. 4. Functional graph of $h_{\langle 2,4\rangle}(y)$ over $\mathbb{F}_{49}^*$ showing each connected component.

6   Conclusion and Future Directions

We have presented a study of parameterized function families derived from the polynomial map $F(X) = (aX^q + bX)(X^q - X)^{n-1}$ over quadratic extensions of finite fields. By leveraging the vector space representation $\mathbb{F}_{q^2} \cong \mathbb{F}_q \times \mathbb{F}_q$ and the dynamical analysis of Brochero and Teixeira [6], we constructed three families with controlled preimage properties.

The key findings are:

- The parameters $\delta_1 = a + b$ and $\delta_2 = b - a$ determine the preimage structure of F. When both are non-zero and $\gcd(n, q - 1) = 1$, the preimage size is uniformly 1 for non-zero elements.
- The shift parameters in Family 2 allow avoidance of the degenerate component containing zero.
- The projection to one coordinate in Family 3 yields a simpler family with optimal universality properties.

Finally, we present directions for future research.

1. **Higher-degree extensions:** The analysis could be extended to $\mathbb{F}_{q^k}$ for $k > 2$, where the vector space structure is richer. In particular, the Frobenius automorphism yields a k-dimensional coordinate representation, and the interplay between the parameters and the preimage structure is expected to produce families with larger key spaces and potentially stronger universality bounds.
2. **Even characteristic field:** Similarly, considering polynomials over a field with even characteristic would be of interest for binary applications.
3. **Other polynomial families:** The functional graphs of polynomials studied in [1,4] may yield similar parameterized families with different properties.
4. **Cycle length distribution:** A more detailed analysis of cycle lengths in the functional graphs of our families, extending the results of [6], would be of interest.
5. **Implementation considerations:** For practical applications, efficient implementation over specific fields (e.g., binary extensions or Mersenne prime fields) merits investigation.

References

1. Aguirre, J.J.R., Lemos, A., Neumann, V.G.L.: On the functional graph of $f(x) = x(x^{q-1} - c)^{q+1}$ over quadratic extensions of finite fields. arXiv:2408.08957 (2024)
2. Beneš, N., Brim, L., Kadlecaj, J.: Exploring attractor bifurcations in boolean networks. BMC Bioinform. **23**, 173 (2022)
3. Bierbrauer, J.: Introduction to Coding Theory. Discrete Mathematics and Its Applications, 2 edn. CRC Press, Taylor & Francis Group, Boca Raton, Florida (2016)
4. Bors, A., Panario, D., Wang, Q.: Functional graphs of generalized cyclotomic mappings of finite fields. Mem. Eur. Math. Soc. **23**, 1–270 (2025)
5. Brochero, F.E., Teixeira, H.R.: On the functional graph of $f(x) = c(x^{q+1} + ax^2)$ over quadratic extensions of finite fields. Finite Fields Appl. **88**, 102180 (2023)
6. Brochero, F.E., Teixeira, H.R.: The dynamics of a function family over quadratic extensions of finite fields. arXiv:2501.09852 (2025)

7. Capaverde, J., Masuda, A.M., Rodrigues, V.M.: Rédei permutations with the same cycle structure. Finite Fields Appl. **81**, 102046 (2022)

8. Carter, J.L., Wegman, M.N.: Universal classes of hash functions. J. Comput. Syst. Sci. **18**(2), 143–154 (1979)

9. Gassert, T.A.: Chebyshev action on finite fields. Discret. Math. **315**, 83–94 (2014)

10. Gravel, C., Panario, D., Teixeira, H.: Cycle structure of discrete functions extended through linear transformations. Submitted to Finite Fields and Their Applications (2025)

11. Gravel, C., Panario, D., Teixeira, H.: The graph of a family of functions over quadratic extensions of finite fields. Discret. Math. **348**, 114500 (2025)

12. Kadelka, C., Wheeler, M., Veliz-Cuba, A., Murrugarra, D., Laubenbacher, R.: Modularity of biological systems: a link between structure and function. J. R. Soc. Interface **20**, 207 (2023)

13. Katz, J., Lindell, Y.: Introduction to Modern Cryptography. Chapman & Hall/CRC Cryptography and Network Security, 3 edn. CRC Press, Taylor & Francis Group, Boca Raton, Florida (2021)

14. Martins, R., Panario, D., Qureshi, C.: A survey on iterations of mappings over finite fields. In: Radon Series on Computational and Applied Mathematics, vol. 23, pp. 135–172. de Gruyter (2019)

15. Mullen, G.L., Panario, D.: Handbook of Finite Fields. Discrete Mathematics and Its Applications. CRC Press (2013)

16. Nyang, D., Song, J.: Fast digital signature scheme based on the quadratic residue problem. Electron. Lett. **33**, 205–206 (1997)

17. Oliveira, J.A., Brochero Martínez, F.E.: Dynamics of polynomial maps over finite fields. Des. Codes Cryptogr. **92**, 1113–1125 (2024)

18. Pollard, J.M.: A Monte Carlo method for factorization. BIT Numer. Math. **15**, 331–334 (1975)

19. Pollard, J.M.: Monte Carlo methods for index computation (mod p). Math. Comput. **32**, 918–924 (1978)

20. Qureshi, C., Panario, D.: Rédei actions on finite fields and multiplication map in cyclic groups. SIAM J. Discret. Math. **29**, 1486–1503 (2015)

21. Qureshi, C., Panario, D.: The graph structure of the chebyshev polynomial over finite fields and applications. Des. Codes Crypt. **87**, 393–416 (2019)

22. Stinson, D.R.: Universal hashing and authentication codes. Des. Codes Crypt. **4**(4), 369–380 (1994)

23. Ugolini, S.: On the iterations of the maps $ax^{2^k} + b$ and $(ax^{2^k} + b)^{-1}$ over finite fields of characteristic two. J. Geom. **112** (2021)

24. Zhang, D., Gao, S., Gao, R., Liu, Z.: Logicgep: boolean networks inference using symbolic regression from time-series transcriptomic profiling data. Briefings Bioinform. **25**(4), bbae286 (2024)

Codegrees and an Exact Triangle Formula for the Norm-One Cayley Graph over $\mathbb{F}_{q^3}$ in Characteristic 2

Francisco-Javier Soto$^{(\boxtimes)}$ [iD]

Universidad Rey Juan Carlos, Madrid, Spain
franciscojavier.soto@urjc.es

Abstract. We consider the Cayley graph on the additive group of $\mathbb{F}_{q^3}$, with $q = 2^m$, whose connection set is the norm-one torus. We show that the number of common neighbours of two distinct vertices depends only on the norm of their difference. For the norm-one class we obtain an explicit codegree formula and, as a consequence, an exact triangle count.

Keywords: finite fields · additive Cayley graphs · norm-one torus · codegrees · triangles

1 Introduction

Counting small subgraphs in Cayley graphs is a concrete way to compare algebraic constructions with random-graph heuristics. Codegrees and triangles are basic statistics in the study of quasirandom and pseudorandom graphs; see Chung–Graham–Wilson [3], Krivelevich–Sudakov [9], and Tao–Vu [17]. On the algebraic side, Cayley graphs over finite fields sit naturally in algebraic combinatorics and association schemes; compare Delsarte [5] and Godsil–Royle [7]. In many cases, norm–trace identities over finite fields give exact counts for small subgraphs.

The Graph. Let $q = 2^m$ with $m \geq 1$, set $k = \mathbb{F}_q$, and let $K = \mathbb{F}_{q^3}$. Write $\mathrm{N} = \mathrm{N}_{K/k}$ and $\mathrm{Tr} = \mathrm{Tr}_{K/k}$ for norm and trace. More explicitly, for $x \in K$,

$$\mathrm{N}(x) = x\, x^q\, x^{q^2} = x^{1+q+q^2}, \qquad \mathrm{Tr}(x) = x + x^q + x^{q^2}.$$

Define

$$T = \{t \in K^* : \mathrm{N}(t) = 1\}. \tag{1}$$

Let Γ be the additive Cayley graph $\mathrm{Cay}(K, T)$, that is, the vertex set is K and distinct vertices x, y are adjacent if and only if $x - y \in T$. Since $\mathrm{char}(K) = 2$ we have $-t = t$ for all $t \in K$, hence $T = -T$ and Γ is undirected. Moreover $0 \notin T$, so Γ is simple. The graph is regular of degree $|T| = q^2 + q + 1$; see Eq. (4). Here and throughout, for a finite set S we write $|S|$ for its cardinality.

L. Batina and F. Özbudak (Eds.): WAIFI 2026, LNCS 16611, pp. 358–365, 2026.
https://doi.org/10.1007/978-3-032-27574-5_22

Furthermore, Γ is connected. If $m = 1$ then $q = 2$ and $K = \mathbb{F}_8$, so $|K^*| = 7$ and hence

$$N(x) = x^{1+q+q^2} = x^7 = 1 \qquad \text{for every } x \in K^*,$$

that is, $T = K^*$. Thus Γ is the complete graph on 8 vertices. For $m \geq 2$, Γ is a generalised Paley graph since $T = (K^*)^{q-1}$ by Lemma 1, and it is connected by [11, Theorem 1.2(1)], because for every proper divisor $a \mid 3m$ we have

$$\frac{2^{3m} - 1}{2^a - 1} > 2^m - 1 = q - 1,$$

so the required divisibility cannot occur.

Since Γ is a generalised Paley graph, it is natural to compare its small-subgraph counts with the cyclotomic approach. A classical example in this area is given by Paley graphs [14], and more generally by generalised Paley graphs and cyclotomic schemes [11]. In that circle of ideas, small-subgraph counts can be expressed through cyclotomic numbers, Jacobi sums, or related character sums; see Storer [16] and Lidl–Niederreiter [10] for background. Explicit evaluations are straightforward when the cyclotomic index is very small, while closed formulas become subtler as the index grows. For related work on clique counts in generalised Paley graphs, see Evans–Pulham–Sheehan [6], where Jacobi sums are used, and Dawsey–McCarthy [4], where finite-field hypergeometric functions are used. For the present graph the connection set has index $q - 1$ in K^*, so a short triangle formula does not seem to follow routinely from cyclotomy.

The equation that appears in the computation of $C(1)$ also sits close to classical problems over finite fields. In our setting, Lemma 2 reduces the count to solutions of $t^{q+1} + t + 1 = 0$, a special case of the family $x^{q+1} + ax + b$ studied by Bluher [2]. More broadly, counting elements with prescribed norm and trace has a substantial literature; see, for example, Moisio [12] and Moisio–Wan [13]. Our aim here is narrower. We isolate the present cubic characteristic-2 Cayley graph and show that its edge codegree and triangle count follow from a short direct argument based on norm and trace.

Norm maps also underpin several extremal constructions in graph theory, notably the norm graphs of Kollár–Rónyai–Szabó [8] and Alon–Rónyai–Szabó [1]. The graph studied here is different in spirit, since it is an additive Cayley graph, but it follows the same general theme that norm constraints can impose strong arithmetic structure on small subgraph counts.

Two choices are built into the definition above, and both affect the statements and the proofs. First, we work in characteristic 2 so that T is automatically symmetric, that is, $T = -T$, and Γ is an undirected graph without modifying the connection set. Over odd characteristic one naturally gets a directed Cayley graph with connection set T, and passing to an undirected graph usually means replacing T by $T \cup (-T)$, which changes codegrees and triangle counts. Second, we focus on the cubic extension as the first case where the intersection problem behind the edge codegree has about q solutions and still allows a closed formula. In the quadratic extension the condition $N(t) = N(1+t) = 1$ collapses to a fixed quadratic equation, so the edge codegree stays bounded as q grows, whereas the total number of triangles still grows with q. In degree 3, the Frobenius orbit has length three and the norm–trace constraints factor cleanly in Lemma 2, reducing the key count to trace-zero elements modulo k^* in Lemma 3. This gives an exact triangle formula with a dependence on the parity of m.

For $\beta \in k^*$ we set

$$C(\beta) = \left|\{t \in K^* : \mathrm{N}(t) = \beta \text{ and } \mathrm{N}(1+t) = \beta\}\right|. \tag{2}$$

Our first point is that the codegree of a pair of vertices depends only on the norm class of the difference.

Theorem 1. *Let $x, y \in K$ with $x \neq y$, and put $\gamma = \mathrm{N}(x - y) \in k^*$. Then the number of common neighbours of x and y in Γ equals $C(\gamma^{-1})$.*

The special value $C(1)$ controls edge codegrees and therefore the number of triangles. We prove an exact formula; see Theorem 2 and Eq. (9),

$$C(1) = \begin{cases} 2q + 2 & \text{if } m \text{ is odd,} \\ 2q & \text{if } m \text{ is even.} \end{cases}$$

As a consequence,

$$\tau(\Gamma) = \frac{q^3(q^2 + q + 1)}{6} C(1), \tag{3}$$

where $\tau(\Gamma)$ denotes the number of triangles in Γ.

The outline is as follows. Section 2 fixes notation and recalls the basic finite-field map onto the norm-one torus. Section 3 proves the codegree reduction. Section 4 evaluates $C(1)$ and deduces the triangle count.

2   The Norm-One Torus

Throughout, $k = \mathbb{F}_q$ and $K = \mathbb{F}_{q^3}$ with $q = 2^m$.

Since K^* is cyclic of order $q^3 - 1 = (q - 1)(q^2 + q + 1)$, the subgroup T in Eq. (1) has size

$$|T| = \frac{q^3 - 1}{q - 1} = q^2 + q + 1. \tag{4}$$

In addition, we will use the map

$$\pi \colon K^* \longrightarrow T, \qquad \pi(x) = x^{q-1}. \tag{5}$$

Lemma 1. *The map π in Eq. (5) is surjective, and its kernel equals k^*. In particular it induces a bijection*

$$K^*/k^* \cong T. \tag{6}$$

Proof. For $x \in K^*$ we have $\mathrm{N}(x^{q-1}) = \mathrm{N}(x)^{q-1} = 1$, so $\pi(K^*) \subseteq T$. Moreover,

$$\ker(\pi) = \{x \in K^* : x^{q-1} = 1\} = k^*.$$

Hence $|\pi(K^*)| = |K^*|/|k^*| = (q^3 - 1)/(q - 1) = |T|$, so π is surjective and induces Eq. (6). This is the finite-field form of Hilbert's Theorem 90; see [15, Chapter X, §1] or [10, Chapter 2]. $\square$

3   Codegrees

We next prove that codegrees are indexed by the norm classes in k^*.

Proof (of Theorem 1). Let $x, y \in K$ with $x \neq y$, and set $w = x - y \in K^*$. A vertex $z \in K$ is a common neighbour of x and y if and only if

$$z - x \in T \quad \text{and} \quad z - y \in T.$$

Put $u = z - x$ and $v = z - y$. Then $u, v \in T$ and, since $\operatorname{char}(K) = 2$,

$$u + v = (z - x) + (z - y) = x - y = w. \tag{7}$$

Conversely, any pair $(u, v) \in T^2$ satisfying Eq. (7) gives a common neighbour $z = x + u$.

Thus the codegree equals the number of $u \in T$ such that $w - u \in T$. Since $w \neq 0$, the map $u \mapsto t = u/w$ is a bijection between such u and elements $t \in K^*$ with $u = wt$. Let $\gamma = \mathrm{N}(w) \in k^*$. Then

$$\mathrm{N}(u) = 1 \iff \mathrm{N}(t) = \gamma^{-1}.$$

Furthermore, with $u = wt$ we have $w - u = w(1 - t)$, and in characteristic 2 we have $1 - t = 1 + t$. Hence

$$\mathrm{N}(w - u) = \mathrm{N}(w(1 + t)) = \gamma \, \mathrm{N}(1 + t),$$

so

$$w - u \in \mathrm{T} \iff \mathrm{N}(1 + t) = \gamma^{-1}.$$

We conclude that

$$\left| \{ z \in K : z \sim x \text{ and } z \sim y \} \right| = \left| \{ t \in K^* : \mathrm{N}(t) = \gamma^{-1}, \ \mathrm{N}(1 + t) = \gamma^{-1} \} \right|$$
$$= C(\gamma^{-1}).$$

which is exactly the stated formula, and the theorem follows. $\square$

4   Special Value and Triangle Count

We now study the special class $\beta = 1$, which controls edge codegrees.

Set

$$S(1) = \{ t \in K^* : \mathrm{N}(t) = 1 \text{ and } \mathrm{N}(1 + t) = 1 \}.$$

Lemma 2. *For $t \in K^*$ we have $t \in S(1)$ if and only if*

$$\left(t^{q+1} + t + 1 \right)\left(t^{q+1} + t^q + 1 \right) = 0. \tag{8}$$

Proof. Write $t_0 = t$, $t_1 = t^q$, and $t_2 = t^{q^2}$. The conditions $N(t) = 1$ and $N(1+t) = 1$ are

$$t_0 t_1 t_2 = 1 \qquad \text{and} \qquad (1 + t_0)(1 + t_1)(1 + t_2) = 1.$$

Expanding the second equality in characteristic 2 and using $t_0 t_1 t_2 = 1$, we get

$$(t_0 + t_1 + t_2) + (t_0 t_1 + t_1 t_2 + t_2 t_0) = 1.$$

Eliminating $t_2 = (t_0 t_1)^{-1}$ and multiplying by $t_0 t_1$ gives

$$t_0^2 t_1^2 + t_0^2 t_1 + t_0 t_1^2 + t_0 t_1 + t_0 + t_1 + 1 = 0.$$

The left-hand side factors as

$$(t_0 t_1 + t_0 + 1)(t_0 t_1 + t_1 + 1).$$

Substituting back $t_0 = t$ and $t_1 = t^q$ gives Eq. (8).

Conversely, assume Eq. (8). If $t^{q+1} + t + 1 = 0$, then $t \neq 0, 1$ and

$$t^q = \frac{t + 1}{t} \qquad \text{and} \qquad t^{q^2} = \frac{t^q + 1}{t^q} = \frac{1}{t + 1}.$$

Hence

$$N(t) = t \cdot t^q \cdot t^{q^2} = 1,$$

and also

$$N(1 + t) = (1 + t)(1 + t^q)(1 + t^{q^2}) = (1 + t) \cdot \frac{1}{t} \cdot \frac{t}{t + 1} = 1.$$

Thus $t \in S(1)$.

If instead $t^{q+1} + t^q + 1 = 0$, then $t \neq 0, 1$ and

$$t^q = \frac{1}{t + 1} \qquad \text{and} \qquad t^{q^2} = \frac{1}{t^q + 1} = \frac{t + 1}{t},$$

and the same calculation gives $N(t) = N(1 + t) = 1$. Therefore $t \in S(1)$, and the equivalence follows, finishing the argument. $\qquad\square$

Let

$$A = \{t \in K^* : t^{q+1} + t + 1 = 0\}, \qquad A^{-1} = \{a^{-1} : a \in A\}.$$

By Lemma 2, we have $S(1) = A \cup A^{-1}$, since inversion sends the equation $t^{q+1} + t + 1 = 0$ to $t^{q+1} + t^q + 1 = 0$.

Lemma 3. *We have* $|A| = q + 1$.

Proof. First note that any $t \in A$ lies in T, that is, $\mathrm{N}(t) = 1$. From $t^{q+1} + t + 1 = 0$ we get $t^q = (t+1)/t$ and $t^{q^2} = 1/(t+1)$, so $\mathrm{N}(t) = t \cdot t^q \cdot t^{q^2} = 1$.

Consider the k-linear subspace

$$V = \{x \in K : \mathrm{Tr}(x) = 0\}.$$

Since $\mathrm{Tr} \colon K \to k$ is a nonzero k-linear map, in fact $\mathrm{Tr}(1) = 1 + 1 + 1 = 1 \neq 0$, and $[K : k] = 3$, we have $\dim_k(V) = 2$, so $|V \setminus \{0\}| = q^2 - 1$.

Let $\pi \colon K^* \to T$ be the map in Eq. (5). For $x \in K^*$ set $t = \pi(x) = x^{q-1}$. Then

$$t^{q+1} + t + 1 = x^{q^2 - 1} + x^{q-1} + 1 = \frac{x^{q^2} + x^q + x}{x} = \frac{\mathrm{Tr}(x)}{x}.$$

Hence $t^{q+1} + t + 1 = 0$ if and only if $\mathrm{Tr}(x) = 0$, that is, $x \in V \setminus \{0\}$.

Moreover, if $x \in V \setminus \{0\}$ and $a \in k^*$, then $\mathrm{Tr}(ax) = a\,\mathrm{Tr}(x) = 0$ and $(ax)^{q-1} = x^{q-1}$, so each fibre of π meets $V \setminus \{0\}$ in exactly $|k^*| = q - 1$ elements. By Lemma 1, all fibres of π have size $q - 1$, and therefore

$$|A| = \frac{|V \setminus \{0\}|}{q - 1} = \frac{q^2 - 1}{q - 1} = q + 1,$$

as desired. $\qquad\square$

Lemma 4. *The intersection $A \cap A^{-1}$ has cardinality 0 if m is odd and 2 if m is even.*

Proof. If $t \in A \cap A^{-1}$, then t satisfies both $t^{q+1} + t + 1 = 0$ and $t^{q+1} + t^q + 1 = 0$. Adding the two equations gives $t^q + t = 0$, hence $t \in k^*$. Substituting into $t^{q+1} + t + 1 = 0$ gives

$$t^2 + t + 1 = 0 \quad \text{in } k.$$

This quadratic is equivalent to $t^3 = 1$ with $t \neq 1$, so it has solutions in k exactly when k^* contains an element of order 3, that is, when $3 \mid (2^m - 1)$, which is equivalent to m being even. In that case there are exactly two solutions, and they are distinct since $(X^2 + X + 1)' = 1$ in characteristic 2. For the standard criterion on roots of unity in finite fields, see, for example, [10, Chapter 2]. $\quad\square$

Theorem 2. *Let $q = 2^m$. Then*

$$C(1) = \begin{cases} 2q + 2 & \text{if } m \text{ is odd,} \\ 2q & \text{if } m \text{ is even.} \end{cases} \tag{9}$$

Proof. By Lemma 2, we have $S(1) = A \cup A^{-1}$, and inversion is a bijection $A \to A^{-1}$. Hence

$$C(1) = |S(1)| = |A| + |A^{-1}| - |A \cap A^{-1}| = 2\,|A| - |A \cap A^{-1}|.$$

Now Lemma 3 gives $|A| = q + 1$, and Lemma 4 gives the intersection size. Substituting these values gives Eq. (9), and the claim holds. $\qquad\square$

Corollary 1. *Let $\tau(\Gamma)$ be the number of triangles in Γ. Then*

$$\tau(\Gamma) = \frac{q^3(q^2 + q + 1)}{6} \, C(1) = \begin{cases} \dfrac{q^3(q^2 + q + 1)(q + 1)}{3} & \text{if } m \text{ is odd,} \\[2ex] \dfrac{q^3(q^2 + q + 1)q}{3} & \text{if } m \text{ is even.} \end{cases} \tag{10}$$

Proof. Every edge of Γ has difference in T, so its norm is 1. By Theorem 1, every edge has exactly $C(1)$ common neighbours.

The number of edges is

$$|E(\Gamma)| = \frac{|K| \cdot |T|}{2} = \frac{q^3(q^2 + q + 1)}{2}.$$

Each triangle is counted once at each of its three edges, hence

$$\tau(\Gamma) = \frac{|E(\Gamma)|\, C(1)}{3} = \frac{q^3(q^2 + q + 1)}{6} \, C(1).$$

Substituting Eq. (9) gives Eq. (10), and the corollary follows. $\qquad\square$

Remark 1 (A check). For $q = 2$ we have $K = \mathbb{F}_8$ and $T = K^*$, so Γ is the complete graph on 8 vertices. Therefore every edge has codegree 6, in agreement with Eq. (9) for $m = 1$, since $C(1) = 2q + 2 = 6$.

5   Conclusion

The norm-one Cayley graph over a cubic extension in characteristic 2 has a structured codegree distribution that already gives an exact triangle formula. The reduction in Theorem 1 shows that codegrees are indexed by norm classes in $\mathbb{F}_q^*$, and the norm-one class can be evaluated explicitly through Theorem 2. This leads directly to the exact triangle count in Corollary 1.

A natural next step is to study the other norm classes $\beta \in \mathbb{F}_q^*$ in Eq. (2) and to ask whether similarly explicit expressions exist for $C(\beta)$, or for further subgraph counts. In this note we focus on triangles as the first case where the count can be written in closed form within this framework, using only elementary norm–trace manipulations over finite fields.

Acknowledgments. The author acknowledges support from the "PREDOCT2022-006" program at Universidad Rey Juan Carlos. Also, he was partially supported by grant PID2023-151238OA-I00, financed by MICIU/AEI/10.13039/501100011033 and by EU, ERDF. The author thanks the anonymous referees for their careful reading and constructive comments, which improved the manuscript.

Disclosure of Interests. The author has no competing interests to declare that are relevant to the content of this article.

References

1. Alon, N., Rónyai, L., Szabó, T.: Norm-graphs: variations and applications. J. Combin. Theory Ser. B **76**(2), 280–290 (1999). https://doi.org/10.1006/jctb.1999.1906
2. Bluher, A.W.: On $x^{q+1}+ax+b$. Finite Fields Appl. **10**(3), 285–305 (2004). https://doi.org/10.1016/j.ffa.2003.08.004
3. Chung, F.R.K., Graham, R.L., Wilson, R.M.: Quasi-random graphs. Combinatorica **9**(4), 345–362 (1989). https://doi.org/10.1007/BF02125347
4. Dawsey, M.L., McCarthy, D.: Generalized Paley graphs and their complete subgraphs of orders three and four. Res. Math. Sci. **8**(2), 1–23 (2021). https://doi.org/10.1007/s40687-021-00254-7
5. Delsarte, P.: An Algebraic Approach to the Association Schemes of Coding Theory. No. 10 in Philips Research Reports Supplements, N.V. Philips' Gloeilampenfabrieken, Eindhoven (1973)
6. Evans, R.J., Pulham, J.R., Sheehan, J.: On the number of complete subgraphs contained in certain graphs. J. Combin. Theory Ser. B **30**(3), 364–371 (1981). https://doi.org/10.1016/0095-8956(81)90054-X
7. Godsil, C., Royle, G.: Algebraic Graph Theory, Graduate Texts in Mathematics, vol. 207. Springer, New York (2001). https://doi.org/10.1007/978-1-4613-0163-9
8. Kollár, J., Rónyai, L., Szabó, T.: Norm-graphs and bipartite turán numbers. Combinatorica **16**(3), 399–406 (1996). https://doi.org/10.1007/BF01261323
9. Krivelevich, M., Sudakov, B.: Pseudo-random graphs. In: Győri, E., Katona, G.O.H., Lovász, L., Fleiner, T. (eds.) More Sets, Graphs and Numbers, Bolyai Society Mathematical Studies, vol. 15, pp. 199–262. Springer, Cham (2006). https://doi.org/10.1007/978-3-540-32439-3_10
10. Lidl, R., Niederreiter, H.: Finite Fields, Encyclopedia of Mathematics and its Applications, vol. 20. Cambridge University Press, Cambridge, 2nd edn. (1997). https://doi.org/10.1017/CBO9780511525926
11. Lim, T.K., Praeger, C.E.: On generalised Paley graphs and their automorphism groups. Mich. Math. J. **58**(1), 293–308 (2009). https://doi.org/10.1307/mmj/1242071694
12. Moisio, M.: Kloosterman sums, elliptic curves, and irreducible polynomials with prescribed trace and norm. Acta Arith **132**(4), 329–350 (2008). https://doi.org/10.4064/aa132-4-3
13. Moisio, M., Wan, D.: On Katz's bound for the number of elements with given trace and norm. J. Reine Angew. Math. **638**, 69–74 (2010). https://doi.org/10.1515/crelle.2010.002
14. Paley, R.E.A.C.: On orthogonal matrices. J. Math. Phys. **12**(1–4), 311–320 (1933). https://doi.org/10.1002/sapm1933121311
15. Serre, J.-P.: Local Fields. GTM, vol. 67. Springer, New York (1979). https://doi.org/10.1007/978-1-4757-5673-9
16. Storer, T.: Cyclotomy and Difference Sets, Lectures in Advanced Mathematics, vol. 2. Markham Publishing Company, Chicago (1967)
17. Tao, T., Vu, V.H.: Additive Combinatorics, Cambridge Studies in Advanced Mathematics, vol. 105. Cambridge University Press, Cambridge (2006). https://doi.org/10.1017/CBO9780511755149

Explicit Bounds on the Existence Probability of Random Multivariate Quadratic Systems over Finite Fields

Michiya Iwata[(✉)], Ryomei Sugai, Kosuke Sakata, and Tsuyoshi Takagi

Department of Mathematical Informatics, The University of Tokyo, Tokyo, Japan
`iwata-michiya07@g.ecc.u-tokyo.ac.jp`

Abstract. The security of multivariate public-key cryptography, a major approach to post-quantum cryptography, is based on the computational hardness of solving systems of multivariate quadratic equations over finite fields (the MQ problem). The MQ problem consists of solving a system of quadratic equations over a finite field of size q, with n variables and m polynomials. The existence probability of solutions to the MQ problem plays a central role in analyzing the security of multivariate cryptography. However, explicit bounds for fixed parameters (q, n, m) have not been sufficiently studied. In this work, we evaluate the existence probability for randomly generated MQ systems with fixed parameters by analyzing the coefficient space arising from the MQ system. Using the inclusion–exclusion principle, we obtain a lower bound (approximately 0.625) and an upper bound (approximately 0.667) on the existence probability, focusing on the case $m = n$. We also derive upper and lower bounds on the probability that the number of solutions is exactly one in the case $m = n$. Finally, we analyze the existence probability of solutions to the MQ problem in the case $m \neq n$.

Keywords: post-quantum cryptography · multivariate public-key cryptography · MQ problem

1 Introduction

Multivariate public-key cryptography is a well-established approach to post-quantum cryptography, originating with the proposal of Matsumoto and Imai in 1988 [15]. Representative signature schemes include UOV [11], MAYO [3], and QR-UOV [6].

The MQ problem (Multivariate Quadratic Problem) is the problem of finding solutions to a system of m quadratic polynomials in n variables over a finite field $\mathbb{F}_q$ of size q. The security of multivariate public-key cryptography is based on the computational hardness of the MQ problem. Various algorithms for solving the MQ problem have been proposed, including Gröbner basis methods [5] and hybrid methods [2]. On the other hand, a satisfactory theoretical understanding of the solution existence probability remains limited.

L. Batina and F. Özbudak (Eds.): WAIFI 2026, LNCS 16611, pp. 366–383, 2026.
https://doi.org/10.1007/978-3-032-27574-5_23

Various classical theories are known for the number of solutions to polynomial systems over finite fields. When the number of variables is large compared with the degrees, results of Warning [17] and refinements by Ax and Katz [1,10] provide nontrivial information on the number of zeros. For a fixed system, evaluations via character sums [18] and algebraic-geometric estimates of rational points [12] are also available. For a single polynomial, the Schwartz–Zippel lemma [16,19] gives an upper bound on the probability that a random evaluation yields zero. These existing studies do not necessarily fit the choice of parameters assumed in the MQ problem, and they do not provide the distributional behavior of MQ problems.

Among probabilistic approaches, inclusion–exclusion has played a central role. For a single *univariate* polynomial of fixed degree, the number of polynomials with no zeros can be computed explicitly by the inclusion–exclusion principle [14]. More recently, Fusco and Bach [7] derived the *asymptotic* distribution of the number of solutions for systems of multivariate polynomials of degree at least two, using inequalities based on inclusion–exclusion.

Our Contributions. We provide explicit finite-parameter bounds on the solution existence probability P_{exist} for random systems of m quadratic equations in n variables over $\mathbb{F}_q$, extending the inclusion–exclusion analysis for univariate polynomials [14] to multivariate quadratic polynomial systems over finite fields.

– In the case $m = n$, we derive bounds $R_{\text{lower}}^{(2)}$ and $R_{\text{upper}}^{(3)}$ (Theorem 1), and an improved lower bound $R_{\text{lower}}^{(4)}$ by analyzing rank drops (Theorem 2).
– We analyze the solution-count distribution by computing the expectation and the variance, and we bound the probability that the MQ system has exactly one solution when $m = n$ (Theorem 4).
– We provide explicit bounds on P_{exist} in the overdetermined case $m > n$ and the underdetermined case $m < n$ (Theorems 5 and 6).
– As an application to cryptanalysis, we revisit the analysis of Hashimoto's algorithm by replacing the commonly used heuristic assumption on P_{exist} with our provable bounds (Sect. 7).

These results provide theoretical foundations for security analyses in multivariate public-key cryptography and offer further insight into the probabilistic behavior of random quadratic systems over finite fields.

Organization of the Paper. Section 2 defines the solution existence probability P_{exist} for the MQ problem, and reviews related work. Section 3 presents our estimation framework based on inclusion–exclusion inequalities and introduces the basic bounds. Section 4 improves the lower bound by analyzing rank drops through affine invariance. Section 5 studies the distribution of the number of solutions and derives upper and lower bounds on the probability that the MQ system has exactly one solution. Section 6 extends the existence-probability analysis to the case $m \neq n$. Finally, Sect. 7 discusses an application to cryptanalysis.

2 The MQ Problem and the Existence Probability

In this section, we define the MQ problem, introduce the solution existence probability P_{exist}, and review related work.

2.1 The Solution-Existence Probability for the MQ Problem

The MQ problem (Multivariate Quadratic Problem) can be defined as follows.

Definition 1 (MQ problem). *The* MQ *problem over a finite field $\mathbb{F}_q$ is the following. Let $n, m \in \mathbb{N}$, and let $f_1, \ldots, f_m \in \mathbb{F}_q[x_1, \ldots, x_n]$ be m quadratic polynomials in n variables $x_1, \ldots, x_n \in \mathbb{F}_q$. The problem is to find $x_1, \ldots, x_n \in \mathbb{F}_q$ satisfying $f_1 = \cdots = f_m = 0$.*

More explicitly, each f_k can be written as follows.

$$
\begin{cases}
f_1(x_1, \ldots, x_n) = \sum_{1 \le i \le j \le n} a_{ij}^{(1)} x_i x_j + \sum_{1 \le i \le n} b_i^{(1)} x_i + c^{(1)} = 0, \\
\qquad\qquad \vdots \\
f_m(x_1, \ldots, x_n) = \sum_{1 \le i \le j \le n} a_{ij}^{(m)} x_i x_j + \sum_{1 \le i \le n} b_i^{(m)} x_i + c^{(m)} = 0.
\end{cases}
\tag{1}
$$

The system is called underdetermined if $n > m$ and overdetermined if $m > n$.
For the subsequent discussion, we introduce notation and definitions.

- We represent an n-tuple of variables by a vector $\boldsymbol{x} = (x_1, \ldots, x_n) \in \mathbb{F}_q^n$.
- Each polynomial $f_k \in \mathbb{F}_q[x_1, \ldots, x_n]$ in an MQ system has $N = \frac{(n+2)(n+1)}{2}$ coefficients. Writing $f_k(x_1, \ldots, x_n) = \sum_{1 \le i \le j \le n} a_{ij}^{(k)} x_i x_j + \sum_{1 \le i \le n} b_i^{(k)} x_i + c^{(k)}$, we define the column vector obtained by listing all coefficients as $\boldsymbol{\theta}^{(k)} = \left(a_{11}^{(k)}, a_{12}^{(k)}, \ldots, a_{nn}^{(k)}, b_1^{(k)}, \ldots, b_n^{(k)}, c^{(k)}\right)^\top \in \mathbb{F}_q^N$. Here, $a_{ij}^{(k)}$ $(1 \le i \le j \le n)$ are ordered lexicographically with respect to the pair (i, j).
- We denote by $\boldsymbol{\theta} = \left(\boldsymbol{\theta}^{(1)}, \boldsymbol{\theta}^{(2)}, \ldots, \boldsymbol{\theta}^{(m)}\right)^\top \in \mathbb{F}_q^{mN}$ the column vector obtained by listing all mN coefficients of $f_1, \ldots, f_m$, and call it the coefficient vector. The tuple $(f_1, \ldots, f_m)$ is in one-to-one correspondence with $\boldsymbol{\theta} \in \mathbb{F}_q^{mN}$.
- For $\boldsymbol{\theta} \in \mathbb{F}_q^{mN}$, we let $f^{(\boldsymbol{\theta})}$ denote the tuple $(f_1, \ldots, f_m)$ determined by $\boldsymbol{\theta}$, and we write the system $f_1(\boldsymbol{x}) = \cdots = f_m(\boldsymbol{x}) = 0$ as $f^{(\boldsymbol{\theta})}(\boldsymbol{x}) = 0$.

Definition 2 (Existence probability of solutions to the MQ problem). *Define the probability space $(\Omega, \mathcal{F}, P)$ by $\Omega = \mathbb{F}_q^{mN}$, $\mathcal{F} = 2^\Omega$, and $P(A) = \frac{|A|}{|\Omega|}$ for $A \subset \Omega$. The existence probability of solutions to the MQ problem is the probability P_{exist} that, when a coefficient vector $\boldsymbol{\theta} \in \mathbb{F}_q^{mN}$ is chosen uniformly at random, the system $f^{(\boldsymbol{\theta})}(\boldsymbol{x}) = 0$ has at least one solution $\boldsymbol{x} \in \mathbb{F}_q^n$, defined by*

$$
P_{\text{exist}} = P\left(\{\boldsymbol{\theta} \in \mathbb{F}_q^{mN} \mid \exists \boldsymbol{x} \in \mathbb{F}_q^n, \, f^{(\boldsymbol{\theta})}(\boldsymbol{x}) = 0\}\right).
$$

2.2 Probability Evaluations and the Inclusion–Exclusion Principle

For a single univariate polynomial, the number of monic polynomials with a root can be computed explicitly via inclusion–exclusion [14]. For multivariate systems, Fusco and Bach [7] derived asymptotic estimates for the probability of having no solution using inclusion–exclusion inequalities.

The Case of a Single Polynomial in One Variable [14] For $x \in \mathbb{F}_q$, let $W_x^d \subset \mathbb{F}_q[x]$ be

$$W_x^d = \{ f \mid f(x) = 0, \ f \text{ is monic}, \ \deg f = d \}.$$

Then the number of monic degree-d polynomials having a root is $\left| \bigcup_{x \in \mathbb{F}_q} W_x^d \right|$. By the inclusion–exclusion principle,

$$\left| \bigcup_{x \in \mathbb{F}_q} W_x^d \right| = \sum_{k=1}^{q} \sum_{x_1 < \cdots < x_k} (-1)^{k+1} \left| W_{x_1}^d \cap \cdots \cap W_{x_k}^d \right|$$

holds. When $k > d$, since a monic degree-d polynomial $f(x) \in \mathbb{F}_q[x]$ has at most d roots, we have $\left| W_{x_1}^d \cap \cdots \cap W_{x_k}^d \right| = 0$. When $k \leq d$, if a monic degree-d polynomial $f(x) \in \mathbb{F}_q[x]$ has k roots $x_1, \ldots, x_k$, then it can be written as $f(x) = (x - x_1) \ldots (x - x_k)\phi(x)$ using an arbitrary monic polynomial $\phi(x)$ of degree $d - k$, and hence $\left| W_{x_1}^d \cap \cdots \cap W_{x_k}^d \right| = q^{d-k}$. Therefore, the number of monic degree-d polynomials having a root is given by $\sum_{k=1}^{d} (-1)^{k+1} \binom{q}{k} q^{d-k}$. Then, the existence probability of solutions to a single monic degree-d univariate polynomial is $\sum_{k=1}^{d} (-1)^{k+1} \binom{q}{k} q^{-k}$. We aim to extend this work to MQ systems.

The Case of Asymptotic Distributions of the Number of Solutions [7]. Let $d \geq 2$ and let p be a prime. For a system with m equations in n variables, let $\alpha = m - n$. For each sufficiently large n, choose $f_1, \ldots, f_{n+\alpha} \in \mathbb{F}_p[x_1, \ldots, x_n]$ uniformly at random among degree-d polynomials. Then, as $n \to \infty$, the probability that the system $f_1 = \cdots = f_{n+\alpha} = 0$ has no solution converges to $\mathrm{e}^{-p^{-\alpha}}$. Moreover, the number of solutions in $\mathbb{F}_p^n$ converges in distribution to the Poisson law with parameter $\lambda = p^{-\alpha}$.

In particular, when $\alpha = 0$, i.e., when $m = n$, the probability of having no solution is asymptotically e^{-1}. In the proofs of these theorems, one sets

$$I_x = \begin{cases} 1 & (f_1(x) = \cdots = f_{n+\alpha}(x) = 0), \\ 0 & (\text{otherwise}), \end{cases}$$ and uses the fact that the probability

of having no solution can be written as

$$\mathbb{E}\left[\prod_x (1 - I_x) \right] = \sum_{k=0}^{p^n} (-1)^k \sum_{x^{(1)} < \cdots < x^{(k)}} \mathbb{E}\left[\prod I_{x^{(i)}} \right].$$

For sufficiently large n, one uses that $\sum_{x^{(1)} < \cdots < x^{(k)}} \mathbb{E}\left[\prod I_{x^{(i)}} \right] \approx \frac{p^{-\alpha k}}{k!}$, and from inclusion–exclusion inequalities one obtains, for odd K,

$$-\frac{p^{-\alpha K}}{K!}(1 + o(1)) \leq \mathbb{E}\left[\prod_x (1 - I_x) \right] - \sum_{k=0}^{K-1} (-1)^k \frac{p^{-\alpha k}}{k!} \leq \frac{p^{-\alpha(K+1)}}{(K+1)!}(1 + o(1)).$$

Hence, as $n \to \infty$, the probability of having no solution converges to $e^{-p^{-\alpha}}$.

This prior work focuses on the asymptotic behavior of the solution-count distribution, and does not provide an evaluation for fixed parameters. Therefore, in this work we aim to give explicit upper and lower bounds on the existence probability of solutions for a fixed n.

2.3   Related Work

Let p be the characteristic of the finite field $\mathbb{F}_q$. Refining Chevalley's theorem [4], Warning proved that, when $n > \sum_i \deg f_i$, the number of solutions to $f_1 = \cdots = f_m = 0$ satisfies (number of solutions) $\equiv 0 \mod p$ [17]. Later, Ax and Katz extended this theorem [1,10].

On the other hand, it is known that the number of solutions can be expressed using character sums. Using a nontrivial additive character ψ, the number of solutions can be written as $q^{-m} \sum_{y \in \mathbb{F}_q^m} \sum_{x \in \mathbb{F}_q^n} \psi\left(\sum_{i=1}^m y_i f_i(x)\right)$ [13,18]. Moreover, from algebraic-geometric methods, for an algebraic irreducible variety V in projective space P^n with $\dim V = r$ and $\deg V = d$, one can obtain an estimate of the form $|(\text{number of rational points of } V) - q^r| \le (d-1)(d-2)q^{r-\frac{1}{2}} + \Lambda q^{r-1}$, where Λ is a constant depending on (n, r, d) [12].

For randomly chosen $f_1, \ldots, f_m \in U \subseteq \mathbb{F}_q[x_1, \ldots, x_n]$, the expected number of solutions is q^{n-m}, and if every function $\mathbb{F}_q^n \to \mathbb{F}_q$ can be represented by a polynomial in U, the number of solutions follows $\mathrm{Bin}(q^n, q^{-m})$.

For a single polynomial, Schwartz–Zippel lemma states that for any nonzero $f \in \mathbb{F}_q[x_1, \ldots, x_n]$ and uniformly random $r_1, \ldots, r_n \in \mathbb{F}_q$, $\Pr[f(r_1, \ldots, r_n) = 0] \le \frac{\deg f}{q}$ [16,19].

3   A Proposal for Estimating the Existence Probability of Solutions to the MQ Problem

In this section, we derive an estimate of the existence probability of solutions to the MQ problem by using inequalities from the inclusion–exclusion principle.

3.1   Inclusion–Exclusion Inequalities and the Proposed Method for Probability Estimation

In the proposed method, we extend the inclusion–exclusion analysis for a single univariate polynomial to MQ systems. The choice of order on $\mathbb{F}_q$ and the induced order $\mathbb{F}_q^n$ is arbitrary, as long as it is fixed throughout the argument. For convenience, we assume that the order on $\mathbb{F}_q^n$ is given by $x < y \iff \exists j \in \{1, \ldots, n\}$ $x_1 = y_1, \ldots, x_{j-1} = y_{j-1}, x_j < y_j$.

Definition 3 (Definitions of Z_x and S_k). *For a given* $x = (x_1, \ldots, x_n) \in \mathbb{F}_q^n$, $k = 1, \ldots, q^n$ *let* $Z_x \subseteq \mathbb{F}_q^{mN}$ *and* S_k *be defined by*

$$Z_x = \{\theta \in \mathbb{F}_q^{mN} \mid (f^{(\theta)}(x)) = 0\}, \quad S_k = \sum_{x^{(1)} < \cdots < x^{(k)}} |Z_{x^{(1)}} \cap \cdots \cap Z_{x^{(k)}}|. \quad (2)$$

By the inclusion–exclusion principle, $\left|\bigcup_{x \in \mathbb{F}_q^n} Z_x\right| = \sum_{k=1}^{q^n}(-1)^{k+1} S_k$ follows. As an inclusion–exclusion inequality, truncating the sum $\sum_{k=1}^{q^n}(-1)^{k+1} S_k$ after an even number of terms gives a lower bound for $\left|\bigcup_{x \in \mathbb{F}_q^n} Z_x\right|$, and truncating it after an odd number of terms gives an upper bound, by the classical Bonferroni inequalities [8]. Therefore,

$$\sum_{k=1}^{K:\,\text{even}}(-1)^{k+1} S_k \leq \left|\bigcup_{x \in \mathbb{F}_q^n} Z_x\right| \leq \sum_{k=1}^{K':\,\text{odd}}(-1)^{k+1} S_k.$$

From the definition of Z_x, the set $\bigcup_{x \in \mathbb{F}_q^n} Z_x$ is the set of coefficients of MQ systems that have at least one solution. Therefore,

$$P_{\text{exist}} = \frac{\left|\bigcup_{x \in \mathbb{F}_q^n} Z_x\right|}{\left|\mathbb{F}_q^{mN}\right|} = \frac{\sum_{k=1}^{q^n}(-1)^{k+1} S_k}{q^{mN}}$$

holds. Here, let

$$R_{\text{lower}}^{(K)} = \sum_{k=1}^{K:\,\text{even}}(-1)^{k+1} \frac{S_k}{q^{mN}}, \quad R_{\text{upper}}^{(K')} = \sum_{k=1}^{K':\,\text{odd}}(-1)^{k+1} \frac{S_k}{q^{mN}}. \tag{3}$$

Then, by the inclusion–exclusion inequalities, the existence probability can be evaluated as

$$R_{\text{lower}}^{(K)} \leq P_{\text{exist}} \leq R_{\text{upper}}^{(K')}.$$

In the case of a single univariate polynomial with $m = n = 1$ discussed in Sect. 2.2, we were able to compute $S_k = \sum_{x^{(1)} < \cdots < x^{(k)}} |Z_{x^{(1)}} \cap \cdots \cap Z_{x^{(k)}}|$ for any k. However, for general n, m, while S_k can be computed for small k, it becomes challenging as k grows. Therefore, in what follows, we aim to derive the lower bound $R_{\text{lower}}^{(K)}$ and the upper bound $R_{\text{upper}}^{(K')}$ by computing S_k for small k.

3.2   A Lower Bound $R_{\text{lower}}^{(2)}$ and an Upper Bound $R_{\text{upper}}^{(3)}$ for the Existence Probability

In practice, by computing S_1, S_2, S_3 in Definition 3, we derive $R_{\text{lower}}^{(2)}$ and $R_{\text{upper}}^{(3)}$ for the case $m = n$, by taking $K = 2$ and $K' = 3$ in (3).

As preparation for evaluating $|Z_{x^{(1)}} \cap \cdots \cap Z_{x^{(k)}}|$, we define,

$$\Phi_x = \left(x_1^2, x_1 x_2, \ldots, x_n^2, x_1, \ldots, x_n, 1\right) \in \mathbb{F}_q^{1 \times N}$$

for $x \in \mathbb{F}_q^n$. Here, the quadratic terms $x_i x_j$ are arranged for $1 \leq i \leq j \leq n$ in lexicographic order with respect to the index pair (i, j). Using Φ_x, we can write $f_k(x) = \Phi_x \theta^{(k)}$, and so on.

Proposition 1. $S_1 = \sum_x |Z_x|$ *satisfies* $S_1 = q^{mN-m+n}$. *In particular, when* $m = n$, *we have* $S_1 = q^{nN}$.

Proof. Fix $\boldsymbol{x} \in \mathbb{F}_q^n$, and define a linear map from the coefficient space $\mathbb{F}_q^{mN}$ to $\mathbb{F}_q^m$ by $F_1 : \boldsymbol{\theta} \longmapsto \left(f_1(\boldsymbol{x}), \dots, f_m(\boldsymbol{x})\right)^{\top}$. In matrix form, F_1 is represented by $A_1 = \mathrm{diag}\left(\Phi_{\boldsymbol{x}}, \dots, \Phi_{\boldsymbol{x}}\right) \in \mathbb{F}_q^{m \times mN}$. Each row $\Phi_{\boldsymbol{x}}$ of A_1 contains the constant term 1, and hence $\mathrm{rank}(A_1) = m$. Since $|Z_{\boldsymbol{x}}|$ equals the number of $\boldsymbol{\theta}$ such that $F_1(\boldsymbol{\theta}) = 0$, the dimension theorem implies $\dim \mathrm{Ker}\, F_1 = mN - m$, and $|Z_{\boldsymbol{x}}| = q^{mN-m}$. Thus,

$$S_1 = \sum_{\boldsymbol{x} \in \mathbb{F}_q^n} |Z_{\boldsymbol{x}}| = q^n \cdot q^{mN-m} = q^{n+m(N-1)}.$$

In particular, when $m = n$, we obtain $S_1 = q^{nN}$.

Proposition 2. $S_2 = \sum_{\boldsymbol{x} < \boldsymbol{y}} |Z_{\boldsymbol{x}} \cap Z_{\boldsymbol{y}}|$ *satisfies* $S_2 = \dfrac{q^{n+m(N-2)}(q^n - 1)}{2}$. *In particular, when* $m = n$, *we have* $S_2 = \frac{q^{nN}(1-q^{-n})}{2}$.

Proof. Fix distinct $\boldsymbol{x}, \boldsymbol{y} \in \mathbb{F}_q^n$, and define a linear map from the coefficient space $\mathbb{F}_q^{mN}$ to $\mathbb{F}_q^{2m}$ by $F_2 : \boldsymbol{\theta} \longmapsto \left(f_1(\boldsymbol{x}), f_1(\boldsymbol{y}), \dots, f_m(\boldsymbol{x}), f_m(\boldsymbol{y})\right)^{\top}$. In matrix form, F_2 is represented by $A_2 = \mathrm{diag}\left(B_2, \dots, B_2\right) \in \mathbb{F}_q^{2m \times mN}$, $B_2 = \begin{pmatrix} \Phi_{\boldsymbol{x}} \\ \Phi_{\boldsymbol{y}} \end{pmatrix} \in \mathbb{F}_q^{2 \times N}$. Hence, $\mathrm{rank}(A_2) = m \cdot \mathrm{rank}(B_2)$.

Let $\alpha, \beta \in \mathbb{F}_q$ and assume $\alpha\Phi_{\boldsymbol{x}} + \beta\Phi_{\boldsymbol{y}} = 0$. From the constant-term component we obtain $\alpha + \beta = 0$. From the linear-term components we obtain $\alpha(x_i - y_i) = 0$ for all i. Since $\boldsymbol{x} \neq \boldsymbol{y}$, there exists j such that $x_j \neq y_j$, which implies $\alpha = 0$, and hence $\beta = 0$. Then, $\Phi_{\boldsymbol{x}}$ and $\Phi_{\boldsymbol{y}}$ are linearly independent, and thus $\mathrm{rank}(B_2) = 2$. Consequently, $\mathrm{rank}(A_2) = 2m$, and by the dimension theorem $\dim \mathrm{Ker}\, F_2 = mN - 2m$. Hence, $|Z_{\boldsymbol{x}} \cap Z_{\boldsymbol{y}}| = q^{mN-2m}$. Therefore,

$$S_2 = \sum_{\boldsymbol{x} < \boldsymbol{y}} |Z_{\boldsymbol{x}} \cap Z_{\boldsymbol{y}}| = \binom{q^n}{2} q^{mN-2m} = \frac{q^{n+m(N-2)}(q^n - 1)}{2}.$$

In particular, when $m = n$, we obtain $S_2 = \frac{q^{nN}(1-q^{-n})}{2}$.

Proposition 3. $S_3 = \displaystyle\sum_{\boldsymbol{x} < \boldsymbol{y} < \boldsymbol{z}} |Z_{\boldsymbol{x}} \cap Z_{\boldsymbol{y}} \cap Z_{\boldsymbol{z}}|$ *satisfies* $S_3 = \frac{q^{n+m(N-3)}(q^n-1)(q^n-2)}{6}$. *In particular, when* $m = n$, $S_3 = \frac{q^{nN}(1-3q^{-n}+2q^{-2n})}{6}$.

Proof. Fix distinct $\boldsymbol{x}, \boldsymbol{y}, \boldsymbol{z} \in \mathbb{F}_q^n$, and define a linear map $F_3 : \mathbb{F}_q^{mN} \to \mathbb{F}_q^{3m}$ by $\boldsymbol{\theta} \longmapsto \left(f_1(\boldsymbol{x}), f_1(\boldsymbol{y}), f_1(\boldsymbol{z}), \dots, f_m(\boldsymbol{x}), f_m(\boldsymbol{y}), f_m(\boldsymbol{z})\right)^{\top}$. In matrix form, F_3 is represented by $A_3 = \mathrm{diag}\left(B_3, \dots, B_3\right) \in \mathbb{F}_q^{3m \times mN}$, $B_3 = \begin{pmatrix} \Phi_{\boldsymbol{x}} \\ \Phi_{\boldsymbol{y}} \\ \Phi_{\boldsymbol{z}} \end{pmatrix} \in \mathbb{F}_q^{3 \times N}$. Hence $\mathrm{rank}(A_3) = m \cdot \mathrm{rank}(B_3)$.

Below, let $\alpha, \beta, \gamma \in \mathbb{F}_q$ and show that if $\alpha\Phi_{\boldsymbol{x}} + \beta\Phi_{\boldsymbol{y}} + \gamma\Phi_{\boldsymbol{z}} = 0$, then $\alpha = \beta = \gamma = 0$.

   (i) Consider the case where $(x_1, \ldots, x_n, 1), (y_1, \ldots, y_n, 1), (z_1, \ldots, z_n, 1) \in \mathbb{F}_q^{n+1}$ are linearly independent over $\mathbb{F}_q^{n+1}$. Looking at the constant-term and linear-term components of $\alpha \Phi_x + \beta \Phi_y + \gamma \Phi_z = 0$, it follows that $\alpha = \beta = \gamma = 0$. Hence Φ_x, Φ_y, Φ_z are linearly independent. Since the case distinction in (i) exhausts all possibilities when $q = 2$, we assume $q \geq 3$ in what follows.

   (ii) Consider the case where $(x_1, \ldots, x_n, 1), (y_1, \ldots, y_n, 1), (z_1, \ldots, z_n, 1) \in \mathbb{F}_q^{n+1}$ are not linearly independent over $\mathbb{F}_q^{n+1}$. x, y, z lie on the same line in $\mathbb{F}_q^n$. Thus we can write $x = \tau_x v + a$, $y = \tau_y v + a$, $z = \tau_z v + a$, where $\tau_x, \tau_y, \tau_z \in \mathbb{F}_q$ are pairwise distinct, $v \in \mathbb{F}_q^n \setminus \{0\}$, and $a \in \mathbb{F}_q^n$. For $\alpha \Phi_x + \beta \Phi_y + \gamma \Phi_z = 0$, combining the relation from constant, linear, and quadratic terms, we have

$$\begin{pmatrix} 1 & 1 & 1 \\ \tau_x & \tau_y & \tau_z \\ \tau_x^2 & \tau_y^2 & \tau_z^2 \end{pmatrix} \begin{pmatrix} \alpha \\ \beta \\ \gamma \end{pmatrix} = 0. \text{ However, } \det \begin{pmatrix} 1 & 1 & 1 \\ \tau_x & \tau_y & \tau_z \\ \tau_x^2 & \tau_y^2 & \tau_z^2 \end{pmatrix} = (\tau_x - \tau_y)(\tau_y - \tau_z)(\tau_z - \tau_x),$$

and since τ_x, τ_y, τ_z are pairwise distinct when $x < y < z$, this determinant is nonzero. Then, $\alpha = \beta = \gamma = 0$, and hence Φ_x, Φ_y, Φ_z are linearly independent.

   From the above, in both (i) and (ii) we have $\mathrm{rank}(B_3) = 3$, and hence $\mathrm{rank}(A_3) = 3m$. Therefore, by the dimension theorem, $\dim \mathrm{Ker}\, F_3 = mN - 3m$, and thus $|Z_x \cap Z_y \cap Z_z| = q^{mN-3m}$. Consequently,

$$S_3 = \sum_{x<y<z} |Z_x \cap Z_y \cap Z_z| = \binom{q^n}{3} q^{mN-3m} = \frac{q^{n+m(N-3)}(q^n - 1)(q^n - 2)}{6}.$$

In particular, when $m = n$, we obtain $S_3 = \frac{q^{nN}(1 - 3q^{-n} + 2q^{-2n})}{6}$.

   When $m = n$, Propositions 1, 2, and 3 yield the following lower and upper bounds $R_{\mathrm{lower}}^{(2)}$ and $R_{\mathrm{upper}}^{(3)}$ for the existence probability:

$$R_{\mathrm{lower}}^{(2)} = \sum_{k=1}^{2} (-1)^{k+1} \frac{S_k}{q^{mN}} = \frac{1}{2}\left(1 + \frac{1}{q^n}\right),$$

$$R_{\mathrm{upper}}^{(3)} = \sum_{k=1}^{3} (-1)^{k+1} \frac{S_k}{q^{mN}} = \frac{2}{3}\left(1 + \frac{1}{2q^{2n}}\right).$$

We thus obtain the following theorem.

Theorem 1 (Lower and upper bounds on the existence probability for the MQ problem with $m = n$). *When $m = n$, the existence probability P_{exist} of solutions for the MQ problem satisfies*

$$\frac{1}{2}\left(1 + \frac{1}{q^n}\right) \leq P_{\mathrm{exist}} \leq \frac{2}{3}\left(1 + \frac{1}{2q^{2n}}\right).$$

   Since $\frac{1}{2}\left(1 + \frac{1}{q^n}\right) \geq \frac{1}{2}$ and, for $q \geq 2$ and $n \geq 1$, $\frac{2}{3}\left(1 + \frac{1}{2q^{2n}}\right) \leq \frac{2}{3}\left(1 + \frac{1}{8}\right) = \frac{3}{4}$, the following corollary holds.

Corollary 1. *When $m = n$, the existence probability P_{exist} of solutions for the MQ problem satisfies* $\dfrac{1}{2} \leq P_{\mathrm{exist}} \leq \dfrac{3}{4}$.

4   Improved Lower Bound $R_{\mathrm{lower}}^{(4)}$ for the Existence Probability

In this section, we compute $R_{\mathrm{lower}}^{(4)}$ when $m = n$, i.e., the quantity in (3) with $K = 4$, by evaluating S_4 in Definition 3.

For $k \geq 4$, in evaluating S_k we must take into account that there exist tuples $\boldsymbol{x}^{(1)}, \ldots, \boldsymbol{x}^{(k)}$ for which a rank drop occurs in $B_k = \begin{pmatrix} \Phi_{\boldsymbol{x}^{(1)}} \\ \vdots \\ \Phi_{\boldsymbol{x}^{(k)}} \end{pmatrix} \in \mathbb{F}_q^{k \times N}$. To study the rank drop of B_k, we first confirm properties satisfied by $\Phi_{\boldsymbol{x}}$ and B_k.

Lemma 1 (Invariance of $\Phi_{\boldsymbol{x}}$ under affine transformations). *Let $T : \mathbb{F}_q^n \to \mathbb{F}_q^n$ be an affine transformation. If $\boldsymbol{x}, \boldsymbol{x}' \in \mathbb{F}_q^n$ satisfy $\boldsymbol{x} = T(\boldsymbol{x}')$, then there exists an invertible matrix $P_T \in \mathbb{F}_q^{N \times N}$ such that $\Phi_{\boldsymbol{x}} = \Phi_{\boldsymbol{x}'} P_T$.*

Proof. An affine transformation T can be written as $\boldsymbol{x} = T(\boldsymbol{x}') = A\boldsymbol{x}' + \boldsymbol{c}$ for a nonsingular matrix $A \in \mathbb{F}_q^{n \times n}$ and a vector $\boldsymbol{c} \in \mathbb{F}_q^n$. Each component of $\Phi_{\boldsymbol{x}}$ can be expressed in terms of A, $\boldsymbol{c}$, and $\boldsymbol{x}'$ as follows:

$$1 = 1 \quad \text{(constant term)},$$

$$x_i = \sum_k A_{ik} x_k' + c_i \quad \text{(linear terms)},$$

$$x_i x_j = \sum_{k,\ell} A_{ik} A_{j\ell} x_k' x_\ell' + \sum_s (A_{is} c_j + A_{js} c_i) x_s' + c_i c_j \quad \text{(quadratic terms)}.$$

By arranging the corresponding coefficients into columns, we can construct a matrix $P_T \in \mathbb{F}_q^{N \times N}$ such that $\Phi_{\boldsymbol{x}} = \Phi_{\boldsymbol{x}'} P_T$.

Moreover, for affine transformations S and T, consider their composition $S \circ T$. Then $\Phi_{S \circ T(\boldsymbol{x})} = \Phi_{T(\boldsymbol{x})} P_S = \Phi_{\boldsymbol{x}} P_T P_S = \Phi_{\boldsymbol{x}} P_{S \circ T}$, and hence $P_T P_S = P_{S \circ T}$. Taking $S = T^{-1}$ yields $P_T P_S = P_T P_{T^{-1}} = P_{T^{-1} \circ T} = P_{I_n} = I_N$, and thus $P_T^{-1} = P_{T^{-1}}$. Therefore, P_T is invertible.

Corollary 2. *Let $T : \mathbb{F}_q^n \to \mathbb{F}_q^n$ be an affine transformation, and suppose that $\boldsymbol{x}^{(i)} = T(\boldsymbol{x}'^{(i)})$ holds for $(i = 1, \ldots, k)$. If $\alpha_i \in \mathbb{F}_q$ and $\sum_{i=1}^k \alpha_i \Phi_{\boldsymbol{x}'^{(i)}} = 0$, then $\sum_{i=1}^k \alpha_i \Phi_{\boldsymbol{x}'^{(i)}} = 0$.*

Proof. By Lemma 1, $\sum_{i=1}^k \alpha_i \Phi_{\boldsymbol{x}^{(i)}} = \sum_{i=1}^k \alpha_i \Phi_{\boldsymbol{x}'^{(i)}} P_T = \left(\sum_{i=1}^k \alpha_i \Phi_{\boldsymbol{x}'^{(i)}} \right) P_T = 0$. Since P_T is invertible, the claim follows.

Corollary 3. *Let $\boldsymbol{x}^{(1)}, \ldots, \boldsymbol{x}^{(k)} \in \mathbb{F}_q^n$ be distinct points. Let $T : \mathbb{F}_q^n \to \mathbb{F}_q^n$ be an affine transformation, and suppose that $\boldsymbol{x}'^{(i)} = T(\boldsymbol{x}'^{(i)})$ holds for $(i = 1, \ldots, k)$. If we define*

$$B_k = \begin{pmatrix} \Phi_{\boldsymbol{x}'^{(1)}} \\ \cdots \\ \Phi_{\boldsymbol{x}'^{(k)}} \end{pmatrix} \in \mathbb{F}_q^{k \times N}, \quad B_k' = \begin{pmatrix} \Phi_{\boldsymbol{x}'^{(1)}} \\ \cdots \\ \Phi_{\boldsymbol{x}'^{(k)}} \end{pmatrix} \in \mathbb{F}_q^{k \times N},$$

then $\mathrm{rank}(B_k) = \mathrm{rank}(B_k')$.

Proof. By Lemma 1, there exists an invertible matrix P_T such that $B_k = B'_k P_T$. Right multiplication by an invertible matrix only changes a basis of the column space, and hence $\mathrm{rank}(B_k) = \mathrm{rank}(B'_k P_T) = \mathrm{rank}(B'_k)$.

4.1 Derivation of S_4 and Improvement of the Lower Bound

Using the affine invariance derived in the previous subsection, we compute S_4.

Proposition 4. $S_4 = \sum_{x<y<z<w} |Z_x \cap Z_y \cap Z_z \cap Z_w|$ *satisfies*

$$S_4 = \left(\binom{q^n}{4} + \binom{q}{4} \frac{q^n - 1}{q - 1} q^{n-1}(q^m - 1) \right) q^{mN-4m}.$$

In particular, when $m = n$, $S_4 = \frac{q^{nN}}{24} \left(1 + \frac{q(q-5)}{q^n} + \frac{-2q^2+10q-1}{q^{2n}} + \frac{q(q-5)}{q^{3n}} \right)$.

Proof. Fix distinct points x, y, z, w, and define the map from $\mathbb{F}_q^{mN}$ to $\mathbb{F}_q^{4m}$ by $F_4 : \boldsymbol{\theta} \mapsto \left(f_1(x), \ldots, f_1(w), \ldots, f_m(x), \ldots, f_m(w) \right)^{\top}$. The map F_4 can be represented by the matrix $A_4 = \mathrm{diag}(B_4, \ldots, B_4) \in \mathbb{F}_q^{4m \times mN}$, whose block diagonal consists of $B_4 = (\Phi_x, \Phi_y, \Phi_z, \Phi_w)^{\top} \in \mathbb{F}_q^{4 \times N}$. Hence, $\mathrm{rank}(A_4) = m \times \mathrm{rank}(B_4)$. Let $\alpha, \beta, \gamma, \delta \in \mathbb{F}_q$ and suppose that $\alpha\Phi_x + \beta\Phi_y + \gamma\Phi_z + \delta\Phi_w = 0$.

 (i) Let $n \geq 2$. Consider the case where x, y, z, w are such that no three of them are collinear in $\mathbb{F}_q^n$. By Corollary 3, we may assume $x = 0$, $y = e_1$, and $z = e_2$. Comparing components in the identity $\alpha\Phi_x + \beta\Phi_y + \gamma\Phi_z + \delta\Phi_w = 0$, we first obtain, for the quadratic components, $\beta + \delta w_1 w_1 = 0$, $\gamma + \delta w_2 w_2 = 0$ and $\delta w_i w_j = 0$. Since no three points are collinear, there exist indices $\{i, j\}$ other than $\{1,1\}$ and $\{2,2\}$ such that $w_i w_j \neq 0$, and hence $\delta = 0$ follows. Then, $\beta = 0$ and $\gamma = 0$, and from the constant-term component $\alpha + \beta + \gamma + \delta = 0$ we obtain $\alpha = 0$. Thus $\Phi_x, \Phi_y, \Phi_z, \Phi_w$ are linearly independent, and hence $\mathrm{rank}(B_4) = 4$. Note that when $q = 2$, any three distinct points are not collinear, so the case distinction is complete with (i). In what follows, we assume $q \geq 3$.

 (ii) Let $n \geq 2$. Consider the case where, among x, y, z, w in $\mathbb{F}_q^n$ for $n \geq 2$, three points are collinear and the remaining point is not on that line. We may assume without loss of generality that x, y, z are collinear. By Corollary 3, we may assume $x = \tau_x e_1$, $y = \tau_y e_1$, $z = \tau_z e_1$. Then, under $\alpha\Phi_x + \beta\Phi_y + \gamma\Phi_z + \delta\Phi_w = 0$, we have $\alpha\tau_x + \beta\tau_y + \gamma\tau_z + \delta w_1 = 0$ and $\delta w_j = 0$ for $j \geq 2$. Since w is not on the same line as the other three points, there exist indices $j \geq 2$ such that $w_j \neq 0$, and hence $\delta = 0$. Combining the constant-term component $\alpha + \beta + \gamma = 0$ and the quadratic-term component $\alpha\tau_x^2 + \beta\tau_y^2 + \gamma\tau_z^2 = 0$, we

obtain $\begin{pmatrix} 1 & 1 & 1 \\ \tau_x & \tau_y & \tau_z \\ \tau_x^2 & \tau_y^2 & \tau_z^2 \end{pmatrix} \begin{pmatrix} \alpha \\ \beta \\ \gamma \end{pmatrix} = 0$, and $\alpha = \beta = \gamma = 0$ follows. Thus $\Phi_x, \Phi_y, \Phi_z, \Phi_w$ are linearly independent, and hence $\mathrm{rank}(B_4) = 4$. Note that when $q = 3$, four distinct points cannot all lie on the same line, so the case distinction is complete with (i) and (ii). In what follows, we assume $q > 3$.

 (iii) Let $n \geq 1$. Consider the case where all four points x, y, z, w are collinear in $\mathbb{F}_q^n$. When $n = 1$ we only need to consider this case. By Corollary 3, we may assume $\boldsymbol{x} = \tau_x \boldsymbol{e}_1$, $\boldsymbol{y} = \tau_y \boldsymbol{e}_1$, $\boldsymbol{z} = \tau_z \boldsymbol{e}_1$, $\boldsymbol{w} = \tau_w \boldsymbol{e}_1$. In this case, the only nonzero column vectors of B_4 are $(\tau_x^2, \tau_y^2, \tau_z^2, \tau_w^2)^\top$, $(\tau_x, \tau_y, \tau_z, \tau_w)^\top$, $(1, 1, 1, 1)^\top$, so $\mathrm{rank}(B_4) \leq 3$. Since the minor determinants are nonzero, we have $\mathrm{rank}(B_4) = 3$.

In cases **(i)** and **(ii)**, by the dimension theorem, we have $\mathrm{rank}(A_4) = 4m$, and thus $\dim \mathrm{Ker}\, F_4 = mN - 4m$. Therefore, $|Z_x \cap Z_y \cap Z_z \cap Z_w| = q^{mN - 4m}$.

In case **(iii)**, by the dimension theorem, we have $\mathrm{rank}(A_4) = 3m$, and thus $|Z_x \cap Z_y \cap Z_z \cap Z_w| = q^{mN - 3m}$. As for the number of quadruples yielding case **(iii)**, the number of distinct lines in $\mathbb{F}_q^n$ is $\frac{q^n - 1}{q - 1} \cdot \frac{q^n}{q}$, and hence the number of tuples $(\boldsymbol{x}, \boldsymbol{y}, \boldsymbol{z}, \boldsymbol{w})$ with $\boldsymbol{x} < \boldsymbol{y} < \boldsymbol{z} < \boldsymbol{w}$ lying on the same line is $\binom{q}{4} \frac{q^n - 1}{q - 1} q^{n-1}$. Therefore,

$$S_4 = \sum |Z_x \cap Z_y \cap Z_z \cap Z_w| = \left(\binom{q^n}{4} + \binom{q}{4} \frac{q^n - 1}{q - 1} q^{n-1}(q^m - 1) \right) q^{mN - 4m}.$$

Note that this formula also holds for $n = 1$ and for $q \leq 3$. Moreover, when $m = n$, S_4 is given by

$$S_4 = \frac{q^{nN}}{24} \left(1 + \frac{q(q - 5)}{q^n} + \frac{-2q^2 + 10q - 1}{q^{2n}} + \frac{q(q - 5)}{q^{3n}} \right).$$

As seen above, in order to compute S_k for general k, it is necessary to take into account rank drops of B_k. Since this requires understanding the k-tuples of points in $\mathbb{F}_q^n$ for which B_k loses rank, it is not easy to derive S_k in general.

From Propositions 1, 2, 3, and 4, we obtain the lower bound $R_{\mathrm{lower}}^{(4)}$ by evaluating the partial sum $\frac{1}{q^{mN}} (S_1 - S_2 + S_3 - S_4)$.

Theorem 2 (A lower bound on the existence probability for the MQ problem with $m = n$). *When $m = n$, a lower bound $R_{\mathrm{lower}}^{(4)}$ on the existence probability for the MQ problem is given by*

$$\frac{5}{8} \left(1 - \frac{q(q - 5)}{15q^n} + \frac{(2q^2 - 10q + 9)}{15q^{2n}} - \frac{q(q - 5)}{15q^{3n}} \right).$$

Corollary 4 ($R_{\mathrm{lower}}^{(4)}$ improves $R_{\mathrm{lower}}^{(2)}$ for $n \geq 2$). *For $n \geq 2$, for the MQ problem with $m = n$, the lower bounds $R_{\mathrm{lower}}^{(2)}$ and $R_{\mathrm{lower}}^{(4)}$ satisfy $R_{\mathrm{lower}}^{(4)} > R_{\mathrm{lower}}^{(2)}$.*

Corollary 5 (Improved constant lower bounds on P_{exist} for $n \geq 2$). *Assume that $m = n$ and $n \geq 2$. Then*

$$P_{\mathrm{exist}} > \frac{7}{12}.$$

Proof. By Theorem 2, $P_{\mathrm{exist}} \geq R_{\mathrm{lower}}^{(4)}$. Hence it suffices to show that $R_{\mathrm{lower}}^{(4)} > \frac{7}{12}$. From the explicit formula for $R_{\mathrm{lower}}^{(4)}$, we obtain

$$24 \left(R_{\mathrm{lower}}^{(4)} - \frac{7}{12} \right) = 1 - \frac{q(q - 5)}{q^n} + \frac{2q^2 - 10q + 9}{q^{2n}} - \frac{q(q - 5)}{q^{3n}}.$$

Set $E_{q,n} := 1 - \frac{q(q-5)}{q^n} + \frac{2q^2 - 10q + 9}{q^{2n}} - \frac{q(q-5)}{q^{3n}}$. It is enough to prove that $E_{q,n} > 0$.

If $q = 2$ or $q = 3$, then $q(q-5) < 0$ and $2q^2 - 10q + 9 = -3$, so $E_{q,n} = 1 + \frac{q(5-q)}{q^n} - \frac{3}{q^{2n}} + \frac{q(5-q)}{q^{3n}} \geq 1 - \frac{3}{q^{2n}} > 0$. If $q = 4$, then $E_{4,n} = 1 + \frac{4}{4^n} + \frac{1}{4^{2n}} + \frac{4}{4^{3n}} > 0$. If $q = 5$, then $E_{5,n} = 1 + \frac{9}{5^{2n}} > 0$.

Assume now that $q \geq 6$, and put $x = q^{-n}$. Since $n \geq 2$, we have $0 < x \leq q^{-2}$. Define

$$F(x) := 1 - q(q-5)x + (2q^2 - 10q + 9)x^2 - q(q-5)x^3.$$

Then $E_{q,n} = F(q^{-n})$, and $F'(x) = -q(q-5) + 2(2q^2 - 10q + 9)x - 3q(q-5)x^2$. For $0 < x \leq q^{-2}$, we have

$$F'(x) \leq -q(q-5) + \frac{2(2q^2 - 10q + 9)}{q^2} = -q(q-5) + 4 - \frac{20}{q} + \frac{18}{q^2} < 0.$$

Hence F is decreasing on $(0, q^{-2}]$, and therefore,

$$\begin{aligned}
E_{q,n} = F(q^{-n}) \geq F(q^{-2}) &= 1 - \frac{q(q-5)}{q^2} + \frac{2q^2 - 10q + 9}{q^4} - \frac{q(q-5)}{q^6} \\
&= \frac{5}{q} + \frac{2q^2 - 10q + 9}{q^4} - \frac{q-5}{q^5} > \frac{5}{q} - \frac{q-5}{q^5} > 0.
\end{aligned}$$

Thus $E_{q,n} > 0$ for all $q \geq 6$ as well.

5   On the Distribution of the Number of Solutions of MQ Systems

In this section, we investigate properties of the distribution of the number of solutions of MQ systems.

5.1   Expectation and Variance of the Number of Solutions

Using the results on S_1 and S_2 obtained in the previous sections, we can prove the following theorem.

Let $Y(\boldsymbol{\theta})$ be the random variable representing the number of solutions of the MQ system determined by the coefficient vector $\boldsymbol{\theta} \in \Omega$. That is, we define

$$Y(\boldsymbol{\theta}) = \left| \left\{ \boldsymbol{x} \in \mathbb{F}_q^n \mid (f^{(\boldsymbol{\theta})}(\boldsymbol{x})) = 0 \right\} \right|.$$

Theorem 3 (Expectation and variance of the number of solutions). *The expectation $\mathbb{E}[Y(\boldsymbol{\theta})]$ and the variance $\mathrm{Var}[Y(\boldsymbol{\theta})]$ are given by*

$$\mathbb{E}[Y(\boldsymbol{\theta})] = q^{n-m}, \quad \mathrm{Var}[Y(\boldsymbol{\theta})] = q^{n-m} - q^{n-2m}.$$

In particular, when $m = n$, we have $\mathbb{E}[Y(\boldsymbol{\theta})] = 1$ and $\mathrm{Var}[Y(\boldsymbol{\theta})] = 1 - \frac{1}{q^n}$.

Proof. For a fixed $x \in \mathbb{F}_q^n$, define $I_x^\theta = \begin{cases} 1 & (f^{(\theta)}(x)) = 0) \\ 0 & (f^{(\theta)}(x)) \neq 0) \end{cases}$. By the definition of

expectation, $\mathbb{E}[Y(\theta)] = \sum_{\theta \in \Omega} Y(\theta) \frac{1}{|\Omega|} = \frac{1}{|\Omega|} \sum_{\theta \in \Omega} \sum_{x \in \mathbb{F}_q^n} I_x^\theta$. By exchanging the order of the finite sums and using that $\sum_{\theta \in \Omega} I_x^\theta = |Z_x| = q^{mN-m}$, which follows from the computation of S_1 in the previous section, we obtain

$$\mathbb{E}[Y(\theta)] = \frac{1}{|\Omega|} \sum_{x \in \mathbb{F}_q^n} \sum_{\theta \in \Omega} I_x^\theta = \frac{1}{q^{mN}} \sum_{x \in \mathbb{F}_q^n} q^{mN-m} = \frac{1}{q^{mN}} q^n q^{mN-m} = q^{n-m}.$$

The variance can also be computed using sums of indicator variables. First,

$$\mathbb{E}[Y(\theta)^2] = \sum_{\theta \in \Omega} Y(\theta)^2 \frac{1}{|\Omega|} = \frac{1}{|\Omega|} \sum_{\theta \in \Omega} \left(\sum_{x \in \mathbb{F}_q^n} I_x^\theta + \sum_{x \neq y} I_x^\theta I_y^\theta \right),$$

where we used $I_x^\theta I_x^\theta = I_x^\theta$. By exchanging the order of the finite sums and using the computation of S_2 in the previous section, which implies $|Z_x \cap Z_y| = \sum_{\theta \in \Omega} I_x^\theta I_y^\theta = q^{mN-2m}$, we obtain

$$\mathbb{E}[Y(\theta)^2] = \frac{1}{|\Omega|} \left(\sum_{x \in \mathbb{F}_q^n} |Z_x| + \sum_{x \neq y} |Z_x \cap Z_y| \right) = q^{n-m} + q^{2(n-m)} - q^{n-2m}.$$

Therefore, $\mathrm{Var}[Y(\theta)] = \mathbb{E}[Y(\theta)^2] - \mathbb{E}[Y(\theta)]^2 = q^{n-m} - q^{n-2m}$.

5.2   On the Probability that the Solution Count Equals Exactly 1

In this subsection, we study the distribution of the number of solutions to the MQ problem. By deriving a representation in terms of S_k, we obtain upper and lower bounds on the probability that the MQ system has exactly one solution. Throughout this section, we use the convention $\binom{a}{b} = 0$ if $a < 0$ or $a < b$.

Lemma 2 (Representation of the solution-count distribution via S_k)). *Fix q and assume $m = n$. Let $\theta \in \mathbb{F}_q^{mN}$ be uniformly random. For $i \in \mathbb{N}$, the probability that the number of solutions satisfies $Y(\theta) = i$ is given by*

$$\Pr[Y(\theta) = i] = \sum_{k=i}^{q^n} \frac{S_k}{q^{mN}} \binom{k}{i} (-1)^{k-i},$$

where $S_k = \sum_{x^{(1)} < \cdots < x^{(k)}} |Z_{x^{(1)}} \cap \cdots \cap Z_{x^{(k)}}|$.

Proof. Define the probability generating function $g(t) = \mathbb{E}[t^{Y(\theta)}]$. Recall $Y(\theta) = \sum_{x \in \mathbb{F}_q^n} I_x^\theta$, then $g(t) = \mathbb{E}\left[\prod_{x \in \mathbb{F}_q^n} t^{I_x^\theta} \right] = \mathbb{E}\left[\prod_{x \in \mathbb{F}_q^n} (1 + (t-1)I_x^\theta) \right]$. Therefore,

$$g(t) = \mathbb{E}\left[\sum_{S \subseteq \mathbb{F}_q^n} (t-1)^{|S|} \prod_{x \in S} I_x^\theta \right] = \sum_{k=0}^{q^n} (t-1)^k \, \mathbb{E}\left[\sum_{\substack{S \subseteq \mathbb{F}_q^n \\ |S|=k}} \prod_{x \in S} I_x^\theta \right].$$

The inner sum counts k-subsets of the solution set, hence $\sum_{\substack{S \subset \mathbb{F}_q^n \\ |S|=k}} \prod_{x \in S} I_x^{\theta} = \binom{Y(\theta)}{k}$. Therefore,

$$g(t) = \sum_{k=0}^{q^n} (t-1)^k \, \mathbb{E}\left[\binom{Y(\theta)}{k}\right] = \sum_{i=0}^{q^n} t^i \sum_{k=i}^{q^n} \mathbb{E}\left[\binom{Y(\theta)}{k}\right] \binom{k}{i} (-1)^{k-i}.$$

On the other hand, by definition of the probability generating function, $g(t) = \sum_{i=0}^{q^n} \Pr[Y(\theta) = i] \, t^i$, so comparing coefficients of t^i yields

$$\Pr[Y(\theta) = i] = \sum_{k=i}^{q^n} \mathbb{E}\left[\binom{Y(\theta)}{k}\right] \binom{k}{i} (-1)^{k-i}.$$

Since $S_k = \sum_{x_1 < \cdots < x_k} \sum_{\theta \in \mathbb{F}_q^{mN}} \mathbb{1}\{\theta \in Z_{x_1} \cap \cdots \cap Z_{x_k}\} = \sum_{\theta \in \mathbb{F}_q^{mN}} \binom{Y(\theta)}{k}$, we have $\mathbb{E}\left[\binom{Y(\theta)}{k}\right] = \frac{S_k}{q^{mN}}$. Substituting this completes the proof.

Lemma 3. *Let* $(M \in \mathbb{N}_0)$ *and* $(i \in \mathbb{N})$. *Then, for* $Y := Y(\theta)$,

$$\sum_{k=i+M+1}^{q^n} \mathbb{E}\left[\binom{Y}{k}\right] \binom{k}{i} (-1)^{k-i} = (-1)^{M+1} \mathbb{E}\left[\binom{Y}{i}\binom{Y-i-1}{M}\right].$$

Proof. By linearity of expectation,

$$\sum_{k=i+M+1}^{q^n} \mathbb{E}\left[\binom{Y}{k}\right] \binom{k}{i} (-1)^{k-i} = \mathbb{E}\left[\sum_{k=i+M+1}^{Y} \binom{Y}{k}\binom{k}{i} (-1)^{k-i}\right].$$

Put $k = i + j$, the inner sum equals $\binom{Y}{i} \sum_{j=M+1}^{Y-i} (-1)^j \binom{Y-i}{j}$. Then we have $\sum_{j=M+1}^{Y-i} (-1)^j \binom{Y-i}{j} = -\sum_{j=0}^{M} (-1)^j \binom{Y-i}{j}$. Using $\sum_{j=0}^{M} (-1)^j \binom{r}{j} = (-1)^M \binom{r-1}{M}$ with $r = Y - i$, the inner sum becomes $(-1)^{M+1} \binom{Y}{i}\binom{Y-i-1}{M}$. Taking expectation yields the claim.

Lemma 4 (Inclusion–exclusion inequalities for the solution-count distribution). *Fix q and choose the coefficient vector $\theta \in \mathbb{F}_q^{mN}$ uniformly at random. For $i \in \mathbb{N}$, let K be even and K' be odd. Then, using S_k, we have*

$$\sum_{k=i}^{i+K'} \frac{S_k}{q^{mN}} \binom{k}{i} (-1)^{k-i} \leq \Pr[Y(\theta) = i] \leq \sum_{k=i}^{i+K} \frac{S_k}{q^{mN}} \binom{k}{i} (-1)^{k-i}.$$

Proof. By Lemma 2 and Lemma 3, for $M \in \mathbb{N}_0$,

$$\Pr[Y(\theta) = i] - \sum_{k=i}^{i+M} \frac{S_k}{q^{mN}} \binom{k}{i} (-1)^{k-i} = \sum_{k=i+M+1}^{q^n} \mathbb{E}\left[\binom{Y}{k}\right] \binom{k}{i} (-1)^{k-i}$$

$$= (-1)^{M+1} \mathbb{E}\left[\binom{Y}{i}\binom{Y-i-1}{M}\right].$$

Hence, if M is even $(M = K)$, then $\Pr[Y(\boldsymbol{\theta}) = i] - \sum_{k=i}^{i+K} \frac{S_k}{q^{mN}} \binom{k}{i}(-1)^{k-i} \leq 0$, and if M is odd $(M = K')$, then $\Pr[Y(\boldsymbol{\theta}) = i] - \sum_{k=i}^{i+K'} \frac{S_k}{q^{mN}} \binom{k}{i}(-1)^{k-i} \geq 0$. This proves the desired inequalities.

Using the above lemmas, we can derive the following bounds on the probability that the MQ system has exactly one solution.

Theorem 4. *Fix q and assume $m = n$. Choose the coefficient vector $\boldsymbol{\theta} \in \mathbb{F}_q^{mN}$ uniformly at random. Then the probability that the number of solutions satisfies $Y(\boldsymbol{\theta}) = 1$ obeys*

$$\frac{1}{3}\left(1 - \frac{q^2 - 5q + 3}{2q^n} + \frac{2q^2 - 10q + 7}{2q^{2n}} - \frac{q(q-5)}{2q^{3n}}\right) \leq \Pr[Y(\boldsymbol{\theta}) = 1] \leq \frac{1}{2}\left(1 - \frac{1}{q^n} + \frac{2}{q^{2n}}\right).$$

Proof. By the inequalities in Lemma 4,

$$\frac{1}{q^{mN}}(S_1 - 2S_2 + 3S_3 - 4S_4) \leq \Pr[Y(\boldsymbol{\theta}) = 1] \leq \frac{1}{q^{mN}}(S_1 - 2S_2 + 3S_3).$$

When $m = n$, we use $S_1 = q^{nN}$, $S_2 = \frac{q^{nN}(1-q^{-n})}{2}$, $S_3 = \frac{q^{nN}(1-3q^{-n}+2q^{-2n})}{6}$, and $S_4 = \frac{q^{nN}}{24}\left(1 + \frac{q(q-5)}{q^n} + \frac{-2q^2+10q-1}{q^{2n}} + \frac{q(q-5)}{q^{3n}}\right)$. Substituting the expressions, we obtain the desired expression.

6   On the Existence Probability for MQ Problems with $m \neq n$

In this section, we evaluate the existence probability for MQ systems with $m \neq n$. For this evaluation, we use the quantities S_k obtained in Propositions 1, 2, and 3, as well as the expectation and variance of the solution count.

6.1   Existence Probability for Overdetermined MQ Systems $(m > n)$

Theorem 5. *Consider an overdetermined MQ instance with $m > n$. Let $k = m - n$. Then the existence probability P_{exist} satisfies*

$$\frac{3}{4}q^{-k} \leq P_{\text{exist}} \leq q^{-k}.$$

Proof. By Proposition 1, 2, we have $S_1 = q^{mN-m+n}$ and $S_2 = \frac{q^{n+m(N-2)}(q^n-1)}{2}$. Hence,

$$P_{\text{exist}} \geq R_{\text{lower}}^{(2)} = \frac{1}{q^{mN}}(S_1 - S_2) = q^{n-m} - \frac{q^{n-2m}(q^n-1)}{2} \geq q^{-k}\left(1 - \frac{q^{-k}}{2}\right).$$

Since $m > n$ implies $k > 0$, we have $q^{-k} \leq 2^{-1}$, and thus $q^{-k}\left(1 - \frac{q^{-k}}{2}\right) \geq q^{-k}\left(1 - \frac{2^{-1}}{2}\right)$. Therefore, $P_{\text{exist}} \geq q^{-k}\left(1 - \frac{q^{-k}}{2}\right) \geq q^{-k}\left(1 - \frac{1}{4}\right) = \frac{3}{4}q^{-k}$.

Let Y denote the number of common zeros of the given MQ system over $\mathbb{F}_q^n$. By Theorem 3, $\mathbb{E}[Y] = q^{n-m} = q^{-k}$. Since $P_{\text{exist}} = \Pr[Y \geq 1]$, Markov's inequality yields $P_{\text{exist}} = \Pr[Y \geq 1] \leq \mathbb{E}[Y] = q^{-k}$. This proves the claim.

6.2 Existence Probability for Underdetermined MQ Systems ($n > m$)

Theorem 6. *Consider an underdetermined MQ instance with $m < n$. Let $k = m - n$. Then the existence probability P_{exist} satisfies*

$$P_{\text{exist}} \geq \frac{1}{1 + q^k - q^{-n}}.$$

In particular,

$$P_{\text{exist}} \geq \frac{1}{1 + q^k}.$$

Proof. In the underdetermined case $m < n$, we have $k < 0$. By Theorem 3, the expectation and variance of the solution count $Y(\boldsymbol{\theta})$ satisfy $\mathbb{E}[Y(\boldsymbol{\theta})] = q^{n-m}$ and $\text{Var}[Y(\boldsymbol{\theta})] = q^{n-m} - q^{n-2m}$. Using the Cauchy–Schwarz inequality,

$$(\mathbb{E}[Y])^2 = \left(\mathbb{E}\big[Y \cdot \mathbf{1}_{Y>0}\big]\right)^2 \leq \mathbb{E}[Y^2]\Pr[Y > 0],$$

we obtain

$$\Pr[Y > 0] \geq \frac{(\mathbb{E}[Y])^2}{\text{Var}[Y] + (\mathbb{E}[Y])^2}.$$

Therefore,

$$P_{\text{exist}} = \Pr[Y > 0] \geq \frac{(q^{n-m})^2}{q^{n-m} - q^{n-2m} + (q^{n-m})^2} = \frac{1}{1 + q^k - q^{-n}}.$$

In particular, $\frac{1}{1+q^k-q^{-n}} \geq \frac{1}{1+q^k}$.

7 Applications

In this section, we illustrate how the solution existence probability of the MQ problem informs the evaluation of cryptographic attack methods.

As an efficient attack method for an underdetermined MQ system with n variables and m polynomials, Hashimoto's algorithm [9] is known. Overall, Hashimoto's algorithm consists of Steps (1)–(5), and Steps (3)–(5) are repeated until a solution is obtained.

Let a, k be nonnegative integers satisfying $a + k \leq m$, and partition the n variables into the following triple (X, Y, Z):

$$\underbrace{x_1, \ldots, x_{m-a-k+1}}_{X}, \underbrace{x_{m-a-k+2}, \ldots, x_{m-k}}_{Y}, \underbrace{x_{m-k+1}, \ldots, x_n}_{Z}.$$

(1), (2) Apply an appropriate linear transformation to the given MQ system.
(3) For Z, choose a candidate Z from the solution space of the system of linear equations obtained by the transformation, and substitute it.
(4) For Y, solve an MQ system with $a - 1$ variables and $a - 1$ polynomials, and substitute the obtained solution.
(5) For X, solve an overdetermined MQ system with $m - a - k$ variables and $m - a$ polynomials. If no solution is obtained, return to Step (3).

In evaluating such an algorithm, the following assumption is made.

Hypothesis 1. *For an* MQ *instance over* $\mathbb{F}_q$ *consisting of n variables and m polynomials, when $n \leq m$, the solution existence probability P_{exist} is assumed to satisfy $P_{\text{exist}} = q^{n-m}$. In particular, when $m = n$, it is assumed that $P_{\text{exist}} = 1$.*

Therefore, for Hashimoto's algorithm, it is evaluated that a solution is obtained in Step (4) with probability 1, and a solution is obtained in Step (5) with probability q^{-k}.

However, according to the theoretical estimates obtained in this work, when $n < m$, the solution existence probability P_{exist} satisfies $\frac{3}{4}q^{n-m} \leq P_{\text{exist}} \leq q^{n-m}$ by Theorem 5, and when $m = n$, we have $\frac{7}{12} \leq P_{\text{exist}} \leq \frac{3}{4} < 1$ by the Corollary 1 and Corollary 5. Hence, for Hashimoto's algorithm, if each MQ system solved in each step can be regarded as randomly given, then Step (4) yields a solution with probability between $\frac{7}{12}$ and $\frac{3}{4}$, and Step (5) yields a solution with probability between $\frac{3}{4}q^{-k}$ and q^{-k}. We note that the above discussion concerns the interpretation of Step (4) in which one solution of the $n = m$ MQ subsystem is computed and substituted into Step (5).

8   Conclusion

In this paper, we investigated the existence probability of solutions to the MQ problem, which underpins the security of multivariate public-key cryptography. By evaluating the image space of a linear map from the coefficient space to the polynomials defining an MQ system and applying the inclusion–exclusion principle, we derived explicit upper and lower bounds on the existence probability for MQ systems.

As future work, it is desirable to determine an explicit value of the existence probability for arbitrary m and n. In this study, we derived only upper and lower bounds on the existence probability. To compute the existence probability via the inclusion–exclusion principle, it is necessary to explicitly evaluate S_k for arbitrary k. However, in this study we were only able to provide explicit evaluations for $k = 1, 2, 3, 4$. The difficulty lies in analyzing the rank drop of the matrix B_k that characterizes S_k for general k, and this remains an important topic for future investigation.

Acknowledgments. This work was supported by JST K Program Grant Number JPMJKP24U2, Japan.

References

1. Ax, J.: Zeroes of polynomials over finite fields. Am. J. Math. **86**(2), 255–261 (1964)
2. Bettale, L., Faugère, J.C., Perret, L.: Hybrid approach for solving multivariate systems over finite fields. J. Math. Cryptol. **3**(3), 177–197 (2009)
3. Beullens, W.: Mayo: Practical post-quantum signatures from oil-and-vinegar maps. In: Selected Areas in Cryptography (SAC 2021), pp. 355–376. (2021)

4. Chevalley, C.: Démonstration d'une hypothèse de M. Artin. Abhandlungen aus dem Mathematischen Seminar der Universität Hamburg **11**, 73–75 (1935)
5. Faugère, J.C.: A new efficient algorithm for computing Gröbner bases without reduction to zero (F5). In: Proceedings of the International Symposium on Symbolic and Algebraic Computation (ISSAC '02), pp. 75–83 (2002)
6. Furue, H., Ikematsu, Y., Kiyomura, Y., Takagi, T.: A new variant of unbalanced oil and vinegar using quotient ring: QR-UOV. In: Advances in Cryptology - ASIACRYPT 2021, Part IV, pp. 187–217 (2021)
7. Fusco, G., Bach, E.: Phase Transition of Multivariate Polynomial Systems. In: Cai, J.-Y., Cooper, S.B., Zhu, H. (eds.) TAMC 2007. LNCS, vol. 4484, pp. 632–645. Springer, Heidelberg (2007). https://doi.org/10.1007/978-3-540-72504-6_58
8. Galambos, J., Simonelli, I.: Bonferroni-Type Inequalities with Applications. Springer, New York (1996)
9. Hashimoto, Y.: An improvement of algorithms to solve under-defined systems of multivariate quadratic equations. JSIAM Lett. **15**, 53–56 (2023)
10. Katz, N.M.: On a theorem of Ax. Am. J. Math. **93**(2), 485–499 (1971)
11. Kipnis, A., Patarin, J., Goubin, L.: Unbalanced oil and vinegar signature schemes. In: Advances in Cryptology - EUROCRYPT '99, pp. 206–222 (1999)
12. Lang, S., Weil, A.: Number of points of varieties in finite fields. Am. J. Math. **76**(4), 819–827 (1954)
13. Leep, D.B., Schueller, L.M.: Zeros of a pair of quadratic forms defined over a finite field. Finite Fields Appl. **5**(2), 157–176 (1999)
14. Leont'ev, V.K.: Roots of random polynomials over a finite field. Math. Notes **80**, 300–304 (2006)
15. Matsumoto, T., Imai, H.: Public quadratic polynomial-tuples for efficient signature-verification and message-encryption. In: Advances in Cryptology - EUROCRYPT '88, pp. 419–453 (1988)
16. Schwartz, J.T.: Fast probabilistic algorithms for verification of polynomial identities. J. ACM **27**(4), 701–717 (1980)
17. Warning, E.: Bemerkung zur vorstehenden Arbeit von Herrn Chevalley. Abh. Math. Semin. Univ. Hambg. **11**, 76–83 (1935)
18. Weil, A.: Numbers of solutions of equations in finite fields. Bull. Am. Math. Soc. **55**(5), 497–508 (1949)
19. Zippel, R.: Probabilistic algorithms for sparse polynomials. In: Proceedings of the International Symposium on Symbolic and Algebraic Manipulation (EUROSAM '79), pp. 216–226 (1979)

Automorphism Groups of Hyperelliptic Curves of 2-Rank Zero

Kohtaro Yamaguchi and Shushi Harashita[✉]

Graduate School of Environment and Information Sciences, Yokohama National University, 79-7 Tokiwadai, Hodogaya-ku, Yokohama 240-8501, Japan
`yamaguchi-kohtaro-vx@ynu.jp, harasita@ynu.ac.jp`

Abstract. In this paper, we determine the reduced automorphism groups of hyperelliptic curves of a small genus in characteristic 2, when they are of 2-rank 0. Such a curve is an Artin-Schreier curve defined in the form $y^2 - y = f(x)$ for a polynomial $f(x)$. After we clarify semidirect-product structures of the automorphism groups for an arbitrary genus, we derive the detailed group structures for the reduced automorphism groups of the curves of a small genus, through computations using the computational algebra system Magma. With these experiments, we formulate two conjectures, which are analogues for our curves of the Oort conjecture on automorphism groups of generic principally polarized supersingular abelian varieties.

1 Introduction

In this paper, we investigate the order and the group structure of automorphism group of hyperelliptic curves in characteristic 2 of small genus which is of 2-rank 0 or is supersingular. This research is motivated in part by Oort's conjecture, which predicted that the automorphism group of generic supersingular principally polarized abelian variety of dimension $g \geq 2$ is $\{\pm 1\} \simeq \mathbb{Z}_2$, where $\mathbb{Z}_m$ denotes the cyclic group of order m. As an analogue of this conjecture, one may ask whether any generic 2-rank 0 (resp. supersingular) hyperelliptic curve in characteristic 2 of genus ≥ 2 is isomorphic to $\mathbb{Z}_2$. Our theorems answers this question for small genera, which lead us to formulate refined conjectures. Although Oort's conjecture has been resolved affirmatively recently ([6] for $g = 3$, [7] for even g and [15] for general g) unless $g = 2, 3$ and $p = 2$, the purpose of this paper is to highlight the interest of studying the case of a class of principally polarized abelian varieties that is, in one sense, narrower and, in another sense, broader than original Oort's case.

The classification of the autormorphism groups of curves contributes to the classification of curves. The representative prior studies are as follows. In [9, Section 8], Igusa classified the automorphism groups of curves of genus two into seven types. The case of genus-three hyperelliptic curves was done by Lercier and Ritzenthaler [10, 3.1]. The aim of this paper can be regarded as obtaining similar results when we restrict ourselves to the case of hyperelliptic curves in characteristic $p = 2$ which is of p-rank zero or supersingular. Here the p-rank of a curve C over an algebraically closed field k in characteristic p is the logarithm with base p of the cardinality of k-valued points of the p-kernel of the Jacobian variety of C. We say that a curve C is supersingular if its Jacobian variety is supersingular, i.e., is isogenous to a product of supersingular elliptic curves. The p-rank of a supersingular curve is known to be zero, but the fact that C has p-rank 0 does not imply that C is supersingular if the genus of C is greater than or equal to 3.

Let K be an algebraically closed field of characteristic 2. We study the following Artin-Schreier curve

$$C : y^2 - y = f(x),$$

where

$$f(x) := x(x^{2n} + a_{n-1}x^{2(n-1)} + \cdots + a_1 x^2 + a_0)$$

with $a_i \in K$. As we review in Sect. 2.1, it is known that such curves C are of 2-rank 0 and conversely any hyperelliptic curve of 2-rank 0 over K is realized in this way. Consider a coordinate-change of the form

$$\begin{cases} x \mapsto X := \alpha x + \beta, \\ y \mapsto Y := y + \gamma_n x^n + \gamma_{n-1} x^{n-1} + \cdots + \gamma_1 x + \gamma_0 \end{cases} \tag{1-1}$$

for $\alpha, \beta, \gamma_0, \ldots, \gamma_n \in K$ with $\alpha \neq 0$. If $Y^2 - Y = f(X)$ holds, then this coordinate-change defines an automorphism of C. Conversely, we shall see in Proposition 1 that any automorphism of C is realized in this way. We denote this automorphism by

$$(\alpha, \beta; \gamma_n, \ldots, \gamma_0).$$

Let $\mathrm{Aut}(C)$ be the automorphism group of C. The automorphism σ defined by

$$\sigma : \begin{cases} x \mapsto x, \\ y \mapsto y + 1, \end{cases}$$

i.e., $\sigma = (1, 0; 0, \ldots, 0, 1)$ is the hyperelliptic involution, since $C/\langle \sigma \rangle$ is isomorphic to $\mathbb{P}^1$. The group $\mathrm{Aut}(C)/\langle \sigma \rangle$ is called *the reduced automorphism group*. We write it as $\mathrm{RA}(C)$.

For an automorphism $(\alpha, \beta; \gamma_n, \ldots, \gamma_0)$ of C, we obtain $\alpha^{2n+1} = 1$ by examining the leading term of $Y^2 - Y = f(X)$. We have the following two homomorphisms

$$\rho_n : \quad \mathrm{Aut}(C) \longrightarrow \mu_{2n+1}$$

sending $(\alpha, \beta; \gamma_n, \ldots, \gamma_0)$ to α, where μ_{2n+1} is the group of $(2n+1)$-th roots of unity, and

$$\tau_n : \quad \mathrm{Aut}(C) \longrightarrow M_n$$

sending $(\alpha, \beta; \gamma_n, \ldots, \gamma_0)$ to (α, β), where M_n is the group of affine transformations $(u, v) : x \mapsto ux + v$ for $u, v \in K$.

The main results of this paper are Theorems 1, 2, and 3 below. The first theorem reveals the group extension structure of $\mathrm{Aut}(C)$.

Theorem 1. (1) *We have* $\mathrm{Ker}\tau_n = \langle\sigma\rangle$ *and therefore*

$$\mathrm{RA}(C) := \mathrm{Aut}(C)/\langle\sigma\rangle \simeq \mathrm{Im}\tau_n.$$

(2) *Put* $U_n := \left\{ \begin{pmatrix} 1 & \beta \\ 0 & 1 \end{pmatrix} \;\middle|\; (1,\beta) \in \mathrm{Im}\tau_n \right\}$. *Then we have*

$$\mathrm{Im}\tau_n \simeq \left\{ \begin{pmatrix} \alpha & \beta \\ 0 & 1 \end{pmatrix} \;\middle|\; (\alpha,\beta) \in \mathrm{Im}\tau_n \right\} \simeq U_n \rtimes \mathbb{Z}_{\#\mathrm{Im}\rho_n}.$$

Moreover, since U_n is an elementary abelian 2-group, there exists a non-negative integer ℓ such that $U_n \simeq \mathbb{Z}_2^\ell$; hence

$$\mathrm{RA}(C) \simeq \mathbb{Z}_2^\ell \rtimes \mathbb{Z}_{\#\mathrm{Im}\rho_n}.$$

(3) *If n is odd with $n \geq 3$, then we have $\mathrm{Ker}\rho_n = \langle\sigma\rangle$, equivalently $\#U_n = 1$.*
(4) *We have*
$$\mathrm{Aut}(C) \simeq \mathrm{Ker}\rho_n \rtimes \mathbb{Z}_{\#\mathrm{Im}\rho_n}.$$

In particular, by (3), if n is odd with $n \geq 3$, then we have $\mathrm{Aut}(C) \simeq \mathbb{Z}_2 \rtimes \mathbb{Z}_{\#\mathrm{Im}\rho_n}$.

As a result, we found that these groups are isomorphic to groups constructed as direct products and semidirect products of several cyclic groups. In the subsequent, Theorem 2, we compute, using the computer algebra system Magma, a Gröbner basis of the ideal generated by the conditions under which the coordinate transformation (1-1) becomes an automorphism. For $n = 1,\ldots,6$, this yields the detailed group structure and defining conditions of the reduced automorphism group $\mathrm{RA}(C)$, as well as the order of the automorphism group $\mathrm{Aut}(C)$ (cf. [16]).

Theorem 2. *For $n = 1,\ldots,6$, we have the table of the group structure of $\mathrm{RA}(C)$ and the order of $\mathrm{Aut}(C)$.*

n	group structure of RA(C)	#Aut(C)
1	$\mathbb{Z}_2^2 \rtimes \mathbb{Z}_3 \simeq A_4$	24
2	$\mathbb{Z}_2^4$ if $a_1 \neq 0$	32
	$\mathbb{Z}_2^4 \rtimes \mathbb{Z}_5$ if $a_1 = 0$	160
3	$\mathbb{Z}_1$ if $a_2^8 + a_2 + a_1^4 \neq 0$ or $a_2^{12} + a_2 a_1^2 + a_0^4 \neq 0$	2
	$\mathbb{Z}_7$ if $a_2^8 + a_2 + a_1^4 = a_2^{12} + a_2 a_1^2 + a_0^4 = 0$	14
4	$\mathbb{Z}_1$ if $a_3 \neq 0$	2
	$\mathbb{Z}_2^6$ if $a_3 = 0$ and $a_2 \neq 0$	128
	$\mathbb{Z}_2^6 \rtimes \mathbb{Z}_3$ if $a_3 = a_2 = 0$ and $a_1 \neq 0$	384
	$\mathbb{Z}_2^6 \rtimes \mathbb{Z}_9$ if $a_3 = a_2 = a_1 = 0$	1152
5	$\mathbb{Z}_1$ if $a_3 \neq 0$ or $a_4 + a_2^4 \neq 0$ or $a_4^8 + a_1^2 \neq 0$ or $a_4^{40} + a_4^{18} + a_4 a_2^2 + a_0^8 \neq 0$	2
	$\mathbb{Z}_{11}$ if $a_3 = a_4 + a_2^4 = a_4^8 + a_1^2 = a_4^{40} + a_4^{18} + a_4 a_2^2 + a_0^8 = 0$	22
6	$\mathbb{Z}_1$ if either $a_5(a_3^2 + a_5^6) = 0$, $a_5 a_3^4 + a_5^{13} + 1 \neq 0$, or $t \neq 0$ is true, and at least of a_5, a_3, $a_4^4 + a_2^2$, $a_4 + a_1^{16}$, $a_4^{24} + a_4^4 a_1^8 + a_4 a_1^4 + a_0^8$ is not 0	2
	$\mathbb{Z}_2$ if $a_5(a_3^2 + a_5^6) \neq 0$ and $a_5 a_3^4 + a_5^{13} + 1 = t = 0$, and at least of a_5, a_3, $a_4^4 + a_2^2$, $a_4 + a_1^{16}$, $a_4^{24} + a_4^4 a_1^8 + a_4 a_1^4 + a_0^8$ is not 0	4
	$\mathbb{Z}_{13}$ if $a_5 = a_3 = a_4^4 + a_2^2 = a_4 + a_1^{16} = a_4^{24} + a_4^4 a_1^8 + a_4 a_1^4 + a_0^8 = 0$	26

Here

$$t = a_1^8 + a_1^4 a_3^{24} + a_1^4 a_5^{72} + a_1^4 a_5^{46} + a_1^4 a_5^{20} + a_2^8 a_5^8 + a_2^2 a_3^{28} + a_2^2 a_5^{84} + a_2^2 a_5^{71}$$

$$+ a_2^2 a_5^{58} + a_2^2 a_5^{45} + a_2^2 a_5^{32} + a_2^2 a_5^{19} + a_2^2 a_5^6 + a_3^{30} a_4 + a_3^{22} + a_3^4 a_4 + a_3^2 a_4 a_5^{84}$$

$$+ a_3^2 a_4 a_5^{71} + a_3^2 a_4 a_5^{58} + a_3^2 a_4 a_5^{45} + a_3^2 a_4 a_5^{32} + a_3^2 a_4 a_5^{19} + a_3^2 a_4 a_5^6 + a_3^2 a_5^{60}$$

$$+ a_3^2 a_5^{47} + a_3^2 a_5^8 + a_4^8 a_5^{24} + a_4 a_5^{12} + a_5^{40} + a_5^{14}. \tag{1-2}$$

The case of genus 1 is well-known. Since a genus-2 curve is supersingular if and only if it is of p-rank 0 (cf. [8, Lem. 1.1 (i)]), the result for genus 2 gives a counter-example of original Oort conjecture, but this has already been proved by Ibukiyama [5] in a different way. In the theorem, we have compiled these cases into a single table so that the reader can compare them.

By this theorem, generic hyperelliptic curves of 2-rank 0 (in characteristic 2) is $\{\pm 1\}$ have the automorphism group $\{\pm 1\}$ (an analogue of Oort conjecture is true) for $2 < g \leq 6$. It would be natural to expect.

Conjecture 1. *The automorphism group of a generic hyperelliptic curve of genus g and 2-rank 0 in characteristic 2 is $\{\pm 1\}$ if $g > 2$.*

Finally we study supersingular curves in characteristic 2 of genus ≤ 9. Scholten-Zhu [14] gave a defining equation of such a curve as in the table below. We call them Artin-Schreier curves of Scholten-Zhu type.

genus g	Artin-Schreier curves of Scholten-Zhu type C_g^{SZ}
1	$y^2 - y = x^3$
2	$y^2 - y = x^5 + c_3 x^3$
3	none
4	$y^2 - y = x^9 + c_5 x^5 + c_3 x^3$
5	$y^2 - y = x^{11} + c_3 x^3 + c_1 x$
6	$y^2 - y = x^{13} + c_3 x^3 + c_1 x$
7	none
8	$y^2 - y = x^{17} + c_9 x^9 + c_5 x^5 + c_3 x^3$
9	$y^2 - y = x^{19} + c^8 x^9 + c^3 x$

Remark that there is no supersingular hyperelliptic curve in characteristic 2 if the genus is of the form $2^k - 1$ for some integer $k \geq 2$, see [13, Thm. 1.2].

We computed the automorphism groups of Artin-Schreier curves of Scholten-Zhu type:

Theorem 3. *The group structure of the reduced automorphism group* $\mathrm{RA}(C_g^{SZ})$ *and the order of the automorphism group* $\mathrm{Aut}(C_g^{SZ})$, *under the coordinate transformation* $(*)$ *of the curve* C_g^{SZ} *are as follows.*

genus g	group structure of $\mathrm{RA}(C_g^{SZ})$	$\#\mathrm{Aut}(C_g^{SZ})$
1	$\mathbb{Z}_2^2 \rtimes \mathbb{Z}_3 \simeq A_4$	24
2	$\mathbb{Z}_2^4$ if $c_3 \neq 0$	32
	$\mathbb{Z}_2^4 \rtimes \mathbb{Z}_5$ if $c_3 = 0$	160
4	$\mathbb{Z}_2^6$ if $c_5 \neq 0$	128
	$\mathbb{Z}_2^6 \rtimes \mathbb{Z}_3$ if $c_5 = 0$ and $c_3 \neq 0$	384
	$\mathbb{Z}_2^6 \rtimes \mathbb{Z}_9$ if $c_5 = c_3 = 0$	1152
5	$\mathbb{Z}_1$ if $c_3 \neq 0$ or $c_1 \neq 0$	2
	$\mathbb{Z}_{11}$ if $c_3 = c_1 = 0$	22
6	$\mathbb{Z}_1$ if $c_3 \neq 0$ or $c_1 \neq 0$	2
	$\mathbb{Z}_{13}$ if $c_3 = c_1 = 0$	26
8	$\mathbb{Z}_2^8$ if $c_9 \neq 0$ or $c_5 \neq 0$ or $c_3 \neq 0$	512
	$\mathbb{Z}_2^8 \rtimes \mathbb{Z}_{17}$ if $c_9 = c_5 = c_3 = 0$	8704
9	$\mathbb{Z}_1$ if $c \neq 0$	2
	$\mathbb{Z}_{19}$ if $c = 0$	38

A naive hyperelliptic analogue of Oort conjecture would predict that any generic supersingular hyperelliptic curve of genus ≥ 2 has the automorphism group $\{\pm 1\}$. However from this theorem this is false in characteristic 2 if g is $2, 4, 8$. A reasonable refinement of the above conjecture would be the following:

Conjecture 2. *The automorphism group of a generic supersingular hyperelliptic curve of genus g in characteristic 2 is $\{\pm 1\}$ if g is not a power of 2. More strongly one may expect that this would hold even if we replace "if" by "if and only if".*

This paper is organized as follows. In Sect. 2, we review a description of hyperelliptic curves in characteristic 2 which are of 2-rank 0, and a basic fact on their automorphism groups. In Sect. 3, we prove the three theorems in Introduction. In Sect. 4, we present a summary of this paper and discuss the issues to be addressed in future work.

2   Preliminaries

In this section, we review the basics of hyperelliptic curves, which are of 2-rank 0, and their automorphism groups.

2.1   Hyperelliptic Curves of 2-Rank Zero

Let K be an algebraically closed field of characteristic 2. Let C be a hyperelliptic curve over K. The function field of C is a quadratic extension of the field $K(x)$ of rational functions, say $K(x)[y]$ with $y^2 + a(x)y = b(x)$ for $a(x), b(x) \in K(x)^{\times}$. By replacing y by $a(x)y$, we see that C is isomorphic to an Artin-Schreier curve

$$y^2 - y = f(x)$$

for $f(x) \in K(x)$. Assume that the 2-rank of C is zero. Then it follows from [11, Lem. 2.6] that $f(x)$ has a single pole on the projective line $\mathbb{P}^1$. By a coordinate change, one may assume that the pole is at the point P_{∞} of infinity, in which case $f(x) \in K[x]$. By replacing y by $y + h$ for some polynomial $h \in K[x]$, we may assume that $f(x)$ does not contain any even-degree terms. Hence we conclude that every hyperelliptic curve of 2-rank 0 is realized as

$$y^2 - y = x(x^{2n} + a_{n-1}x^{2(n-1)} + \cdots + a_1 x^2 + a_0) \qquad (2\text{-}3)$$

for $a_i \in K$. Conversely any nonsingular curve defined as above is hyperelliptic curves of genus n and 2-rank zero (cf. [11, Lem. 2.6]).

2.2   Automorphisms

Let C be a hyperelliptic curves of 2-rank 0, say defined as in (2-3).

Proposition 1. *The automorphisms of the curve C are only those induced by coordinate transformations (1-1).*

Proof. Since the morphism $\pi : C \to \mathbb{P}^1$ of degree two is unique up to isomorphism by [4, Chap. IV, Prop. 5.3], for an automorphism ψ of C there exists an automorphism φ of $\mathbb{P}^1$ that makes the diagram commute

$$
\begin{array}{ccc}
C & \xrightarrow{\ \psi\ } & C \\
\downarrow{\scriptstyle \pi} & & \downarrow{\scriptstyle \pi} \\
\mathbb{P}^1 & \xrightarrow{\ \varphi\ } & \mathbb{P}^1.
\end{array}
$$

Since the point P_∞ at infinity of $\mathbb{P}^1$ is the unique branch point of π, the point P_∞ is not moved by φ. Hence ψ induces an automorphism of $C \backslash \pi^{-1}(P_\infty)$. Let x and X be the usual x-coordinate of $\mathbb{P}^1$ of the source and the target of φ. Since φ stabilizes P_∞, φ sends x to $\alpha X + \beta$. Write $f(x) := x(x^{2n} + a_{n-1}x^{2(n-1)} + \cdots + a_1 x^2 + a_0)$. The coordinate ring of $C \backslash \pi^{-1}(P_\infty)$ is

$$
K[x, y]/(y^2 - y - f(x)).
$$

The associated isomorphism ψ^* of coordinate rings

$$
\psi^* : K[x, y]/(y^2 - y - f(x)) \longrightarrow K[X, Y]/(Y^2 - Y - f(X))
$$

sends x to $\alpha X + \beta$ and y to $h_1 Y + h_2$ for some $h_1, h_2 \in K[X]$, $h_1 \neq 0$. It sends $y^2 - y - f(x)$ to

$$
(h_1 Y + h_2)^2 - (h_1 Y + h_2) - f(\alpha X + \beta) = (h_1^2 - h_1)Y + h_1^2 f(X) - f(\alpha X + \beta) + h_2^2 - h_2
$$

in $K[X, Y]/(Y^2 - Y - f(X))$, which is zero. If $h_1^2 = h_1$ did not hold, then Y would be written only by X, which contradicts that π is of degree 2. Hence we have $h_1 = 1$ as $h_1 \neq 0$. It is clear that the degree of h_2 has to be less than or equal to n.

3   Proof

In this section, we prove the theorems presented in Sect. 1. The notation introduced in Sect. 1 will be used freely throughout this section.

3.1   Proof of Theorem 1

Recall that we study a nonsingular curve of the form

$$
C : y^2 - y = f(x) := x(x^{2n} + a_{n-1}x^{2(n-1)} + \cdots + a_1 x^2 + a_0)
$$

for $a_i \in K$ and that an element of $\mathrm{Aut}(C)$ is of the form

$$
(\alpha, \beta; \gamma_n, \ldots, \gamma_0)
$$

for some $\alpha, \beta, \gamma_0, \ldots, \gamma_n \in K$ which denotes a coordinate-change

$$\begin{cases} x \mapsto X := \alpha x + \beta, \\ y \mapsto Y := y + \gamma_n x^n + \gamma_{n-1} x^{n-1} + \cdots + \gamma_1 x + \gamma_0 \end{cases} \tag{3-4}$$

and satisfies $Y^2 - Y = f(X)$. We now begin the proof of Theorem 1.

Proof. (1) What we want to show is that

$$\mathrm{Ker}\tau_n = \{(\alpha, \beta; \gamma_n, \ldots, \gamma_0) \in \mathrm{Aut}(C) \mid (\alpha, \beta) = (1, 0)\}$$

is generated by the hyperelliptic involution σ. Recall $\sigma = (1, 0; 0, \ldots, 0, 1)$. We see that $(\alpha, \beta) = (1, 0)$ implies $\gamma_n = \cdots = \gamma_1 = 0$ and $\gamma_0 = 0, 1$, by examining the x^{2i}-coefficient of $Y^2 - Y = f(X)$ with X and Y as in (3-4) for $i = 0, \ldots, n$. Hence we have $\mathrm{Ker}\tau_n = \langle \sigma \rangle \simeq \mathbb{Z}_2$ and therefore $\mathrm{RA}(C) = \mathrm{Aut}(C)/\mathrm{Ker}\tau_n \simeq \mathrm{Im}\tau_n$.

(2) For two elements $(\alpha, \beta), (A, B) \in \mathrm{Im}\tau_n$, their product in $\mathrm{Im}\tau_n$ is given by

$$(\alpha, \beta) \cdot (A, B) = (\alpha A, \alpha B + \beta).$$

The first isomorphism $\mathrm{Im}\tau_n \simeq \left\{ \begin{pmatrix} \alpha & \beta \\ 0 & 1 \end{pmatrix} \;\middle|\; (\alpha, \beta) \in \mathrm{Im}\tau_n \right\}$ is obtained by comparing with the product of the associated matrices. To show the second isomorphism, it suffices to show that $U_n := \left\{ \begin{pmatrix} 1 & \beta \\ 0 & 1 \end{pmatrix} \;\middle|\; (1, \beta) \in \mathrm{Im}\tau_n \right\}$ is a normal subgroup of $\mathrm{Im}\tau_n$, that a complement of U_n in $\mathrm{Im}\tau_n$ is isomorphic to $\mathbb{Z}_{\#\mathrm{Im}\rho_n}$, and that a complement of U_n exists in $\mathrm{Im}\tau_n$. We consider the group homomorphism

$$\mathrm{Im}\tau_n \to \mu_{2n+1}$$

sending $\begin{pmatrix} \alpha & \beta \\ 0 & 1 \end{pmatrix}$ to α. Since the kernel of this homomorphism is U_n, we have $U_n \lhd \mathrm{Im}\tau_n$. Also, since the image of this homomorphism is a cyclic group of order $\#\mathrm{Im}\rho_n$, by the fundamental theorem on homomorphisms we have $\mathrm{Im}\tau_n/U_n \simeq \mathbb{Z}_{\#\mathrm{Im}\rho_n}$. Here, the complement of U_n in $\mathrm{Im}\tau_n$ is isomorphic to $\mathrm{Im}\tau_n/U_n$. Finally, we show that U_n has a complement in $\mathrm{Im}\tau_n$. From the Schur-Zassenhaus Lemma (cf. [12, Chap. 7, Thm.7.41]), it is sufficient to show that the orders of U_n and $\mathrm{Im}\tau_n/U_n$ are coprime. For any $(1, \beta) \in \mathrm{Im}\tau_n$, since $(1, \beta) \cdot (1, \beta) = (1, 0) = e$ holds, the order of any nontrivial element of U_n is 2. Therefore, U_n is either the trivial group $\{e\}$ or an elementary abelian 2-group, and its order is found to be a power of 2. On the other hand, since the order of $\mathrm{Im}\tau_n/U_n \simeq \mathbb{Z}_{\#\mathrm{Im}\rho_n}$ is a divisor of $2n+1$, that is, an odd number, the orders of U_n and its complement in $\mathrm{Im}\tau_n$ are coprime. From the above, the second isomorphism was shown. From the proof so far and Theorem 1 (1), there exists a non-negative integer ℓ such that $\mathrm{RA}(C) \simeq \mathbb{Z}_2^\ell \rtimes \mathbb{Z}_{\#\mathrm{Im}\rho_n}$.

(3) We determine

$$\mathrm{Ker}\rho_n = \{(\alpha, \beta; \gamma_n, \ldots, \gamma_0) \mid \alpha = 1\},$$

when n is odd with $n \geq 3$. The x^ℓ-coefficient $c(\ell)$ of the right hand side of the equation $Y^2 - Y = \sum_{\ell=0}^{2n+1} c(\ell)X^\ell$ obtained by applying the coordinate transformation (3-4) to the curve C is

$$c(\ell) = \binom{2n+1}{\ell}\alpha^\ell \beta^{2n+1-\ell} + \sum_{i=\lfloor \frac{\ell}{2}\rfloor}^{n-1} a_i \binom{2i+1}{\ell}\alpha^\ell \beta^{2i+1-\ell} + \gamma_{\frac{\ell}{2}}^2 + \gamma_\ell,$$

where we put $\gamma_\ell = 0$ for $\ell > n$ and $\gamma_{\frac{\ell}{2}} = 0$ if $\frac{\ell}{2} \notin \mathbb{N}$. The following equation is a necessary and sufficient condition for the coordinate transformation (3-4) to be an automorphism.

$$c(\ell) = \begin{cases} 1 & (\ell = 2n+1) \\ 0 & (\ell : \text{even}) \\ a_{\frac{\ell-1}{2}} & (\ell : \text{odd and } \ell \neq 2n+1) \end{cases} \tag{3-5}$$

Let $n \geq 3$ be an odd integer. To consider the kernel of ρ_n, we assume $\alpha = 1$. Since

$$c(2n-1) = \binom{2n+1}{2n-1}\beta^2 + a_{n-1}\binom{2n-1}{2n-1}\beta^0 = \beta^2 + a_{n-1},$$

the Eq. (3-5) reads $\beta^2 + a_{n-1} = a_{n-1}$. Hence we obtain $\beta = 0$. Also, we have $\gamma_n = 0$ and $\gamma_0 = 0, 1$ from $c(2n) = \beta + \gamma_n^2 = 0$ and $c(0) = \gamma_0^2 + \gamma_0 = 0$. Let m be an integer with $1 \leq m \leq n$. By $\beta = 0$, we have

$$c(2m-1) = \binom{2n+1}{2m-1}\beta^{2(n-m+1)} + \sum_{i=m-1}^{n-1} a_i \binom{2i+1}{2m-1}\beta^{2(i-m+1)} + \gamma_{2m-1}$$

$$= a_{m-1} + \gamma_{2m-1}.$$

The Eq. (3-5) reads $a_{m-1} + \gamma_{2m-1} = a_{m-1}$ and therefore we have $\gamma_{2m-1} = 0$. Finally, we show that $\gamma_\ell = 0$ holds for any even number $\ell \geq 2$. From $\beta = 0$, when $\ell \geq 2$ is even,

$$c(\ell) = \gamma_{\frac{\ell}{2}}^2 + \gamma_\ell = 0$$

holds. By substituting $\ell = 2$, we see that $\gamma_2 = 0$, and for even ℓ such that $4 \leq l \leq 2n$, it follows inductively that $\gamma_\ell = 0$. From the above, we have

$$\mathrm{Ker}\rho_n = \{(1,0;0,\ldots,0,0),(1,0;0,\ldots,0,1)\} = \langle \sigma \rangle.$$

(4) Since $\mathrm{Aut}(C)/\mathrm{Ker}\rho_n \simeq \mathrm{Im}\rho_n$, if $\#\mathrm{Ker}\rho_n$ and $\#\mathrm{Im}\rho_n$ are coprime, then $\mathrm{Aut}(C) \simeq \mathrm{Ker}\rho_n \rtimes \mathrm{Im}\rho_n$ that we aim to prove here follows from the Schur-Zassenhaus Lemma. Since $\mathrm{Im}\rho_n$ is a subgroup of the finite cyclic group μ_{2n+1}, we have $\mathrm{Im}\rho_n \simeq \mathbb{Z}_{\#\mathrm{Im}\rho_n}$. It remains to show that $\#\mathrm{Ker}\rho_n$ and $\#\mathrm{Im}\rho_n$ are coprime. By Theorem 1 (1), (2), there exists a non-negative integer ℓ such that

$$\#\mathrm{Aut}(C) = \#\mathrm{Ker}\tau_n \cdot \#\mathrm{Im}\tau_n = \#\mathbb{Z}_2 \cdot \#U_n \cdot \#\mathbb{Z}_{\#\mathrm{Im}\rho_n} = 2 \cdot 2^\ell \cdot \#\mathrm{Im}\rho_n$$

holds. Therefore we have $\#\mathrm{Ker}\rho_n = \#\mathrm{Aut}(C)/\#\mathrm{Im}\rho_n = 2^{\ell+1}$, but on the other hand, $\#\mathrm{Im}\rho_n$ is an odd number because it is a divisor of $2n+1$. Hence $\#\mathrm{Ker}\rho_n$ and $\#\mathrm{Im}\rho_n$ are coprime.

3.2 Proof of Theorem 2

In this subsection, we prove Theorem 2, namely, for $n \leq 6$ we determine the group structure of the reduced automorphism group $\mathrm{RA}(C)$ of

$$C: \quad y^2 - y = x(x^{2n} + a_{n-1}x^{2(n-1)} + \cdots + a_1 x^2 + a_0)$$

for $a_i \in K$. For the proof, we used the computational software Magma (cf. [1]).

Proof. From the expression for the coefficients of the curve used in the proof of Theorem 1 (3), $c(\ell)$, it can be seen that once α and β are determined, $\gamma_n, \ldots, \gamma_1$ are determined uniquely, and γ_0 is determined in two ways. From this fact and Theorem 1 (2), in order to determine the group structure of the reduced automorphism group $\mathrm{RA}(C)$, it is sufficient to find the number of values that α and β can take. For each $n = 1, \ldots, 6$, we derived the number of possible values of α and of β (that is, the order of $\mathrm{Im}\rho_n$ and of U_n), as well as their conditional formulas, using the computational software Magma (cf. [1]). More precisely, we used Magma to find the Gröbner basis of the ideal generated by polynomials that are necessary and sufficient for the coordinate transformation of the curve C to be an automorphism, and analyzed the obtained output. Considering the fiber structure $\mathrm{Im}\tau_n \to \mathrm{Im}\rho_n$: $(\alpha, \beta) \mapsto \alpha$, we used a lexicographic order satisfying $\beta \succ \alpha$ as the monomial order to compute the Gröbner basis above.

Algorithm 1. Computing Gröbner basis

Input: A natural number $n \in \mathbb{N}$
Output: The Gröbner basis G of the ideal generated by polynomials that are necessary
  and sufficient for the coordinate transformation of the curve to be an automorphism
1: Set the polynomial $F := y^2 - y - x(x^{2n} + \sum_{i=0}^{n-1} a_i x^{2i})$ and the coordinate transformation $\varphi := (x \mapsto \alpha x + \beta, \ y \mapsto y + \sum_{i=0}^{n} \gamma_i x^i)$
2: Compute $H := F - \varphi(F)$
3: Set $L := \emptyset$ and the set M of monomials contained in the polynomial H
4: **for** the monomial m in M **do**
5:   $L := L \cup \{\text{the coefficient of } m\}$
6: **end for**
7: Set the ideal I generated by L
8: Compute the Gröbner basis G of I with respect to the lexicographic order $\gamma_n \succ$
  $\cdots \succ \gamma_0 \succ \beta \succ \alpha \succ a_0 \succ \cdots \succ a_{n-1}$
9: **return** G

If we put $\alpha = 1$, i.e., we replace "$L := \emptyset$" by "$L := \{\alpha - 1\}$" in the third line of Algorithm 1, the algorithm provides a Gröbner basis determining U_n.

Here, as an example, the steps of the proof are described concretely for $n = 3$. For the curve

$$C : y^2 - y = x(x^6 + a_2 x^4 + a_1 x^2 + a_0)$$

over K and its coordinate transformation

$$\begin{cases} x \mapsto \alpha x + \beta, \\ y \mapsto y + \gamma_3 x^3 + \gamma_2 x^2 + \gamma_1 x + \gamma_0, \end{cases}$$

the necessary and sufficient condition for this coordinate transformation to be an automorphism is that the following holds.

$$\begin{cases} \alpha^7 = 1 \\ \alpha^6 \beta + \gamma_3^2 = 0 \\ \alpha^5 \beta^2 + a_2 \alpha^5 = a_2 \\ \alpha^4 \beta^3 + a_2 \alpha^4 \beta + \gamma_2^2 = 0 \\ \alpha^3 \beta^4 + a_1 \alpha^3 + \gamma_3 = a_1 \\ \alpha^2 \beta^5 + a_1 \alpha^2 \beta + \gamma_1^2 + \gamma_2 = 0 \\ \alpha \beta^6 + a_2 \alpha \beta^4 + a_1 \alpha \beta^2 + a_0 \alpha + \gamma_1 = a_0 \\ \beta^7 + a_2 \beta^5 + a_1 \beta^3 + a_0 \beta + \gamma_0^2 + \gamma_0 = 0 \end{cases}$$

By computing the Gröbner basis of the ideal generated by this polynomials using Magma, we obtain the following output.

$$\begin{cases} \gamma_3 + \alpha^3 a_1 + \alpha^3 a_2^2 + a_1 + a_2^2 = 0 \\ \gamma_2 + \alpha^3 a_1^3 + \alpha^3 a_1^2 a_2^2 + \alpha^3 a_1 a_2^4 + \alpha^3 a_2^6 \\ \quad + \alpha^2 a_0^2 + \alpha^2 a_1^3 + \alpha^2 a_1 a_2^4 + \alpha^2 a_2^6 + a_0^2 + a_1^2 a_2^2 = 0 \\ \gamma_1 + \alpha^3 a_1 a_2 + \alpha^3 a_2^3 + \alpha a_0 + \alpha a_1 a_2 + a_0 + a_2^3 = 0 \\ \gamma_0^2 + \gamma_0 + \alpha^6 a_1^2 a_2^3 + \alpha^6 a_2^7 + \alpha^3 a_1^3 a_2 + \alpha^3 a_1^2 a_2^3 + \alpha^3 a_1 a_2^5 + \alpha^3 a_2^7 + \alpha^2 a_1^3 a_2 \\ \quad + \alpha^2 a_1^2 a_2^3 + \alpha^2 a_1 a_2^5 + \alpha^2 a_2^7 + \alpha a_0 a_1^2 + \alpha a_0 a_2^4 + \alpha a_1^3 a_2 + \alpha a_1 a_2^5 + a_0 a_1^2 \\ \quad + a_0 a_2^4 + a_1^3 a_2 + a_1^2 a_2^3 + a_1 a_2^5 + a_2^7 = 0 \\ \beta + \alpha a_1^2 + \alpha a_2^4 + a_1^2 + a_2^4 = 0 \\ \alpha^7 + 1 = 0 \\ \alpha a_0^4 + \alpha a_1^2 a_2 + \alpha a_2^{12} + a_0^4 + a_1^2 a_2 + a_2^{12} = 0 \\ \alpha a_1^4 + \alpha a_2^8 + \alpha a_2 + a_1^4 + a_2^8 + a_2 = 0 \end{cases}$$

We have $\alpha^7 = 1$ from the 6th equation. Also, from 7th and 8th equation, if both $a_2^8 + a_2 + a_1^4 = 0$ and $a_2^{12} + a_2 a_1^2 + a_0^4 = 0$ hold, then $\alpha^7 = 1$; otherwise, $\alpha = 1$ is obtained. This allows us to determine the number of values α can take. Next, assuming $\alpha = 1$, we find the number of values β can take. We have $\beta = 0$ by substituting $\alpha = 1$ for the 5th equation, so the number of values that β can take is 1. In this way, we analyzed the output (Gröbner basis) obtained with Magma and proved Theorem 2 by determining the order of $\mathrm{Im}\rho_n$ and of U_n. The source code used is available on [16]. For each $n = 1, \ldots, 6$, the details are described below.

(1) <u>The case of $n = 1$</u>: By substituting $\ell = 3$ for $c(\ell)$, we obtain $\alpha^3 = 1$. Since the formula concerning α is only this formula, $\#\mathrm{Im}\rho_1 = 3$ always holds. Next,

let $\alpha = 1$, and show that $U_n \simeq \mathbb{Z}_2^2$, that is, β takes 4 different values. From the calculation results of Magma , it follows that $\beta^4 + \beta = 0$, and since this equation does not have a multiple root, it can be seen that β takes on 4 different values.

(2) $\underline{\text{The case of } n = 2:}$ By substituting $\ell = 5$ for $c(\ell)$, we obtain $\alpha^5 = 1$. Also, we obtain $a_1(\alpha + 1) = 0$ from the calculation results of Magma. Since $\mathrm{Im}\rho_2$ is a cyclic group of order 1 or 5, when $a_1 \neq 0$ holds, we have $\#\mathrm{Im}\rho_2 = 1$, and when $a_1 = 0$ holds, we have $\#\mathrm{Im}\rho_2 = 5$. Next, let $\alpha = 1$, and show that $U_n \simeq \mathbb{Z}_2^4$, that is, β takes 16 different values. From the calculation results of Magma, it follows that

$$\beta^{16} + a_1^4 \beta^8 + a_1^2 \beta^2 + \beta = 0$$

and since this equation does not have a multiple root, it can be seen that β takes on 16 different values.

(3) $\underline{\text{The case of } n = 3:}$ By substituting $\ell = 7$ for $c(\ell)$, we obtain $\alpha^7 = 1$. Also, we obtain

$$\begin{cases} (a_2^8 + a_2 + a_1^4)(\alpha + 1) = 0 \\ (a_2^{12} + a_2 a_1^2 + a_0^4)(\alpha + 1) = 0 \end{cases}$$

from the calculation results of Magma. Since $\mathrm{Im}\rho_3$ is a cyclic group of order 1 or 7, when $a_2^8 + a_2 + a_1^4 = a_2^{12} + a_2 a_1^2 + a_0^4 = 0$ holds, we have $\#\mathrm{Im}\rho_3 = 7$, and otherwise, we have $\#\mathrm{Im}\rho_3 = 1$. Next, we show that $U_n = \{e\}$. From the proof of Theorem 1 (3), when n is an odd number greater than or equal to 3, if $\alpha = 1$, then $\beta = 0$. Therefore $U_n = \{e\}$ holds.

(4) $\underline{\text{The case of } n = 4:}$ By substituting $\ell = 9$ for $c(\ell)$, we obtain $\alpha^9 = 1$. Also, we obtain

$$\begin{cases} a_3(\alpha + 1) = 0 \\ a_2(\alpha + 1) = 0 \\ a_1^2(\alpha^3 + 1) = 0 \end{cases}$$

from the calculation results of Magma. Since $\mathrm{Im}\rho_4$ is a cyclic group of order $1, 3$, or 9, when either $a_3 \neq 0$ or $a_2 \neq 0$ holds, we have $\#\mathrm{Im}\rho_4 = 1$, when $a_3 = a_2 = 0$ and $a_1 \neq 0$ holds, we have $\#\mathrm{Im}\rho_4 = 3$, and when $a_3 = a_2 = a_1 = 0$ holds, we have $\#\mathrm{Im}\rho_4 = 9$. Next, let $\alpha = 1$, and show that $U_n = \{e\}$ when $a_3 \neq 0$ holds, and that $U_n \simeq \mathbb{Z}_2^6$ when $a_3 = 0$ holds.

(i) $\underline{\text{The case of } a_3 \neq 0:}$ From the calculation results of Magma, we obtain $a_3\beta = 0$. Since $a_3 \neq 0$, $\beta = 0$ holds, so we have $U_n = \{e\}$.

(ii) $\underline{\text{The case of } a_3 = 0:}$ From the calculation results of Magma, it follows that

$$\beta^{64} + a_2^8 \beta^{32} + a_1^8 \beta^{16} + a_1^4 \beta^4 + a_2^2 \beta^2 + \beta = 0$$

and since this equation does not have a multiple root, it can be seen that β takes on 64 different values. Therefore $U_n \simeq \mathbb{Z}_2^6$.

(5) $\underline{\text{The case of } n = 5:}$ By substituting $\ell = 11$ for $c(\ell)$, we obtain $\alpha^{11} = 1$. Also, we obtain

$$\begin{cases} a_3(\alpha + 1) = 0 \\ (a_4 + a_2^4)(\alpha + 1) = 0 \\ (a_4^8 + a_1^2)(\alpha + 1) = 0 \\ (a_4^{40} + a_4^{18} + a_4 a_2^2 + a_0^8)(\alpha + 1) = 0 \end{cases}$$

from the calculation results of Magma. Since $\mathrm{Im}\rho_5$ is a cyclic group of order 1 or 11, when $a_3 = a_4 + a_2^4 = a_4^8 + a_1^2 = a_4^{40} + a_4^{18} + a_4 a_2^2 + a_0^8 = 0$ holds, we have $\#\mathrm{Im}\rho_5 = 11$, and otherwise, we have $\#\mathrm{Im}\rho_5 = 1$. Next, we show that $U_n = \{e\}$. From the proof of Theorem 1 (3), when n is an odd number greater than or equal to 3, if $\alpha = 1$, then $\beta = 0$. Therefore $U_n = \{e\}$ holds.

(6) <u>The case of $n = 6$:</u> By substituting $\ell = 13$ for $c(\ell)$, we obtain $\alpha^{13} = 1$. Also, we obtain $a_5(\alpha + 1) = 0, a_3(\alpha + 1) = 0$ from the calculation result of Magma, where we used the grevlex order to terminate the computation. Since $\mathrm{Im}\rho_6$ is a cyclic group of order 1 or 13, when $a_5 = a_3 = 0$ does not hold, we have $\#\mathrm{Im}\rho_6 = 1$. By performing the calculation with $a_5 = a_3 = 0$, we obtain the following three equations.

$$\begin{cases} (a_4^4 + a_2^2)(\alpha + 1) = 0, \\ (a_4 + a_1^{16})(\alpha + 1) = 0, \\ (a_4^{24} + a_4^4 a_1^8 + a_4 a_1^4 + a_0^8)(\alpha + 1) = 0. \end{cases}$$

From the considerations so far, it can be seen that the necessary and sufficient condition for $\#\mathrm{Im}\rho_6 = 13$ is $a_5 = a_3 = a_4^4 + a_2^2 = a_4 + a_1^{16} = a_4^{24} + a_4^4 a_1^8 + a_4 a_1^4 + a_0^8 = 0$. Next, we consider the group structure of U_n. From the calculation results of Magma with $\alpha = 1$, we obtain the following three equations.

$$\begin{cases} (a_5 a_3^4 + a_5^{13} + 1)\beta = 0, \\ t\beta = 0, \\ \beta^2 + a_5(a_3^2 + a_5^6)\beta = 0, \end{cases}$$

where t is as in (1-2). When the necessary condition $a_5 = 0$ for $\#\mathrm{Im}\rho_6 = 13$ is satisfied, from the third equation, $\beta = 0$, that is, $U_n = \{e\}$ holds. Therefore, if $\#\mathrm{Im}\rho_6 = 13$ holds, the group structure of $\mathrm{RA}(C)$ is determined as $\mathbb{Z}_{13}$. On the other hand, when $\#\mathrm{Im}\rho_6 = 1$ holds (that is, when at least one of $a_5, a_3, a_4^4 + a_2^2, a_4 + a_1^{16}, a_4^{24} + a_4^4 a_1^8 + a_4 a_1^4 + a_0^8$ is nonzero), from the three equations above, if $a_5(a_3^2 + a_5^6) \neq 0$ and $a_5 a_3^4 + a_5^{13} + 1 = t = 0$ hold, then β takes two possible values, so we have $U_n \simeq \mathbb{Z}_2$, otherwise $\beta = 0$, so we have $U_n = \{e\}$.

Regarding the order of the automorphism group $\mathrm{Aut}(C)$, since $\#\mathrm{Aut}(C) = 2 \cdot \#\mathrm{RA}(C) = 2 \cdot \#U_n \cdot \#\mathrm{Im}\rho_n$ holds, the order of $\mathrm{Aut}(C)$ can be immediately determined from the group structure of $\mathrm{RA}(C)$.

3.3   Proof of Theorem 3

In this subsection, we prove Theorem 3. That is, we determine $\mathrm{RA}(C_g^{SZ})$ of Artin-Schreier curves C_g^{SZ} of Scholten-Zhu type.

Proof. In the case of $g = 1, \dots, 6$, it immediately follows from Theorem 2. Using a method similar to the proof of Theorem 2, we show the case $g = 8$ and 9.

(1) The case of $g = 8$: By substituting $\ell = 17$ for $c(\ell)$, we obtain $\alpha^{17} = 1$. Also, we obtain

$$\begin{cases} c_9(\alpha + 1) = 0 \\ c_5^2(\alpha + 1) = 0 \\ c_3^4(\alpha + 1) = 0 \end{cases}$$

from the calculation results of Magma. Since $\mathrm{Im}\rho_8$ is a cyclic group of order 1 or 17, when $c_9 = c_5 = c_3 = 0$ holds, we have $\#\mathrm{Im}\rho_8 = 17$, and otherwise, we have $\#\mathrm{Im}\rho_8 = 1$. Next, let $\alpha = 1$, and show that $U_n \simeq \mathbb{Z}_2^8$, that is, β takes 256 different values. From the calculation results of Magma, it follows that

$$\beta^{256} + c_9^{16}\beta^{128} + c_5^{16}\beta^{64} + c_3^{16}\beta^{32} + c_3^8\beta^8 + c_5^4\beta^4 + c_9^2\beta^2 + \beta = 0$$

and since this equation does not have a multiple root, it can be seen that β takes on 256 different values.

(2) The case of $g = 9$: By substituting $\ell = 19$ for $c(\ell)$, we obtain $\alpha^{19} = 1$. Also, we obtain $c^{32}(\alpha + 1) = 0$ from the calculation results of Magma. Since $\mathrm{Im}\rho_9$ is a cyclic group of order 1 or 19, when $c \neq 0$ holds, we have $\#\mathrm{Im}\rho_9 = 1$, and when $c = 0$ holds, we have $\#\mathrm{Im}\rho_9 = 19$. Next, we show that $U_n = \{e\}$. From the proof of Theorem 1 (3), when n is an odd number greater than or equal to 3, if $\alpha = 1$, then $\beta = 0$. Therefore $U_n = \{e\}$ holds.

4   Concluding Remarks

In this paper, we considered the nonsingular curve over an algebraically closed field K of characteristic 2 defined by

$$C : y^2 - y = x(x^{2n} + a_{n-1}x^{2(n-1)} + \cdots + a_1x^2 + a_0)$$

for $a_i \in K$. These are hyperelliptic curves of 2-rank 0. We examined the group structure and the order of the automorphism group $\mathrm{Aut}(C)$, as well as the reduced automorphism group $\mathrm{RA}(C)$. We were able to determine the group structure of $\mathrm{RA}(C)$ when $n \leq 6$, but for bigger n we were unable to determine it due to computational difficulties. Of course, by improving the algorithm, enhancing the computational environment, and investing a substantial amount of time, it may be possible to obtain similar results for slightly larger values of n. Based on these experiments, we formulated analogues of Oort's conjecture for these families of curves, see Conjectures 1 and 2 in Introduction.

As a future work, we would like to have a theoretical approach toward general results on the group structure of $\mathrm{Aut}(C)$ as well as the detailed group structure of $\mathrm{RA}(C)$ for any n. The remaining problem for odd n is to classify the image of ρ_n. When n is even, in addition to the above issues, determining the structure of the kernel also becomes a major problem. However when n is a power of 2, say $n = 2^m$, as far as our experiments indicate, only the following two types of groups appeared as U_n.

$$U_n \simeq \begin{cases} \mathbb{Z}_1 & \text{for specific } (a_i) \text{ chosen in our experiments,} \\ \mathbb{Z}_2^{2(m+1)} & \text{if } a_i = 0 \text{ unless } i \text{ is a power of 2,} \end{cases}$$

where we note $\mathrm{RA}(C) \simeq U_n \rtimes \mathbb{Z}_{\#\mathrm{Im}\rho_n}$ by Theorem 1. However, it is uncertain whether one can assert that only these two cases occur, and we also regard this as an interesting problem, much like Conjectures 1 and 2.

Acknowledgments. The first author would like to express his deepest gratitude to Professor Harashita, his supervisor, for the generous guidance and valuable advice provided throughout the preparation of this paper. The authors are also grateful to the members of the Harashita Laboratory for their insightful comments during seminars and discussions. They also thank the referees for their comments and remarks.

References

1. Bosma, W., Cannon, J., Playoust, C.: The MAGMA algebra system. I. The user language. J. Symb. Comput. **24**(3-4), 235–265 (1997)
2. van der Geer, G., van der Vlugt, M.: Reed-Muller codes and supersingular curves. I. Compos. Math. **84**(3), 333–367 (1992)
3. van der Geer, G., van der Vlugt, M.: On the existence of supersingular curves of given genus. J. Reine Angew. Math. **458**, 53–62 (1994)
4. Hartshorne, R.: Algebraic Geometry. Graduate Texts in Mathematics, vol. 52. Springer (1997)
5. Ibukiyama, T.: Principal polarizations of supersingular abelian surfaces. J. Math. Soc. Japan **72**(4), 1161–1180 (2020)
6. Karemaker, V., Yobuko, F., Yu, C.-F.: Mass formula and Oort's conjecture for supersingular abelian threefolds. Adv. Math. **386**, Article ID 107812 (2021)
7. Karemaker, V., Yu, C.-F.: Supersingular Ekedahl-Oort strata and Oort's conjecture, arXiv:2406.19748
8. Ibukiyama, T., Katsura, T., Oort, F.: Supersingular curves of genus two and class numbers. Compos. Math. **57**, 127–152 (1986)
9. Igusa, J.: Arithmetic variety of moduli for genus two. Ann. Math. **2**(72), 612–649 (1960)
10. Lercier, R., Ritzenthaler, C.: Hyperelliptic curves and their invariants: geometric, arithmetic and algorithmic aspects. J. Algebra **372**, 595–636 (2012)
11. Pries, R., Zhu, H.J.: The p-rank stratification of Artin-Schreier curves. Ann. Inst. Fourier **62**(2), 707–726 (2012)
12. Rotman, J.J.: An Introduction to the Theory of Groups. Graduate Texts in Mathematics, vol. 148. Springer (1995)
13. Scholten, J., Zhu, H.J.: Hyperelliptic curves in characteristic 2. Int. Math. Res. Not. **2002**(17), 905–917 (2002)
14. Scholten, J., Zhu, H.J.: Families of supersingular curves in characteristic 2. Math. Res. Lett. **9**(5–6), 639–650 (2002)
15. Viehmann, E.: Oort's conjecture on automorphisms of generic supersingular abelian varieties, arXiv:2603.06033
16. https://github.com/Kohtaro-Yamaguchi/code_Magma.git

Author Index

L. Batina and F. Özbudak (Eds.): WAIFI 2026, LNCS 16611, pp. 399–400, 2026.
https://doi.org/10.1007/978-3-032-27574-5